High-Temperature Superconducting Materials Science and Engineering

New Concepts and Technology

PERGAMON TITLES OF RELATED INTEREST

Books

BEVER
Encyclopedia of Materials Science and Engineering
8 Volume Set

BIANCONI & MARCELLI
High T_c Superconductors

CAHN
Encyclopedia of Materials Science and Engineering
Supplementary Volumes 1, 2 and 3

EVETTS
Concise Encyclopedia of Magnetic and Superconducting Materials

KROTO, FISCHER & COX
The Fullerenes

ROSE-INNES & RHODERICK
Introduction to Superconductivity

Journals

Applied Superconductivity

Journal of Physics and Chemistry of Solids

Progress in Crystal Growth and Characterization of Materials

Solid State Communications

High-Temperature Superconducting Materials Science and Engineering

New Concepts and Technology

Editor: **Donglu Shi**
University of Cincinnati, USA

PERGAMON

UK   Elsevier Science Ltd, The Boulevard, Langford Lane, Kidlington, Oxford OX5 1GB, UK

USA   Elsevier Science Inc., 660 White Plains Road, Tarrytown, New York 10591-5153, USA

JAPAN Elsevier Science Japan, Tsunashima Building Annex, 3-20-12 Yushima, Bunkyo-ku, Tokyo 113, Japan

First Edition 1995

Library of Congress Cataloging in Publication Data
High-temperature superconducting materials science and engineering : new concepts and technology / edited by Donglu Shi.
p. cm.
Includes index.
1. High-temperature superconductivity. 2. High-temperature superconductors.
I. Shi, Donglu.
QC611.98.H54H5425   1994
537.6'23—dc20   94-5255

British Library Cataloguing in Publication Data
A catalogue record for this book is available from the British Library

ISBN 9780080421513

Transferred to digital print 2008

Printed and bound in Great Britain by
CPI Antony Rowe, Chippenham and Eastbourne

Contents

8   Bulk Processing and Characterization of YBa$_2$Cu$_3$O$_{6+x}$ 345
        P. McGinn

**9   Processing Bi-Based High-T$_c$ Superconducting Tapes, Wires,
        and Thick Films for Conductor Applications   383**
        E. E. Hellstrom

Preface

We intend this book to provide an up-to-date introduction to the fascinating field of high-temperature superconductivity. The focus of the book is on the basic concepts and recent developments in the field, particularly in the areas of vortex-state properties, structure, synthesis and processing, phase equilibrium, defects characterization, thin films, conductor development, and applications. Most of the theoretical issues are addressed in a straightforward manner so that technical nonspecialists and university students can benefit from the information. Furthermore, many physical concepts in superconductivity are explained in light of current theories.

This book is written for a large readership including university students and researchers from diverse backgrounds such as physics, materials science, engineering and chemistry. Both undergraduate and graduate students will find the book a valuable reference not only on superconductivity, but also on materials-related topics, including structure and phase diagrams of complex systems, novel ceramic and composite processing methods, thin film fabrication, and new materials-characterization techniques. In particular, this book gives a detailed introduction to experimental methods at low temperatures and high magnetic fields. Thus, it can serve as a comprehensive introduction to researchers in electromagnetic ceramics in general, and can also be used as a graduate-level text in superconductivity.

The book devotes two chapters (Chapters 2 and 6) to cryogenic systems and low-temperature measurements. Detailed experimental procedures for both transport and magnetic measurements are presented at a level suitable for people with no previous training in these areas. The book systematically introduces cryogenics and low-temperature techniques specifically for characterizing superconducting properties. Furthermore, the book includes numerous developments in magnetic property measurements of high-T_c superconductors by using SQUIDs, ac susceptometers and vibrating sample magnetometers. Recent developments in determining important superconducting parameters such as T_c, H_{c2}, and J_c are presented in detail, providing essential information for researchers in the field. Chapters 2 and 6 are also valuable to beginners and technical nonspecialists in the study of superconductivity and in low-temperature measurements.

Chapter 4 deals with vortex properties such as flux pinning, the Bean critical state, and flux creep. Although these concepts were introduced long ago, new vortex phenomena have been recently observed in high-T_c superconductors, requiring new and modified physical models to interpret them. For example, dynamic models including collective creep and Josephson coupling of bilayers of CuO plates have been developed to solve the flux motion problem. More fundamentally, flux avalanches observed near the critical state cannot be explained by traditional flux-creep models: Chapter 4 gives a detailed introduction to a new theory—self-organized criticality—for treating such vortex motion problems.

Chapters 6, 7 and 8 should be of most interest to researchers in materials processing. These chapters give detailed descriptions of many traditional and novel methods in synthesis and manufacture of ceramic materials with tailored microstructures and desired forms for practical applications. Recent research results in phase reactions, phase diagrams and heat treatment are presented for several high-T_c systems. In addition, Chapter 5 presents new characterization techniques using transmission electron microscopy, including the electron energy-loss spectrum method. With these highly advanced techniques, defect structures including tweeds, twin boundaries, grain boundaries, stacking faults and dislocations in high-T_c ceramics are studied in detail.

In Chapter 3, the crystal structure of various perovskites is introduced for most of the high-T_c superconductors. The effects of structural changes on superconductivity and oxygen-ordering behavior are analyzed and illustrated by means of various experiments.

The last chapter concerns one of the most exciting aspects of superconductivity: applications. Although industrial and scientific applications of conventional superconductors have been well established for several decades, it is a challenging task to use high-T_c oxides at liquid nitrogen temperature. Microstructure control, reproducibility and grain boundary weak links are still challenges to be overcome. Nevertheless, the potential applications of high-T_c materials are certainly recognized in many areas such as power transmission lines, magnetic levitation and nuclear magnetic resonance. The field is young, offering researchers exciting opportunities to make a major contribution to science and technology.

Acknowledgments

We thank all the authors for their enthusiasm, effort, cooperation and excellent contribution in their area of expertise. In particular, we acknowledge the superb technical editing work by Dr Gail Pieper of Argonne National Laboratory. It is impossible to even imagine the completion of this book without her participation in copyediting, proofreading, computer formatting and organizing this book. We are also grateful to Dr Roger Poeppel and Dr Harold Myron of Argonne National Laboratory for their valuable suggestions during the production of this book.

Acronyms

A-H model	Ambegaokar-Halperin model
BCS theory	Bardeen, Cooper and Schrieffer theory
CBED	Convergent beam electron diffraction
CEBAF	Continuous Electron Beam Accelerator Facility
CFC	Continuous-flow cryostat
CGR	Carbon glass resistance
CSL mode	Coincidence site lattice model
CCSL model	Constrained coincidence site lattice model
CMOS	Complementary metal oxide semiconductor
CSL	Coincidence site lattice
CVD	Chemical vapor disposition
DSC lattice vectors	Displacement-shift-complete lattice vectors
EDX	Energy dispersive x-ray spectroscopy
EELS	Electron energy-loss spectroscopy
EXAFS	Extended x-ray absorption fine structure
FENIX	Fusion Engineering International Experimental Magnet Facility
FLL	Flux line lattice
GBD	Grain boundary dislocation
G-L	Ginzburg-Landau
grp	Glass-reinforced plastic
HAGB	High-angle grain boundary
HEMT	High-electron mobility transistors
HREM	High-resolution electron microscopy
HRTEM	High-resolution transmission electron microscopy
HTF	High-temperature ferroelectrics
HTS	High-temperature superconductor
IR	Irreversibility line
ITER	International Thermonuclear Experimental Reactor
LAGB	Low-angle grain boundary
LCT	Large Coil Task
LHC	Large Hadron Collider
LTS	Low-temperature superconductor

MHD	Magnetohydrodynamics
MOCVD	Metal-organic chemical vapor deposition
MOD	Metal-organic deposition
MPMG	Melt-powder-melt-growth
MRI	Magnetic resonance imaging
MTG	Melt-texture growth
NMR	Nuclear magnetic resonance
OPIT	Oxide powder-in-tube
OSHA	Occupational Safety and Health Administration
PDMG	Platinum-doped melt-growth
PIT	Powder-in-tube
PLD	Pulsed laser deposition
PMP	Powder-melt process
PRT	Platinum resistance thermometer
QMG	Quench-and-melt-growth
RPC	Resonant pinning center
SBGP model	Sommerfeld-Bloch-Gruneisen-Peierls model
SFFT	Superconducting flux flow transistor
SIS	Superconductor-insulator-superconductor
SLMG	Solid-liquid-melt-growth
SMES	Superconducting Magnetic Energy Storage
SNS	Superconductor-normal metal-superconductor
SOC	Self-organized criticality
SQUID	Superconducting quantum interference device
SSC	Superconducting Super Collider
SUM	Structural unit model
TAFF	Thermally assisted flux flow
TEM	Transmission electron microscopy
T-J	Tinkham-Josephson
VLSI	Very large scale integration
VSM	Vibrating sample magnetometry
VTI	Variable-temperature insert
YSZ	Yttria-stabilized zirconia

Contributors

E. Gregory, IGC Advanced Superconductors Inc., 1875 Thomaston Ave., Waterbury, CT 06704, USA

E. E. Hellstrom, Department of Materials Science & Engineering, University of Wisconsin, 1509 University Ave., Madison, WI 53706, USA

J. D. Hettinger, Materials Science Division, Argonne National Laboratory, Argonne, IL 60439, USA

F. Izumi, National Institute for Research in Inorganic Materials, 1-1 Namiki, Tsukuba, Ibaraki 305, Japan

R. Jenkins, Clarendon Laboratory, University of Oxford, Parks Road, Oxford OX1 3PU, UK

H. Jones, Clarendon Laboratory, University of Oxford, Parks Road, Oxford OX1 3PU, UK

P. J. McGinn, Center for Materials Science & Engineering, Department of Electrical Engineering, University of Notre Dame, Notre Dame, IN 46556, USA

J. C. Phillips, AT&T Bell Laboratories, 600 Mountain Ave., Murray Hill, NJ 07974, USA

J. M. Phillips, AT&T Bell Laboratories, 600 Mountain Ave., Murray Hill, NJ 07976, USA

S. Sengupta, Materials Science Division, Argonne National Laboratory, Argonne, IL 60439; Department of Electrical Engineering, University of Notre Dame, Notre Dame, IN 46556, USA

D. Shi, Department of Materials Science and Engineering, University of Cincinnati, Cincinnati, OH 45221-0012, USA

D. G. Steel, Materials Science Division, Argonne National Laboratory, Argonne, IL 60439, USA

Z. J. J. Stekly, IGC Advanced Superconductors Inc., 1875 Thomaston Ave., Waterbury, CT 06704, USA

E. Takayama-Muromachi, National Institute for Research in Inorganic Materials, 1-1 Namikik, Tsukuba, Ibaraki 305, Japan

Y. Zhu, Materials Science Division, Brookhaven National Laboratory, Upton, NY 11973, USA

Credits

Support from the following organizations is acknowledged:

Chapter 2: U.S. Department of Energy, Divisions of Basic Energy Sciences-Materials Sciences, Conservation and Renewable Energy-Advanced Utility Concepts-Superconducting Technology Program (JDH), under contract #W-31-109-ENG-38 and the National Science Foundation Office of Science and Technology Centers under contract DMR 91-20000 (DGS).

Chapter 4: U.S. Department of Energy, Basic Energy Sciences-Materials Sciences, under Contract No. W-31-109-ENG-38. and the Midwest Superconductivity Consortium (DOE Contract DE-FG02-90ER45427).

Chapter 5: U.S. Department of Energy, Division of Materials Sciences, Office of Basic Energy Sciences under Contract No. DE-AC02-76CH00016.

Chapter 9: Advanced Research Projects Agency (N000014-90-J-4115) and the Electric Power Research Institute (RP8009-05).

1

High-Temperature Superconductivity in the Layered Cuprates: An Overview

J. C. Phillips

1.1 Crystal Chemistry—Structure and Function: Why are Layered Cuprates so Special?

In the past seven years, more than 35,000 research papers have appeared on the subject of high-temperature superconductivity, with special emphasis on the layered cuprates. The latter have evolved as layered multinary oxides with pseudoperovskite structures from the cubic (Ba, Pb, Bi) oxide perovskite superconductors, which in turn evolved from the cubic Chevrel chalcogenide cluster compounds. Their crystal chemistry is just as different from that of intermetallic compounds (such as Nb_3Sn) as their transition temperatures (about ten times larger), and it is clear that their novel crystal chemistry is the origin of their novel superconductive properties. One of the central themes of this chapter (and indeed of other chapters as well) is this close relationship, which has many ramifications, not only conceptually, but also technologically, for processing these materials to obtain desirable properties.

The simplest and most general way to analyze crystal chemistry is to discuss size and electronegativity differences together with average valences. The latter present no problem for any number of elements, but there are many different definitions of the former, even for binary cases, and until quite recently all seemed to have virtues in some cases and weaknesses in others. It was, moreover, not clear that the binary definitions of differences could be extended to multinary cases. Rabe et al., however, have recently shown that a recipe does exist that organizes the entire crystallographic data base with a success level higher than 95% for simple, common binary compounds. When this recipe is applied to complex novel compounds, diagrams of the sort shown in Figure 1.1 emerge [1].

When we examine Figure 1.1, we see that three kinds of complex, marginally stable materials form small islands on the diagram. The three kinds are stable quasi-crystals, high-T_c (> 500 K) ferroelectrics (HTF), and high-T_c (> 10 K) superconductors (HTS). Because so many ferroelectrics have the $BaTiO_3$ cubic perovskite structure, we are not surprised to see that the HTS oxide island is adjacent to (but does not overlap) the HTF island. On the opposite side of the HTF oxide island is the HTS chalcogenide (Chevrel) island. Thus, while the average electronegative difference $|\Delta X|$ between cations and an-

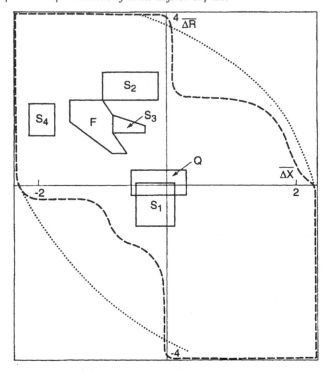

Fig. 1.1. A schematic summary of the domains of marginally unstable lattices with special physical properties, adopted from [1] in terms of average multinary size $\overline{\Delta R}$ and electronegativity $\overline{\Delta X}$ differences. The dashed boundary denotes the region spanned by binary alloys, the dotted boundary that spanned by known ternary and multinary alloys. Here Q denotes stable quasi-crystals, and F denotes stable ferroelectrics with $T_c >$ 500 K. There are four superconductive domains S_i with $T_c > 10$ K. Domain S_1 includes primarily compounds such as Nb_3Sn with the A15 structure. Domain S_2 includes binary compounds, primarily NbC and NbN (rock salt structure). Domain S_3 is the Chevrel compounds, while S_4 is primarily the layered cuprates. Note how F straddles S_3 and S_4.

ions is much larger in the oxides than in the chalcogenides, the average size difference, $\overline{\Delta R}$ is much the same. This average size difference $\overline{\Delta R}$ is about half that found in binary HTS such as NbN. This may be the reason why the anionic (oxide and chalcogenide) HTS can form in spite of the packing problems characteristic of perovskite and related structures.

What do we mean when we say that there are packing problems in the perovskite structure ABO_3? We mean that there is only one lattice parameter (the cubic lattice constant [a]) which can be adjusted to reduce strain energies, but there are two bond lengths, A-O and B-O, which would like to have certain "natural" (or binary) values. (In the crystal chemical literature these "natural" lattice constants $<a_i>$ are sometimes called prototypical.) Similarly, for the layered pseudoperovskites one can speak of prototypical lattice constants $<b_i>$ for each layer. The condition for the formation of layered cuprates is that the prototypical planar square lattice constants $<b_i>$ be nearly equal to [b] for CuO_2 planes for all layers i, even in parent compounds that contain no CuO_2 planes, such as $Bi_4Ti_3O_{12}$ and $Bi_2Mo(W)O_6$. In fact, through knowledge of [b] in the parent compounds, one would be led to discover the Bi- and Tl-based cuprates [1].

The actual values of [a] or [b] would appear to be a compromise between the values $<a_i>$ or $<b_i>$, but this is not the whole story. Interplanar strain energies can be reduced by the presence of a certain concentration of native defects in each plane i, such as anionic vacancies and interstitials, which may be partially aggregated or ordered. The concentration and spatial distribution of such defects can be optimized for superconductive properties by suitable processing, as described elsewhere in this book. In the case of the cuprates, this processing is especially easy because of the high oxygen mobilities, second only to a few materials (such as ZrO_2) used commercially as oxygen getters in high-temperature glass processing ($T \gtrsim 10^{3\circ}C$). The high oxygen mobilities suggest small activation energies for diffusion, and these are related to the high densities of oxygen vacancies and/or interstitials, together with the resonant states associated with these electronic defects. We thus believe that it is no accident that the normal-state and superconductive properties of the layered cuprates are so sensitive to processing. Any theory of these properties should recognize and discuss the role of such defects if it is to be considered more than a logical tautology.

We pause at this/point to discuss the extent to which electrically active defects can be observed in complex crystals. There have been a great many diffraction studies of cuprates, and many defects have been observed, either directly or indirectly, from the static broadening of diffraction patterns and the large R values which imply a high defect concentration. In such complex materials one can expect to find many defects, only a few of which are electrically active and which are associated directly with superconductivity. For this reason, only a few diffraction experiments have succeeded in obtaining direct evidence relating defects to superconductivity. These have involved very carefully prepared samples where phase separation into ordered and disordered regions has taken place [2,3]. As shown in Figure 1.2, channeling experiments have also revealed structural changes at T_c not observable by neutron diffraction [4,5]. In any case, it is important to recognize that although direct structural evidence for superconductive coupling to defects is difficult to obtain, this problem arises because in such complex materials the background "noise" level is high (large R values). There have been some cases in which this problem has been overcome, and these carry much more weight than the many routine structural studies that have failed to identify such coupling as a result of poor sample quality or inadequate resolution.

We see, then, that part of the practical answer to what makes the cuprates so special is prosaic high oxygen mobilities, not mysterious and magical many-electron interactions. In many respects the high oxygen mobilities of the cuprates represent one of nature's felicitous coincidences. (Another felicitous coincidence, crack-free SiO_2 on Si, is the basis of integrated circuits and the microelectronics industry.) There is, however, yet another felicitous coincidence, and this is the nature of the resonant states that pin E_F and enormously enhance T_c. We discuss this microscopic problem later. First, however, we discuss the normal-state transport properties of the cuprates and explain microstructurally why these are so different from normal metals. The same microstructural model enables us to understand many aspects of $\rho(T,H)$ for T near T_c. After discussing the chemistry of resonant pinning centers, we collect all these novel concepts to discuss the fundamental issue, which is why T_c is so high in the cuprates compared with normal metals.

1.2 Normal-State Percolative Transport in Ferroelastic Anionic Metals

Some ten years ago most condensed matter physicists believed that no large qualitative differences in electronic transport existed between simple nearly free electron s-p metallic crystals with one or a few atoms per unit cell and complex multinary metals involving three or more elements and many atoms per unit cell. In all cases it was supposed (Sommerfeld model) that electrical currents were carried by electronic wave packets moving ballistically with Fermi group velocities v_F, mean scattering times, τ, and mean free paths, $l = v_F\tau$. The temperature dependence of the electrical conductivity could

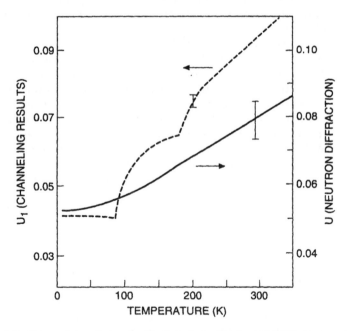

Fig. 1.2. Comparison of atomic vibrational amplitudes in YBCO measured by conventional neutron diffraction Debye-Waller factors and by channeling, which is sensitive to out-of-plane displacements. A distinctive anomaly at $T_c \sim 90$ K is observed by channeling, but not by diffraction. The figure is adapted from [4].

be explained (Bloch-Grüneisen), as in simple s-p metals, primarily by phase space considerations, augmented in the case of transition metals by increased scattering due to Peierls Umklapp (lattice recoil) processes [6]. The only differences to be expected for complex metals would be those associated with larger unit cells and more complex energy band structures.

Before proceeding further, we should mention that there were already hints of anomalous metallic behavior in low-dimensional organic polymers (such as TTF-TCNQ) and layer compounds (such as $NbSe_2$) which were extensively studied [7] in the 1970s. Most of these anomalies were associated with collective electronic phases (charge density waves) which characteristically appear in conjunction with one- or two-dimensional structures. What makes the anomalies in the quasi-crystal and superconductive high-T_c novel metals so striking is that they do not seem to be collective. The conductivities, for example, are linear either in temperature T or in T^{-1}, a property that is strongly suggestive of some phase-space mechanisms that affect single particles. Ordinary phase space mechanisms in d dimensions, however, give rise either to T^{-2} or T^n behavior, with n ~ Θ ~ d, so that linearity in T seems almost impossible to achieve. This is because the usual arguments are based on Fermi liquid theory, which has worked well for elemental and binary metals. We will see, however, that an explanation of electronic transport in novel ternary metals requires a fundamentally different approach.

The Fermi liquid description of the broad trends in the temperature dependence of the electrical resistivities $\rho(T)$ of elemental metals (such as the alkalies and alkaline earths, the noble metals, Al and Pb) is remarkably successful, as shown [8] by de Haas and his coworkers at Leiden in the 1930s. The electronic momentum and energy are dissipated by electron-phonon scattering. The rate at which this occurs is closely related to the lattice thermal conductivity and specific heat, as described by the Bloch-Grüneisen formula, which contains the lattice Debye θ as a characteristic temperature. In this formula $\rho(T)$ is linear for $T \gtrsim \theta$, freezes out for $T \lesssim \theta$ with a linear region $0.2 \lesssim T/\theta \lesssim 0.5$, and finally goes to zero like T^α with $\alpha = 3(5)$ with (without) Umklapp scattering. As the Dutch workers noted, the theory fails only to account for a resistance minimum at $T = T_m$ which shifts to lower temperatures for purer metals and which occurs at $T_m \gtrsim 10$ K for the purest materials. The minimum was ascribed to internal (magnetic) degrees of freedom of the impurities, and a full perturbative theory of this effect was developed by J. Kondo [9] and others some three decades later. The correctness of the wave-packet description is demonstrated by good quantitative agreement in the elemental metals between the values of θ derived by fitting $\rho(T)$ and the lattice specific heat [6].

The temperature dependencies of the resistivities of quasi-crystals and HTS do not resemble at all those of normal metals. Annealing at constant composition has almost no effect on $\rho(T)$ in normal metals, but small changes in processing can drastically affect the functional form of $\rho(T)$ in these ternary materials. In what follows we quote examples of what we believe to be the "ideal" behavior of $\rho(T)$ in novel metals. We will justify our choices both phenomenologically and theoretically, but it is important to recognize that

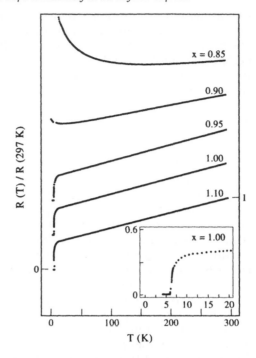

Fig. 1.3. The planar resistivity $\rho_{||}(T)$ in $(Bi_{2-x}Sr_x)_2CuO_{6+d}$ as a function of x, from the semiconductive to the linear regime, from [10], as x is varied.

the range of behavior in novel metals is far greater than in normal metals. This point by itself already tells us, without detailed knowledge of the functional forms of $\rho(T)$ that the usual normal-metal picture of electronic wave packets being scattered by vibrational wave packets and isolated, randomly distributed impurities, is inadequate for novel metals.

The most famous feature of $\rho(T)$ in ternary novel metals is its linear behavior in the HTS cuprates. This behavior was observed initially in the planar resistivity of polycrystalline samples of $(La,Sr)_2CuO_4$ and $YBa_2Cu_3O_{7-x}$, where the conduction takes place primarily in CuO_2 planes. There this behavior is masked at low temperatures by the superconductive phase transitions, but in $(Bi_{2-x}Sr_x)_2CuO_{6+\delta}$ (where T_c is low because the CuO_2 planes are widely separated by semiconductive layers) the planar resistivity $\rho_{||}$ is linear [10] in T from 7 K to 700 K, a result (Figure 1.3) that cannot be explained in the normal metal context by any reasonable vibrational spectrum.

A close correlation exists between the linearity of $\rho_{||}(T)$ and the superconductive transition temperature T_c. At the maximum T_c one has n = 1 in the relation $\rho_{||}(T) = \rho_0 + aT^n$, but as T_c decreases n increases to n = 2, for example [11] in $Tl_2Ba_2CuO_{6+\delta}$ as δ is increased from ~0.0 to ~0.1, as shown in Figure 1.4. Some researchers have suggested that this effect is associated for n = 1 with Fermi surface nesting, but this correlation seems extremely unlikely. Such nesting would give rise to charge density waves which would reduce $N(E_F)$

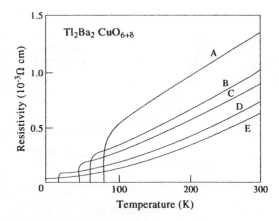

Fig. 1.4.   The resistivity of $Tl_2Ba_2CuO_{6+\delta}$ as δ is varied, from the linear regime to the superlinear regime, from [11]. In both Figures 1.3 and 1.4, $T_c(X) = T_c^{max}$ when $\alpha(X)$ in $\rho_{||}(T)$ proportional to T^{α} is such that $\alpha = 1.0$.

and decrease T_c rather than increase it. Also, the nesting condition is a geometrical accident, but the correlation between high T_c and $n = 1$ is generally observed in all high-T_c superconductors.

When we analyze transport (and even thermal) data on oxides, we must know that these materials as prepared are generally well removed from equilibrium. With HTS it is customary to process the samples until they exhibit narrow superconductive transitions, ΔT_c, with transition temperatures, T_c, close to the maximum reported by other workers. Such samples may still be quite inhomogeneous, however, as measured for instance by Meissner volume filling factors well below unity. (Note that a filling factor of unity means only that there are superconductive filaments on a scale of the penetration depth $\lambda \sim 1000$ A and that, with coherence lengths $\xi_0 \lesssim 100$ A, this is merely a necessary but still far from sufficient condition for true homogeneity.) What this means is that $\rho(T)$ in the normal state can be far more sensitive to departures from equilibrium and homogeneity than T_c and ΔT_c. Strong correlations can be observed, however, between the Meissner filling factors and linearity in $\rho_{ab}(T)$. For example, $(La_{1-x}Sr_x)_2CuO_4$ undergoes an orthorhombic-tetragonal phase transition in the *a-b* plane at a temperature T_t which goes to zero near $x = 0.1$. By annealing polycrystalline samples at constant composition and high temperatures for long times (a month or more), Takagi et al. found [12] that the Meissner volume increased (decreased) in the orthorhombic (tetragonal) phases, in such a way as to suggest that the second-order structural transition was producing a first-order change in electronic properties. This included not only the superconductive phase transition but also the linearity of $\rho_{ab}(T)$, which became T^{α} with $\alpha \sim 2$ in the tetragonal phase while retaining $\alpha = 1$ in the orthorhombic phase. Behavior of this type never occurs in normal metals described by Fermi liquid theory.

A characteristic feature of the layered cuprates is that the resistivity tensor

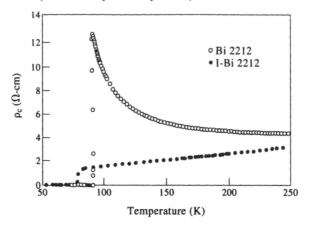

Fig. 1.5. Intercalation of iodine into $Bi_2Sr_2CaCu_2O_{6+\delta}$ makes $\rho_{||}(T)$ also linear in T, as shown in [13]. The dependence of $\rho_\perp$ on film thickness and Bi-Pb alloying is discussed in [14,15].

is very anisotropic, with the c-axis resistivity, ρ_c, being two to three orders of magnitude larger than the planar resistivity ρ_{ab}. Because the layers are alternately metallic and semiconductive, this anisotropy is not surprising, but $\rho_c(T)$ has turned out to be as interesting as $\rho_{ab}(T)$. This was not apparent in the early powder data, which showed a gradual change from "metallic" ($d\rho/dT > 0$) to "semiconductive" ($d\rho/dT < 0$) as T decreased from ~θ to ~θ/3. While $\rho_c(T)$ usually exhibits a sharp (broad) metal-semiconductor transition in sufficiently homogeneous single-crysal (powder) cuprates, much more dramatic changes can be achieved by modifying the semiconductive barriers between the metallic CuO_2 planes. The compound $Bi_2Sr_2CaCu_2O_{6+\delta}$ has two CuO_2 planes per unit cell and is micaceous; that is, it cleaves readily normal to the c-axis. Xiang et al. [13] showed that iodine layers could be inserted epitaxially in the semiconductive barriers and that this had the remarkable consequence [12] of converting $\rho_c(T)$ from semiconductive ($d\rho_c/dT < 0$) to novel metallic ($d\rho_c/dT$ = const. > 0), as shown in Figure 1.5.

More recent work has shown that the intercalability of I is a result of internal stress generated by interlayer misfit [14]. When this internal stress is (largely) removed (by suitable cation alloying [14] or by oxygen overdoping [15] or by growing very thin films [15]), $\rho_c(T)$ becomes linear at almost the same optimal composition as $\rho_{||}(T)$.

All these results clearly differ dramatically from those predicted by conventional wave-packet Sommerfeld-Bloch-Gruneisen-Peierls (SBGP) theory or its closely related modern version, Fermi liquid theory, which at best can reduce α in $\rho(T) \sim T^\alpha$ from α ~ 3 - 5 (SBGP) to α = 2 (Landau's electron-electron scattering mechanism [16], which is usually valid only for excitation energies $\hbar\omega$ well above kθ). Some equal but more subtle surprises, however, are contained in Figure 1.5. The intercalated layers are well removed from the conductive CuO_2 planes, and the magnitude of ρ_c is not much changed by intercalation, since it remains several orders of magnitude larger than ρ_{ab}. Nevertheless, as Figure 1.5 shows, the functional form changes drastically.

At this point we can say that mean field theories fail to describe the data because they omit the internal degrees of freedom connected with impurities or defects. To some extent this explanation resembles the Leiden one that connected the resistivity minimum in elemental metals to magnetic impurities [8]. Here, however, the anomaly refers not merely to a small correction $\Delta\rho(T)$ to $\rho(T)$ at low temperatures, but instead to the entire scattering process of the current carrying particles whose character can no longer be described in terms of perturbed wave packets. This is not so surprising if we remember that at $x = 0$, $(La_{1-x} Sr_x)_2 CuO_4$ is an antiferromagnetic insulator. The Sr not only renders the material metallic, but it may also act as a defect that breaks crystal symmetry by occupying a displaced La site and by rearranging the O atoms in its nearest neighbor coordination shell. Because of crystalline complexity, it is difficult to confirm by diffraction experiments whether such behavior can occur, especially if that behavior is enhanced at defects gettered by domain walls. A number of correlated anomalies were found by Tan et al., however, in extended X-ray absorption fine structure (EXAFS) studies of the Sr environment, which suggest that the Sr impurities do not merely replace La but also destroy the local crystal symmetry of nearby oxygen ions [17].

In fact, given the proximity of high-T_c cuprates to the metal-insulator

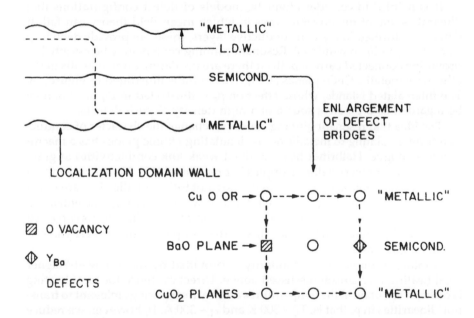

Fig. 1.6. Metallic percolation paths in a layered cuprate pass alternately from CuO_2 planes through resonant pinning centers (which are locally metallic) in semiconductive planes to other CuO_2 or CuO_{1-x} planes, as illustrated here for $YBa_2Cu_3O_{7-x}$ and in [18].

transition, it is easy to explain the linearity of $\rho(T)$ by modifying Landau's argument. In the present case most of the electronic states, even at $E = E_F$, are localized, and only a few are extended, carry current in the normal state, and carry supercurrents for T just below T_c. These extended states scatter mainly against localized states. The final state combinations into which they scatter, and the corresponding values of α, are (extended, extended, 2), as in a conventional metal; (extended, localized, 1), and (localized, localized, 0). The last term is the largest, but as it is temperature-independent, it is regarded as part of the background resistance. The (extended, extended, 2) term is the smallest, because there are so few extended states. Thus $\rho(T)$ is dominated by the (extended, localized, $\alpha = 1$) term [18].

The replacement of Landau's T^2 factor in $\rho(T)$ by the linear T dependence due to the dominance of extended→localized scattering in quantum percolation theory is a new effect that depends explicitly on Fermi statistics. It is analogous to other effects, however, that are similar in many respects. In particular, the quantum-mechanical effect is analogous to the classical accumulation of charge and large voltages near metal-semiconductor interfaces which gives rise to local heating and a quasi-linear dependence of the conductivity on frequency, known as the Maxwell-Wagner effect [18]. (The fact that this classical quasi-linear dependence becomes exactly linear in the presence of Fermi rather than Maxwell-Boltzmann statistics merely reflects the sharpening effects of Fermi statistics.)

It is helpful to consider atomistic models of defect configurations that illustrate some of the possible ways in which mean field theory can fail. If defects are formed in the semiconductive layers, these can provide deep trap states that pin E_F to within $k\theta$. Resonant pinning centers may be essential to providing connected current paths if there are insulating domain walls in the otherwise metallic CuO_2 planes [17,3]. This point is illustrated in Figure 1.6. The intercalated islands, whose effect on ρ_c is illustrated in Figure 1.5, may be regarded as a possible source of a high density of such defects.

The idea that resonant pinning states can change the character of conduction from insulating to metallic across insulating atomic planes has a macroscopic analogue. Halbritter has examined weak-link conductivities at grain boundaries in polycrystalline samples in great detail [19]. He has shown that while the grain boundary conductivities are not fully metallic, they are much larger than one would expect if the interfaces were insulating. The enhanced conductivities are attributed to nanobridges where the conductivity is enhanced by localized pinning states at E_F. This is an example of self-similarity at different length scales.

Another example of self-similarity is provided by the Maxwell-Wagner local heating mechanism discussed above. Direct evidence for local heating is difficult to obtain at the temperatures and barrier spacings relevant to transport linearities in ρ, that is, $T_L \sim 300$ K and $\ell_L \sim 300$ Å. If, however, we reduce T_L to $T_V \sim 0.3$ K $\sim 10^{-4}$ T_L and increase ℓ_L to $\ell_V \sim 0.3$ mm $\sim 10^4$ ℓ_L, we are in a macroscopic regime where local heating is demonstrable. This is, in fact, the regime where so-called vortex tunneling (an elegant mathematical concept) had been supposed to occur in the context of a residual magnetic relaxation rate that did not go to zero, as it should have done had it been thermally activated, but instead saturated below T_V. This saturation is prosaically explained [19] by local heating, just as local heating prosaically explains the linearity of $\rho(T)$ in T.

Some HTS perovskites are known in the alloy family (Ba, K) (Pb, Sb, Bi) O_3 with T_cs as high as 30 K. These materials have many properties similar to those of the cuprates. The most striking shared simple property is that they may undergo an abrupt first-order electronic transition (actually a metal-insulator transition) at a critical composition where a weak second-order lattice distortion (cubic-rhombohedral) begins. Again this transition cannot be explained by a mean-field or Fermi liquid energy band model based on the observed alloy structures (e.g., in $Ba_{1-x}K_xBiO_3$). The transition, however, can be understood if we assume that, in the metallic phase, insulating domain wall fragments coalesce at the critical composition to render the sample insulating. The observed small lattice distortions which grow slowly in insulating phase are caused by the anisotropic internal surface tension of this domain wall network.

The normal-state conductivities in this family exhibit a remarkable series of anomalies when they are measured in superconductive tunneling experiments of Sharifi et al. [20]. The tunneling experiments involve the metallic perovskite, which is separated from a normal metallic electrode by an insulating tunneling barrier. A remarkable feature of these experiments is that they show that their insulating tunneling barriers are provided by a *native* insulating phase at the surface of the perovskite, which is presumably associated with surface composition gradients. This is the only case in which such behavior has been convincingly demonstrated. For a series of alloys the normal state tunneling conductance, dI/dV, is linear in voltage up to 0.2 V, which in itself is unusual. Even more striking are two linear correlations: first, between $\sigma(0)$ and $d\sigma/dV$ in the normal state, and second, between T_c and $d\sigma/dV$, as shown in Figure 1.7. The first correlation, which describes the tunneling barriers, shows that they are characteristic of the perovskite alloy. The second correlation is truly unprecedented, because it correlates the superconductive transition in the metallic perovskite with the transport character of the insulating perovskite barrier.

1.3 Field-Dependent Resistive Transition and Tinkham-Josephson Networks: Microscopic Percolation

If one wishes to understand the microscopic mechanisms responsible for superconductivity it is important to study properties near and just below T_c. Bardeen, Cooper, and Schrieffer stressed this point in their classic paper [21] on weak-coupling superconductivity in elemental metals such as Al, Sn, and Pb. In particular, they emphasized that the way the energy gap develops below T_c and the way it determines the energy-dependent pairing amplitude of quasi-particles can be tested by studying anomalies generated by coherence factors in such properties as nuclear resonance and ultrasonic attenuation. These anomalies are largely absent in high-temperature superconductors, the simplest explanation for their absence being that we are in the strong-coupling limit [21,22].

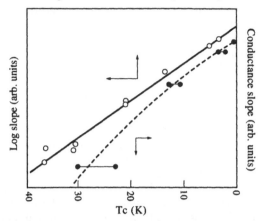

Fig. 1.7. The zero bias conductance and the conductance slope of perovskite superconductor tunneling barriers both correlate linearly with T_c in perovskite superconductors [19], which suggests that T_c itself is determined by microscopic internal barriers and resonant pinning centers in those barriers.

In place of the coherence factors characteristic of the BCS theory, one always finds broadened transitions in which $\rho_s(T,H)$ drops from $\rho_n(T)$ to zero over a temperature range ΔT. This broadened transition should be studied in the middle range $0.1\,\rho_n \lesssim \rho_s \lesssim 0.9\,\rho_n$, as the edges of the range reflect too strongly sample inhomogeneities. In the early days most workers thought that $\Delta T > 0$ reflected only inhomogeneities in T_c characteristic of granular samples, but more recent experiments have shown that $\Delta T > 0$ even in twin-free, nominally single-crystal samples. One can argue that this means that even the latter must be inhomogeneous, but this in turn means that the inhomogeneities are unavoidable and may not be incidental. In fact, there is a general tendency in disordered, inhomogeneous systems towards self-similarity at different length scales. Thus inhomogeneities of granular samples on a scale of μm may parallel those seen on a scale of nm, that is, the microscopic scale of the coherence length which determines the magnitude of T_c.

At present, only one quantitative theory describes $\rho_s(T,H)$ quantitatively in the middle range $0.1\,\rho_n < \rho_s < 0.9\,\rho_n$. This is the Tinkham model of microscopic Josephson networks, which was developed as an extension of the phenomenological model [23] of thermally activated fluxon motion near T_c. The Tinkham-Josephson network model [24] explains ΔT in terms of phase slippage in a network of heavily damped current-driven Josephson microjunctions. With the magnetic field normal to these junctions (which lie primarily in the a-b planes) one has cylindrical symmetry and a characteristic field-dependent Fraunhofer diffraction pattern. The effect of this pattern is to convolute the Boltzman factor associated with thermal activation, exp (-U/

kT), where U = U(T,T$_c$,H), into a Bessel function [25]. With suitable scaling arguments for U(T,T$_c$,H), Tinkham found that these Bessel functions gave an excellent one-parameter fit to ρ_s(T, H) over the entire middle range for granular samples of YBa$_2$Cu$_3$O$_{7-x}$. In this range the fluxons behave like a gas or liquid moving in the potential U; at lower temperatures the liquid freezes into a polycrystalline Abrikosov lattice in twin-free single crystals of YBCO, or in less ideal samples of YBCO or other materials, into a vortex glass.

One of the interesting questions is the effect of material inhomogeneities on thermally activated fluxon losses. By analogy with diffusive losses in ordinary crystals, we expect that the most mobile fluxons will be those pinned to lower-dimensionality extended defects, such as two-dimensional grain boundaries of Abrikosov fluxon crystallites or even one-dimensional partial edge dislocations generated by stacking faults in the materials themselves. Most samples of BSCCO exhibit a high density of stacking faults and partial edge dislocations in the cation sublattices which are observable by electron diffraction [26]. One of the characteristic features of BSCCO is that the width, ΔT(H), is much greater than for YBCO, and this width shows a different dependence on magnetic field in BSCCO than in YBCO.

At this point, something quite surprising happens, and it provides strong support for the Tinkham-Josephson (T-J) microjunction model. One can use the Boltzmann factor exp [-U/kT] to analyze ρ(T, H) instead of its Bessel function convolute because it is more convenient due to the ready availability of semi-logarithmic graph paper. When this is done for BSCCO, one finds [27,28] that

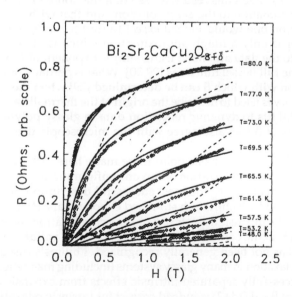

Fig. 1.8. Comparison between the predictions of the Tinkham-Josephson model for ρ(T,H) with H fixed and β = 1 (dashed lines) and β = 1/3 (solid lines), courtesy of H. A. Blackstead (private communication).

the data are best described by a stretched exponential, exp $[-(U/kT)^\beta]$, with $\beta = 0.5$, while the same analysis for YBCO gave $\beta = 1.0$. The value $\beta = 0.5$ can be justified [27,28] by analogy with thermally activated motion of edge dislocations via formation of double kinks; the energy of the latter is proportional to $H^{-1/2}$. Alternatively, one can perform a more careful fitting using the Bessel functions required by the T-J model. In this case one finds [29,30] for YBCO that $\beta = 1.0$ while for BSCCO the best fit is obtained with $\beta = 0.33(1)$. This fit is shown in Figure 1.8.

The difference between the two values of β obtained for BSCCO reflects different physical assumptions regarding the nature of vortex relaxation. In the simple exponential model one focuses on the motion of pairs of vortices independently bound to partial edge dislocations. In the full T-J model with Bessel functions one can explain $\beta = 0.33(1)$ in a different way. The intermediate-time relaxation of nonequilibrium electronic and molecular systems is often described better by a stretched exponential function than by any other function. (This is a very old result; it was first discovered by Kohlrausch in 1847.) Phillips has recently shown [31] in an extensive analysis of stretched exponential relaxation in electronic and molecular systems that in the absence of contact problems or other extrinsic factors, the value of β is given by $d/(d+2)$, where d is the effective dimensionality of the relaxing system. This is a very general result, independent of pair approximations, which in the present case requires only that there be sinks to which the vortices can diffuse. (Such sinks are provided by nonsuperconductive regions of the sample toward which flux is internally driven. This can be described as an internal Meissner effect.) Given this result, we see that the double kink model is only a rough approximation to the exact solution, and that with d = 1 at a partial edge dislocation one would indeed *expect* to find $\beta = 1/3$, not $\beta = 1/2$, and that the latter is only a rough approximation which is obtained when less detailed fits to the data are made using simple exponential functions [27,28] rather than the full Bessel functions [29,30]. What is perhaps most interesting is the precision with which β can be determined [29] when the proper microscopic T-J theory is used to describe the origin of the thermally activated losses. Equally careful fits to dynamic relaxation data in glassy polymers [32] have yielded values of β that also agree with the microscopic theory to within a few percent.

As an aside, there is a basic point to be made here. All measurements of fluxon motion concern relaxation of fluxon configurations. Quite generally in glassy systems near $T = T_g$ the most successful description of relaxation kinetics is provided by Kohlrausch stretched exponentials. Particularly in electrical systems the intrinsic aspects of Kohlrausch relaxation can be masked by extrinsic effects at the contacts, and for this reason the validity of the formula $\beta = d/(d+2)$ has not been generally recognized. Phillips's discussion [31] of Kohlrausch relaxation in many glassy systems (including magnetic spin glasses), however, successfully separated intrinsic effects from extrinsic ones, and he concluded that $\beta = d/(d+2)$ is indeed correct for intrinsic relaxation. The kink pair model, on the other hand, is valuable in its own way, because it provides material suitable for pleasant bedtime reading, although the database used to justify it is negligible compared with that described by Kohlrausch relaxation.

The microscopic T-J model is supported by a growing body of direct evidence for the presence of internal Josephson junctions [33,34]. These occur in BSCCO but not in YBCO. Because BSCCO is micaceous while YBCO is not, these junction effects may be prosaically explained as the result of large-scale weak links such as would be produced by extensive stacking faults. Related anomalies are seen in the Meissner magnetization [35] when H makes a small (6°) angle with the c-axis and H = 10 oersteds. Similar behavior has been observed in polycrystalline samples when flux starts to penetrate along grain boundaries. It is argued [35] that this mechanism cannot explain the same effect in single crystals, but this argument is too credulent. There can easily be a low density of extensive stacking faults in the "single crystals" that is not observed by diffraction. For small angles away from the c-axis and small H, the spacing between the stacking faults can be quite large, and this would not be detected by diffraction. For much larger fields, however, where $\rho(T,H)$ has been measured and fitted by the Tinkham-Josephson model, the spacing of the weak links must be much smaller. What we have again is an example of self-similarity on two quite different length scales which differ by factors of order 10^4.

1.4 Resonant Pinning Centers and Interlayer Coupling: Microscopic Formulae for T_c and E_g

The earliest data on layered cuprates raised serious problems concerning the connection between interlayer coupling and high-temperature superconductivity. On the one hand, T_c was found to increase as the number of CuO_2 layers per unit cell increased. On the other hand, measurements of the c-axis resistivity in the normal state, $\rho_\perp(T)$, consistently showed semiconductive behavior with $d\rho_\perp(T)/dT < 0$ at low T. At the same time, attempts to calculate T_c from the energy bands of the ideal crystal, using methods that had been quite successful for elemental metals (Al, Sn, Pb) and intermetallic compounds (Nb_3Sn) generally could not be stretched to yield values higher than about 30 K. Finally, as we noted earlier, homogeneous models that purported to yield $\rho_{||}(T)$ proportional to T produced at best an effective $N(E_F) = N_b(E_F)Z(E_F)$, which was constant, not peaked, as one would expect from $T_c \sim 100$ K.

Any theory that faces so many difficulties must be based on one or more false assumptions, and there may also be serious gaps in the experimental data. Very early, Phillips pointed out [36] that in view of the complexity of the crystal chemistry of the cuprates as well as their alternation of metallic and semiconductive planes, which in effect form an infinite series of Schottky barriers, it was extremely unlikely that electronic structure would be correctly described by energy bands based on the ideal crystal structure alone. On the contrary, just as with Schottky barriers, one could expect to find a high density of defects in the semiconductive layers. Some of the electronic states associated with these defects could reasonably pin the Fermi energy, so that their contribution $N_r(E_F)$ to the total $N(E_F)$ would be several times larger than that of the band states $N_b(E_F)$. For $(La,Sr)_2CuO_{4+\delta}$ these resonant pinning states could be the dopant cations $Sr_{[La]}$, but more generally they could also be oxygen vacancies or interstitials as well.

Such resonant pinning centers have many attractive features. They will almost surely be distributed inhomogeneously, and they will tend to concentrate near atomic or electronic domain walls. They thus generate percolation paths for coherent extended states as well as a high concentration of incoherent localized states. As we mentioned, such a combination of localized and extended states easily explains [3] why $\rho_{||}(T)$ is proportional to T. Because $N_r(E_F) >> N_b(E_F)$, we can understand why T_c is so large. Moreover, the occasional reports [37] of superconductive effects (such as abrupt drops in $\rho(T)$ or as Meissner effects with $f_M \lesssim 10^{-2}$) at temperatures well above T_c are also easily explained by gettering of RPC by extended defects. Only two problems remain. We need to understand why $\rho_\perp(T)$ is so different from $\rho_{||}(T)$, and we need some way of characteristizing the interaction between Cooper pairs and resonant pinning centers (RPCs) which will enable us to estimate T_c and E_g, the two basic quantities predicted by the BCS theory for lower-temperature elemental and intermetallic superconductors.

If the electronic paths are indeed percolative and pass coherently from one layer to the next via resonant pinning centers, then $\rho_\perp(T)$ and $\rho_{||}(T)$ should exhibit the same temperature dependence. If the concentration of resonant pinning centers is low, and electron scattering by such centers is strong, the magnitude of $\rho_\perp(T)$ may be much larger than that of $\rho_{||}(T)$, but their ratio should be nearly constant. In fact, several recent experiments have shown that when the stress resulting from misfit of interlayer prototypical lattice constants (see the introduction to this chapter) is reduced, either by alloying one of the cations other than Cu, or by intercalation of layers (such as I) which relieve the misfit by island formation of the intercalated layer, or merely by growing an ultrathin film, the semiconductive character of $\rho_\perp(T)$ is reduced or even disappears, and $\rho_\perp(T)/\rho_{||}(T)$ does become virtually constant [13–15]. The semiconductive behavior observed in the earlier experiments was due to insulating planes such as one might expect from stacking faults, and these same insulating planes may, in extreme cases (such as micaceous BSCCO), create the internal Josephson junctions discussed above.

Before we proceed, we consider a generic problem with resonant pinning centers. In complex molecules a large value of $N(E_F)$, that is, a high degree of electronic degeneracy at $E = E_F$ is often excluded because such a situation generates Jahn-Teller distortions, which reduce the energy of the molecule by opening an electronic energy gap as the degenerate states form bonding and antibonding pairs. In fact, before the discovery of high-temperature superconductivity in the cuprates, it was generally supposed that such Jahn-Teller distortions would always occur to a considerable extent in transition metal oxides. This supposition was consistent with the poor electrical conductivity of these materials, some of which (such as NiO) were predicted by band theory to be good metals in their ideal crystal structures.

What was needed by theory at this point were some simple examples which could explain why the Jahn-Teller effect did not reduce $N_r(E_F)$ in the cuprates. Of course, molecular chemistry contains many such examples, especially in the planar aromatic hydrocarbons, where π bonding stabilizes the resonant π states against σ Jahn-Teller distortions. For this reason the layered cuprate structure based on stable CuO_2 planes is extremely suggestive. Recently, simple examples have been obtained that meet these needs. We mention the anti–

Jahn-Teller effect in layered Bi_2Te_3:Bi (which contains Te vacancies analogous to oxygen vacancies in the cuprates and which shows an increase in carrier density of about a factor of two between room temperature and low temperatures) [38]. This is exactly the opposite of what one would ordinarily expect in the context of the Jahn-Teller effect, and thus this behavior was termed the anti–Jahn-Teller effect [38]. Phillips explained it in the context of screening of internal electric fields by RPC. With decreasing T the resonant width narrows, and the density of current carriers increases, in turn reducing the enthalpy further. Note that this mechanism is effective only in predominantly ionic metals where the density of free carriers is not too high, as the latter in normal metals already screen the internal fields.

One would expect that the anti–Jahn-Teller effect would increase in strength as we pass from X = Te to X = S and finally to X = O. The recently discovered layered ternary sulfide $BaCo_{1-x}Ni_xS_{2-y}$ illustrates this trend [39]. This material exhibits two phases, one semiconductive and antiferromagnetic, and one metallic. Normally one would expect the metallic phase to be the high-temperature phase where the antiferromagnetic spin density wave energy gap has collapsed. Here, however, the metallic phase is the low-temperature phase, and the resistivity of this phase decreases as the concentration y of S vacancies increases. This shows that a resonant pinning state is associated with the S vacancies and that this state [38] is primarily responsible for the metallic properties, which disappear as y → 0.

Considering the behavior of the Te and S examples [37,39], we can now see that in layered oxides it is possible that RPC can be stable at temperatures up to some formation temperature of order 500°C, where anion disorder sets in. This seems to be the case for the layered cuprate HTS.

How would we go about identifying resonant pinning centers? As we have indicated, sophisticated techniques like EXAFS have yielded suggestive results [17], but even these sophisticated methods do not provide direct information on the concentration of such centers and their contribution to the

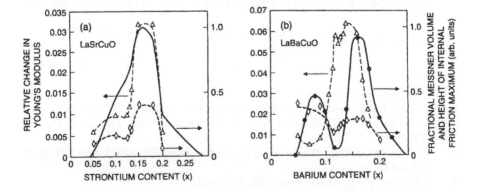

Fig. 1.9. Comparison between the fractional Meissner volume $f_M(x)$ and Young's modulus changes and internal friction peak height [40,45].

electron-boson coupling which is needed to produce HTS. The technique that has proved most successful may well seem surprising: it is the very unsophisticated technique of mechanical energy dissipation at frequencies of Au 10^{-3} s^{-1} or less. The loss peaks, or internal friction peaks as they are called, were explained by Debye with a simple oscillator model that shows that the peaks occur when $\omega\tau = 1$. Here τ is the RPC relaxation time, which is thermally activated, $\tau = \tau_0 \exp(-E_A/kT)$, where E_A is an activation energy that is usually similar to an optical phonon energy. In fact such losses have been studied in perovskite ferroelectrics where the typical value is $E_A \sim 0.06$ eV, quite comparable to an optical phonon energy.

The strengths, $S_I(x)$, show a dependence on composition in $La_{2-x}Sr_xCuO_4$, which is remarkably similar to that of the Meissner filling factor $f_M(x)$, as shown in Figure 1.9 (for a detailed discussion of the data, see Refs. 40–45). The important point is that the data suggest that in the cuprates T_c^{max}, the maximum value of T_c obtained in a series of alloys, is proportional to E_A, although E_A itself is nearly independent of composition. This is what one would expect from quantum percolation theory, but it is quite different from what is predicted by any homogeneous model known to us.

From study of this data, it has been suggested that T_c^{max} is given by

$$T_c^{max} = 2kt\theta_D E_A/W,$$

where $\theta \sim 500$ cm^{-1} is an average CuO_2 phonon energy, E_A is the interlayer resonant pinning center activation energy, and $W \sim 2$ eV is the CuO_2 planar valence band width (or average density of states) near E_F. This equation is of great interest because it predicts that $E_A \sim 0.25 - 0.3$ eV in a material, such as the new Hg compounds, with $T_c^{max} \sim 165$ K. Such an energy should be easy to measure, since it lies well outside the range of one-phonon energies.

1.5 References

1.   K. M. Rabe, J. C. Phillips, P. Villars, and I. D. Brown, Phys. Rev. B **45**, 7650 (1992).
2.   M. Marezio, H. Takagi, and B. Batlogg, Physica C **185**, 873 (1991).
3.   J. C. Phillips, Phys. Rev. B **46**, 8542 (1992).
4.   R. P. Sharma, F. J. Rotella, J. D. Jorgensen, and L. E. Rehn, Physica C **174**, 409 (1991).
5.   J. C. Phillips, Phys. Rev. B **42**, 6795 (1990).
6.   A. J. Dekker, *Solid State Physics*, Macmillan, London (1960), Chap. 11.
7.   A. N. Bloch, R. B. Weisman, and C. M. Varma, Phys. Rev. Lett. **28**, 753 (1972).
8.   W. J. de Haas, J. de Boer and G. J. vanden Berg, Physica 1, **609** (1934); 4, 683 (1937).
9.   J. Kondo, Prog. Theor. Phys. **32**, 37 (1964).
10.   G. Xiao, M. Z. Cieplak, and C. L. Chien, Phys. Rev. B **38**, 11824 (1988).
11.   Y. Kubo, Y. Shimakawa, T. Manako, and H. Igarashi, Phys. Rev. B **43**, 7875 (1991).

12. H. Takagi, R. J. Cava, M. Marezio, B. Batlogg, J. J. Krajewski, W. F. Peck, Jr., P. Bordet, and D. E. Cox, Phys. Rev. Lett. **68**, 3777 (1992).
13. X.-D. Xiang et al., Phys. Rev. Lett. **68**, 530 (1992).
14. L. Forro, Phys. Lett. A **179**, 140 (1993).
15. F. X. Régi, J. Schneck, H. Savary, R. Mellet, and C. Daguet, Supercond. **1**, 627 (1993).
16. N. W. Ashcroft and N. D. Mermin, *Solid State Physics*, Saunders, Philadelphia (1976), 346.
17. Z. Tan, M. E. Filipkowski, J. I. Budnick, E. K. Heller, D. L. Brewe, B. L. Chamberland, C. E. Bouldin, J. C. Woicik, and D. Shi, Phys. Rev. Lett. **64**, 2715 (1990).
18. J. C. Dyre, Phys. Rev. B **48**, 12511 (1993).
19. H. Halbritter, Phys. Rev. B **48**, 9735 (1993); A. Gerber and J. J. M. Franse, Phys. Rev. Lett. **71**, 1895 (1993).
20. F. Sharifi, A. Pangellis, and R. C. Dynes, Phys. Rev. Lett. **67**, 509 (1991).
21. J. Bardeen, L. N. Cooper and J. R. Schrieffer, Phys. Rev. **108**, 1175 (1957); P. B. Allen, Nature **349**, 396 (1991).
22. A. A. Golubov, M. R. Trunin, and S. V. Shulga, Physica C **213**, 139 (1993); M. L. Horbach and W. Vansaarlos, Phys. Rev. B **46**, 432 (1992); N. Miyakawa, Y. Shiina, T. Kaneko, and N. Tsuda, J. Phys. Soc. Jap. **62**, 2445 (1993).
23. Y. Yeshurun and A. P. Malozemoff, Phys. Rev. Letters **60**, 2202 (1988).
24. M. Tinkham, Phys. Rev. Letters **61**, 1658 (1988); M. Tinkham and C. J. Lobb, Solid State Physics **42**, 91 (1989); K. H. Lee and D. Stroud, Phys. Rev. B **46**, 5699 (1992).
25. V. Ambegaokar and B. I. Halperin, Phys. Rev. Lett **22**, 1364 (1969).
26. O. Eibl, Physica C **168**, 249 (1990). See also Z. L. Wang, A. Goyal, and D. M. Kroeger, Phys. Rev. B **47**, 5373 (1993).
27. J. T. Kucera, T. P. Orlando, G. Virshup, and J. N. Eckstein, Phys. Rev. B **46**, 11004 (1992).
28. H. Yamasaki, K. Endo, S. Kosaka, M. Umeda, S. Yoshida, and K. Kajimura, Phys. Rev. Lett. **70**, 3331 (1993).
29. H. A. Blackstead, D. B. Pulling, D. G. Kiefer, M. Sankararaman, and H. Sato, Phys. Lett. A **170**, 130 (1992); H. A. Blackstead, Supercond. Sci. Tech. **6**, 579 (1993).
30. T. W. Krause, A. Shi and W. R. Datar, Physica C **205**, 99 (1993).
31. J. C. Phillips, J. Non-Cryst. Sol., in press (1994).
32. D. J. Plazek and K. L. Ngai, Macromol. **24**, 1222 (1991).
33. R. Kleiner, F. Steinmeyer, G. Kunkel, and P. Müller, Phys. Rev. Lett. **68**, 2394 (1992).
34. K. E. Gray and D. H. Kim, Phys. Rev. Lett. **70**, 1693 (1993).
35. N. Nakamara, G. D. Gu, and N. Koshizuka, Phys. Rev. Lett. **71**, 915 (1993).
36. J. C. Phillips, Phys. Rev. Lett. **59**, 1856 (1987); *Physics of High-T_c Superconductors*, Academic Press, Boston (1989).
37. V. N. Moorthy, S. K. Agarwal, B. V. Kumaraswamy, P. K. Dutta, V. P. S. Awana, P. Maruthikumar, and A. V. Narlikar, J. Phys. Cond. Mat. **2**, 8543 (1990).
38. G. A. Thomas, D. H. Rapkin, R. B. Van Dover, L. F. Mattheiss, W. A. Sunder, L. F. Schneemeyer, and J. V. Waszczak, Phys. Rev. B **46**, 1553 (1992); J. C.

Phillips, Phys. Rev. B **47**, 11615 (1993).

39. L. S. Martinson, J. W. Schweitzer, and N. C. Baenziger, Phys. Rev. Lett. **71**, 125 (1993).

40. M. Gazda, B. Kusz, R. Barczynski, O. Gzowski, L. Murawski, I. Davoli, and S. Stizza, Physica C **207**, 300 (1993).; P. Roth and E. Hegenbarth, Ferroelectrics **79**, 323 (1988); Ferroel. Lett. **10**, 33 (1989).

41. Y. Mi, R. Schaller, H. Berger, W. Benoit, and S. Sathish, Physica C **172**, 407 (1991).

42. G. A. Thomas, D. H. Rapkine, S. L. Cooper, S.-W. Cheong, A. S. Cooper, L. F. Schneemeyer, and J. V. Waszczak, Phys. Rev. B **45**, 2474 (1992). 43. M. Gazda, B. Kusz, and R. J. Barczynski, Sol. State Comm. **83**, 793 (1992).

44. O. M. Nes, K. Fossheim, N. Motohira, and K. Kitazawa, Physica C **185**, 1391 (1991).

45. J. C. Phillips, Physica C, in press (1994).

1.6 Recommended Reading for Chapter 1

1. *Physics of High-T$_c$ Superconductors*, J. C. Phillips (Academic Press, Boston, 1989).

2

Properties of High-Temperature Superconductors: New Considerations in Measurements

J. D. Hettinger and D. G. Steel

This chapter is written for technical nonspecialists who wish to make measurements of properties of high-temperature superconductors, for the purpose of improving material quality or for a particular application. The chapter is somewhat tutorial in style to give the reader a means to understand the technical aspects of making a measurement and decide on the appropriate technique for the material property of interest. The benefits of certain measurements over others are pointed out. We also make special attempts to indicate the importance of material quality in the interpretation of results.

The properties of high-temperature superconductors have caused several changes to the techniques of superconducting measurements. The high transition temperatures and large anisotropies in these materials have provided many new insights into the dynamics of magnetic vortices. Dissipation at grain boundaries or other weak links in materials in conventional superconductors is correctly thought of as extrinsic. Conventional superconductors typically have larger volumes of superconducting coherence; and, as a result, small imperfections in the material microstructure are nondetrimental. In high-temperature superconductors, this is not the case. Any defect in a unit cell can modify the structure over at least the range of coherence, causing a region of reduced superconductivity. For this reason, grain boundaries in these new superconductors are of great importance since very few applications exist for the materials in single crystalline form.

The structure of this chapter is as follows. The first section introduces fundamental superconducting parameters and nomenclature that will be used throughout the chapter. The second section deals with cryogenic systems, including particular cryogens, the use of different thermometers, and the designs of low-temperature cryostats. The following section details how certain measurements are made. Here we deal with electrical transport, magnetic susceptibility, and magnetization. All of these measurements are related in that they measure dissipation. The level of dissipation in each measurement is discussed, allowing the results of each measurement to be connected with the others through a current-voltage curve. In addition, a few comments will be made regarding sample quality. Finally, we discuss the results of measurements of these fundamental properties. Transport properties consider both inter- and intragranular behavior. Susceptibility measurements focus on the connection to electrical resistivity through a frequency-dependent skin depth. Magnetization results are discussed with a view toward the determination of critical current densities as well as upper critical magnetic fields.

2.1 Background

Since this chapter describes extensive measurements on high-temperature superconductors, we first review some of the fundamental properties of superconductivity. This section outlines relevant parameters that describe superconductivity, such as fundamental length and energy scales, and summarizes properties of low transition temperature materials. It is not the purpose of this chapter to describe the properties and parameters in great detail; for additional information the reader is referred to specialized texts on superconductivity [1–3].

2.1.1 Parameters Describing Superconductivity

When a superconducting material is cooled from high temperature, it initially behaves as a normal metal with finite resistance until at a certain temperature, known as T_c, the resistance falls to zero. This defines the temperature of a thermodynamic transition to the superconducting state, in which electrons are paired. The details of the quantum interactions within the superconducting state may be simplified by a macroscopic quantum wavefunction, ψ, where $n_s = \psi\psi* = |\psi|^2$ gives the density of superconducting electron pairs. This quantity ψ is also known as the complex superconducting order parameter and is the basis of the phenomenological Ginzburg-Landau (G-L) formalism for superconductors. The full treatment of the G-L theory appears in many general texts on superconductivity, for example, the book by Tinkham [1].

Two characteristic lengths arise from this theory: the magnetic penetration length, λ, over which spatial variations in the local magnetic field occur, and the superconducting coherence length, ξ, the length scale over which spatial variations of the order parameter take place, that is, the spatial extent of the superconducting electron pair (Cooper pair). The relative sizes of λ and ξ distinguish two different classes of superconductor. Materials with $\lambda < \xi/\sqrt{2}$ are known as type I superconductors and lose their superconductivity completely when a magnetic field penetrates. The case of $\lambda > \xi/\sqrt{2}$ is known as type II superconductivity, and it is energetically favorable to form normal regions within the superconductor above a certain field, known as the lower critical field, H_{c1}. This is called the *mixed state* or vortex state. An applied magnetic field will be screened by the superconducting regions of the sample, but will pass through filamentary normal regions (vortices), where the order parameter will be reduced to zero. The size scale of the coherence length and the penetration length are shown in Figure 2.1. It is required that the wavefunction introduced above be single-valued at all points within the superconductor, which leads to quantization of the magnetic flux through each vortex. The value of the flux quantum may be shown to be $\phi_0 = 2.07 \times$

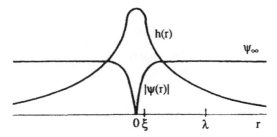

Fig. 2.1. Radial variations of the order parameter and the local magnetic field at a vortex, showing the characteristic length scales of ξ and λ described in the text. Notice that the order parameter is zero at the vortex core center. See [1].

10^{-15} Webers [1]. The radius of the normal core of the vortex line is the coherence length, ξ, since that sets the length scale for variations in the order parameter. To screen the magnetic field within the flux line from the more strongly superconducting region outside and self-consistently generate the flux inside, shielding currents flow around the vortex, with a characteristic length scale of λ. The currents interact with those from other vortices and give rise to a mutual repulsive interaction between vortices.

When one minimizes the free energy as a function of lattice configuration, the vortices in the mixed state are arranged in a hexagonal pattern known as the *Abrikosov lattice* [4,5]. The expected hexagonal pattern has been observed in many type II superconductors, and high-temperature superconductors are no exception [6]. The lattice spacing, a_v, is determined by the applied magnetic field and is given by

$$a_v \sim \sqrt{\frac{\phi_0}{B}} \, . \tag{2.1}$$

This shows that an ensemble of vortices provides an interesting system where the density of lattice points may be tuned by varying the magnetic field. The phase diagram for a conventional (low-T_c) type II superconductor is shown in Figure 2.2. As stated earlier, at the lower critical field it becomes energetically more favorable to admit vortices than to maintain the Meissner state; there is a second-order phase transition to the mixed state at this field. As the field is increased further, the order parameter is progressively reduced; and at another value of the field, the order parameter becomes zero with a second-order phase transition to the normal state. This is the upper critical field, $H_{c2}(T)$. It is given by the equation

$$H_{c2} = \frac{\phi_0}{2\pi\xi^2}, \tag{2.2}$$

which may be qualitatively understood from the destruction of the superconducting state because of the overlap of vortex cores.

Having so far considered the static vortex structures, we now turn to examine

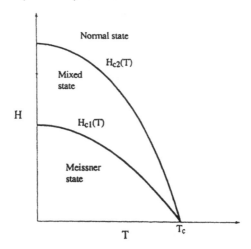

Fig. 2.2. Phase diagram for conventional type II superconductor.

vortex motion. Consider a transport current of density J flowing through a superconductor in the mixed state. Since magnetic flux passes though each vortex core, there will be a Lorentz force acting on each vortex given by $(J \times \phi_0 z)l_\phi$. where l_ϕ is the vortex length. The effect of this force is that flux lines will tend to move transverse to the current, and in moving they will induce an electric field parallel to the current which acts as a resistive voltage [7]. (This dissipation may be seen either in terms of electrodynamics or from the voltage produced by the changing phase due to the moving flux line.) This is a very important result. Although the Cooper pairs exist, the resistance is no longer zero, due to the motion of the vortices.

Within real superconducting materials, chemical and physical defects may reduce the local superconducting condensation energy, and exert forces on the vortices, keeping them pinned at the defect locations. As long as the vortices do not move, zero resistance will be maintained. Once the Lorentz forces exceed the pinning forces, however, the vortices will move, and there will be dissipation. Vortex dynamics separate into at least two different regimes. If the vortices move freely and pinning is not important, the resulting dissipation is known as *flux flow*. Thermal activation can depin the vortices from their wells, leading to another regime of behavior, namely, *flux creep*. The details of how these apply in the case of high-T_c superconductors will be discussed in the next section of this chapter.

2.1.2 Low-T_c Superconductors

Researchers are often asked what is new about the high-temperature superconductors, other than their elevated transition temperatures. In some ways these superconductors display fundamental properties very similar to conventional superconductors, but in other ways they are quite different. In low-T_c materials, pinning is usually large and typical thermal energies are compara-

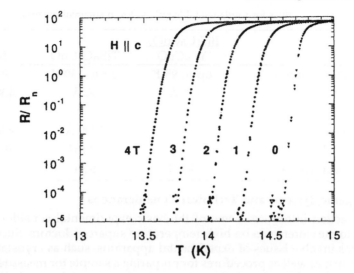

Fig. 2.3. Resistance vs temperature for a thin film of niobium nitride in different magnetic fields applied perpendicular to the film. Notice the continuous suppression of the temperature of the transition as the magentic field is increased, while the slope of the transition is changed only slightly. This is in contrast to the behavior of the high-temperature superconductors (discussed later in this chapter).

tively small so that the effects of thermal activation across pinning barriers are small. High-T_c materials, by contrast, have lower pinning energies due to small normal cores and activation can therefore be highly significant. Figure 2.3 shows that the effect of a magnetic field on the resistive transitions in a low-T_c material is to depress the transition temperature as a consequence of the upper critical field. In a high-transition temperature superconductor, the transition is broadened by magnetic field rather than being depressed, and this is a consequence of the dissipation arising from vortex motion. This difference has important consequences even for something as simple as the transition temperature. Although T_c is easy to define for the low-T_c systems as the zero-resistance temperature, this is not applicable for the smeared transitions of high-T_c superconductors. Many of the differences between the old and new superconductors may be viewed from the coherence of the superconducting wavefunction introduced earlier in this section. That length is much shorter in the case of the high-T_c materials, leading to the reduction of the pinning energy and the importance of grain boundaries in these materials, as will be discussed later. The high-T_c materials are also highly anisotropic, consisting of many copper-oxide planes. This property leads to an even shorter range of coherence perpendicular to the layers. Table 2.1 gives approximate values for these fundamental length scales in high-temperature and low-temperature superconductors [1,8]. (The notations a, b, and c refer to the directions with respect to the crystallographic axes: a and b are in the copper-oxide planes, while c is perpendicular to them.)

Table 2.1. Values for fundamental length scales in superconductors.

	BiSrCaCuO/ YBaCuO	TlBaCaCuO	Nb
Transition temperature, T_c	up to 95 K	up to 130 K	9 K
Coherence length, ξ_{ab}	15 Å	30 Å	400 Å
Coherence length, ξ_c	2 Å	1 Å	—
Penetration length, λ_{ab}	1500Å	2000 Å	400 Å
Lower critical field, H_{c1}	10 mT	10 mT	0.1 T
Upper critical field, H_{c2}	300 T	60 T	0.3 T

2.2 Cryogenic Systems and Technical Considerations

This section introduces some of the basic experimental considerations relevant to measurements on high-temperature superconductors. Such considerations involve issues of experimental apparatus, such as cryostats and thermometry, as well as procedures for preparing a sample for measurement, for example, attaching electrical contacts and using photolithography (which will be discussed in the next section). Within the limited space available, we can provide only a brief introduction to each subject, but more extensive reviews and discussions are available. For a general reference on low-temperature experimental techniques, the reader is referred to the texts by White and Richardson [9,10].

2.2.1 Cryogens

The title "high-temperature superconductors" refers to a class of materials with high transition temperatures relative to the superconducting materials known before 1986 [11]. Despite the title, no material has yet been confirmed to be superconducting above about 130 K. The implication is that to achieve superconducting behavior, one must cool the sample significantly below room temperature. Such cooling may be achieved by using a mechanical cooling system such as a refrigerator or by direct cooling with cold liquids, known as *cryogens*.

Simple closed-cycle refrigerators can produce temperatures down to about 8 K and provide an easy way of performing low-temperature measurements. The disadvantage of most refrigerators is that the geometry of their construction, which is necessary for the cooling operation, is not well suited to many of the different types of measurements performed. For this reason, most experiments on high-temperature superconductors are made using cryogenic liquids.

The most common cryogens used are liquid nitrogen (which has a boiling point of 77 K) and liquid helium (which boils at 4.2 K). This section will briefly describe the use, storage, and safeguards of these cryogens. Since the transition temperatures of the high-T_c materials are often above 77 K, liquid nitrogen may provide sufficiently low temperatures for many experiments. However, for measurements on samples with lower transition temperatures or at high magnetic fields (when the resistive transition can be extended to significantly lower temperatures), it will be necessary to use liquid helium for cooling samples.

Liquid nitrogen is produced by liquefying air, which makes it relatively inexpensive. Cost is usually what determines the storage conditions for cryogenic liquids, so for the case of nitrogen, an acceptable boil-off rate of liquid from a dewar is obtained for a container with a single vacuum jacket or a single vacuum jacket with limited superinsulation (aluminized mylar). The liquid is usually removed from such a storage container by using a tube extended to the bottom of the dewar, which is maintained at an overpressure of a few pounds per square inch. The production and storage of liquid helium are quite different. Since there is little helium in the atmosphere, it is obtained from naturally occurring underground reservoirs of gas. Compared with nitrogen, then, the production of liquid helium is significantly more costly, a fact that partly motivates the additional measures taken for storage. Older helium storage dewars comprise concentric layers of metal vacuum walls and an outer jacket filled with liquid nitrogen to reduce the helium boil-off rate. More modern storage dewars require no nitrogen reservoir and instead have layers of superinsulation in vacuum to prevent heat from reaching the stored cryogen. Liquid helium is removed from storage dewars in a similar way to nitrogen, with an overpressure to provide the pressure differential that drives the liquid through the tube. The tube used is a special "transfer line", however, which consists of two concentric tubes separated by a vacuum space to

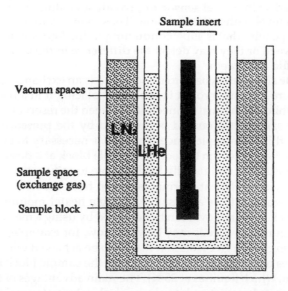

Fig. 2.4. Schematic of an exchange gas cryostat used for variable temperature studies of high-temperature superconductors. The sample insert is immersed in a reservoir of liquid helium.

reduce the influx of heat from thermal conduction and convection.

Several different safety hazards are associated with the use of cryogenic fluids. The most obvious, of course, comes from the extremely cold temperatures of the cryogenic liquids, for which certain standard procedures are recommended to ensure safety. Gloves should be worn but should not be too tight on the hands that they can trap cold liquid on the surface of the skin. The eyes and face should be protected with safety glasses and a face mask. Safety shoes and loosely fitting clothes will also prevent dangerous exposure of the skin to the cold liquid, should there be a spill. Another major danger is asphyxiation. Since the gas/liquid volume expansion ratio is so large (about 700 in the case of nitrogen and about 760 for helium), a small amount of liquid can displace a large volume of air and thus reduce the oxygen concentration in the laboratory. To protect against this, one should use cryogens in well-ventilated workspaces. In addition, the large expansion ratio can lead to high pressures being developed, and thus adequate pressure relief devices must be employed on all cryogenic systems. For complete details on the safe handling of cryogenic fluids, the reader should refer to specialist safety publications, such as those published by the Occupational Safety and Health Administration (OSHA).

2.2.2 Cryostats

This section discusses two of the most important types of liquid helium cryostats available for measurements on high-temperature superconducting materials. Certain aspects are common to different types of cryostat, such as being equipped with a level sensor to provide a continuous readout of the liquid helium level within the reservoir. These sensors may be based on resistance, using a wire that is superconducting in the liquid helium and resistive state above it, or they may detect the difference in the dielectric constant of helium liquid and vapor.

Perhaps the simplest cryostat to understand is an exchange gas system, as shown in Figure 2.4. The cryostat is cooled by placing it in a reservoir of liquid helium, while the thermal connection between the reservoir and the block on which the sample is mounted is provided by the presence of some exchange gas within the sample space. This gas is necessary to provide a thermal link to the liquid helium to cool the sample block at a desired rate. Once at low temperatures, it will usually be desirable to reduce the amount of gas in order to provide sufficient thermal isolation from the helium and allow a range of temperatures to be accessed by using a heater mounted directly on the sample block. Such removal can be achieved by using a mechanical pump, although the pumping speeds may be very slow, for example, when helium is used as the exchange gas. Isolation may also be achieved with an exchange gas that freezes out at low temperatures once the sample block is cooled (for example, neon, which liquifies at 25 K). The main advantages of the exchange gas cryostat are its low cost and its mechanical stability for making measurements.

A second type of cryostat is the helium gas flow system, in which the sample is heated or cooled by flowing helium vapor [12]. This type of cryostat is

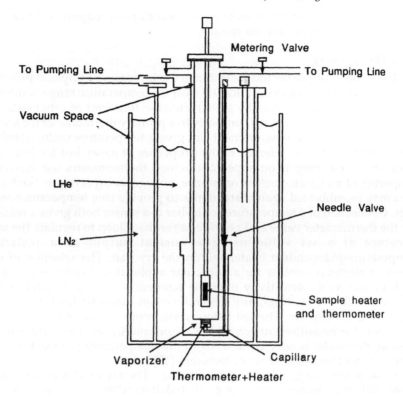

Fig. 2.5. Schematic of a gas-flow cryostat. This type of cryostat offers the flexibility of varying the temperature from 1.4 K to 300 K.

shown in Figure 2.5.

For measurements between about 4 K and room temperature, liquid helium from a reservoir flows through a needle valve and a capillary tube to a resistive heater (vaporizer). It then flows vertically up through the sample space and out to atmosphere through a flow valve. The temperature of the vapor, and thus of the sample, is controlled by the heater. A thermometer at the heater allows the vapor temperature to be monitored, and by connecting to a controller, the temperature of the flowing vapor may be regulated. In this mode of operation, the sample is at a pressure close to one atmosphere, with the flow through the capillary driven by the excess pressure in the helium reservoir and controlled by the needle valve.

As an alternative, the sample space may be connected to a roughing pump to draw the liquid through the capillary tube, in which case the liquid helium reservoir pressure is close to one atmosphere, rather than overpressurized. To operate at lower temperatures down to 1.4 K, liquid helium is allowed through the capillary tube into the sample space, where it is pumped to produce evaporative cooling. This type of cryostat has the advantages of accessing a very wide temperature range and giving good thermal coupling between the flowing helium vapor and the sample.

2.2.3 *Thermometry*

The basic principle behind thermometry is to measure a physical property of a material that changes significantly over the temperature range of interest but is independent (or may be a characteristic function) of other external parameters. By calibrating that property to a primary thermometer (or knowing a calibration), the measurement can give a temperature scale. Absolute temperature scales are referred to the properties of gases; but for low temperature measurements on superconductors, thermometers use electrical properties of materials, such as resistance, voltage, or capacitance. Such sensors may be calibrated against standards to provide true temperature readings. Connection of a temperature controller to a sensor both gives a readout for the thermometer value and provides a feedback loop to regulate the temperature at a set value using a current output to a resistive, temperature-independent heater within the cryostat. The selection of one thermometer over another for a particular application should be based on such factors as the sensitivity over the temperature range of interest, the reproducibility of readings, any dependence on magnetic field, the sensor size and geometry, etc. The list below gives some details of relevant thermometers for measurements on high-T_c superconductors. For further information, the reader is referred to texts on low-temperature physics [9] or literature from thermometer manufacturers [13].

Diode sensors (e.g., silicon, germanium): The forward bias voltage is measured for a constant current (e.g., 10 μA). It increases monotonically with decreasing temperature. These sensors can be used over a large temperature range, but readings are strongly dependent on the orientation of the sensor with respect to magnetic field and also vary significantly with magnetic field magnitude at low temperatures.

Carbon glass resistance (CGR) sensors: The resistance is measured by using a four-terminal configuration. It increases monotonically with decreasing temperature. These sensors have the advantage of a small magnetic field dependence, and readings are independent of sensor orientation with respect to field. These are probably the most commonly used thermometers in investigations of high-temperature superconducting properties.

Capacitance sensors: The capacitance of the sensor is measured using a two-terminal electrical measurement. It has a nonmonotonic dependence on temperature. These sensors have poor reproducibility over time and thermal cycling, but show negligible dependence on magnetic field. Thus, they are suitable as control thermometers at high fields when there may be appreciable magnetoresistance in the setpoint thermometer.

Platinum resistance thermometer (PRT): The resistance is measured by using a four-terminal configuration. It decreases monotonically with decreasing temperature. These thermometers have poor sensitivity below about 40 K, but above this region their magnetic field dependence is low.

2.2.4 Magnetic Fields

Several different types of magnet may be used to produce laboratory magnetic fields for low-temperature measurements on high-T_c superconductors. For small fields (up to about 1 tesla) an electromagnet may be used. This consists of two large coils of wire, which are usually enclosed in jackets for water cooling to prevent excessive heating at the highest currents and, therefore, the highest fields. Within each coil there is often an iron pole-piece to concentrate the field in the sample region between the coils. Power for the magnet comes from a standard dc power supply.

To produce higher fields up to 19 tesla, a superconducting magnet may be used. The ability of a superconducting wire to carry a large current without dissipation makes it highly suitable for use as a magnet winding. Superconducting magnets have the additional advantage that they can be placed in a persistent mode in which no current is supplied from an external power supply and a very stable field is produced. Such magnets can be in the geometry of a solenoid to provide an axial field or split coils for a transverse field. They are wound from superconducting wire (usually NbTi in commercial systems) which is cooled to below its transition temperature by liquid helium. The current for the magnet is supplied from a power supply that allows both the magnitude of the current and the ramp rate to be controlled. If the magnet current is changed too quickly and a large emf is induced in the coils, it is possible to "quench" the magnet, meaning that the wire loses superconductivity and heat is generated. This process causes the rapid loss of helium liquid from the dewar, becauses the magnet dissipates energy which is taken up by the helium. Quenching can also occur if the magnet is operated when the helium level is too low. Thus, it is important when running a superconducting magnet to carefully monitor the amount of liquid helium present in the dewar, as measured by the level sensor within the cryostat.

Certain hazards are associated with the operation of a high-field magnet. The field itself can erase the information on the magnetic strip on credit cards, affect heart pacemakers, and draw in metallic objects that are left physically too close. All these hazards can be reduced by proper signs and ensuring a clear area around the magnet. The electrical power being supplied to the magnet also poses a possible hazard.

The sections that follow show how these techniques of low-temperature measurement may be used to determine properties of the high-temperature superconductors.

2.3 Measurements of Properties of High-Temperature Superconducting Materials

Many different types of measurements have been used extensively to study the properties of high-temperature superconductors in magnetic fields. In most of those techniques, the basic principle is the same. A dc magnetic field is used to produce flux lines within the superconducting material, and the response of the vortices to some electromagnetic probe is then examined. In the case of electrical transport, the probe is a transport current. The response of the superconductor is measured from the voltage that is produced by the motion of flux lines. In the case of the screening response, the probe is a small ac magnetic field. The corresponding response function is often termed *ac susceptibility;* however, that description applies only for the case of a true volume response, which may not always be the case. These techniques can be used to provide information about the dynamics of the vortices, such as their thermally activated motion, as well as possible phase transitions. They are highly versatile methods, with a number of parameters that may be adjusted in the measurements. Changing the dc magnetic field allows the vortex separation (density) to be varied, while the strength of the ac drive field in a susceptibility experiment allows the vortex structure to be probed with a range of amplitudes. The frequency of the ac field may be changed over a wide range, and this corresponds to examining the flux line dynamics over different time scales. The temperature of the sample may affect the magnetic response, and this too may be varied within a cryogenic environment. Hence, with all the parameters outlined here, the techniques to be discussed in this section offer great potential for an investigation of the properties of the mixed state of high-temperature superconductors.

2.3.1 Electrical Transport

Measurements of the resistivity in high-temperature superconducting systems are conceptually very simple. When studying the resistivity as a function of magnetic field, temperature, current density, or crystallographic direction, however, several factors must be considered for the measurement to give reliable results. This subsection will introduce some of these considerations, referring frequently to the preceding section discussing cryogenic systems.

Shown in Figure 2.6 is a schematic of the measurement electronics for an ac or dc measurement of the resistivity. The current supply can be either a dc constant current supply (battery with a limiting resistor) or a constant voltage oscillator with a current limiting resistance in series. This resistor is much larger than the resistance of the sample under consideration, thereby creating a constant current ac supply. In the case where a dc supply is used, if at each temperature or magnetic field the current is reversed to eliminate any offsets from thermal voltages, one can expect to reach voltage levels of approximately 30–50 nV using a Keithley 182 sensitive voltmeter. For the ac measurement, using a standard lock-in amplifier with a PAR 1900 transformer giving a gain of 100 (if the experiment is performed at the frequency of 19 Hz and the total resistance including lead resistance is not greater than 1 kΩ), one can expect to measure a minimum dissipation of approximately 1 nV.

Because of the small resistance levels one needs to measure in superconducting samples, it is essential to use a four-point measurement technique. By *four-point measurement*, we mean that each current lead and each voltage lead is a distinct, nonoverlapping electrical contact. In a two-terminal measurement, the current and voltage probes share a contact. In this case the total resistance measured is

$$R_{total} = R_{sample} + R_{contacts}. \tag{2.3}$$

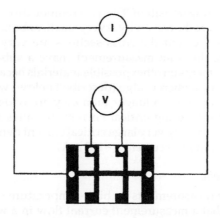

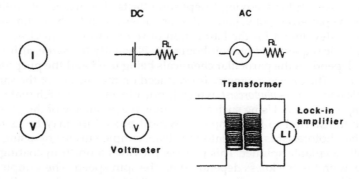

Fig. 2.6. Schematic of the four-point resistance measurement. For the dc case a voltmeter is used, while in the ac measurement, a transformer and lock-in amplifier provide more sensitivity. The sample represented in the schematic shows a typical thin-film microbridge configuration which may be produced lithographically.

As a result, if there is a 1-Ω contact resistance, a 1-mΩ sample resistance is hardly measurable. In a four-terminal measurement, the contacts at the voltage probes become part of the lead; and since the voltmeter has a large input impedance, no current will flow in that direction. Hence, one can measure only the potential difference between the voltage contacts.

If measurements are made on high-quality materials with large cross-sections or very short lengths, the measurement sensitivities listed above may not be adequate to study the resistivity as it varies over several orders of magnitude near the superconducting transition. This is a common difficulty when measuring single crystals where the resistivity is very small, or bulk samples where the resistivity may be larger but the cross sections are very large. For simple geometrical reasons, thin-film measurements have a substantial advantage over similar measurements on other possible materials because of the ability to lithographically define a microbridge (described below) which allows one to create structures that are much longer than they are wide. This property is nearly essential for measuring critical current densities using standard dc transport techniques, making very large critical current densities ($\sim 10^6 A/cm^2$) measurable with very modest currents ($\sim 100\,mA$).

2.3.1.1 Photolithography

In order to make measurements on high-temperature superconductors, it is often important that a measurement current flow in a well-defined geometry. This means using photolithographic patterning of the sample. The goal is to produce geometries of superconducting material that define some particular physical property for measurement or that allow novel effects to be studied. For example, a simple narrow wire allows an accurate and local determination of the transport critical current, while overlapping layers may provide a system for studying Josephson effects. The use of photolithography for the patterning of superconductors owes much to the techniques that have been developed for the fabrication of semiconductor devices.

The basic sequence of photolithography is usually the same, although the details depend on the particular chemicals being used and the purpose of the patterning. The device geometry is produced on the surface of the sample in a chemical known as *photoresist*, which is not affected by an etch that removes the sample material. Using such a technique, the geometry of the photoresist may be transferred to the sample. A typical procedure may be as follows. First, the photoresist is spun onto the clean sample surface by using a commercially available spinner. This process provides a uniform coating on the surface, with the thickness dependent on the spin speed. The sample is next baked at low temperatures (maybe around 80°C) to dry and partially harden the resist. Following this step, a contact or projection mask is exposed by using a bright light source that operates in a region of the spectrum where the photoresist undergoes polymerization or polymer bonds are broken. The mask allows a particular image to be exposed on the resist. Contact masks are placed on the surface of the sample, while projection masks are used in optical systems that reduce the size of the image onto the sample. Following exposure, the sample is immersed in a chemical developer, which removes the photoresist from regions where it was exposed to light (for positive resist) or from areas where it was not exposed to light (for negative resist), leaving photoresist on the sample displaying the design of the mask. The sample may then be baked again to fully harden the resist, after which it is ready for etching to define the device structure in the sample material. The regions where the photoresist remains are protected from the etch. Following completion of the etch through the unprotected regions, the sample may be rinsed in solvent to remove all remaining areas of resist.

2.3.1.2 Electrical Contacts

For electrical transport measurements, it is important to maintain low contact resistance so that there is little dissipation in the contacts. Attaching measurement leads to high-temperature superconductors has been problematic because of the poor quality of sample surface layers. Several different techniques have been used to achieve low-resistance electrical contacts to both single crystals and thin films. The simplest method involves direct contact of the measurement wire to the surface of the sample. This may be done using an electrically conductive paint or paste (usually containing silver) which binds the wire to the surface, and, once dry, holds down to cryogenic temperatures. The wire may also be secured by using a pressed pad of indium metal, which is soft and adheres well to the sample surface, or indium solder. Although these contact techniques may be used directly to the surface of the sample, often the surface must be prepared first. For example, surface layers can be degraded by contact with moisture in the air, and etching can remove these layers to expose high-quality conducting layers beneath. An example of such an etch that has been used for high-temperature superconducting materials is bromine in ethanol [14]. Metallic contacts may be deposited on the surface of the sample to protect the high-quality surface after etching and to prevent further degradation. Metals such as silver and gold can be easily deposited by using sputtering or thermal evaporation and with shadow masks to define the contact pad configuration. Low-temperature annealing can further improve the contact resistance, although diffusion of the metal atoms into the superconductor must be considered. Limited diffusion may be advantageous, but if it is excessive, degradation of the superconducting properties can result.

2.3.1.3 Sample Holder/Wiring

Although some very low temperature refrigeration systems (dilution refrigerators) require warming to room temperature to modify the sample configuration, most cryostats used for studies of high-temperature superconductivity are performed in systems that remain cold between experimental runs. These cryostats are typically configured with a sample insert that extends from the room-temperature environment to the very low temperature cryogenic environment. Both the exchange gas system and the continuous flow system that were discussed earlier in this chapter may use this type of sample insert. This insert will vary a great deal depending on the measurement being performed. The sample probe is often configured with more than one thermometer if the sample insert will be used over a large range of temperatures and in an applied magnetic field.

The sample holder has electrical wiring for the thermometer and sample that goes to room temperature along the length of the sample insert. These electrical leads are typically twisted in pairs to eliminate problems associated with magnetic pickup for low-level voltage measurements. The leads must be thermally anchored at several fixed points along the length of the sample insert. It is also very important to thermally anchor the leads for both the sample and the thermometers to the sample holder, which is constructed of a material with a very large thermal conductivity such as copper. This is particularly relevant when measurements are performed on thin-film samples of high-temperature superconductors that may be fabricated on substrates with very poor thermal conductivities, effectively thermally isolating the film from the "constant temperature reservoir." In this case, as in the case of an encapsulated thermometer, much of the cooling of the material takes place through the electrical leads connected to the sample or sensor. For this reason, it is very important to have the leads at the same temperature as the sample block. This requirement ensures that the primary heating and cooling of the sample are done by the sample block (where the thermometer is located) and not by the gas in the sample space (whether it is exchange gas or flowing vapor), as the latter can lead to a nonuniform temperature profile over the sample.

Regardless of the type of cryostat—closed-cycle refrigerator, gas flow, or exchange gas—good thermal contact, such as between the sample and sample block, is one of the most important considerations. The typical adhesive used for making thermal contact is GE 7031 varnish. This material has a thermal conductivity that is roughly 4000 times smaller than that of copper, but its major advantage is that small amounts provide a very strong bond. Therefore, the conduction path can be kept short. Thin layers of grease or epoxy may also be used for thermal contact.

Closely related to the issues regarding thermal contact are concerns regarding equilibrium measurements. Although good thermal contact allows the entire sample holder "system" to reach thermal equilibrium quickly, it can never be perfect so cooling of the sample holder must be performed in a controlled manner. The thermal conductivities are a decreasing function of temperature so that the thermal relaxation times are longer at low temperatures. Therefore, to test whether the cooling rate is slow enough at all temperatures, it is sufficient to perform a cooling and warming cycle at the lowest temperature that will be used and to evaluate the reproducibility of the results. Of course, this test must be performed on a nonhysteretic property.

Finally, as materials processing is modified to achieve the largest critical currents, it is becoming apparent that one must evaluate more than a critical current based on a single voltage criterion. Specifically, to make an accurate determination of the process responsible for limiting the capacity of these materials to carry a dissipationless current, one must evaluate the full voltage/current characteristic and attempt to quantitatively evaluate the nonlinear portion of the curve. Even though good electrical contact has been made to the material and a lithographically defined microbridge has made this type of measurement possible, it is important to limit the power dissipation in the microbridge. A good rule-of-thumb is 0.01 to 0.1 mW but is very dependent upon the thermal contact achieved in the experimental setup and the cryostat design.

We now consider favorable sample geometries for some specific transport measurements. When measuring the linear resistivity as a function of temperature, one should perform the experiment at a constant current density.

Therefore, to achieve the maximum measurement sensitivity in terms of resistance, the sample geometry should be as long as possible. Cross-sectional area in such a measurement is irrelevant as long as the cross section is not so large that the total measurement current causes heating at the contacts. For a single crystal, a typical cross-sectional area might be 6×10^{-4} cm^2. To reach a current density of 10 A/cm^2, one needs to apply a current of 6 mA. Therefore, the total contact resistance should be below 3Ω. Within these limitations, whether one is measuring a bulk polycrystalline sample, a single crystal, or a thin film, no system has a decided advantage over the others. Suppose, however, that one wishes to evaluate a physical property as a function of current density. Then it is clear that a small cross-sectional area is advantageous. Suppose that one has a contact resistance of 0.1 Ω and that it is important to measure to a current density of 10^5A/cm^2. In a single crystal of the cross section mentioned before, a current of 60 A would be required, corresponding to a power dissipated at the contacts of 360 W! Thin-film samples with a thickness of 3000 Å and widths of order 100 µm are lithographically simple to define. The corresponding cross-sectional area in the sample would require a current of 30 mA, which gives a power dissipation of less than 0.1 mW.

2.3.2 Screening Response (ac Susceptibility)

The general case of the response to an ac magnetic field is applicable to several different measurements that have been made of high-temperature superconductors. The clearest case is that of the *susceptibility* (or permeability), which may be measured by a multicoil magnetometer [15]. This is the experimental configuration that will be described in most detail in this chapter. Analogous experiments involve the oscillation of a sample within a magnetic field. This is similar because application of an alternating magnetic field to a sample may be equivalent to vibration of the sample within a constant magnetic field (with additional account being taken of possible effects from angular motion). Such experiments have been performed by the damping of vibrating superconductors in various configurations, such as suspended on wires, mounted on reeds performing flexural vibrations, and glued on vibrating tongues.

For example, single crystals of high-temperature superconductors were epoxied to small silicon oscillators, which were driven self-resonantly with the use of a phase-locked loop. The frequency of measurement was essentially fixed at 2 kHz, and the dissipation of the high-Q oscillator was monitored as a function of temperature [16]. In low-frequency torsional oscillator experiments, the sample was suspended on a tungsten wire and released after an initial displacement from equilibrium. The resulting oscillations were monitored with an optoelectronic system, and the dissipation was obtained from the free decay of the angular oscillations [17,18]. In the vibrating reed technique, a reed of the superconductor was clamped at one end and electrostatically driven by an electrode near the free end [19]. In ultrasonic measurements, sound waves with frequencies of a few megahertz were produced by transducers on the ends of the sample, and the attenuation was measured by an echo signal [20–22]. The phonons of the ultrasonic probe couple to the vortex lines through pinning within the sample. One advantage of the ultrasonic probe over magnetic measurements is that the interaction with the flux lines is purely within the bulk—there are not the surface effects that can be associated with magnetic penetration into the material. The dependence on pinning, however, means that this technique may be subject to the details of the disorder within the material.

Measurements of the ac screening response to be presented later in this chapter were made using a conventional first-order magnetic gradiometer [23], although other geometries may be used. Such a device consists of a primary circuit used to drive the sample with an ac magnetic field, and two secondary circuits for pickup of the sample response and cancellation with respect to the background signal. A schematic of this is shown in Figure 2.7. The susceptometer was made in two parts: a hollow cylinder of phenolic material, around which insulated copper wire was wound to form a solenoid, and a second piece of phenolic in the form of a rod, which was sized to fit concentrically within the cylinder. For the region closest to the center, the magnetic field produced by the solenoid was about 40 oersted per amp drive current. Around the inner rod were wound two small coils each of 75 turns of insulated copper wire to act as the pickup (secondary) coils for the magnetometer. The single crystals of superconducting $Bi_2Sr_2CaCu_2O_x$ used as samples for this experiment were placed inside one of the pick-up coils. Since the material structure of $Bi_2Sr_2CaCu_2O_x$ and hence the physical properties are highly anisotropic, it is very important that the orientation of the crystal be well defined with respect to the ac and dc fields. In addition, in designing such a magnetometer, it is important to achieve good coupling between the solenoid that produces the drive field and the pickup coils. For these experiments, this gave a measurable response even for very low driving fields, typically 5 mOe. The driving field amplitude could be increased well above this level to probe larger dis-

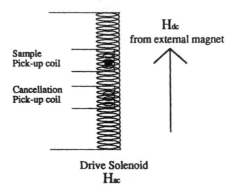

Fig. 2.7. Schematic of an ac susceptometer using primary and secondary coils. A magnetic field is generated by driving the primary (outer) coil with an ac current. The superconducting sample is placed within one of the secondary (pickup) coils. An external dc field generates the vorticies within the superconductor (after [23]).

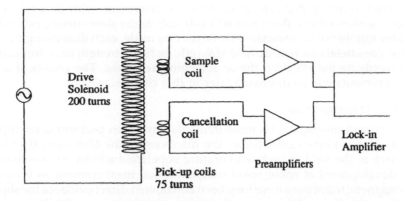

Fig. 2.8. Shematic of the ac susceptibility measurement circuit. Variable gain on the two preamplifiers allows the signals from the two pick-up coils to be nulled against each other, so that the resultant response is due to the screening of the ac magnetic field by the superconducting sample (after [23]).

placements of the vortices. An external dc field from an electromagnet was used to produce the vortices within the sample and gave the ability to tune their density.

For most of the measurements, the solenoid was driven with the sinusoidal output of a lock-in amplifier, taking the direct voltage output and connecting it to a limiting resistor to provide a current source (just as for the electrical tansport measurements discussed earlier). Some of the measurements, generally those made at low currents or at higher frequencies where inductances became important, were performed by taking the drive from an ac-dc current calibrator driven by a function generator. For both techniques, the drive frequency was easily changed. The two pickup coils were each connected to low-noise preamplifiers. The outputs of these preamplifiers were connected in opposition, and the resultant null signal was measured by the lock-in amplifier. Variable gain on the preamplifiers allowed the signal to be properly nulled between the two pickup coils, thereby providing maximum sensitivity to changes of response between the coils arising from the sample contribution. Both in-phase and out-of-phase components of the response were detected by the lock-in amplifier. The measurement circuit for the ac screening response is shown in Figure 2.8.

The susceptometer was located within a variable gas flow cryostat (as described earlier in this chapter), with changes in the temperature of the flowing vapor being followed by the temperature of the sample. Measurements were made by monitoring the sample behavior as the magnetometer was cooled from the normal state through any features in the response. For fixed values of the dc field, ac drive amplitude, and ac drive frequency, the temperature controller for the helium vapor was programmed to reduce slowly the output to the vaporizer in order to give a uniform cooling rate. The temperature of the sample (measured by a resistance thermometer on the sample block) was observed to follow this rate quite closely. Comparison with data taken on warming showed that there was no hysteresis at the slow cooling rate used. A large number of such measurements may be made, each time changing one of the parameters such as dc field strength, ac field strength or ac frequency, to investigate the effect on the ac screening response. The results of such measurements will be discussed later in this chapter.

2.3.3 dc Magnetization

Although magnetization measurements have been performed on super-conducting materials since before the discovery of the Meissner effect [24], research in the field of high-temperature superconductivity has resulted in the development of some novel magnetization measurement techniques. Diamagnetic transitions have long been considered direct evidence for super-conductivity in newly discovered superconducting materials. Distinctions between reversible and irreversible magnetization have been used to understand pinning effects in type II superconductors using Bean's model [25]. In this section, we discuss traditional magnetometry techniques and demonstrate

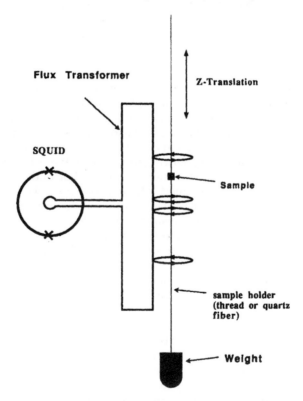

Fig. 2.9. Schematic of the SQUID and flux transformer. The sample is mounted on a fiber and moved throught the transformer coils. The fiber is always within the transformer, and so the measured signal is due to the superconducting sample.

the novel interpretation of results for high-temperature superconductors later in the chapter.

2.3.3.1 SQUID Magnetometry

Superconducting quantum interference devices (SQUIDs) are typically used in the measurement of magnetization for small samples with low magnetizations since their measurement sensitivity exceeds that of any other available technique. A schematic representation of pickup coils that are inductively coupled to a SQUID is shown in Figure 2.9. This type of pickup coil, originally used as a second derivative gradiometer, is used in a SQUID manufactured by Quantum Design Inc., to eliminate background signals from the sample holder that extends over the entire range of the gradiometer. As the sample is moved through the gradiometer coil, the resulting "field gradient" is entirely due to the sample. SQUID magnetometers have been extensively discussed in other books [1,2,26,27], and details will be given here only when appropriate to demonstrate experimental procedures.

Integrated systems providing computer control for temperature, magnetic field, and sample translation are available and can detect magnetic moments as small as 10^{-8} emu. To succeed in measuring materials that test this sensitivity limit requires considerable effort, however. First, one must ensure that the magnetometer is not situated near magnetic materials such as furniture or even structural support for a building, which could result in a magnetic field gradient within the sample space. Second, the suspension of a sample in such an apparatus requires using a low-background sample holder that extends through the pickup coil apparatus. For measurements requiring the largest sensitivity, quartz fibers are recommended, both for their small background signal and the robust support they provide in fairly small volumes [28]. Typically it is adequate to mount a small sample to the quartz fiber using a thin layer of GE 7031 varnish.

Because of the limitations on sample holder design, temperature control in these systems is difficult. It is not useful to construct a sample holder out of copper because of the large background signal this will contribute. Therefore, measurements are always made at fixed temperatures so that the entire apparatus is in thermal equilibrium with the thermometer. This procedure is standard for most magnetization techniques and puts many constraints on the measurements, most importantly making them time consuming. For example, magnetization measurements as a function of temperature using a SQUID may be evaluated, but the temperature must be stabilized at each individual setpoint to generate the overall curve. For this reason it is often most convenient to take data as a function of magnetic field or time, thereby avoiding most of the difficulties of temperature stability, although the temperature can drift over the duration of the magnetic field sweep. Such time constraints can make the interpretation of data difficult because of the time dependence of the magnetic relaxation. This situation will be discussed in more detail later when we compare different measurement techniques.

Although each of the techniques used to measure magnetization require constant temperatures to make accurate measurements, the stability required by the SQUID magnetometer is the most demanding. A Quantum Design SQUID averages over several measurements of the magnetic moment. Although the software package available with the Quantum Design SQUID allows the operator to vary the number of measurements to average over, a single measurement can take 50–60 seconds, assuming that the temperature is stable. If the magnetic field and temperature are stable, for averaging over three measurements, the length of time to collect a single data point is close to three minutes. Therefore, without accounting for possible temperature drifts and magnetic field ramp times, 2.5 hours are needed to collect a data set of 50 data points. Clearly, this requirement puts a significant constraint on the number of data points that may be collected in a reasonable time.

2.3.3.2 *Vibrating Sample Magnetometry (VSM)*

In the vibrating sample magnetometer [29], a vibration mechanism is positioned at the top of the cryostat, which may be as simple as a loudspeaker assembly. This mechanism moves the sample in the applied magnetic field parallel to a set of pickup coils, and an emf is induced in a pickup coil surrounding the sample specimen. Other geometries of the pickup coil configuration are possible [29]. This induced emf is proportional to the magnetization of the sample. Integrated systems based on this type of magnetometer are also commercially available and can measure magnetic moments to ~10^{-6} emu.

The VSM is based on Faraday's law, in which a changing magnetic field induces an emf proportional to the rate of change of magnetic flux. In this regard, this measurement technique is very similar to the ac susceptibility technique described earlier. In the ac technique, a small field ripple was applied to the sample, and the screening of this ripple by the sample was measured. In the VSM technique, the sample is moved in a static uniform field, and the change in the magnetic field due to the sample is measured. In the first case, since one applies a field ripple, the slope of the magnetization versus magnetic field curve (or susceptibility) is determined. In the second case, since one is basically measuring the change in the magnetic field, it is a direct measurement of the magnetization.

The frequency of oscillation is important to the sensitivity of the measurement. Typical measurement frequencies may be around 70 Hz. The pickup in the coils is detected by using lock-in techniques. Therefore, problems caused by cryostat or building vibrations may be eliminated by driving the sample at a frequency that is far away from any extrinsic vibration frequencies.

2.3.3.3 *Faraday Methods*

The Faraday method requires placing a specimen in a magnetic field gradient and determining the resulting force on the material. The force is directly related to the magnetization of the specimen and the magnetic field gradient

$$F = -\frac{dB}{dZ} M \qquad (2.4)$$

and may be measured using a capacitive force transducer [30]. The sample is mounted on the upper plate of a capacitor, where the restoring force of the upper capacitor plate is assumed to be linear as a function of displacement for small total translations. Capacitance may be measured very accurately, so proper calibration of the magnetometer can give an accurate measurement of the magnetization. Measurements made by using this technique typically include a component of the torque, which is given by

$$\tau = \mathbf{M} \times \mathbf{B}, \tag{2.5}$$

because of the force transducer used.

A field gradient is present in any solenoid magnet away from the center of the magnet. Some magnet systems may be supplied with a gradient coil, or one can be constructed. In any case, to check for or eliminate a torque contribution, one can simply use the geometry of the experimental setup. In a typical solenoid magnet, the magnetic field is largest at the center, and the field drops off away from the center. If the force is measured above the magnet center and at an equal distance below the magnet center, the sign of the gradient field changes while the overall size of the field is a constant. Therefore, by subtracting the two signals, one can eliminate the contribution from the torque. Using this technique, one can measure both quantities simultaneously.

This description has thus far excluded several experimental details. First, the sample must be small enough that one can assume a constant field over its size, yet the field gradient must be large enough to exert a significant force on the sample. Neither the force nor the torque should change the position of the sample to the extent that the anisotropy of the sample or the field gradient becomes important. By using feedback circuitry, any net motion of the sample during the measurement may be eliminated. Also, the torque is proportional to the magnetic field, a fact that makes the determination of the magnetization at small fields very difficult. If one is using the magnetic field profile of a solenoid, compensation coils on the magnet may make the measurement difficult. Also, the gradient is usually proportional to magnetic field, thereby introducing measurement problems at small fields. If a gradient coil is used, however, the sample can sit in a magnetic field gradient while being in zero applied field.

For more detailed descriptions of measurements with Faraday techniques, the reader is referred to [30] and references contained within that paper.

2.3.4 Comparison of Techniques

We now attempt to connect the different measurements using a discussion of the effective electric-field criteria of each technique described above. To make comparisons between results obtained through the various measurements, we use Faraday's law to estimate the electric field of each experimental technique. If the measurement is made at static field, the electric field criterion is determined by the relaxation of magnetization. If the magnetic field is ramped at rates larger than the magnetization relaxation rate, the electric field criterion is determined by the magnetic field ramp rate. In either case, the correct expression for the electric field generated by a changing magnetic field is Faraday's law, which may be written

$$\nabla \times E = -\mu_0 \frac{\partial H}{\partial t} , \qquad (2.6)$$

where $B = H - 4\pi M$ in SI units. In cylindrical coordinates this expression becomes

$$E_c(V/m) = \mu_0 \frac{\Delta H}{\Delta t} R = \frac{\Delta B}{\Delta t} R, \qquad (2.7)$$

where R is the appropriate sample radius. Therefore, it is easy to determine the effective electric field criterion if we know the magnetization relaxation rate for a static field, or we use a constant ramp rate when increasing the magnetic field in a dynamic measurement. One can see a definite advantage to measurements made in a ramped magnetic field if a constant electric field criterion is needed.

As was mentioned previously, the SQUID measurement must be made under conditions of constant temperature and magnetic field. Therefore, for this measurement method, E_c is determined by properties of the sample. No reliable methods exist to predict an appropriate value for the magnetization relaxation. (In the Results section, we give an example of a possible technique for estimating this value.)

In the case of an ac susceptibility measurement, the applied magnetic field is always changing since there is an applied field ripple. Typical values of the primary ac magnetic field may vary from 1 mG to several gauss, while the frequencies vary from 100 Hz to 100 kHz. In a VSM experiment where the magnetic field is constantly changing, the change in the magnetization is no longer the correct quantity. The parameter that determines the electric field criterion is the magnetic field ramp rate. In the ac susceptibility measurement, based on the numbers given earlier, the effective ramp rate may vary from 10^{-1} to 10^5 G/s, giving an electric field around 1.2×10^{-7} to 10^{-2} V/m. In the case where the magnetic field is ramped for a VSM or a Faraday balance measurement, a typical ramp rate for a superconducting magnet may be 100 G/s. This corresponds to an E_c of 10^{-5}V/m. These values should be contrasted to those used for determining the critical current in a transport measurement, typically 10^{-4} V/m. The numbers imply that with some knowledge of the current flowing in the material, the E-J curve may be constructed from these magnetic measurements.

2.3.5 Material Quality

Many studies of the fundamental properties of high-temperature superconductors, particularly those published soon after the discovery of these materials, resulted in the possible misinterpretation of experimental data because of the poor quality of samples measured. As research has continued, the different issues involved in such studies have emphasized the importance of an interdisciplinary approach. Measurements of fundamental properties in polycrystalline materials must be accompanied by companion microstructural and quantitative compositional characterizations, thereby providing a closed loop between processing, microstructure and properties. This is true for any type of measurement including electrical transport, susceptibility, and magnetization.

2.3.6 Importance of Material Quality to the Measurement of Superconducting Parameters

Measurements of phase-pure materials are essential to understanding the properties of high-temperature superconducting systems. Such characterizations are of particular importance since the superconducting transitions are thermally smeared and strongly dependent upon characteristic length scales in the superconducting state. This property can even make it difficult to define parameters as conceptually simple as the transition temperature. For example, multiple transitions can make it nearly impossible to distinguish fundamental effects. If two phases of material that have significantly different critical temperatures are present in a sample (e.g., two superconductors from the Tl-O family, $TlBa_2CaCu_2O_7$ and $Tl_2Ba_2CaCu_2O_8$, with T_cs of 80 K and 110 K, respectively), the critical temperatures of the two phases may be obvious. When a magnetic field is applied, however, transitions in the two systems behave differently, and the resistance may go to zero at or near the same temperature for both phases if the applied magnetic field is large enough. In this case, the properties resulting from each phase will be almost impossible to separate.

Earlier, we suggested the importance of compositional analysis. If single-phase material is available, however, characterizations of the microstructural properties of a sample remain important for several reasons. By investigating the microstructure resulting from various processing techniques and by comparing these studies with corresponding transport measurements on the same materials, one can determine the optimum processing conditions for enhancing a desired property. For example, researchers have shown that intergranular critical currents are enhanced by increasing the texturing that exists in polycrystalline materials [31]. When a hydrostatic pressure is applied to a material at high temperatures, very small planar defects are formed within a single grain of a material. These act as very effective pinning centers, increasing the intragranular critical currents [32].

Detailed microstructural characterizations also give important information about growth mechanisms. For example, in a silver-sheathed material, there exists evidence that the crystallites begin their growth near the silver/superconductor interface and this leads to the growth of material with the c-axis perpendicular to this interface [33]. The result is texturing near the interface, which can greatly enhance the intergranular critical currents. This type of growth also reduces the existence of voids in the material, which significantly modify measured parameters.

For thin films, the compatibility of the superconductor with the underlying substrate materials may be very important. For example, all materials containing barium that are fabricated on yttria-stabilized zirconia (YSZ) substrates, whether single crystal or polycrystalline, will develop an interaction layer ($BaZrO_3$) near the substrate-material interface [34]. This can actually be a significant cross section of the overall sample dimension, depending again on the process history. After systematic studies of the development and existence of such an interaction layer are performed, processes may be modified to eliminate the existence of these defects or measurements of the transport properties may be modified by assuming a smaller effective cross-sectional area of the sample.

So far in this subsection we have discussed defects that complicate the interpretation of experimental results. Understanding the properties of polycrystalline materials is very important since wires used in many applications are necessarily polycrystalline. Undertaking a new area of research with only polycrystalline materials is difficult, however, since multiple mechanisms contribute to a particular property. One must first evaluate the properties of an individual crystallite and continue from that to two crystallites with known relative orientations to each other. In this way progress may be made, starting with fundamental properties and adding to them in order to build a composite system. This is the only realistic way to approach an understanding of the complex processes involved in current transport in composite high-temperature superconducting systems.

In conventional superconductors, polycrystalline materials have performed well for applications in a magnetic field. Grain boundaries in these materials represent areas of reduced superconductivity which act as strong pinning centers to prevent vortex motion. In high-temperature superconducting materials, however, the grain boundaries behave as weak links and thus are detrimental to dissipationless current transport. This situation is a result of the extremely short coherence lengths in these materials. The grain boundaries are nearly the same dimension as the coherence length, causing a reduction of the order parameter at grain boundaries. This is in contrast to traditional type II superconductors, as may be seen in Table 2.1, with the range of coherence being significantly larger. Research has shown that the larger the misorientation angle between adjacent grains, the worse the current-carrying capabilities. [35]. These weak-link properties have been exploited for electronic devices [36–39].

2.4 Results

Directly related to applications of high-temperature superconductors is the nature of the vortex state in the materials as a function of temperature, magnetic field, and anisotropy. Many different experiments using various experimental probes have been performed attempting to understand more about the dissipation that occurs as a result of the dynamics of vortices with current flow parallel to the CuO bi- or tri-layers [40–43]. Here, we give several examples of experimental results on high-temperature superconducting systems that differ from those observed in traditional low-temperature superconductors.

2.4.1 Electrical Transport Properties of High-Temperature Superconducting Materials

In the two subsections that immediately follow, we discuss fundamental electrical transport properties of high-temperature superconducting materials. The first subsection deals with single crystal or intragranular properties of materials and the second with the properties related to current transport across grain boundaries. Extrinsic electrical transport properties that are a function of the strength of the intergranular coupling may be obvious in any polycrystalline sample of high-temperature superconducting material and are considered a fundamental property for the purposes of this section.

2.4.1.1 Intragranular Electrical Transport Properties of High-T_c Materials

The materials that are the focus of this section fall naturally into several categories, determined by the material anisotropy. The least anisotropic high-temperature superconducting material mentioned here is $YBa_2Cu_3O_7$. Several reviews have been written on the possible origins of the resistivity below the superconducting transition in this material, and readers are referred to these articles for further information [44,45]. Since $YBa_2Cu_3O_7$ is isotropic (at least in comparison with the other materials discussed here), magnetic vortices in this material are nearly always three-dimensional (3D). All other materials discussed here are much more anisotropic and have in common that the vortices are two-dimensional (2D) over a significant range of temperatures and magnetic fields, and are often referred to as "pancake" vortices within the Cu-O layers. These materials have received far less attention regarding their fundamental properties, although they play a significant part in research efforts aimed at making long lengths of wires for high-current applications. For this reason, as well as the new science they display, these materials are the primary focus of this subsection.

High-quality single crystal samples of the more anisotropic materials, including various phases of TlBaCaCuO, have been difficult to fabricate. The most reliable work (with regard to fundamental properties) has been performed on epitaxial films, which have many of the same properties as single crystals. By *epitaxial* we mean that regardless of the growth mechanism, the crystallographic structure of the resulting material matches that of the underlying substrate.

The most systematic data available in the large body of literature on the vortex state in high-temperature superconductors are from electrical transport studies. Many of these studies were measurements of dissipation in single-crystal-like materials [46,47]. Figure 2.10 shows the resistivity data for an epitaxial film of $Tl_2Ba_2CaCu_2O_8$ taken as a function of temperature in an applied magnetic field directed along the c-axis (perpendicular to the superconducting Cu-O planes). Note that the resistivity is plotted with a logarithmic vertical scale. The resistive transition is greatly broadened as a result of vortex motion. (This should be contrasted with the data of Figure 2.3.) As a measure of the characteristic magnetic field and temperature of the onset of vortex motion (referred to in the literature as "irreversibility behavior") we use a single resistivity criterion ($\rho/\rho_n = 10^{-5}$, where ρ_n is the temperature dependent normal state resistivity), making comparisons between materials with different anisotropies possible [41]. The dashed line drawn in Figure 2.10 indicates this criterion.

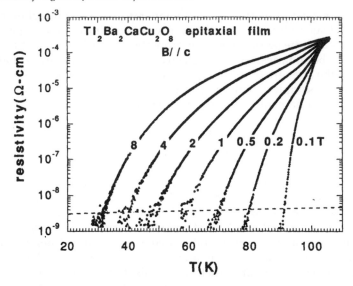

Fig. 2.10. Resistivity vs temperature for a $Tl_2Ba_2CaCu_2O_8$ epitaxial film in a magnetic field applied along the c-axis (perpendicular to the film plane). Notice the suppression of the temperature of the transition as a function of magnetic field. The dashed line indicates a criterion of $\rho/\rho_n = 10^{-5}$.

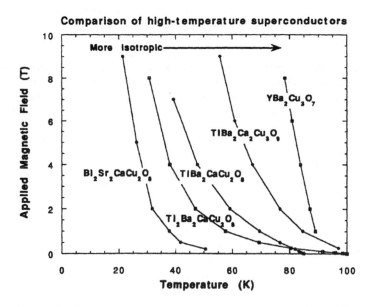

Fig. 2.11. Lines representing the separation of a region of irreversibility (low temperatures) and reversibility (high temperatures) as defined from date such as shown in Figure 2.10 using a $\rho/\rho_n = 10^{-5}$ criterion. Lines connecting the data points are only guides to the eye.

These results, as well as those for other materials, have been compiled and are shown in Figure 2.11. This plot indicates that when a magnetic field is applied, the resistive transition is extended to much lower temperatures than expected for $T_c(H)$. Similar data have been compiled by Kim et al. [41], and these authors have plotted the results in reduced units of $H/H_{c2}(0)$ and $(1-T/T_c)$. When viewed in this way, the data fall naturally into groups determined by the separation between the Cu-O layers in the materials studied. This grouping gives an indication of a possible energy scale that determines the onset of vortex motion. We expect that the onset is related to the Josephson coupling between the layers [48].

To confirm the importance of the Josephson coupling, we refer to the electrical transport in the c-axis direction. C-axis resistivity data presented in the literature for $Bi_2Sr_2CaCu_2O_x$ show a distinct upturn near the critical temperature as the temperature is reduced [49]. This upturn has been shown to be consistent with the freezing out of tunneling quasi-particles, while the reduction in resistivity below the transition temperature is consistent with an increasing Josephson coupling [50].

In a model system consisting of Nb-AlO-Nb Josephson junctions, the magnetic field dependence of the coupling energy was experimentally determined [51] and shown to be consistent with the usual Josephson coupling energy presented by Ambegaokar and Baratoff [52], with some modification to account for the application of the magnetic field. The energy for the Josephson coupling may be written as

$$E_{cj} = \frac{\hbar I_{jc}(T)}{e} = \frac{\pi\hbar\Delta(T)}{2e^2 R_N} \tanh\left(\frac{\Delta(T)}{2k_B T}\right), \tag{2.8}$$

where $I_{cj}(T)$ is the temperature-dependent critical current of the Josephson junction in the absence of thermal fluctuations, R_N is the normal state resistance of a single Josephson junction, and $\Delta(T)$ is the temperature-dependent energy gap.

The modification for magnetic field that must be made to the above expression will enter through the term R_N. The normal state resistance of the junction may be written as $R_N = \frac{\rho_{Nc} s B}{\phi_0}$, where s is the c-axis repeat distance and ρ_{Nc} is the c-axis normal state resistivity and where it is assumed that the effective junction area is determined by the magnetic field dependent area $\frac{\phi_0}{B}$ of a single vortex [41]. The experimental results verify the $1/B$ dependence of the coupling area. Although both theoretical and experimental work on highly anisotropic high-temperature superconducting materials were performed in which this field dependence was assumed [53]. Prior to the investigation of this model system, this was the first direct experimental verification of such a magnetic field dependence [51].

We now address how this coupling energy may lead to important physics that is relevant to applications of highly anisotropic superconductors. Con-

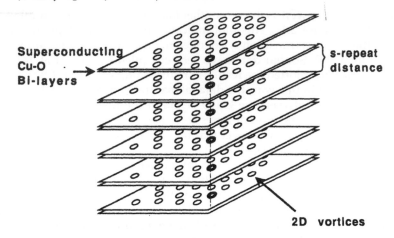

Fig. 2.12. Schematic representation of a high-temperature superconductor as a stack of superconducting planes connected through Josephson coupling in the presence of a magnetic field perpendicular to the layers. The rings represent two-dimensional vortices. Notice that they align in stacks and in a hexagonal lattice within planes in the absence of pinning.

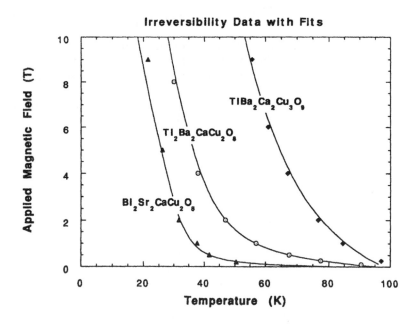

Fig. 2.13. Representative curves extracted from Fig. 2.11 showing fits to the data using the model outlined in the text.

sider as a model of a high-temperature superconducting material a stack of Cu-O planes separated by insulators. This is shown in Figure 2.12 with a magnetic field applied perpendicular to the layers.

If the material were isotropic, the "line tension" of the vortices would be uniform. But because the layers are Josephson coupled, the order parameter between the layers can never be greater than the order parameter within the layers. Therefore, the line tension is weakest in the region between the layers. If we consider the possibility of thermal fluctuations, the probability that an

individual Josephson junction will become decoupled is $\exp[\dfrac{-E_{cj}(B,T)}{k_B T}]$.

Therefore, at any finite temperature, there is a finite probability that the vortex will become disconnected from its nearest neighbor.

After a junction becomes disconnected from its nearest neighbor, pinning within the layer prevents 2D vortex motion. To estimate the temperature and field dependence of the pinning energy, we will use the expression for core pinning [7]

$$E_p = \alpha_p \pi r_d^2 s \ \frac{B_c^2(T)}{2\mu_0} \ (1 - \frac{B}{B_{c2}(T)})^2, \qquad (2.9)$$

where r_d is the defect radius, s is the c-axis repeat distance, B_c is the thermodynamic critical field, and α_p characterizes the pinning strength. If the vortices were each decoupled from their nearest neighbors, the probability of thermal activation would be $\exp(-E_p(B,T)/k_B T)$. Therefore, the overall probability of activated motion of 2D vortices is the product of the probabilities of decoupling from two layers and the probability of overcoming the pinning barrier. This makes the overall energy barrier preventing vortex motion

$$E_{barrier} = 2E_{cj}(B,T) + E_p(B,T). \qquad (2.10)$$

Let us assume a very simple criterion that vortex motion occurs when the energy barrier is of the order of the thermal energy, $E(B,T)_{barrier} = k_B T$ [41]. This criterion is used to fit the data of Figure 2.11, and fits for several materials are shown in Figure 2.13. Samples of very different geometries and microstructures give very similar results for similar phases of materials. These fits are very convincing evidence that Josephson coupling plays a large role in the effectiveness of the pinning in these materials. The results also suggest that one should pursue the use of materials with lower anisotropies in wire fabrication for high-field/high-current applications.

2.4.2 Experimental Studies of Polycrystalline Materials

Large-scale current transport and magnetic applications of high-temperature superconductors necessitate the use of polycrystalline materials, making dissipation at grain boundaries of great importance. Researchers have investigated processes suitable for large-scale production that could possibly render the resulting bulk material textured. Recently a large body of research has been directed at investigating the dissipation at grain boundaries. The first

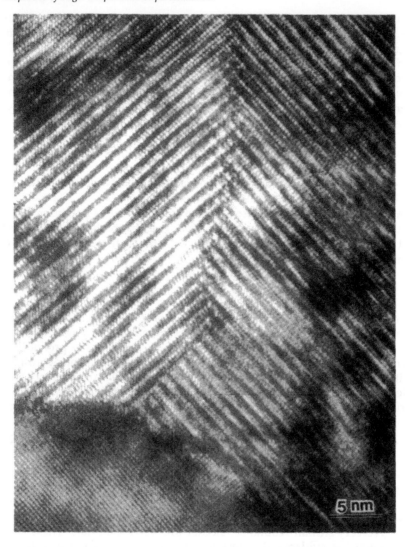

Fig. 2.14. A TEM micrograph showing a grain boundary in a $YBa_2Cu_3O_7$ thin film on a substrate containing many ion milled channels. (Figure courtesy of D. J. Miller).

work was performed on $YBa_2Cu_3O_7$ films grown on bicrystal substrates by Dimos and coworkers [35]. These results indicated that all high-angle grain boundaries (HAGB) behave the same, as weak links with Josephson junction–like properties regardless of whether they are tilt or twist boundaries [35]. This characteristic has been exploited for the fabrication of electronic devices. In fact, much of the work that has followed these initial investigations has been directed toward device fabrication.

Although studies of "single property" systems represent a first step in understanding polycrystalline transport, they are rather unrealistic models of true polycrystalline materials where many grain boundaries, both high and low angle, in parallel and series, are sampled simultaneously. At the other extreme of polycrystalline materials, however, studies suffer from a lack of knowledge about the individual grain-boundary structure and how current may redistribute to minimize the dissipation, particularly in highly anisotropic materials [54,55]. Therefore, it is difficult to determine whether all or a few grain boundaries of a certain type extending across the current path are contributing to the observed dissipation in a transport measurement.

The experimental studies that we discuss here are attempts to look at an experimental regime between single-grain-boundary properties and measurements of randomly textured polycrystalline materials. First, we discuss results from electrical transport measurements on $YBa_2Cu_2O_7$ grown on a substrate consisting of a series of step edges [56]. In what follows, the microstructure of these materials is described and then related to the various transport measurements performed. Then we discuss similar results on a typical polycrystalline textured thick film of $TlBa_2Ca_2Cu_3O_9$ [57], a single thallium layer compound that has been shown to be the best alternative so far to $YBa_2Cu_2O_7$ for high-field and high-temperature applications [41].

Microstructural characterization of a $YBa_2Cu_2O_7$ sample, grown *in situ* by laser ablation on the substrate with a series of ion-milled channels, was performed in order to verify the types of grain boundaries present. Figure 2.14 shows a transmission electron microscope (TEM) image of a typical [100] tilt boundary. The lattice fringes observed in this micrograph represent the direction of the Cu-O planes. This sample contains a series of very well defined grain boundaries that are continuous across the conduction path. The thickness of the film is approximately the depth of the ion-milled channels, or 3000 Å. Each channel creates four grain boundaries of a similar type. The repeat distance is about 13 μm, indicating that for our measurements, where we sample a 1-mm length of the material, we are actually sampling around 300 similar grain boundaries. The regions between the HAGB are epitaxial to the substrate (many different areas of the film were imaged with similar results).

The results to be presented next were obtained on a phase-pure, highly textured, 2.7-μm-thick film grown at General Electric [57]. The average grain size was in the range of 1–5 μm as determined from scanning electron microscopy (SEM). For this sample, it was estimated that there were roughly 400 grain boundaries in series and 40 in parallel, making a determination of the conduction path quite difficult.

In order to verify the textured microstructure and to determine the types of grain boundaries present in materials grown in this manner, TEM studies were also performed on this material. Figure 2.15 shows a TEM image of a few typical grain boundaries in a section of the textured $TlBa_2Ca_2Cu_3O_9$ thick film. This film is similar to those of [58], where a brick-wall structure has been noted. The same reference shows random in-plane misorientations. From the micrograph shown, we too can see the brick-wall structure. We may also observe several different grain boundaries. At point A, the lattice fringes abruptly discontinue at the interface with the adjacent grain. The point denoted as B

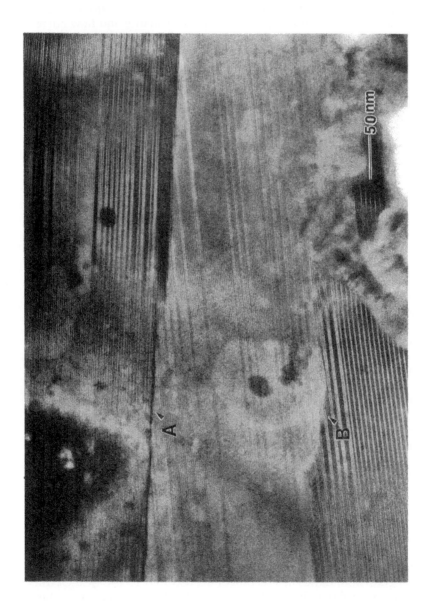

Fig. 2.15. A TEM micrograph showing several grain boundaries in a thick polycrystalline film of $TlBa_2Ca_2Cu_3O_y$. Point A denotes a HAGB, and point B is a LAGB. (Figure courtesy of D. J. Miller).

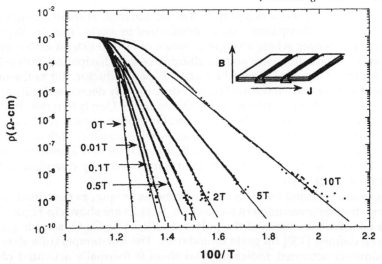

Fig. 2.16. Resistivity vs inverse temperature for a YBa$_2$Cu$_3$O$_x$ film containing many high-angle grain-boundaries in series. The solid lines indicate a low-temperature thermally activated resistivity with a temperature-independent energy barrier. The inset shows the geometry of the measurement (after [56]).

in the graph shows a very low angle grain boundary (LAGB), demonstrating that the sample consists of a distribution of grain boundaries, both high and low angle. Since different types of grain boundary exist in series and in parallel, at small current densities the current path will consist of LAGB with the strongest superconducting properties. The small current densities allow a redistribution of currents so as to avoid HAGB in parallel with the conduc-

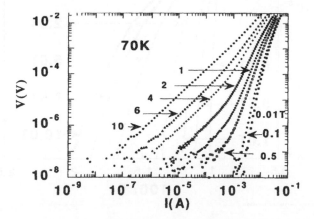

Fig. 2.17. Voltage vs current curves at different magnetic fields for the same sample measured in Fig. 2.16 at a constant temperature of 70 K (after [56]).

tion path. Therefore, even in a polycrystalline material, at small enough current densities, the dissipation may be dominated by vortex motion. Regions could exist, however, where a single or series of HAGB extend across much or the entire conducting path and no alternative low-dissipation path exists.

At zero field, the HAGB may be a good superconductor. But as the magnetic field is increased, the critical current density may decrease significantly. Again, there are two possible ways for this to occur. One is that the effective conduction path is decreasing because a smaller cross section is available after some of the grains in parallel are made inaccessible because of high-dissipation grain boundaries blocking access. The other is that there is a uniform reduction in the critical current of a grain boundary effectively extending across the total conduction path.

Data were collected by using a four-probe technique, as described in the earlier section on measurement techniques. Results are shown in Figure 2.16 for the resistive transitions of the polycrystalline $YBa_2Cu_3O_7$ sample grown with well-defined [100] tilt grain boundaries. The low-temperature data are approximately activated, indicating that there is thermally activated phase slip at the grain boundaries. At low temperatures and current densities, it is assumed that the dissipation is from the junction with the weakest superconducting properties.

Figure 2.17 is a plot of a family of V-I curves at 70 K for the $YBa_2Cu_3O_7$ sample. At larger current levels, one notices a negative curvature. At lower temperatures, a family of E-J curves generated in the same way show only

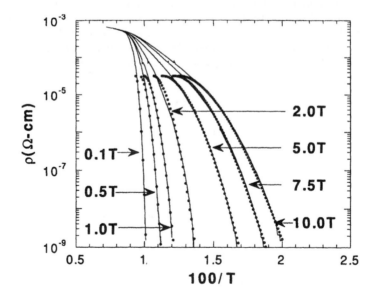

Fig. 2.18. Resistivity vs inverse temperature for a $TlBa_2Ca_2Cu_2O_9$ thick film in an applied magnetic field perpendicular to the film. The symbols are fits to the data using a thermally activated resistivity with a temperatue-independent energy barrier (after [56]).

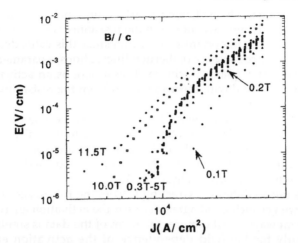

Fig. 2.19. Voltage vs current curves for the same sample as measured in Fig. 2.18 at a constant temperature of 40 K (after [56]).

this negative curvature. In what follows we examine this behavior and compare it with the results for the resistive transitions. This may be an indication of the signature of grain-boundary dissipation and a first step toward understanding whether vortex motion or phase slip at grain boundaries limit the critical current densities in polycrystalline materials. This is an important result to test various processing techniques.

Figure 2.18 shows the resistive transition for polycrystalline $TlBa_2Ca_2Cu_3O_9$. Often, in a polycrystalline system, a double transition is observed, one attributed to the intragranular material and one attributed to the intergranular coupling. Notice that there are not two well-defined resistive transitions for this sample. This indicates that the grain boundaries are dominating the resistivity of the system at all temperatures. The temperature dependence of the activation energy is similar to that observed for polycrystalline $Tl_2Ba_2CaCu_2O_8$ thin films, indicating that the grain boundaries are similar in the two systems.

Figure 2.19 shows a family of E-J curves at 40 K for the $TlBa_2Ca_2Cu_3O_9$ sample. Here we can observe an effect that may be directly attributed to HAGB. At small applied fields and large current densities, the E-J curves shift very noticeably as the field is increased. At slightly higher fields, however, the critical current density, as defined by an electric field criterion, is nearly magnetic field independent. The independence of field extends from 1 T to 4 T at this temperature. Also notice that at this temperature, there always exists negative curvature in the E-J plot when viewed on a log-log scale.

The data in Figure 2.17 are reminiscent of data generated from the Ambegaokar-Halperin (A-H) model [59]. Using this connection, one can fit the general shape of the E-J curve to the A-H model, which requires the choice of a value for the normal state resistance of the junction. In the analysis shown here, a temperature-dependent normal state resistance was used. This was dictated by the data, which showed nearly ohmic behavior at 10 T and at temperatures of 75 K and 78 K, although the resistance was noticeably smaller at 75 K. This is a fairly important assumption since this value determines the critical current, I_c, in the absence of thermal fluctuations, a parameter extracted from the model. This quantity will be shown here as an activation energy which will be compared with those extracted from the resistivity transitions shown in Figure 2.16.

The quality of the fits to the A-H model are not very good over the full range of data collected. This is probably due to nonuniformity over all grain boundaries. Nonetheless, with these deficiencies aside, values for $I_c(B,T)$ are extracted that are related to the activation energies extracted from the R vs T data. The results are shown in Figure 2.20.

Given the uncertainties in extracting I_c from the E-J curves, it is very surprising that we get such good agreement for the activation energy extracted in two different ways. Furthermore, the form of the data is similar to results shown recently for the field dependence of the activation energy for a Nb-AlO-Nb junction, where the coherent area of the Josephson junction is not the physical junction dimension, but a field-dependent area, which at large fields is U(H)~1/H [51].

In Figure 2.19, the E-J curves for the single thallium layer compound $TlBa_2Ca_2Cu_3O_9$ show only a region with "negative curvature" when viewed on a log-log plot. Similar behavior was also noted in $Tl_2Ba_2CaCu_2O_8$ polycrys-

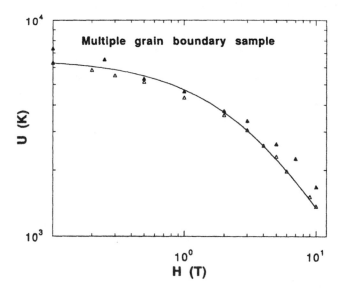

Fig. 2.20. Activation energies extracted from the resistivity vs inverse temperature data shown in Fig. 2.16. Also shown are the results of fitting the data of Fig. 2.17 to the Ambegaokar-Halperin model. The solid line is a fit to the results as described in the text (after [56]).

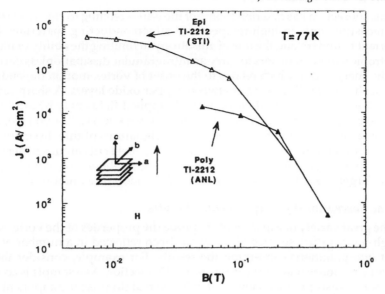

Fig. 2.21. Critical current density, J_c, for two samples of $Tl_2Ba_2CaCu_2O_8$. One is epitaxial and the other polycrystalline. The epitaxial material supports a much larger critical current density at small fields than the polycrystalline material.

talline films. In contrast, epitaxial $Tl_2Ba_2CaCu_2O_8$ films have shown positive curvature in the same region of parameter space. Therefore, this behavior is believed to be a result of the grain boundaries. Similar curvature is also observed for the $YBa_2Cu_3O_7$ sample discussed earlier at lower temperatures. This supports the assertion that the curvature is a result of the grain boundary properties.

The sharp reduction in the measured J_c at small applied fields can be attributed to the destruction of the superconducting properties of HAGB. The field-independent region is a result of the strong superconducting properties of the remaining LAGB as well as substantial pinning in the intragranular materials. Both types of grain boundaries could be observed in Figure 2.15.

Finally, we explicitly demonstrate the role of grain boundaries in limiting the measured critical current densities in high temperature superconductors. Shown in Figure 2.21 are critical current densities for two double Tl-O layer superconductors. One is epitaxial, and the other is polycrystalline. As may be seen directly from the plot, the critical currents are suppressed by an order of magnitude in the polycrystalline sample. However, the magnetic field at which the critical current tends to a small value is nearly the same for the two sample microstructures. At least in this case, the grain boundaries appear much more prevalent at small fields, while the vortex motion may still be dominant at higher magnetic fields. In the $YBa_2Cu_3O_7$ sample discussed earlier, the dissipation is dominated by grain boundaries at all magnetic fields since the coupling between grains is very weak.

This subsection has described some of the issues relating to electrical transport measurements in high-temperature superconducting materials. It is important to understand the role of anisotropy in limiting the ability of highly anisotropic superconductors to carry an intragranular dissipationless current, and the energy scale which relates to the onset of vortex motion depends on the Josephson coupling energy between copper oxide layers. A sharp reduction in the critical current density in small applied fields may be attributed to the existence of grain boundaries acting as weak links. The voltage/current characteristic of the material will show signatures of this dissipation at the grain boundaries. Vortex motion limits the high-field critical current densities in some polycrystalline materials (where the grain-boundary coupling energy is larger than the energy barrier for activated vortex motion).

2.4.3 ac Susceptibility – Experimental Results

The great variety of experiments to probe the properties of the vortex state in high-temperature superconductors has been reflected in a number of different interpretations concerning the results. For example, consider the ac screening response that will be discussed in this section. As a sample is cooled, the *apparent* susceptibility shows a step in the real (in-phase) component and a peak in the imaginary (out-of-phase) component. It was first proposed that these features indicated the onset of superconductivity [60]. Later comparison with hysteresis observed in the dc magnetization suggested identification with an "irreversibility line" [61,62]. More recently, there have been considerations of the vortex dynamics, such as the thermally assisted flux flow (TAFF) [63] and critical state models [64–67]. These involve microscopic details of the vortex behavior and possess empirical parameters. Later in this chapter, it will be seen that the results can be understood in terms of a model

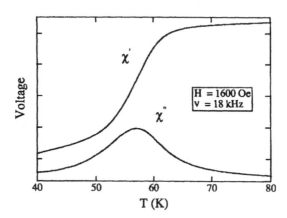

Fig. 2.22. Typical experimental data for the real (in-phase, χ') and imaginary (out-of-phase, χ'') components of the response signal for ac screening response measurements (after [23]).

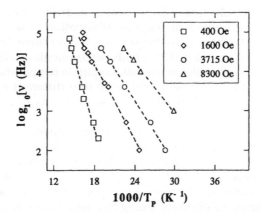

Fig. 2.23. Drive frequency, ν, and temperature, T_p, of the dissipation peak shown on an Arrhenius-like plot. The data were taken at four values of the dc magnetic field, cooling through the peak in the out-of-phase response and determining the peak temperature. Both ac and dc fields were perpendicular to the layers of the crystal. The dashed lines are fits to the data (after [23]).

based on the electromagnetic skin depth for the superconductor treated simply as a material of finite resistance with no account being taken of the microscopic details of the vortex dynamics. For experiments other than susceptibility, there have also been substantial differences in interpreting the results. For the silicon oscillator mentioned earlier, a peak was observed in the dissipation (imaginary component of the response); this was identified as a melting transition of the vortex lattice with the melting temperature defined as the location of the dissipation peak. The justification for a phase transition was claimed to be that below the melting temperature, the bulk modulus of the vortex lattice contributes an additional stiffness to the oscillator, while the response softens as the lattice melts [16]. As will become clear later, the weakness of this experiment was that it was made at a single frequency. For the low-frequency torsional oscillator measurements, the dissipation peak observed was also interpreted as due to a flux-lattice melting transition [17,18], while vibrating reed experiments have been explained in terms of thermally activated depinning (a decoupling of the flux line lattice from the reed) with diffusive flux motion at higher temperatures. With such different explanations, the origin of the behavior has been far from clear.

As described in the section on experimental techniques, the ac screening response measurements were made by monitoring the sample behavior as the susceptometer was cooled down from the normal state through any features in the response. For all the data taken on cooling, the measured response displayed a characteristic steplike transition in the real part (χ') of the output signal and the peak in the imaginary part (χ''). Typical results are shown in Figure 2.22, in this case for a dc field of 1600 Oe, an ac field of 5 mOe, and an ac frequency of 18 kHz. The magnitude of the response signal was typically a few megavolts. Results from a large number of different measurements are shown in Figure 2.23, which displays the dissipation peak temperature, T_p, as a function of different ac drive frequencies, ν, over three decades from 100 Hz to 100 kHz with ac and dc fields both perpendicular to the layers of the crystal. Each point on this plot represents a separate temperature scan at fixed drive frequency and dc magnetic field (with four different values shown in this plot). The temperature of the dissipation peak was obtained from the out-of-phase component of the response.

Notice that Figure 2.23 shows there is strong frequency dependence of the response at the magnetic fields studied, with significant shifts of T_p even at low frequencies. Such dispersion already has major implications for the identification of the peak as being due to a phase transition, such as a melting transition of the vortex lattice. Since a true phase transition is defined in the zero-frequency limit, the observed strong frequency dispersion seriously calls into question the identification of a dissipation peak obtained at fixed frequency as being due to a phase transition. This result contrasts sharply to some interpretations of other results mentioned earlier in this chapter, and exposes the weakness of such measurements at a single frequency when there may be important time scales in the system. For example, simple viscoelastic response within a physical system may lead to a peak in the out-of-phase response at finite frequency, without any need to invoke a melting transition to explain the behavior.

Figure 2.24 shows the dissipation peak temperature as a function of magnetic field, measured at four different frequencies. The strong frequency dispersion between the four sets of data points clearly rules out the lines as phase

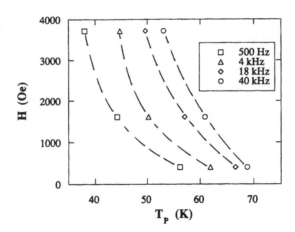

Fig. 2.24. Dissipation peak temperature, T_p, as a function of magnetic field for four different frequencies of measurement. This type of plot has been claimed to show phase boundaries, but the strong frequency dispersion questions identification as due to a phase transition within the field and temperature range studies. These measurements are all within the mixed state—the fields are above the lower critical field boundary $H_{c1}(T)$ and below the upper critical field $H_{c2}(T)$ (after [23]).

boundaries. The discussion to follow will show how the results may be understood.

It was mentioned earlier that one of the advantages of the screening response technique is that the ac drive amplitude may be varied quite easily in order to probe the vortex dynamics with different forces. Despite that simplicity, the results and interpretations for different probe amplitudes have been sources of significant controversy in this field of research. Some authors have claimed to find no amplitude dependence in the screening response [68–70], while others have observed a large temperature reduction in the response curve for increased driving amplitudes [71–74]. Linear response has been analyzed within such models as the TAFF model, while nonlinear response has been viewed in terms of critical state behavior. Typically, nonlinear response has been observed at higher values of the drive amplitude, while the response remained linear at low amplitudes. Similar behavior has been observed with mechanical oscillators, where the oscillation amplitude was reported to have been kept small ≤ 150 nm to ensure linear response [16].

The measurements on $Bi_2Sr_2CaCu_2O_x$ described in this section were made with a driving field of about 5 mOe. Compared with the ac field strengths used in most other susceptibility experiments, such as those referred to above, this is a very low excitation level, and the response was observed to be linear over about two decades above this amplitude. At substantially larger driving fields (≥ 1 Oe), however, the peak did shift to lower temperature, signaling the transition to nonlinear response. It thus appears that for these measurements, amplitude dependence is to be observed only at high values of the driving field. Of course, the change from linear to nonlinear response with ac field amplitude involves sample specific quantities such as the pinning and hence need not be universal. But the nature of the response and the comparison with theory still depend critically on the regime one is in. Given the results seen in this work, it should be reasonable to use the data within amplitude-independent models, without referring to models with complicated critical state profiles, etc.

Geshkenbein et al. [75] showed that the ac magnetic response of a superconductor in the mixed state may be very simply understood from classical electromagnetism (without the need to consider the presence of flux lines in the problem). According to this explanation, the peak in the dissipative part of the response is due simply to a size effect, and occurs when the electromagnetic skin depth, δ, becomes comparable to the sample size, d. As a starting point for this treatment, consider the superconductor as being simply a material of finite resistance, ignoring the vortex behavior that gives rise to that resistance. Since it is the vortex motion that is producing the dissipation within the material, and hence the resistance, the flux motion is contained in this theory as the underlying mechanism, but none of its details will be important. That is the fundamental simplification of this treatment, which proceeds just as for a regular, resistive metal. The derivation of the electromagnetic skin depth and the particular conditions for the penetration of an ac magnetic field may be found in any basic text of electromagnetism [76,77]. The response function, $\chi(\omega)$, is then given by

$$\chi(\omega) = \frac{\int B(x)dx}{\mu_0 H_0 d} - 1, \tag{2.11}$$

which gives real and imaginary components of

$$\chi'(\omega) = \frac{\sinh u + \sin u}{u(\cosh u + \cos u)} - 1 \tag{2.12}$$

$$\chi''(\omega) = \frac{\sinh u - \sin u}{u(\cosh u + \cos u)} \tag{2.13}$$

where

$$u = \frac{d}{\delta} = \left(\frac{d^2 \mu_0 \omega}{2\rho}\right)^{1/2}. \tag{2.14}$$

The maximum in χ'' occurs at $u = 2.254$, and for a given sample size, the peak position is therefore set by the ratio of the frequency to the resistivity.

As pointed out by Geshkenbein et al., the physical interpretation of such behavior is very simple. Considering the peak in χ'', for $\delta \to 0$, the screening is complete: $\chi' \to -1$ and $\chi'' \to 0$. This corresponds to the high-frequency limit; and with the short skin depth, the field cannot penetrate. For $\delta \to \infty$, the field penetrates the sample completely and the out-of-phase signal $\chi'' \to 0$. This is the zero-frequency, or dc, limit; and in that case the skin depth is no longer a relevant parameter for the problem. The peak in the out-of-phase component is then a clear consequence of the variation of the skin depth through the appropriate sample dimension, with the precise position of the peak determined by a geometrical factor of order unity (in this particular slab geometry, that factor is 2.254). Bulk response is obtained in the limit $\omega \to 0$; but if the resistivity is thermally activated so that $\rho \approx \exp(-U/k_B T)$, then the reciprocal of the peak temperature varies as $\ln(\omega)$. This means that the zero-frequency limit applies on a logarithmic scale, so that seemingly low frequencies of, for example, 1 Hz are far from the true limit. The skin-depth treatment of the screening response also shows a need for greater precision in the common description of such an experiment as measuring *susceptibility*. At temperatures above the peak in the imaginary response, the signal is indeed due to ·bulk screening within the material (since the skin depth is comparatively long relative to the size of the sample), and thus the response may be interpreted as due to the bulk susceptibility. Below the peak, however, the response is due to a size effect arising from the skin depth and can no longer be interpreted as due to the true volume susceptibility. Although this may appear to be a question of terminology rather than physics, the importance lies in the origin of the behavior that is measured in this type of experiment.

To test the skin-size effect explanation described above, one must compare, on the one hand, measurements of the electromagnetic screening response and, on the other, determinations of the electrical resistivity.

A small flake of the crystal used for the ac screening response studies was used for direct four-point resistance measurements in magnetic field. Measurements were made using a sample card for electrical transport on the probe of the gas flow dewar, in place of the susceptometer used for screening response. The flake was mounted to the sample card using a thin layer of grease to the copper sample block. The transport measurements were made using an ac technique. Magnetic fields were applied using a water-cooled electromagnet, with a maximum field of about 6000 Oe. With the contacts at the edges of the sample, it was possible to use a van der Pauw technique to obtain the normal state resistivity, which was found to be approximately 600 μΩcm. (The van der Pauw technique provides a way of obtaining the resistivity of a lamellar sample from two four-point measurements of the resistance, together with the sample thickness [78].) The electrical transport was found to be linear and independent of frequency from dc up to several kilohertz. The temperature dependence of the resistance in three different values of the magnetic field is shown on an Arrhenius plot in Figure 2.25.

For constant values of the dc field, the frequency of the dissipation peak and the directly measured in-plane resistivity, ρ, at the same temperature were used to calculate the ratio of the measured sample size, d, to the inferred electromagnetic skin depth, δ, where

$$\delta = \left(\frac{2\rho}{\mu_0\omega}\right)^{1/2}. \qquad (2.15)$$

The ratio d/δ calculated from the values of ρ and ω for each dissipation peak was found to be 1.9±0.2 for magnetic fields of 400, 1600, and 3715 Oe.

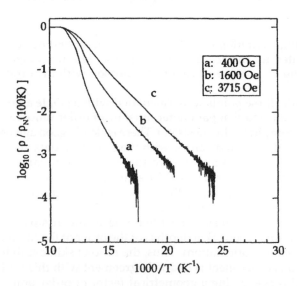

Fig. 2.25. Arrhenius plot of the directly measured resistivity of a crystal for three different magnetic field values. The normal state resistivity, ρ_N, (100 K) was 600 μΩ cm (after [23]).

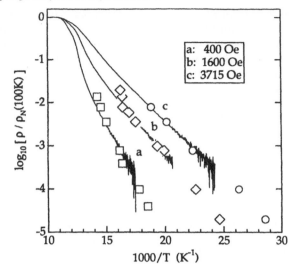

Fig. 2.26. Comparison of the directly measured resistivity of a crystal with inferred values from screening response measurements. Data are shown for three different magnetic field values. The squares, diamonds and circles show the resistivity inferred from ac screening response measurements in fields of 400, 1600 and 3715 Oe, respectively, according to skin-size effects. A single value for the ratio d/δ of 1.9 was used for all the screening response data (after [23]).

Using the single criterion of $d/\delta = 1.9$, the corresponding resistance values were calculated for the ac screening response data from Figure 2.23. These points are shown in Figure 2.26 for comparison with the directly measured resistance.

To reiterate, these points were calculated with a single adjustable parameter for them all, and that parameter must be of order unity for the skin-size explanation to be physically consistent. There is clearly good agreement between the derived and measured values, with the screening data allowing extrapolation to lower temperatures where the resistance became too low to be measured by direct means. The 400 Oe screening response data have a somewhat steeper slope compared with the measured resistivity, although there is still good agreement with the value of $d/\delta = 1.9$.

According to the skin-size effect hypothesis, the dissipation peak occurs when the electromagnetic skin depth is of the order of the sample size (for the appropriate sample orientation). Thus, the ratio of skin depth to sample size obtained in this experiment is in good agreement with this explanation over a substantial range, giving a geometrical factor of order unity. (For a given geometry there will be a factor relating sample size and skin depth at T_p, stated earlier to be 2.25 for a slab geometry with the ac field perpendicular to the c-axis. This factor likely differs for the ac field parallel to the c-axis, as in our case.) The frequency independence of the measured resistance justifies the use of a single resistivity value in the determination of the skin depth. It also shows that the frequency dependence of the dissipation peak does not arise from strong dispersion of the resistivity.

It can be seen in Figures 2.23 and 2.25 that both the dissipation peak frequency and the resistance show approximately exponential dependence on $(1/T)$ at low temperatures. The graphs show similar behavior, emphasizing the connection between the measurements that was formally expressed in the treatment of the skin depth, δ. For both cases, exponential behavior is generally interpreted as thermal activation of vortices across pinning barriers, with a dependence of the form $v_{p}, \rho \approx \exp(-U/k_B T)$ [79–82]. This dependence motivates more detailed correlation of the resistance to the ac screening response within thermal activation models. By so doing, it will better connect the macroscopic response to microscopic parameters. For this connection it is necessary to turn to a discussion of more detailed theories of the behavior of a superconductor in the mixed state in response to an applied ac magnetic field. That situation is quite complicated, as it involves nonlocal effects through the coupling of supercurrent density and vortex displacements and the effects of collective as well as individual pinning. Such a treatment is outside the scope of this discussion; for further details the reader is referred to other publications [83–86]. For the purposes of this chapter, it is sufficient to point out that the microscopic models allow information about the pinning energies within the material to be extracted from the screening response and resitivity data.

This experiment has shown how the relatively simple technique of ac screening response (ac susceptibility) can provide useful information about the properties of the mixed state of high-temperature superconductors. For the particular experimental conditions studied in this experiment, the response was found to be consistent with a model based on the electromagnetic skin depth for the material and independent of the precise microscopic details of the vortex behavior [87]. As stated at the beginning of this section, there are many similar experiments to probe the vortex behavior. Some of the variations have been to measure different high-temperature superconducting material systems, at stronger driving amplitudes in order to probe the critical state, in different orientations to examine the anisotropy, etc. Clearly, these constitute a significant class of experiments in understanding the properties of the high-temperature superconductors.

2.4.4 dc Magnetization Results

2.4.4.1 Measurements of the Upper Critical Magnetic Field, $H_{c2}(T)$

One of the most important fundamental parameters for describing the superconducting state of a material is the upper critical magnetic field, H_{c2}. Conventional techniques for measuring this parameter, however, have not been very successful in high-temperature superconducting materials. As introduced earlier, the application of magnetic field to a high-temperature superconducting material has the effect of severely broadening the resistive transition. A reasonable value for the normal state resistivity is never reached at temperatures substantially below the zero field critical temperature. Therefore, it is very difficult to measure H_{c2} using this technique, since no feature of the data is sharp enough to give a well-defined result. The diamagnetic transition is also systematically reduced without significant broadening in low-temperature superconductors, yielding well-defined $H_{c2}(T)$ values. But in high-temperature superconducting materials, this transition is smeared in a very similar way to the resistive transitions. Therefore, a more detailed analysis is needed.

Close to the superconducting transition where the order parameter is small, the magnetization is expected to vary linearly with magnetic field as [88]

$$-4\pi M = \frac{H_{c2}(T) - H}{1.16(2\kappa^2 - 1)},$$

(2.16)

where κ is a ratio of the fundamental length scales ($\frac{\lambda}{\xi}$) describing the superconducting state. It is possible to fit this expression to a region of the measured magnetization very near $T_c(H)$. Figure 2.27 shows the data and the straight line fits for a commercially available Nb-Ti sample which has a fairly modest T_c of 10 K [89]. Clearly there is good agreement over a fairly substantial range of magnetization. Using the number extracted for H_{c2} from this fitting yields a slope, $dH_{c2}/dT = -3\,T/K$, similar to results present in the literature for this material [90].

To test the relevance of this technique to high-temperature superconducting materials, we show in Figure 2.28 similar analysis to that performed on the NbTi wire for data from a sample of $YBa_2Cu_3O_{7-\delta}$ with a critical temperature of 62 K [91]. Again, good agreement is seen between the expression (2.16) and the data. Extracting the numbers from each of the fits gives an upper

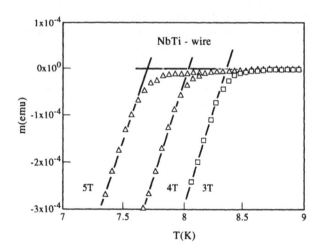

Fig. 2.27. Temperature dependence of the magnetic moment of a sample of NbTi wire measured in three magnetic fields. The solid lines indicate the construction of the phase boundary using Equation (2.16) (after [89]).

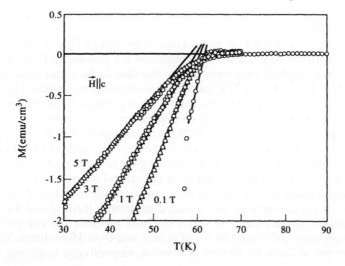

Fig. 2.28. Temperature dependence of the magnetization of a $YBa_2Cu_3O_{7-\delta}$ crystal in an applied magnetic field. Again, the construction of the phase boundary using Equation (2.16) is shown (after [91]).

critical field slope, $dH_{c2}/dT = -2$ T/K.

Similar measurements have been made on many different materials with varying degrees of success. For example, this linear model does not apply to BSCCO single crystals [92,93], where a somewhat different analysis was applied incorporating a logarithmic field dependence of the magnetization [94]. This analysis requires the introduction of several new concepts that will take this discussion far from its intended thrust. Therefore, we refer the readers to [95] for further clarification.

Although the expression (2.16) was derived in 1957, the details of its use in determining the upper critical field were not utilized systematically until the discovery of high-temperature superconducting materials. Therefore, we present it here because it represents a new technique developed for high-temperature superconductors for measuring a fundamental parameter.

2.4.5 Critical Currents (Transport and Magnetization)

Critical current densities have been measured in many different material geometries requiring different measurement techniques. Because of the number of measurements of critical current densities performed using transport and magnetization techniques, it is important to understand the connection between the two, since the extracted values are rarely identical, even for materials processed in very similar ways [96]. In fact, even on the same sample there is no reason to expect that the two techniques will yield identical results, although most investigators think that the results may be directly compared. The work presented in this section, which is an extension of the section comparing measurement techniques, should help in understanding the differences in the J_c results of the two measurements for a homogeneous specimen [97].

Measurements of the J_c of an epitaxial thin film of $Tl_2Ba_2CaCu_2O_8$ determined by both transport and magnetization are presented here. The magnetic field was applied perpendicular to the film surface and measurements were made at 5 K, 40 K, and 77 K for each measurement technique. The samples used for the transport part of the study were patterned into microbridges. The magnetization measurements were performed on another section of the same "chip" in a Quantum Design SQUID magnetometer. The sample was also measured when it was whole, cut into 4 sections, and cut into 16 sections at each of the three temperatures confirming the homogeneity of the sample.

Bean's model [25] predicts that for a cylindrical sample, the hysteresis in the magnetic moment per unit volume is proportional to J_c and the sample radius, R_0. If the currents are not induced uniformly throughout the sample, these relations are not valid. In order to check whether the currents are induced everywhere, the sample is cut into 4 and then 16 sections. The magnetic moment reduces by factors of 2 and 4, respectively, implying that the

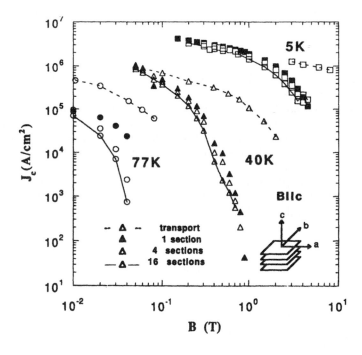

Fig. 2.29. Plot of the J_cs determined from Bean's model for the magnetization measured on three different sample sizes (solid symbols denote 1 section, open symbols denotes 4 sections, and open symbols with solid line denote 16 sections). Also shown is the J_c as determined directly by transport measurements (open symbols with dashed line) (after [97]).

currents induced are not limited by weak links and flow throughout the entire sample: a necessary condition to evaluate R_o in Bean's model.

Having established the validity of Bean's model for these samples, the comparisons of the J_cs obtained by both transport and magnetization may now be presented. The results are shown in Figure 2.29. Notice that the J_cs determined by using the Bean's model for the three different subdivided configurations are nearly identical. J_cs determined by direct transport, however, agree with the magnetization results only at very small magnetic fields and the lowest two temperatures, whereas at higher fields the two results are different by orders of magnitude. When attempting to compare transport and magnetization J_cs, as was mentioned in the section comparing measurement techniques, care must be taken to be aware of the differences in the electric field criterion used to determine the J_c. The electric field criterion which "determines" J_c in the magnetization measurement is not a constant, unlike in the case of a transport measurement. It is a function of the magnetic field, temperature, and the time of the measurement with respect to the stabilized

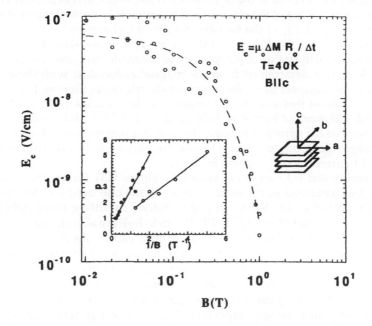

Fig. 2.30. Plot of the effective electric field critreion as a function of magnetic field at 40 K. Inset: Values of p determined by approximating the E-J curve by two points—the transport J_c with the E_c of 10^{-6} V/cm and the magnetization J_c with the effective E_c from Fig. 2.2 (hollow circles). Also shown are the values of p determined by fitting the experimental E-J curves (solid circles). The solid lines are fits of the data to 1/B (after [97]).

applied magnetic field.

Figure 2.30 shows the magnitude and the magnetic field dependence of the effective electric field criterion used in the magnetization measurements at 40 K over the same range of B as Figure 2.1. These values are derived directly from the relaxation over the length of time of the measurement using Faraday's law. The expression in cylindrical coordinates may be approximated as E_c(V/m) $\approx (\Delta B / \Delta t) R_o$. In general, magnetic relaxation values as functions of the parameters listed here are difficult to determine in measurements of the magnetic hysteresis using a SQUID magnetometer. However, the software associated with the magnetometer used here averages over three different values of the magnetic moment, recording each value. The time for each measurement was approximately 50 seconds. Each subsequent measurement is lower than the one before, indicating that this is a good estimate of the magnetic relaxation at that time.

The scatter in the plot can be attributed to variations in the length of time needed for the magnetic field and temperature to reach target values. The electric field criterion decreases when the applied field is increased and the overall magnetic flux gradient is decreased as a result of the reduction in the pinning (or J_c). Because of the nonlinear electric field-current density characteristic, a knowledge of the electric field criterion is important when attempting to compare the J_c of the two measurement techniques.

Assuming that the E-J curve follows a single power-law, $E \sim J^p$, one can then determine the value of p necessary to reconcile the J_c and effective electric field values determined from the magnetization data with those determined from transport using the transport electric field criterion, $E_{cT} = 10^{-4}$ V/m. The results of this analysis are shown as hollow circles in the inset of Figure 2.30. The work of Kim et al. [98], Zeldov et al. [99], and Koch et al. [100] indicates that the values of p determined in this manner are within a reasonable range. However, to confirm this, a comparison with experimental electric field-current density curves is made.

Experimental E-J curves are shown in Figure 2.31 for another film prepared and patterned in the same manner as the one used for the measurements reported so far. Values of p determined by fitting these data in the range of electric field from 10^{-7} to 10^{-6} V/cm to $E \sim J^p$ are also shown in the inset of Figure 2.30 as solid circles. As may be observed in the inset, at low field values where p is large, the log(E) vs log(J) curves cut sharply toward zero dissipation, and the electric field criterion used is of little importance, changing the measured J_c only by a small amount. At larger magnetic fields, however, where the E vs. J curves become closer to ohmic, the electric field criterion becomes increasingly important. This is the range where the effective electric field criterion used in the magnetization measurement differs the most from that used in the transport measurement. This is exactly the explanation of the discrepancies in the J_c values of Figure 2.29. The differences in J_c are due to the differences in electric field criterion, differences that become more significant at larger applied fields where the value of the electric field criterion is the most important.

The inset of Figure 2.30, which shows the estimated exponents as well as

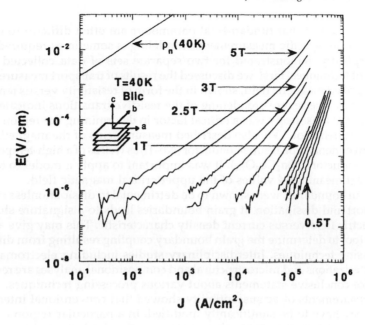

Fig. 2.31. Experimental E-J curves measured for a sample prepared in the same manner as the one used to generate the results of previous figures. The curves between those labeled by magnetic field values are evenly spaced in field (after [97]).

those determined by fitting the low end dissipation of the E vs J data to the function $E \sim J^p$, indicates that the magnetic field dependencies are both roughly $1/B$. The magnitudes are considerably different, however. One would expect that the estimated and measured values of p would be closer if we could measure the dc E-J curve to a lower dissipation level.

The main points of this section are summarized in the following necessary steps to make comparisons of the magnetization and transport critical currents: (1) Bean's model must be satisfied; in other words, the model must be applied when $H \gg H^*$, and it must be verified that $R=R_o$, the total sample radius; (2) the electric field criterion must be determined for each magnetic field and temperature at which a comparison is intended since the criterion is not a constant in the magnetization measurement. This is primarily important in high-temperature superconductors because of the large relaxation effects; and (3) the E-J characteristic must be known since it is not a power law, and any comparison at different electric field criteria requires this information.

2.5 Summary

In this chapter we described basic measurement techniques and gave some background material relevant to understanding the results of various measurements. We discussed specific examples to demonstrate the novel interpretations of several measurements compared to low-temperature superconducting materials.

It was shown that fundamental parameters are often difficult to extract directly from specific measurements. Modeling is sometimes required. This was explicitly demonstrated for two separate sets of data collected using different techniques. First, we discussed the results of transport measurements in an applied magnetic field, shown in the form of resistivity versus temperature curves. The severe broadening of the resistive transitions indicated that the material anisotropy was a critical factor in determining the region of "irreversible behavior." We also described measurements of the magnetization of a conventional type II superconductor (Nb-Ti) as well as of a high-temperature superconductor ($YBa_2Cu_3O_{7-\delta}$). It was important to apply a model to be able to extract meaningful values of the upper critical magnetic field.

Grain boundaries were shown to be detrimental to dissipationless current transport and dissipation at grain boundaries leads to a signature shape of the electric field versus current density characteristic. This may give a diagnostic tool to determine the grain boundary coupling resulting from different processing techniques. Interdisciplinary studies including electromagnetic characterizations and microstructural and compositional analyses are required to make conclusive statements about various processing techniques.

Measurements of ac susceptibility showed that conventional interpretations may have to be significantly modified. In a particular region of magnetic field and temperature where vortex motion leads to dissipation, the susceptibility results may be interpreted as in the case of a normal metal in an oscillating magnetic field. These results were shown to be analogous to direct electrical transport measurements.

Specific criteria were presented for the application of Bean's model to extract a reliable value of the critical current density. We compared the various techniques introduced in this chapter by considering the dissipation level sampled by each measurement. In the case of the SQUID magnetometer results, the magnetic relaxation, and therefore the effective electric field at which the measurement was performed, was not a constant. Thus, care must be taken when comparing critical currents using such an effective electric field, which depends on magnetic field and temperature.

High-temperature superconductivity has provided unanticipated detail in many physical properties. These properties are often complex and require systematic measurement of high quality materials to determine the nature of the physical processes. We hope that the discussions presented here give some guidelines to those new to the field of high-temperature superconductivity for measuring and interpreting experimental data.

2.6 References

1.   M. Tinkham, *Introduction to Superconductivity*, McGraw-Hill, New York (1974).
2.   T. Van Duzer and C. W. Turner, *Principles of Superconductive Devices and Circuits*, Elsevier, New York (1981).
3.   *Physics of High-Temperature Superconductors*, ed. S. Maekawa and M. Sato, Springer-Verlag, Berlin (1992).

4.   A. A. Abrikosov, Sov. Phys. JETP **5**, 1174 (1957).
5.   W. .H. Kleiner, L. M. Roth, and S. H. Autler, Phys. Rev. A **133**, 1226 (1964).
6.   G. J. Dolan, F. Holtzberg, C. Field, and T. R. Dinger, Phys. Rev. Lett. **62**, 2184 (1989).
7.   P. W. Anderson, Phys. Rev. Lett. **9**, 309 (1962).
8.   B. Batlogg, Physics Today **44**, 44 (1991).
9.   G. K. White, *Experimental Techniques in Low-Temperature Physics*, Oxford University Press, Oxford, UK (1987).
10. R. C. Richardson and E. N. Smith, *Experimental Techniques in Condensed Matter Physics at Low Temperatures*, Addison Wesley, Redwood City, Calif. (1988).
11. J. G. Bednorz and K. A. Muller, Z. Phys. B **64**, 189 (1986).
12. Operating Instructions for Detachable Tail Research Dewars, Janis Research Company.
13. Product Catalog, LakeShore Cryotronics, Westerville, Ohio.
14. R. P. Vasquez, B. D. Hunt, and M. C. Foote, Appl. Phys. Lett. **53**, 2692 (1988).
15. Q. Y. Chen et al., in *Magnetic Susceptibility of Superconductors and Other Spin Systems*, ed. R. Hein, T. Francavilla, and D. Liebenberg, Plenum, New York (1992).
16. P. L. Gammel, L. F. Schneemeyer, J. V. Waszczak, and D. J. Bishop, Phys. Rev. Lett. **61**, 1666 (1988).
17. R. G. Beck, D. E. Farrell, J. P. Rice, D. M. Ginsberg, and V. G. Kogan, Phys. Rev. Lett. **68**, 1594 (1992).
18. D. E. Farrell, J. P. Rice, and D. M. Ginsberg, Phys. Rev. Lett. **67**, 1165 (1991).
19. P. Esquinazi, J. Low Temp. Phys. **85**, 139 (1991).
20. J. Pankert, G. Marbach, A. Comberg, P. Lemmens, P. Froning, and S. Ewert, Phys. Rev. Lett. **65**, 3052 (1990).
21. P. Lemmens, P. Froning, S. Ewert, J. Pankert, G. Marbach, and A. Comberg, Physica C **174**, 289 (1991).
22. Y. Horie, T. Miyazaki, and T. Fukami, Physica C **175**, 93 (1991).
23. D. G. Steel and J. M. Graybeal, Phys. Rev. B **45**, 12643 (1992); D.G. Steel, Ph.D. thesis, Massachusetts Institute of Technology, Cambridge, Mass. (1993).
24. W. Meissner and R. Ochsenfeld, Naturwissenschaften **21**, 787 (1933).
25. C. P. Bean, Rev. Mod. Phys. **36**, 31 (1964).
26. T. P. Orlando and K. A. Delin, *Foundations of Applied Superconductivity*, Addison-Wesley Publishing Company, Reading, Mass. (1991).
27. S. T. Ruggiero and D. Rudman, *Superconducting Devices*, Academic Press, New York (1990).
28. U. Welp, private communication (1994).
29. S. Foner, Rev. Sci. Instrum. **30**, 548 (1959).
30. A. G. Swanson, Y. P. Ma, J. S. Brooks, R. M. Markiewicz, and N. Miura, Rev. Sci. Instrum. **61**, 848 (1990).
31. S. Jin, R. B. van Dover, T. H. Tiefel, and J. E. Graebner, Appl. Phys. Lett. **58**, 868 (1991).
32. D. J. Miller, S. Sengupta, J. D. Hettinger, D. Shi, K. E. Gray, A. S. Nash,

and K. C. Goretta, Appl. Phys. Lett. **61**, 2823 (1992).

33. D. J. Miller, private communication (1994).
34. J. Hu, private communication (1994).
35. D. Dimos, P. Chaudhari, and J. Mannhart, Phys. Rev. **B41**, 4038 (1990).
36. K. Char, M. S. Clelough, S. M. Garrison, N. Newman, and G. Zaharchuk, Appl. Phys. Lett. **59**, 733 (1991).
37. C. L. Jia, B. Kabius, K. Urban, K. Herrmann, C. J. Cui, J. Schubert, W. Zander, A. I. Braginski, and C. Heiden, Physica C **175**, 545 (1991).
38. J. P. Hong, T. W. Kim, H. R. Fetterman, A. H. Cardona and L. C. Bourne, Appl. Phys. Lett. **59**, 991 (1991).
39. D. S. Ginley, J. F. Kwak, E. L. Venturini, B. Morosin, and R. J. Baughman, Physica C **140**, 42 (1989).
40. J. D. Hettinger, A. G. Swanson, W. J. Skocpol, J. S. Brooks, J. M. Graybeal, P. M. Mankiewich, R. E. Howard, B. L. Straughn, and E. G. Burkhardt, Phys. Rev. Lett. **62**, 2044 (1989).
41. D. H. Kim, K. E. Gray, R. T. Kampwirth, J. C. Smith, D. S. Richeson, T. J. Marks, J. H. Kang, J. Talvacchio, and M. Eddy, Physica C **177**, 431 (1991).
42. T. M. Palstra, B. Batlogg, L. F. Schneemeyer, and J. V. Waszczak, Phys. Rev. Lett. **61**, 1662 (1988).
43. P. L. Gammel, L. F. Schneemeyer, and D. J. Bishop, Phys. Rev. Lett. **66**, 953 (1991).
44. D. J. Bishop, P. L. Gammel, D. A. Huse, and C. A. Murray, Science **255**, 165 (1992).
45. D. J. Bishop, P. L. Gammel, and D. A. Huse, Scientific American **268**, 48 (1993).
46. W. K. Kwok, U. Welp, G. W. Crabtree, K. G. Vandervoort, R. Hulscher, and J. Z. Liu, Phys. Rev. Lett. **64**, 966 (1990).
47. Y. Iye, S. Nakamura, and T. Tamagai, Physica C **159**, 433 (1989).
48. B. D. Josephson, Phys. Lett. **1**, 251 (1962).
49. G. Briceno, M. F. Crommie, and A. Zettl, Phys. Rev. Lett. **66**, 2164 (1991).
50. K. E. Gray and D. H. Kim, Phys. Rev. Lett. **70**, 1693 (1993).
51. D. H. Kim, K. E. Gray, and J. H. Kang, Phys. Rev. B **45**, 7563 (1992).
52. V. Ambegaokar and A. Baratoff, Phys. Rev. Lett. **10**, 486(1963).
53. J. R. Clem, preprint (1993).
54. A. C. Wright, T. K. Xia, and A. Erbil, , Phys. Rev. B **45**, 5607 (1992).
55. J. W. Ekin, H. R. Hart, Jr., and A. R. Gaddipati, J. Appl. Phys. **68**, 2285(1990).
56. J. D. Hettinger, D. H. Kim, D. J. Miller, J. G. Hu, K. E. Gray, J. E. Sharping, K. Daly, C. Pettiette-Hall, J. E. Tkaczyk, and J. Deluca, IEEE Trans. Appl. Supercond. **3**, 1211(1993).
57. J. A. DeLuca, M. F. Garbauskas, R. B. Bolon, J. G. McMullen, W. E. Balz, and P. L. Karas, J. Mater. Res. **6**, 1415(1991).
58. D. J. Miller, J. G. Hu, J. D. Hettinger, K. E. Gray, J. E. Tkaczyk, J. DeLuca, P. L. Karas, and M. F. Garbauskas, Appl. Phys. Lett. **63**, 556 (1993).
59. V. Ambegaokar and B.I. Halperin, Phys. Rev. Lett. **22**, 1364 (1969).
60. T. K. Worthington, W. J. Gallagher, D. L. Kaiser, F. H. Holzberg, and T. R. Dinger, Physica C **153**, 32 (1988).
61. K. Heine, J. Tenbrink, and M. Thoner, Appl. Phys. Lett. **55**, 2441 (1989).

62. A. P. Malozemoff, T. K. Worthington, Y. Yeshurun, F. Holtzberg, and P. H. Kes, Phys. Rev. B **38**, 7203 (1988).

63. J. van den Berg, C. J. van der Beek, P. H. Kes, J. A. Mydosh, M. J. V. Menken, and A. A. Menovsky, Supercond. Sci. Technol. **1**, 249 (1989).

64. J. Z Sun, M. J. Scharen, L. C. Bourne and J. R. Schrieffer, Phys. Rev. B **44**, 5275 (1991).

65. R. B. Flippen, T. R. Askew, M. S. Osofsky, Physica C **201**, 391 (1992).

66. W. Xing, B. Heinrich, J. Chrzanowski, J. C. Irwin, H. Zhou, A. Cragg, and A. A. Fife, Physica C **205**, 311 (1993).

67. J. Wang, H. S. Gamchi, K. N. R. Taylor, G. J. Russell, and Y. Yue, Physica C **205**, 363 (1993).

68. J. van den Berg, C. J. van der Beek, P. H. Kes, J. A. Mydosh, M. J. V. Menken, and A. A. Menovsky, Supercond. Sci. Technol. **1**, 249 (1989).

69. D. G. Steel and J. M Graybeal, Physica C **189**, 2227 (1991).

70. P. H. Kes, J. Aarts, J. van den Berg, C. J. van der Beek, and J. A. Mydosh, Supercond. Sci. Technol. **1**, 242 (1989).

71. Ch. Heinzel, Ch. Neumann, and P. Ziemann, Europhys. Lett. **13**, 531 (1990).

72. J. H. P. M. Emmen, G. M. Stollman, and W. J. M. De Jonge, Physica C **169**, 418 (1990).

73. C. J. van der Beek, M. Essers, P. H. Kes, M. J. V Menken, and A. A. Menovsky, Supercond. Sci. Technol. **5**, 260 (1992).

74. C. J. van der Beek, Ph.D. thesis, Leiden University, Leiden, The Netherlands (1992).

75. V. B. Geshkenbein, V. M Vinokur, and R. Fehrenbacher, Phys. Rev. B **43**, 3748 (1991).

76. L. D. Landau and E. M. Lifshitz, *Course of Theoretical Physics Vol. 8: Electrodynamics of Continuous Media*, Pergamon, New York (1960).

77. J. D. Jackson, *Classical Electromagnetism*, Wiley, New York (1975).

78. L. J. van der Pauw, Philips Res. Repts. **13**, 1 (1958).

79. T. T. M. Palstra, B. Batlogg, R. B. van Dover, L. F. Schneemeyer, and J. V. Waszczak, Phys. Rev. B **41**, 6621 (1990).

80. J. H. P. M. Emmen, V. A. M. Brabers, and W. J. M. De Jonge, Physica C **176**, 137 (1991).

81. M. Inui, P. B. Littlewood, and S. N. Coppersmith, Phys. Rev. Lett. **63**, 2421 (1989).

82. P. L. Gammel, J. Appl. Phys. **67**, 4676 (1990).

83. E. H. Brandt, Phys. Rev. Lett. **67**, 2219 (1991).

84. M. W. Coffey and J. R. Clem, Phys. Rev. Lett. **67**, 386 (1991).

85. M. W. Coffey and J. R. Clem, Phys. Rev. B **45**, 9872 (1992).

86. M. W. Coffey and J. R. Clem, Phys. Rev. B **45**, 10527 (1992).

87. S. Sammarappuli, A. Schilling, M. A. Chernikov, H. R. Ott and Th. Wolf, Physica C **201**, 159 (1992).

88. A. A. Abrikosov, Zh. Eksp. Teor. Fiz. **32**, 1442 (1957).

89. U. Welp, personal communication (1994).

90. M. Suenaga, A. K. Ghosh, Y. Xu, and D. O. Welch, Phys. Rev. Lett. **66**, 177 (1991).

91. K. G. Vandervoort, U. Welp, J. E. Kessler, H. Claus, G. W. Crabtree, W. K.

Kwok, A. Umezawa, B. W. Veal, J. W. Downey, and A. P. Paulikas, Phys.Rev. B. **43**, 13042 (1991).

92. W. Kritscha, F. M. Sauerzopf, H. W. Weber, G. W. Crabtree, Y. C. Chang, and P. Z. Jiang, Physica C **178**, 59 (1991).

93. P. H. Kes, C. J. van der Beek, M. P. Maley, M. E. McHenry, and D. A. Huse, Phys. Rev. Lett. **67**, 2383 (1991).

94. U. Welp. S. Fleshler, W. K. Kwok, K. G. Vandervoort, J. W. Downey, B. W. Veal, and G. W. Crabtree, in *Physical Phenomena at High Magnetic Fields*, ed. E. Manousakis, Addison-Wesley, Redwood City, Calif. (1991).

95. Z. Hao, J. R. Clem , M. W. McElfresh, L. Civale, A.P. Malozemoff, and F. Holtzberg, Phys. Rev. B **43**, 2844 (1991); Z. Hao and J. R. Clem, Phys. Rev. Lett. 67, 2371 (1991).

96. K. Wantanabe, T. Matsushita, N. Kobayashi, H. Kawabe, E. Aoyagi, K. Hiraga, H. Yamane, H. Kurosawa, T. Hirai, and Y. Muto, Appl. Phys. Lett. **56**, 1490 (1990).

97. J. D. Hettinger, D. H. Kim, K. E. Gray, U. Welp, R. T. Kampwirth, and M. Eddy, App. Phys. Lett. **60**, 2153 (1992).

98. D. H. Kim, K. E. Gray, R. T. Kampwirth, and D. M. McKay, Phys. Rev. B 39, 6249 (1990).

99. E. Zeldov, N. M. Amer, G. Koren, A. Gupta, M. W. McElfresh, and R. J. Gambino, Appl. Phys. Lett. **56**, 680 (1990).

100. R. H. Koch, V. Foglietti, W. J. Gallagher, G. Koren, A. Gupta, and M. P. A. Fisher, Phys. Rev. Lett. **63**, 1511 (1989).

2.7 Recommended Readings for Chapter 2

General texts on superconductivity:

1. *Introduction to Superconductivity*, M. Tinkham, McGraw-Hill, New York, 1974.

2. *Foundations of Applied Superconductivity*, T. P. Orlando and K. A. Delin, Addison-Wesley Publishing Co., Reading, Mass, 1991.

3. *Principles of Superconductive Devices and Circuits*, T. Van Duzer and C. W. Turner, Elsevier, New York, 1981.

Properties of high-temperature superconductors:

4. *High-Temperature Superconductivity*, J. W. Lynn, ed. Springer-Verlag, New York, 1990.

5. *Physics of High-Temperature Superconductors*, S. Maekawa and M. Sato, eds., Springer-Verlag, New York, 1992.

Low-temperature measurement techniques:

6. *Experimental Techniques in Low-Temperature Physics*, G. K. White, Oxford University Press, Oxford, UK, 1987.

7. *Experimental Techniques in Condensed Matter Physics at Low Temperatures*, R. C. Richardson and E. N. Smith, Addison Wesley, Redwood City, Calif., 1988.

ac Susceptibility measurements:

8. *Magnetic Susceptibility of Superconductors and Other Spin Systems*, R. Hein, T. Francavilla, and D. Liebenberg, eds., Plenum, New York, 1992.

3

Crystal Structures and Phase Equilibria

F. Izumi and E. Takayama-Muromachi

3.1 Structural Features of Superconducting Copper Oxides

Many superconducting oxides have appeared since the discovery of high-temperature superconductivity in $(La_{1-x}Ba_x)_2CuO_4$ in 1986. An enormous amount of data on physical and chemical properties, crystal and defect structures, and phase relations has already been reported in the literature. These oxides are most interesting from the viewpoint of not only solid-state physics but crystal chemistry owing to the wide variety of their crystal structures.

The average and local structures of these copper oxides were investigated by neutron powder diffraction, single-crystal x-ray diffraction, high-resolution transmission electron microscopy (HRTEM), extended x-ray absorption fine structure (EXAFS), and so on. In particular, neutron powder diffraction, combined with Rietveld analysis, has been used most widely to refine the structural parameters of new superconductors and to clarify relationships between defects and physical properties. It is usually difficult to grow single crystals of superconductors large enough to collect their neutron diffraction data, whereas polycrystals can be prepared easily in most cases but often contain impurities. Even in such cases, the structure of the main phase can be refined by using the multiphase capability if the structures of the impurities are known and their contents are relatively small. In addition, HRTEM provides us with direct structural information with which initial structural models for Rietveld analysis can be constructed. Therefore, the average structures of superconductors can be determined by using only polycrystalline samples, without solving phase problems.

3.1.1 Coordination and Bonding Behavior

Physical and structural information indicates that superconductivity is achieved in copper oxides only when their crystal structures, chemical compositions, and carrier concentrations satisfy the following prerequisites:
1. Cu atoms form coordination polyhedra of O atoms, that is, $[CuO_4]$ square planes (Figure 3.1a), $[CuO_5]$ quadrangular pyramids (Figure 3.1b), and $[CuO_6]$ octahedra (Figure 3.1c), which are joined together by sharing corners to give two-dimensional $^{IV}CuO_2$, $^{V}CuO_2$, and $^{VI}CuO_2$ sheets, respectively. (A roman numeral at the upper left of each element will be used to designate its coordination number throught this chapter. Thus, $^{VIII}Ca^{2+}$ denotes the eight-coordinated Ca^{2+} ion.)

2. The two-dimensional CuO_2 sheets are at the electronic hearts of the copper oxides and essential for their superconductivity.
3. Carriers responsible for superconductivity are holes (p-type) or electrons (n-type) doped into the CuO_2 conduction sheets.
4. The superconducting transition temperature, T_c, reaches the maximum on doping an appropriate concentration of carriers, that is, 0.15–0.2 per Cu atom, into the CuO_2 sheets.
5. Hole-doped superconductors have $^VCuO_2$ or $^{VI}CuO_2$ sheets.
6. Electron-doped superconductors always contain $^{IV}CuO_2$ sheets.

The c-axis is usually set perpendicular to the CuO_2 sheet even if other settings of axes are possible. For example, the space group of the low-temperature orthorhombic form for $(La_{1-x}Sr_x)_2CuO_4$ is selected to be not standard Cmca but Bmab, where the a-, b-, and c-axes in Cmca correspond to a-, č-, and b-axes, respectively. This conventional manner of setting axes will be also adopted in this chapter.

The formula of CuO_2 means that this two-dimensional sheet is composed of Cu and O atoms in the amount-of-substance ratio of 1 to 2, as shown in Figure 3.2. These Cu and O atoms are not necessarily located on the same (flat) plane in the sense that their z coordinates may slightly differ from each other. Therefore, "CuO_2 sheet" is preferred to "CuO_2 plane" in what follows.

The O atom that forms a Cu–O bond parallel to the c-axis is often referred to as "apical oxygen." We adopt this conventional usage in this chapter. The term "apical" is not very appropriate, however, because O atoms on the CuO_2 sheet also occupy "apical" positions (corners) in the coordination polyhedra of Cu.

Each Cu atom on the CuO_2 sheet has four firm Cu–O bonds on this sheet. Their Cu–O interatomic distances of 0.189–0.197 nm indicate a large degree of covalency. On the other hand, bonds between Cu and apical oxygen are much longer, ranging in length from 0.22 to 0.29 nm. In particular, axial Cu–O bonds in most Hg- and Tl-containing superconductors are extremely long; apical oxygen is far nearer to Hg or Tl than to Cu. Such coordination behavior is typical of the Jahn-Teller distortion expected for a d^9 ion. In fact, high-energy spectroscopy, etc., revealed that holes occupy the 2p orbital of O rather than the 3d orbital of Cu. Each hole may as well be represented as $(Cu–O)^+$ in view of the overlap of Cu 3d and O 2p orbitals.

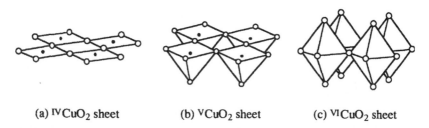

(a) $^{IV}CuO_2$ sheet   (b) $^VCuO_2$ sheet   (c) $^{VI}CuO_2$ sheet

Figure 3.1. (a) $^{IV}CuO_2$, (b) $^VCuO_2$, and (c) $^{VI}CuO_2$ sheets contained in superconducting oxides. Filled circle: Cu, open circle: O.

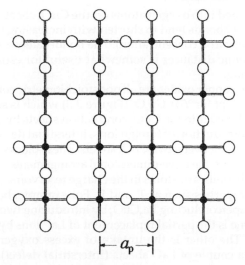

Figure 3.2. Configuration of Cu (filled circle) and O (open circle) on the CuO_2 sheet. Double the Cu–O bond length (0.378–0.394 nm) is approximately equal to the unit-cell dimension of the perovskite-type compound, a_p.

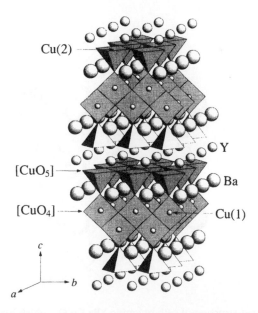

Figure 3.3. Crystal structure of orthorhombic $YBa_2Cu_4O_8$ containing $[CuO_5]$ pyramids and $[CuO_4]$ square planes. Double chains of $[CuO_4]$ planes run parallel with the b-axis $(a < b < c)$.

Holes may be doped into oxygen atoms on the CuO_2 sheet, apical oxygen, or both of them. Cu–O bonds tend to shorten with increasing hole concentration (or decreasing electron concentration). The bond-valence-sum calculated from Cu–O interatomic distances is somewhat useful for estimating the degree of carrier doping.

Superconducting copper oxides are all nonstoichiometric compounds and/ or solid solutions except for $YBa_2Cu_4O_8$ (Figure 3.3) which is self-doped with holes. Carriers are doped into the CuO_2 conduction sheets by charge transfer resulting from the introduction of foreign ions, interstitial defects, and vacancies in structural blocks between two CuO_2 sheets. These blocks are often called "charge reservoirs." Thus, the three-dimensional arrangements, oxidation states, and occupation probabilities of atoms in the charge reservoirs, markedly affect superconducting properties such as T_c and J_c. For example, hole carriers can be doped into nonsuperconducting La_2CuO_4 by introducing two different types of lattice defects. One is the partial replacement of La^{3+} ions by Sr^{2+} ions (substitutional defect). The other is the uptake of excess oxygen at interstitial positions between a couple of LaO sheets (interstitial defect).

For high-temperature superconductors, a purely ionic picture is not strictly valid, and considerable covalent character must be conceded. Nevertheless, they obey principles for building up of ionic crystals rather faithfully, and various methods of estimating structural stabilities or properties such as Pauling's rule, effective ionic radii, bond-valence sums, and Madelung energies are applicable to them without serious errors.

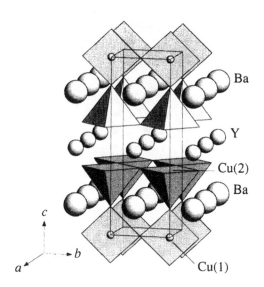

Figure 3.4. Idealized structure of orthorhombic $YBa_2Cu_3O_7$, which has two Cu sites: Cu(1) in a one-dimensional chain of $[CuO_4]$ square planes and Cu(2) within $[CuO_5]$ pyramids. Cu(2) atoms form the two-dimensional CuO_2 sheets. In reality, the O site on the z = O plane is slightly deficient.

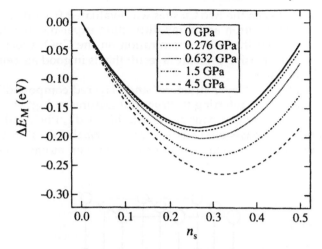

Figure 3.5. Relationship between the relative change in Madelung energy, DE_M, and the number of holes per Cu(2), n_s, at five different pressures in $YBa_2Cu_4O_8$ [1].

Two kinds of cations with comparable effective ionic radii, r, can occupy the same site randomly without changing the structures of copper oxides, for example, $^{IX}La^{3+}$ (r = 0.1216 nm) and $^{IX}Sr^{2+}$ (r = 0.131 nm) in $(La_{1-x}Sr_x)_2CuO_4$. Unless the radii of substituting and substituted ions differ by more than 15%, a wide range of substitution may be expected at room temperature. Higher temperatures and/or higher pressures permit a somewhat greater tolerance. On the other hand, cations with larger ionic radii generally prefer occupying sites with larger coordination numbers. Two kinds of cations are perfectly ordered at two sites with different chemical environments provided the difference in ionic radii between them is very large, as in $^{VIII}Y^{3+}$ (r = 0.1019 nm) and $^{X}Ba^{2+}$ (r = 0.152 nm) in $YBa_2Cu_4O_8$ (Figure 3.3) and $YBa_2Cu_3O_7$ (Figure 3.4).

If coordination numbers suitable for alkaline-earth, rare-earth, Cu, and other ions become too large, and/or oxidation states of Cu become too high in order to maintain electrical neutrality, one or more oxygen positions lack partially or perfectly to give stable structures. In $YBa_2Cu_3O_7$ (Figure 3.4), $^{VIII}Y^{3+}$ and $^{X}Ba^{2+}$ ions occupy two different sites corresponding to the A site in the P-type compound. Cu ions are situated at two sites corresponding to the B site: four-coordinated Cu(1) in the $[CuO_4]$ chain and five-coordinated Cu(2) on the CuO_2 sheet. In other words, O atoms are removed in such a way that the coordination numbers suitable for these three cations are satisfied while keeping a reasonable oxidation state of Cu.

Madelung-energy calculations under an approximation of 100% ionic character are sometimes used to estimate the distribution of carriers in crystal lattices. For example, this method was applied to explain the very large pressure coefficient of T_c, $dT_c/dp = 5.5\,K/GPa$, in $YBa_2Cu_4O_8$ with a T_c of approximately 80 K. This superconductor has two Cu sites: Cu(1) in the double chain of $[CuO_4]$ square planes and Cu(2) on the CuO_2 sheet (Figure 3.3). The distribution of holes between the two Cu sites was evaluated from its crystal data obtained by *in situ* neutron and x-ray diffraction experiments (Figure 3.5). Figure 3.5 shows that the hole concentration on the CuO_2 sheet increases appreciably with increasing pressure, a result that is in good agreement with the large dT_c/dp value.

The technical terms "layer structure" and "layered compound" have often been misused when referring to structural features of high-T_c superconductors. Typical layered compounds are graphite, CdI_2, PbO, and micas. In these compounds, interlayer bindings (e.g., van der Waals forces) are far weaker than bonds within layers, which often enables foreign atoms to be interca-

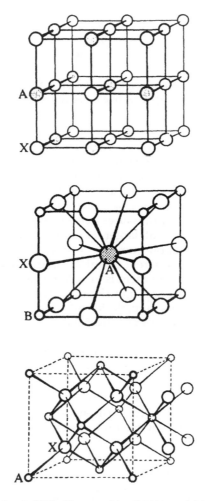

Figure 3.6. (*a*) Rock-salt (AX), (*b*) perovskite (ABX_3), and (*c*) fluorite (AX) type compounds. A and B are cations, and X is an anion (halide or oxide ion) [3] (by permission of Oxford University Press).

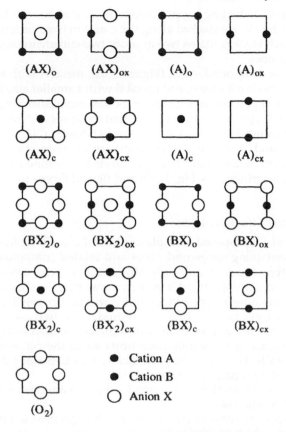

Figure 3.7. Meshes to describe the structures of superconducting copper oxides and related compounds (adapted from [4]).

lated between the layers. Strictly speaking, the above two terms should be applied for only Bi-containing superconductors, where the binding between two BiO sheets is weak enough to intercalate I_2 between them, in marked contrast with all the other superconductors.

3.1.2 Three Types of Structural Blocks

Crystal structures of most superconductors containing Cu are regarded as consisting of two-dimensional blocks (slabs) related to three popular structural types illustrated in Figure 3.6: rock-salt (R), perovskite (P), and fluorite (F) types. Other superconductors contain structural blocks associated with pyroxenes and carbonates. The structures of superconducting oxides are stabilized by the introduction of lattice defects, for example, formation of solid solutions through the occupation of one site by two or more cations and that of oxygen deficiency in R- and/or P-type structural blocks.

Next, we outline the method of constructing structures for superconductors. Most superconducting copper oxides have structures where R-, P-, F-, and other type blocks are stacked along the c-axis in fixed orders and periods. In addition, each block contains two or more two-dimensional sheets such as CuO_2 and BiO ones.

In the P-type compound, ABX_3 (Figure 3.6a), metal A with a larger size is coordinated to twelve X atoms, and metal B with a smaller size to six X atoms octahedrally. In superconducting copper oxides, metal sites corresponding to the B site in the P-type compound are occupied by atoms with relatively smaller sizes, that is, in almost all cases, Cu, and rarely Nb in $Sr_2(Nd_{1-x}Ce_x)_2NbCu_2O_{10}$. On the other hand, A sites in P-type units can accommodate alkaline-earth metals, rare-earth metals, and Pb, R-type units by alkaline-earth metals (Ca, Sr, and Ba), rare-earth metals, Hg, Tl, Pb and Bi, and F-type units by rare-earth metals.

3.1.2.1 Minimum Structural Units

Santoro et al. [2] proposed a simple method of describing the (ideal) structures of Cu-containing superconductors and related compounds composed of P- and R-type blocks using only alphanumeric characters. Izumi and Takayama-Muromachi modified it in such a way that the structures of superconductors containing F-type blocks and partially occupied O sites can be represented as well. In their revised method, each two-dimensional sheet perpendicular to the c-axis is divided into meshes shown in Figures 3.7 and 3.8, and a sequence of these minimum units along the c-axis is represented one-dimensionally. The unit to be periodically repeated along the [001] direction is enclosed by a pair of brackets.

The names of the meshes in Figures 3.7 and 3.8 are derived according to the following simple rules:
1. The number of atoms included in one mesh is given as a form of a chemical formula, which is enclosed by parentheses.
2. Subscripts o and c indicate that a cation, A or B, in a mesh is situated at its origin and center, respectively.
3. Subscript x means that a mesh is translated by half the period along the [100] direction (y for translation along the [010] direction). For example,

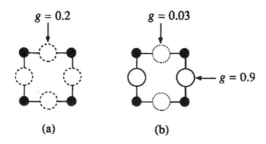

Figure 3.8. Two meshes where oxygen sites are partially deficient: (a) $(CuO_{2x0.2})_o$ mesh and (b) $(CuO_{0.9+0.03})_o$ mesh. Symbol g denotes the occupation factor.

$(BX)_{ox}$ results from the translation of $(BX)_o$ by $a/2$.

4. In case a site is partially deficient, (number of equivalent sites contained in the mesh) x (occupation factor) is given as a subscript (Figure 3.8a). Subscript 1x may be omitted. When two or more anion sites are present in the mesh, the sum of their occupancies is described (Figure 3.8b).

5. The coordination number of each cation may be indicated by a roman numeral at its upper left, for example, $({}^VCuO_2)_o$ and $({}^{IX}LaO)_c$.

Cations A and B in Figures 3.7 and 3.8 occupy positions corresponding to the A and B sites in the P-type compound ABX_3 (Figure 3.6b). The two-dimensional CuO_2 sheets are made up of $(CuO_2)_o$ or $(CuO_2)_c$ meshes (B = Cu and X = O). That is, it is P-type blocks that contain the CuO_2 conduction sheets. In the P-type compound, $({}^{VI}BX_2)_o$ and $({}^{XII}AX)_c$ meshes are stacked alternately: $[({}^{VI}BX_2)_o({}^{XII}AX)_c]$. The structure shown in Figure 3.5b partially includes the $(AX)_c$ mesh common to the R-type structure (Figure 3.6a). In fact, the R-type structure is described as $[({}^{VI}AX)_c({}^{VI}AX)_o]$ using the two meshes in Figure 3.7. The twelve-coordinated A cation in the P-type compound, however, has a chemical environment quite different from the six-coordinated one in the R-type compound.

Two-dimensional sheets with compositions of AX or BX_2 result from the full occupation of the X anion sites in Figures 3.7 and 3.8. On the other hand, anion-deficient sheets, A or BX, are formed when X atoms are deleted from AX and BX_2 sheets, respectively. Meshes containing only anions, (X_2), are found in the structure of the F-type compound AX_2 (Figure 3.6c): $[({}^{VIII}A)_o(X_2)({}^{VIII}A)_c(X_2)]$. In this structure, every A atom is surrounded by eight X atoms arranged at the corners of a cube, and every X by four A arranged tetrahedrally.

Superconducting copper oxides always contain groups of meshes related to the P-type structure: $(AO_m)_c(CuO_2)_o(A'O_n)_c$ or $(AO_m)_o(CuO_2)_c(A'O_n)_c$, where A = A' or A ≠ A'. O atoms may be completely removed from the AO and A'O sheets (m = 0 and/or n = 0), or two or more cations are not distributed randomly but ordered, so that the coordination form and number of each cation are both satisfied.

The classification given in Figures 3.7 and 3.8 is devised so that we can easily understand the principle of constructing superconductors from the basic components, idealizing their structures. In reality, atoms that are strictly not located on the same plane are often contained in one mesh, and those displaced from the origin, middle point of the edge, or center are located at their ideal positions.

A group of two-dimensional sheets composed of meshes that exhibit R-, P-, and F-type mode of connection will hereafter be referred to as a "unit."

3.1.2.2 Methods of Joining P-, R-, and F-type Units

Joints of two different units fall into the following four types:

1. Two P-type units are simply connected with each other without sharing any sheet (abbreviated as P-P).
2. An AO sheet is shared by P- and R-type units (P = R).
3. One corner-sharing chain of $[CuO_4]$ square planes shares edges with another $[CuO_4]$ chain to form a double one-dimensional chain (P/P).

4. P- and F-type units share a sheet containing only cations in eight-coordination (P=F). Every cation on this sheet is coordinated to four O^{2-} ions on the O_2 sheet and to four O^{2-} ions on the CuO_2 sheet.

The periodic unit along the c-axis is enclosed by a pair of brackets using the above expressions in a similar way to mesh groups, for example, [F = P - P =].

Because the structure of $YBa_2Cu_3O_7$ has an infinite P-type unit, no joint is contained in it. In such a case, the infinite unit is expressed as P_∞. $YBa_2Cu_4O_8$ (orthorhombic, Ammm, a < b < c), which is structurally very similar to $YBa_2Cu_3O_7$ as shown in Figures 3.3 and 3.4 contains crystallographic shears where two P-type units with a thickness of c/2 are joined together in the P/P manner through displacement by half the b period. Thus, the c/2 value in $YBa_2Cu_4O_8$ is larger than that in $YBa_2Cu_3O_7$ by the difference in widths of double and single chains of $[CuO_4]$ square planes.

3.1.2.3 Pyroxene and Carbonate Blocks

Up to 1991, the structures of superconducting copper oxides had been regarded as the stacks of blocks closely related to P-, R-, and F-type structures along the [001] direction. This principle of structural construction was broken through by the appearance of new compounds containing novel structural blocks: $(Y_{1-x}Ca_x)Sr_2GaCu_2O_7$ (Figure 3.9) and a family of superconductors containing nonspherical anions such as CO_3^{2-}.

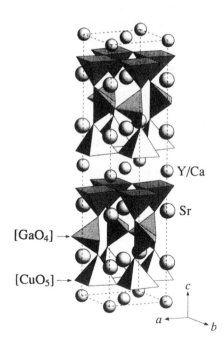

Figure 3.9. Body-centered orthorhombic unit cell ($b < a < c$) of $(Y_{1-x}Ca_x)Sr_2GaCu_2O_7$.

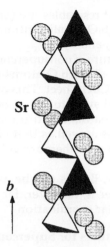

Figure 3.10. $(GaO_3^{3-})_\infty$ chain in $(Y_{1-x}Ca_x)Sr_2GaCu_2O_7$.

In $(Y_{1-x}Ca_x)Sr_2GaCu_2O_7$, Sr_2GaO_3 slabs between two CuO_2 sheets contain $(GaO_3^{3-})_\infty$ zigzag chains of $[GaO_4]$ tetrahedra that run parallel with the b-axis by sharing two corners of each tetrahedron (Figure 3.10). Two remaining O atoms in $[GaO_4]$ occupy apical positions of Cu on the conduction sheet. The Sr_2GaO_3 block cannot be related to any of the P-, R-, and F-type structures.

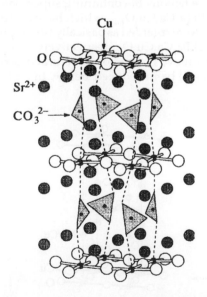

Figure 3.11. Crystal structure of $Sr_2CuO_2CO_3$. Broken lines approximately parallel to the c-axis are weak bonds between Cu atoms on the CuO_2 sheets and O atoms in CO_3^{2-} ions.

The $(GaO_3^{3-})_\infty$ chain somewhat resembles the $(SiO_3^{2-})_\infty$ chains of $[SiO_4]$ tetrahedra in pyroxenes such as $MgSiO_3$. Similar infinite chains also lie in the crystal lattices of $CuGeO_3$, CrO_3, SO_3, and $Pb(PO_3)_2$.

Figure 3.11 shows the structure of nonsuperconducting $Sr_2CuO_2CO_3$, where Cu has an oxidation state of +2. Two O atoms in the CO_3^{2-} ion occupy the apical positions of Cu on the CuO_2 sheet. This oxide carbonate can be made superconducting by (1) partial substitution of BO_3^{3-} ions for CO_3^{2-} ions as in $Sr_2CuO_2(CO_3)_{1-x}(BO_3)_x$ and (2) partial substitution of Cu atoms (plus O atoms bonded to them) for CO_3^{2-} ions as well as that of Ba^{2+} ions for Sr^{2+} ions as in $(Ba_{1-x}Sr_x)_2Cu_{1+y}O_{2+2y+z}(CO_3)_{1-y}$. Both of them are unique methods of doping holes into the CuO_2 sheet.

As will be described later, B sites can be assigned to Ga and C atoms high-handedly. Such a manner is, however, far from reasonable in view of their tetrahedral and triangular coordination.

3.2 Descriptions of the Structures for Superconducting Copper Oxides

3.2.1 Selection of Representative Copper Oxides

Yvon and François [5] summarized the crystal structures of superconducting copper oxides that had been discovered till the beginning of 1989. They selected the following twelve compounds as representative superconductors whose structures were analyzed in detail by single-crystal x-ray diffraction and neutron powder diffraction: $(Nd_{1-x}Ce_x)_2CuO_4$, $(La_{1-x}Sr_x)_2CuO_4$, $(Nd_{1-x}Sr_x)(Nd_{1-y}Ce_y)CuO_4$, $YBa_2Cu_3O_7$, $YBa_2Cu_4O_8$, $Y_2Ba_4Cu_7O_{14+z}$, $Pb_2Sr_2(Y_{1-x}Ca_x)Cu_3O_8$, $TlBa_2CaCu_2O_7$, $TlBa_2Ca_2Cu_3O_9$, $Tl_2Ba_2CuO_6$, $Tl_2Ba_2CaCu_2O_8$, and $Tl_2Ba_2Ca_2Cu_3O_{10}$. Three famous Bi-containing superconductors, $Bi_2Sr_2CuO_{6+z}$, $Bi_2Sr_2CaCu_2O_{8+z}$ and $Bi_2Sr_2Ca_2Cu_3O_{10+z}$, which have complex incommensurate modulated structures, were regarded as basically isomorphous with $Tl_2Ba_2CuO_6$, $Tl_2Ba_2CaCu_2O_8$, and $Tl_2Ba_2Ca_2Cu_3O_{10}$, respectively.

On the basis of the above method of structural description, the structures of the twelve oxides will be introduced below. In addition, we will discuss

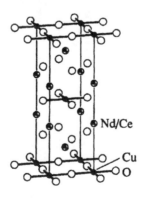

Figure 3.12. Electron-doped superconductor, $(Nd_{1-x}Ce_x)_2CuO_4$.

1) $(Nd_{1-x}Ce_x)_2CuO_4 \equiv M'_2CuO_4$: $[(^{IV}CuO_2)_o(^{VIII}M')_o(O_2)(^{VIII}M')_o(^{IV}CuO_2)_c(^{VIII}M')_o(O_2)(^{VIII}M')_c]$ [P=F=P=F=]

2) $(La_{1-x}Sr_x)_2CuO_4 \equiv M_2CuO_4$: $[(^{VI}CuO_2)_o(^{IX}MO)_c(^{IX}MO)_o(^{VI}CuO_2)_c(^{IX}MO)_o(^{IX}MO)_c]$ [P-P-]

3) $(Nd_{1-x}Sr_x)(Nd_{1-y}Ce_y)CuO_4 \equiv MM'CuO_4$: $[(O_2)_o(^{VIII}M')_c(^{IV}CuO_2)_o(^{VIII}M')_c(^{IX}MO)_o(^{VI}CuO_2)_o]$ [F=P-P=]

4) $YBa_2Cu_3O_7$: $[(^{IV}CuO)_o(^{X}BaO)_c(^{V}CuO_2)_o(^{VIII}Y)_c(^{V}CuO_2)_o(^{X}BaO)_c]$ [P/P/]

5) $YBa_2Cu_4O_8$: $[(^{VIII}Y)_c(^{V}CuO_2)_o(^{X}BaO)_o(^{IV}CuO)_o(^{IV}CuO)_o(^{X}BaO)_{cy}(^{V}CuO_2)_{oy}(^{VIII}Y)_{cy}]$ P_∞

6) $Y_2Ba_4Cu_7O_{14+z}$: $[(CuO_{p+q})_o(BaO)_c(^{V}CuO_2)_o(^{VIII}Y)_c(^{V}CuO_2)_o(^{X}BaO)_{cy}(^{IV}CuO)_o(^{IV}CuO)_o(^{X}BaO)_{cy}(^{V}CuO_2)_{oy}(^{VIII}Y)_{cy}$ $(^{V}CuO_2)_o(BaO)_{cy}(CuO_{p+q})_{oy}(BaO)_{cy}(^{V}CuO_2)_{oy}(^{VIII}Y)_{cy}(^{V}CuO_2)_{oy}(^{X}BaO)_c]$ [P/P/]

7) $Pb_2Sr_2(Y_{1-x}Ca_x)Cu_3O_8 \equiv Pb_2Sr_2M'Cu_3O_8$: $[(^{VIII}M')_o(^{V}CuO_2)_c(^{IX}SrO)_o(^{V}PbO)_c(^{II}Cu)_o(^{V}PbO)_c(^{IX}SrO)_o(^{V}CuO_2)_c]$ [P/P/]

8) $(La_{1-x}Ca_x)_2(Ca_{1-y}La_y)Cu_2O_6 \equiv M_2M'Cu_2O_6$: $[(^{VIII}M')_c(^{V}CuO_2)_o(^{IX}MO)_c(^{IX}MO)_o(^{V}CuO_2)_c(^{IX}MO)_o(^{V}CuO_2)_c]$ [P-P-]

9) $(Ba_{1-x}Nd_x)_2(Nd_{1-y}Ce_y)_2Cu_3O_{8+z}$
$[(CuO_{2x0.46})_o(MO)_c(^{VIII}M')_o(^{VIII}M')_o(O_2)_o(^{VIII}M')_o-(CuO_{2x0.46})_o(MO)_c(^{V}CuO_2)_c(MO)_o(^{V}CuO_2)_c(^{VIII}M')_o(O_2)_o(^{VIII}M')_c(^{V}CuO_2)_o(MO)_c]$ [P-P-]

$M_2M'_2Cu_3O_{8+z}$:
$\equiv$ [P=F=P=F=]

10) $Sr_2(Nd_{1-x}Ce_x)_2NbCu_2O_{10} \equiv NbSr_2M'_2Cu_2O_{10}$: $[(^{V}NbO)_o(^{X}SrO)_c(^{V}CuO_2)_o(^{VIII}M')_c(O_2)(^{VIII}M')_o(^{V}CuO_2)_c(^{X}SrO)_o-(^{V}NbO_2)_c$ $(^{X}SrO)_o(^{V}CuO_2)_c(^{VIII}M')_o(O_2)(^{VIII}M')_c(^{V}CuO_2)_o(^{X}SrO)_c]$ [P=F=P=F=]

11) $TlBa_2CaCu_2O_7$: $[(^{VI}TlO)_o(^{IX}BaO)_c(^{V}CuO_2)_o(^{VIII}Ca)_c(^{V}CuO_2)_o(^{IX}BaO)_c]$ [R=P=]

12) $TlBa_2Ca_2Cu_3O_9$: $[(^{VI}TlO)_o(^{IX}BaO)_c(^{V}CuO_2)_o(^{VIII}Ca)_c(^{V}CuO_2)_o(^{VIII}Ca)_c(^{V}CuO_2)_o(^{IX}BaO)_c]$ [R=P=]

13) $Tl_2Ba_2CuO_6$: $[(^{VI}TlO)_o(^{IX}BaO)_o(^{VI}CuO_2)_c(^{IX}BaO)_o(^{VI}TlO)_o(^{VI}TlO)_c(^{IX}BaO)_o(^{VI}CuO_2)_c(^{IX}BaO)_o(^{VI}TlO)_c]$ [R=P=R=P=]

14) $Tl_2Ba_2CaCu_2O_8$: $[(^{VIII}Ca)_o(^{V}CuO_2)_c(^{IX}BaO)_o(^{VI}TlO)_o(^{VI}TlO)_c(^{IX}Ca)_c(^{V}CuO_2)_o(^{IX}BaO)_c(^{VI}TlO)_o-(^{VI}TlO)_o(^{IX}BaO)_o(^{V}CuO_2)_c]$ [P=R=P=R=]

15) $Tl_2Ba_2Ca_2Cu_3O_{10}$: $[(^{IV}CuO_2)_c(^{VIII}Ca)_o(^{V}CuO_2)_c(^{IX}BaO)_o(^{VI}TlO)_o(^{VI}TlO)_c(^{IX}Ca)_c(^{V}CuO_2)_o-(^{VIII}Ca)_c(^{V}CuO_2)_o(^{IX}BaO)_c(^{IX}BaO)_o(^{VI}TlO)_o$ $(^{VI}TlO)_c(^{IX}BaO)_o(^{V}CuO_2)_c(^{VIII}Ca)_c(^{V}CuO_2)_c]$ [P=R=P=R=]

16) $HgBa_2CuO_4$: $[(^{II}Hg)_o(^{VIII}BaO)_c(^{VI}CuO_2)_o(^{VIII}BaO)_c]$ [R=P=]

17) $HgBa_2CaCu_2O_6$: $[(^{II}Hg)_o(^{VIII}BaO)_c(^{V}CuO_2)_o(^{VIII}Ca)_c(^{V}CuO_2)_o(^{VIII}BaO)_c]$ [R=P=]

18) $HgBa_2Ca_2Cu_3O_8$: $[(^{II}Hg)_o(^{VIII}BaO)_c(^{V}CuO_2)_o(^{VIII}Ca)_c(^{V}CuO_2)_o(^{VIII}Ca)_c(^{V}CuO_2)_o(^{VIII}BaO)_c]$ [P=R=P=R=]

19) $(Sr_{1-x}La_x)CuO_2 \equiv M'CuO_2$: $[(^{IV}CuO_2)_o(^{VIII}M')_c]$ P_∞

20) $(Y_{1-x}Ca_x)Sr_2GaCu_2O_7 \equiv M'Sr_2GaCu_2O_7$: $[(^{VIII}M')_o(^{V}CuO_2)_c(^{VIII}SrO)_o(^{IV}GaO)_c(^{VIII}SrO)_o(^{V}CuO_2)_c(^{VIII}M')_o(^{V}CuO_2)_c-(^{VIII}SrO)_o(^{IV}GaO)_c(^{VIII}SrO)_o$ $(^{V}CuO_2)_c]$ P_∞

21) $Sr_2CuO_2CO_3$: $[(^{IV}CuO_2)_c(^{VIII}SrO)_o(CO)_c(^{VIII}SrO)_o(^{VI}CuO_2)_o(^{VIII}SrO)_c(^{VI}CuO_2)_o(CO)_o(^{VIII}SrO)_c]$ P_∞

Table 3.1. Crystal structures of 21 copper oxides represented by the one-dimensional description method.

nine oxides that were discovered after the work of Yvon and François [5]: $(La_{1-x}Ca_x)_2(Ca_yLa_{1-y})Cu_2O_6$, $(Ba_{1-x}Nd_x)_2(Nd_{1-y}Ce_y)_2Cu_3O_{8+z}$, $Sr_2(Nd_{1-x}Ce_x)_2NbCu_2$ O_{10}, $HgBa_2CuO_4$, $HgBa_2CaCu_2O_6$, $HgBa_2Ca_2Cu_3O_8$, $(Sr_{1-x}La_x)CuO_2$, $(Y_{1-x}Ca_x)$ $Sr_2GaCu_2O_7$, and $Sr_2CuO_2CO_3$ (nonsuperconductor).

Table 3.1 lists the structure of the 21 superconductors represented with the above method. In the following descriptions, their crystal systems, space groups (space group numbers in International Tables, Vol. A), and lattice constants will be outlined. Symbol M denotes a mixed-metal atom whose coordination number is larger than eight, while the coordination number of M' is eight.

3.2.1.1 $(Nd_{1-x}Ce_x)_2CuO_4$

Tetragonal, I4/mmm (No. 139); $a = 0.39469$ nm and $c = 1.20776$ nm for $x = 0.0775$. This structure is often referred to as the T'-type conventionally. $(^{VIII}M')_c(^{IV}CuO_2)_o(^{VIII}M')_c$ and $(^{VIII}M')_o(^{IV}CuO_2)_c(^{VIII}M')_o$ are P-type mesh groups where O atoms are completely removed from (M'O). Because $(^{VIII}M')_c(O_2)(^{VIII}M')_o$ and $(^{VIII}M')_o(O_2)(^{VIII}M')_c$ are F-type mesh groups, M'_2CuO_4 can be regarded as a compound in which P- and F-type units are stacked alternately by sharing the M' sheet (Figure 3.12). Every Cu ion is four-coordinated with no O^{2-} ions at its apical positions; hence, one can dope electrons into the CuO_2 sheet.

3.2.1.2 $(La_{1-x}Sr_x)_2CuO_4$

High-temperature tetragonal (HTT) form (Figure 3.13a): I4/mmm (No. 139); $a = 0.3778$ nm and $c = 1.3237$ nm at 293 K for $x = 0.075$. Low-temperature orthorhombic (LTO) form (Figure 3.13b): Bmab ($\equiv$ Cmca, No. 64), $a = 0.5324$ nm, $b = 0.5348$ nm, and $c = 1.3200$ nm at 7 K for $x = 0.075$. This compound has the K_2NiF_4-type (T-type) structure. $(^{IX}MO)_c(^{VI}CuO_2)_o(^{IX}MO)_c$ and $(^{IX}MO)_o(^{VI}CuO_2)_c(^{IX}MO)_o$ have P-type atomic configurations, and $(^{IX}MO)_c(^{IX}MO)_o$ and $(^{IX}MO)_o(^{IX}MO)_c$ have R-type ones. If we preferentially adopt mesh groups

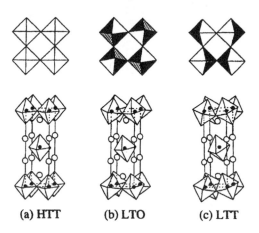

(a) HTT   (b) LTO   (c) LTT

Figure 3.13.   Three polymorphs of the K_2NiF_4-type compound $(La_{1-x}A_x)_2CuO_4$ (A = Ca, Sr, and Ba) [46: (*a*) I4/mmm, (*b*) Bmab, and (*c*) $P4_2$/ncm forms (for only A = Ba).

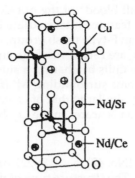

Figure 3.14. T*-type superconductor, $(Nd_{1-x}Sr_x)(Nd_{1-y}Ce_y)CuO_4$.

as large as possible, this compound is considered as a P-P structure where two P-type units are connected with each other. In this way, the P-P joint contains submeshes $(^{IX}MO)_c(^{IX}MO)_o$ and $(^{IX}MO)_o(^{IX}MO)_c$.

3.2.1.3 $(Nd_{1-x}Sr_x)(Nd_{1-y}Ce_y)CuO_4$

Tetragonal, P4/nmm (No. 129); a = 0.38564 nm and c = 1.24846 nm in $(Nd_{0.59}Sr_{0.41})(Nd_{0.73}Ce_{0.27})CuO_{3.92}$. This structure is conventionally called the

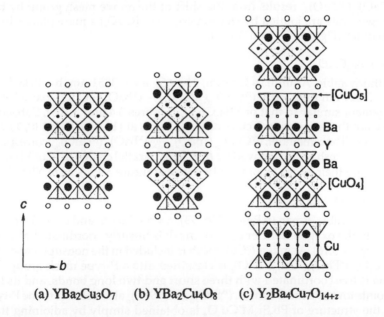

(a) $YBa_2Cu_3O_7$ (b) $YBa_2Cu_4O_8$ (c) $Y_2Ba_4Cu_7O_{14+z}$

Figure 3.15. Projections of the structures for (a) $YBa_2Cu_3O_7$, (b) $YBa_2Cu_4O_8$, and (c) $Y_2Ba_4Cu_7O_{14+z}$ along the [100] direction. Open square: partially occupied O site. (Courtesy of K. Yanagisawa.)

T*-type. MM'CuO$_4$ is a "half-blood" of the Nd$_2$CuO$_4$- and K$_2$NiF$_4$-type compounds, having structural features intermediate between these two (Figure 3.14). $(^{VIII}M')_o(O_2)(^{VIII}M')_c$ is an F-type mesh group, and $(^{VIII}M')_c(^{V}CuO_2)_o(^{IX}MO)_c$ and $(^{VIII}M')_o(^{V}CuO_2)_c(^{IX}MO)_o$ are P-type mesh groups. Each F-type unit is connected with the two P-type units by sharing a sheet containing only M'. The arrangement of eight O atoms surrounding M' (four in the CuO$_2$ sheet and four on the O$_2$ sheet) is very similar to that of O atoms in the F-type structure, but the position of the M' ion is nearer to the O$_2$ sheet than to the CuO$_2$ sheet.

3.2.1.4 YBa$_2$Cu$_3$O$_7$

Orthorhombic, Pmmm (No. 47); a = 0.38227 nm, b = 0.38872 nm, and c = 1.16802 nm in YBa$_2$Cu$_3$O$_{6.93}$. The structure type of this famous superconductor is referred to as the *oxygen-deficient triperovskite*. As this name indicates, this compound is fundamentally classified into the P-type structure and contains two kinds of oxygen-deficient meshes, $(^{IV}CuO)_o$ and $(^{VIII}Y)_c$. Ba and Y atoms are completely ordered at two different A positions to afford a unit cell with a tripled c dimension in comparison with that of the P-type compound (Figures 3.4 and 3.15a).

3.2.1.5 YBa$_2$Cu$_4$O$_8$

Orthorhombic, Ammm ($\equiv$ Cmmm, No. 65); a = 0.38402 nm, b = 0.38708 nm, and c = 2.72309 nm. $(^{IV}CuO)_o(^{X}BaO)_c \cdots (^{X}BaO)_c(^{IV}CuO)_o$ is a P-type mesh group also contained in YBa$_2$Cu$_3$O$_7$. The other mesh group, $(^{IV}CuO)_{oy}(^{X}BaO)_{cy} \cdots (^{X}BaO)_{cy}(^{IV}CuO)_{oy}$ results from the shift of the above mesh group by b/2. These two units are jointed by sharing edges of [CuO$_4$] square planes in the P/P manner (Figures 3.3 and 3.15b).

3.2.1.6 Y$_2$Ba$_4$Cu$_7$O$_{14+z}$

Orthorhombic, Ammm ($\equiv$ Cmmm, No. 65); a = 0.3851 nm, b = 0.3869 nm, and c = 5.029 nm for z = 0.6. $(^{IV}CuO)_o(^{X}BaO)_c \cdots (^{X}BaO)_c(^{IV}CuO)_o$ has a P-type arrangement similar to that of YBa$_2$Cu$_3$O$_7$ (Figures 3.4 and 3.15a). Subscripts p and q are the occupation factors of two O sites at (1/2, 0, 0) and (0, 1/2, 0), respectively. The remainder, $(^{IV}CuO)_o(^{X}BaO)_{cy} \cdots (^{X}BaO)_{cy}(^{IV}CuO)_{oy}$, corresponds to the above mesh groups moved by b/2. Two parallel chains of [CuO$_4$] square planes share their edges to join the two units (Figure 3.15c), as in YBa$_2$Cu$_4$O$_8$.

3.2.1.7 Pb$_2$Sr$_2$(Y$_{1-x}$Ca$_x$)Cu$_3$O$_8$

Orthorhombic, Pman ($\equiv$ Pmna, No. 53); a $\approx$ b $\approx$ 3.8 nm and c $\approx$ 1.57 nm. The Cu$^+$ ion in the oxygen-deficient $(^{II}Cu)_o$ mesh is linearly coordinated to two O atoms (Figure 3.16). A similar $(^{II}Cu)_o$ mesh is included in the nonsuperconductor YBa$_2$Cu$_3$O$_6$. $(^{V}PbO)_c(^{II}Cu)_o(^{V}PbO)_c$ is classified into a P-type mesh group. The Pb^{2+} ion is five coordinated with three short and two long bonds, and its lone pair points towards the Cu sheet. $(^{IX}SrO)_o \cdots (^{IX}SrO)_o$ also belongs to the P-type. Thus, the structure of Pb$_2$Sr$_2$M'Cu$_3$O$_8$ is obtained simply by adjoining these two P-type meshes (Figure 3.16). Of course, the R-type mesh groups, $(^{IX}SrO)_o(^{V}PbO)_c$ and $(^{IX}SrO)_c(^{V}PbO)_{o'}$ are also included, but the P-type mesh groups are preferred which are larger than the R-type ones.

3.2.1.8 $(La_{1-x}Ca_x)_2(Ca_{1-y}La_y)Cu_2O_6$

Tetragonal, I4/mmm (No. 139); a = 0.38160 nm and c = 1.94214 nm in $(La_{0.88}Ca_{0.11})_2(Ca_{0.96}La_{0.04})Cu_2O_6$. $(^{IX}MO)_c(^VCuO_2)_o(^{VIII}M')_c(^VCuO_2)_o(^{IX}MO)_c$ and $(^{IX}MO)_o(^VCuO_2)_c(^{VIII}M')_o(^VCuO_2)_c(^{IX}MO)_o$ are both P-type mesh groups which are simply connected to each other, in a similar way to $(La_{1-x}Sr_x)_2CuO_4$ (Figure 3.17).

3.2.1.9 $(Ba_{1-x}Nd_x)_2(Nd_{1-y}Ce_y)_2Cu_3O_{8+z}$

Tetragonal, I4/mmm (No. 139); a = 0.38747 nm and c = 2.8599 nm in $(Ba_{0.633}Nd_{0.367})_2(Nd_{0.675}Ce_{0.325})_2Cu_3O_{8.91}$. $(^{VIII}M')_c(^VCuO_2)_o$ ⋯⋯ $(^VCuO_2)_o(^{VIII}M')_c$ and $(^{VIII}M')_o(^VCuO_2)_c$ ⋯⋯ $(^VCuO_2)_c(^{VIII}M')_o$ are both P-type mesh groups containing oxygen-deficient meshes (M') and $(CuO_{2×0.46})$. Similar meshes are also present in the solid solutions $Ln(Ba_{1-x}Ln_x)_2Cu_3O_{6+z}$ (Ln = La, Pr, Nd, Sm, and Eu) which are isomorphous with tetragonal $YBa_2Cu_3O_{6+z}$. $(^{VIII}M')_c(O_2)(^{VIII}M')_o$ and $(^{VIII}M')_o(O)_o(^{VIII}M')_c$ are F-type mesh groups. That is, P- and F-type units are stacked alternately by sharing M' sheets in this oxide (Figure 3.18). Every M' cation is surrounded by eight O^{2-} ions arranged at the corners of a distorted cube in a similar way to the T' and T*-type compounds.

3.2.1.10 $Sr_2(Nd_{1-x}Ce_x)_2NbCu_2O_{10}$

Tetragonal, I4/mmm (No. 139); a = 0.3885 nm and c = 2.8864 nm for x = 0.25. $(^{VIII}M')_c(^VCuO_2)_o$ ⋯⋯ $(^VCuO_2)_o(^{VIII}M')_c$ and $(^{VIII}M')_o(^VCuO_2)_c$ ⋯⋯

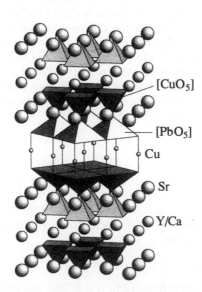

Figure 3.16. Crystal structure of $Pb_2Sr_2(Y_{1-x}Ca_x)Cu_3O_8$ containing [CuO$_5$] and [PbO$_5$] coordination polyhedra.

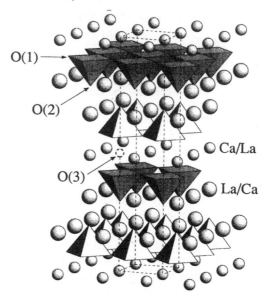

Figure 3.17. Crystal structure of $(La_{1-x}Ca_x)_2(Ca_{1-y}La_y)Cu_2O_6$. The location of an interstitial oxygen defect, O(3), is also indicated.

$(^VCuO_2)_c(^{VIII}M')_o$ are very long P-type mesh groups, with which F-type meshes $(^{VIII}M')_c(O_2)(^{VIII}M')_o$ and $(^{VIII}M')_o(O_2)_o(^{VIII}M')_c$ share $(^{VIII}M')$ meshes (Figure 3.19). This compound has a unique structural feature that Nb, which is a metal other than Cu, fully occupies a B site in the P-type unit. The Nb^{5+} ions readily occupy the B site because the $^{VI}Nb^{5+}$ ion (r = 0.064 nm) has roughly the same size as the Cu^{2+} ions. The concentration of holes on the CuO_2 sheet tends to be decreased when highly charged cations are included in the charge reservoir. Nevertheless, the negative charges of six O^{2-} ions surrounding the Nb^{5+} ion provide the CuO_2 sheet with hole carriers, whose concentration is high enough to make this oxide superconducting.

3.2.1.11 $TlBa_2CaCu_2O_7$

Tetragonal, P4/mmm (No. 123); a = 0.38566 nm and c = 1.2754 nm. $(^{IX}BaO)_c(^{VI}TlO)_o(^{IX}BaO)_c$ has an atomic configuration of the R-type, and $(^{IX}BaO)_c$ ····· $(^{IX}BaO)_c$ has that of the oxygen-deficient P-type. That is, the R- and P-type units are connected with each other *via* the BaO sheet (a). In the P-type unit, the O site on the Ca sheet is 100% vacant owing to the preference of the Ca^{2+} ion for eight-coordination.

3.2.1.12 $TlBa_2Ca_2Cu_3O_9$

Tetragonal, P4/mmm (No. 123); a = 0.3853 nm and c = 1.5913 nm. This superconductor has a structure similar to $TlBa_2CaCu_2O_7$, except for the very long P-type mesh, $(^{IX}BaO)_c(^VCuO_2)_o$ ····· $(^VCuO_2)_o(^{IX}BaO)_c$ (Figure 3.20b).

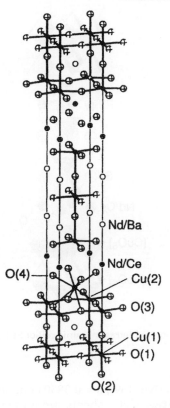

Figure 3.18. Crystal structure of $(Ba_{1-x}Nd_x)_2(Nd_{1-y}Ce_y)_2Cu_3O_{8+z}$. O(1) is a partially occupied O site. Cu(2) and O(3) atoms form the CuO_2 conduction sheets. Eight M'–O bonds are shown for only one M' atom at (1/2, 1/2, 0.204).

3.2.1.13 Ba_2CuO_6

Tetragonal form: I4/mmm (No. 139); a = 0.38625 nm and c = 2.32243 nm in a sample with a T_c of 85 K. Orthorhombic form: Fmmm (No. 69); a ≈ 0.549 nm, b ≈ 0.546 nm, and c ≈ 2.321 nm in a sample with a T_c of 85 K. $(^{IX}BaO)_o(^{VI}CuO_2)_c(^{IX}BaO)_o$ and $(^{IX}BaO)_c(^{VI}CuO_2)_o(^{IX}BaO)_c$ are P-type mesh groups while $(^{IX}BaO)_o(^{VI}TlO)_c(^{VI}TlO)_o(^{IX}BaO)_c$ and $(^{IX}BaO)_c(^{VI}TlO)_o(^{VI}TlO)_c(^{IX}BaO)_o$ are R-type ones. That is, the P- and R-type units alternate with the BaO sheets as interfaces (Figure 3.21a). The R-type unit in this compound is thicker than those in $TlBa_2CaCu_2O_7$ and $TlBa_2Ca_2Cu_3O_9$ because it includes double TlO sheets.

3.2.1.14 $Tl_2Ba_2CaCu_2O_8$

Tetragonal, I4/mmm (No. 139); a = 0.38558 nm and c = 2.92596 nm. This superconductor has a structure where one Ca and one CuO_2 sheets are added to the P-type unit in $Tl_2Ba_2CuO_6$, which results in the increased thickness of this unit (Figure 3.21b).

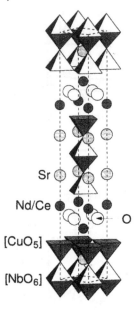

Figure 3.19. Body-centered tetragonal unit cell of $Sr_2(Nd_{1-x}Ce_x)_2NbCu_2O_{10}$.

3.2.1.15 $Tl_2Ba_2Ca_2Cu_3O_{10}$

Tetragonal, I4/mmm (No. 139); a = 0.38503 nm and c = 3.588 nm. In this structure, two Ca and two CuO_2 sheets are added to the P-type unit in $Tl_2Ba_2CuO_6$ (Figure 3.21c).

3.2.1.16 $HgBa_2CuO_4$

Tetragonal, P4/mmm (No. 123); a = 0.38750 nm and c = 0.95132 nm. In Hg-containing superconductors, Hg atoms occupy sites corresponding to the A site in the P-type compound, in a similar way to Tl atoms in Tl-containing superconductors. Then, $(^{VIII}BaO)_c(^{II}Hg)_o(^{VIII}BaO)_c$ has an atomic configuration of the R-type. The O site on the Hg sheet is completely or nearly vacant. There are two short, collinear Hg–O bonds parallel to the c-axis. In addition, weaker Hg–O bonds perpendicular to them result from introduction of excess oxygen into the Hg sheet. Linear complexes employing sd hybrid orbitals in bond formation are very common among Hg^{2+} compounds. $(^{VIII}BaO)_c(^{VI}CuO_2)_o$ $(^{VIII}BaO)_c$ is a P-type mesh group. That is, the R- and P-type units are connected to each other by sharing the BaO sheet (Figure 3.22a).

3.2.1.17 $HgBa_2CaCu_2O_6$

Tetragonal, P4/mmm (No. 123); a = 0.38552 nm and c = 1.26651 nm. This oxide has a structure where one Ca and one CuO_2 sheets are added to the P-type unit in $HgBa_2CuO_4$, which results in the increased thickness of that unit (Figure 3.22b). The structure of $HgBa_2CaCu_2O_6$ can be derived from that of $TlBa_2CaCu_2O_7$ by removing almost all O atoms in the TlO sheet and sub-

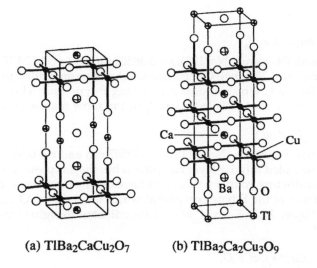

(a) TlBa$_2$CaCu$_2$O$_7$ (b) TlBa$_2$Ca$_2$Cu$_3$O$_9$

Figure 3.20. Superconductors with single TlO sheets: *(a)* TlBa$_2$CaCu$_2$O$_7$ and *(b)* TlBa$_2$Ca$_2$Cu$_3$O$_9$. Cu and O atoms are connected with bonds.

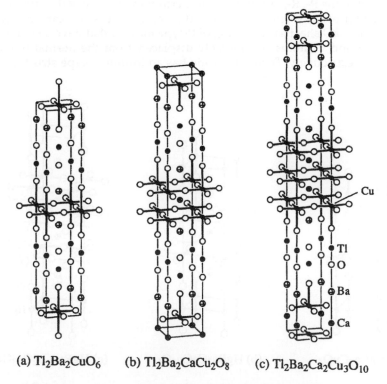

(a) Tl$_2$Ba$_2$CuO$_6$ (b) Tl$_2$Ba$_2$CaCu$_2$O$_8$ (c) Tl$_2$Ba$_2$Ca$_2$Cu$_3$O$_{10}$

Figure 3.21. Superconductors with double TlO sheets: *(a)* Tl$_2$Ba$_2$CuO$_6$, *(b)* Tl$_2$Ba$_2$CaCu$_2$O$_8$, and *(c)* Tl$_2$Ba$_2$Ca$_2$Cu$_3$O$_{10}$.

stituting Hg for Tl.

3.2.1.18 $HgBa_2Ca_2Cu_3O_8$

Tetragonal, P4/mmm (No. 123); a = 0.38501 nm and c = 1.57837 nm. In this structure (Figure 3.22c), two Ca and two CuO_2 sheets are added to the P-type unit of $HgBa_2CuO_4$, which affords a very long P-type unit. This structure is also derived from that of $TlBa_2Ca_2Cu_3O_9$ in the same manner as above.

3.2.1.19 $(Sr_{1-x}La_x)CuO_2$

Tetragonal, P4/mmm (No. 123); a = 0.39507 nm and c = 0.34090 nm for x = 0.1. This so-called infinite-layer compound has the simplest structure of all the superconducting copper oxides that have been ever discovered. It has a structural form of the P-type compound which loses all the O atoms on the AO sheet (Figure 3.23). Each M' atom is coordinated to eight O atoms on the CuO_2 sheet.

3.2.1.20 $(Y_{1-x}Ca_x)Sr_2GaCu_2O_7$

Orthorhombic, I2mb (≡ Ima2, No. 46); a = 0.5474 nm, b = 0.5384 nm, and c = 2.2813 nm. Strictly speaking, the structural block containing Ga should be related not to the P-, R-, or F-type but to pyroxenes, in view of the tetrahedral coordination of Ga (Figures 3.9 and 3.10). Nevertheless, we can coercively look on the GaO sheet as consisting of P-type meshes that have a configuration of Ga and O atoms considerably displaced from the normal B and X positions, respectively. Then, this oxide has an infinite P-type structure.

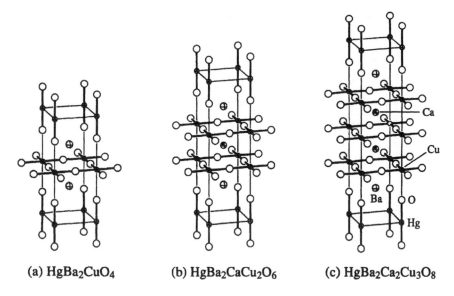

(a) $HgBa_2CuO_4$ (b) $HgBa_2CaCu_2O_6$ (c) $HgBa_2Ca_2Cu_3O_8$

Figure 3.22. Hg-containing superconductors: (a) $HgBa_2CuO_4$, (b) $HgBa_2CaCu_2O_6$, and (c) $HgBa_2Ca_2Cu_3O_8$.

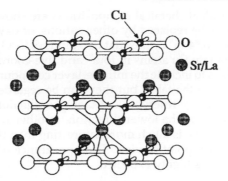

Figure 3.23. Infinite-layer compound, $(Sr_{1-x}La_x)CuO_2 \equiv M'CuO_2$. Eight M'–O bonds are shown for only one M' atom.

3.2.1.21 $Sr_2CuO_2CO_3$

Tetragonal, $I\bar{4}$ (No. 82); a = 0.78045 nm and c = 1.4993 nm. Like La_2CuO_4, this stoichiometric oxide carbonate (Figure 3.11) is not a superconductor per se but a mother phase into which hole carriers need to be injected. It has been selected as the prototype of superconductors containing CO_3^{2-} ions, however, because it is the only stoichiometric copper oxide carbonate with a relatively simple structure. The inclusion of triangular CO_3^{2-} ions in this compound makes it difficult to describe the structure on the basis of only the P-, R-, and F-type units. It is, however, convenient to relate the structure to the P-type because C atoms often share the same sheet with Cu atoms in superconductors containing CO_3^{2-} ions. If the C atom is regarded formally as a B site metal, the structure of $Sr_2CuO_2CO_3$ can be described as an infinite P-type unit. However, three O atoms bonded to each C atom are attracted toward it with C–O bonds as short as about 0.13 nm. They are, therefore, displaced appreciably from the normal anion positions in the P-type compound.

3.2.2 Seven Modes of Stacking Structural Blocks

In summary, the sequences of stacking the R-, P-, and F-type blocks in the above 21 compounds fall into the following seven types: [P-P-], [F=P-P-], [P=F=P=F=], P_∞, [P/P/], [R=P=], and [R=P=R=P=] ≡ [P=R=P=R=]. Thus, structural similarities among various types of superconductors can be easily examined using this classification method. We emphasize again that P-type units are contained in these superconductors without exception. Thus, the structures of superconducting copper oxides can be obtained by modifying those of P-type compounds in a variety of ways.

3.2.3 Combinations of CuO_2 Sheets and Charge Reservoirs

The structures of superconductors are classified in Table 3.2 on the basis of the kinds of CuO_2 conduction sheets (columns) and charge reservoirs separating the CuO_2 sheets (rows). This table lists stacking sequences of two-dimensional sheets in parallel with the c axis using mesh groups.

Representative and ideal chemical compositions are shown in parentheses. Actually, cations may be replaced by other cations, or oxygen contents may be varied. The notation "4+5" means that two kinds of CuO_2 sheets are present that respectively contain Cu atoms in four- and five-coordination. This classification method fails to include the infinite-layer compounds, $(Sr_{1-x}Ln_x)CuO_2$ (Ln = La, Pr, or Nd), in the table, but they can be regarded as consisting of $Sr_{1-x}Ln_x$ sheets as a charge reservoir and CuO_2 conduction sheets.

Extending Table 3.2 to the lower direction, in other words, finding new types of structural blocks, is much more interesting than filling missing cells from a crystal-chemical point of view. The appearance of compounds with novel structural units is highly expected.

3.3 Roles of Charge Reservoirs in Superconductivity

In this section we focus on charge reservoirs and the way in which the superconducting behavior of materials is affected by lattice defects.

3.3.1 Apical Oxygen and Defects in Charge Reservoirs

Apical oxygen is evidently one of the most important members of the charge reservoir. As described above, the bond between Cu on the CuO_2 sheet and apical oxygen is much weaker than that contained in the sheet. Apical oxygen

Table 3.2 Classification of superconducting copper oxides. Ln: lanthanoid metal; A alkaline-earth metal; R: rare-earth metal; $(CuO_2)_2(R_{1-x}A_x)$: $(CuO_2)(R_{1-x}A_x)(CuO_2)$; $(CuO_2)_3Ca_2$: $(CuO_2)(Ca)(CuO_2)(Ca)(CuO_2)$; integer (4, 5, or 6): coordination number of Cu on CuO_2 sheets.

No.	Block	(CuO_2)	$(CuO_2)_2$ $(R_{1-z}M_z)$	$(CuO_2)_3$ Ca_2
1	$(Ln_{1-x}M_xO)(Ln_{1-x}A_xO)$	6	5	
2	$(AO)(CuO)(AO)$		5	
3	$(AO)(CuO)(CuO)(AO)$		5	
4	$(AO)(TlO)(TlO)(AO)$	6	5	4+5
5	$(AO)(BiO)(BiO)(AO)$	6	5	4+5
6	$(AO)(PbO)(Cu)(AO)$		5	
7	$(AO)(PbO)(Cu)(PbO)(AO)$	6	5	
8	$(R)(O_2)(R)$	4, 5		
9	$(AO)(TlO)(AO)$	6	5	4+5
10	$(AO)(Pb_{1-x}Cu_xO)(AO)$	5	5	
11	$(AO)(NbO_2)(AO)$	5		
12	$(AO)(GaO)(AO)$	5	5	
13	$(AO)(Hg)(AO)$	6	5	4+5
14	$(AO)(C_{1-x}Cu_xO_{1-y})(AO)$	6	5	4+5
15	$(AO)(C_{1-x}B_xO)(AO)$	6	4+5	

greatly affects superconducting properties, however, in view of the fact that the coordination number of in-sheet Cu is closely related to the type of carrier. $(Nd_{1-x}Ce_x)_2CuO_4$ (Nd_2CuO_4- or T'-type structure), $(Nd_{1-x}Sr_x)(Nd_{1-y}Ce_y)CuO_4$ (T*-type structure), and $(La_{1-x}Sr_x)_2CuO_4$ (K_2NiF_4- or T-type structure) have analogous metal arrangements but different coordination numbers of Cu, 4, 5, and 6, respectively (Figures 3.12, 3.13, and 3.14). $(Nd_{1-x}Ce_x)_2CuO_4$ is an n-type superconductor, and the other two are p-type ones. Electron-doped superconductivity has been achieved only in the absence of apical oxygen, which is interpreted in terms of the repulsion between the negative charge of apical oxygen and electron carriers on the CuO_2 sheet. By contrast, the presence of apical oxygen stabilizes holes doped into the CuO_2 sheet. The distance between apical oxygen and the CuO_2 sheet is another important factor affecting the electrostatic potential of the carrier. The electrostatic stability of holes on the CuO_2 sheet becomes higher with decreasing distance between apical oxygen and the CuO_2 sheet.

Carrier concentrations on the CuO_2 sheets are changed by introducing various types of defects in charge reservoirs. Incorporation of negative charges (O^{2-} ions) into vacancies or interstitial positions increases the concentration of hole carriers. Substitution of metals with lower oxidation states for those with

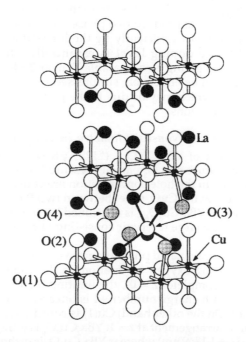

Figure 3.24. Local structure of La_2CuO_{4+z} near an interstitial oxygen defect, O(3). Cu (filled circles) and O (open circles) are connected with white bonds. La and O(3) atoms are connected with black bonds. O(4) atoms, which are displaced from their ideal positions as a result of the uptake of O(3), were shaded to distinguish them from apical O(2) atoms.

higher oxidation states has a similar effect. That is, introduction of oxygen and substitutional defects in charge reservoirs makes it possible to accept electrons from or supply electron to the CuO_2 sheets. Intergrowths are another kind of a defect influencing superconducting properties more or less. Even if impurity phases are hardly detected by x-ray or neutron powder diffraction, intergrowths of foreign blocks are often inserted in superconducting copper oxides. The effects of such planar defects on their superconducting properties, however, have been overlooked in many cases. Relationships between various defects in charge reservoirs and superconductivity in important materials will be described below in connection with their structural properties.

3.3.2 Interstitial Oxygen Defects in La_2CuO_{4+z}

In La_2CuO_{4+z}, O^{2-} ions corresponding to z are incorporated under high oxygen pressure in interstitial positions, O(3), surrounded tetrahedrally by four La^{3+} ions (Figure 3.24).

O(4) atoms around O(3) are pushed aside by about 0.05 nm because of electrostatic repulsion. Such introduction of oxygen defects between two LaO sheets causes p-type superconductivity through transfer of electrons from the CuO_2 sheet to the charge reservoir, that is, two successive LaO sheets. When a sample of La_2CuO_{4+z} ($z \approx 0.05$) is cooled to room temperature, the interstitial oxygen defects segregate to produce two phases with $z \approx 0.08$ and $z \approx 0$. The former is doped with holes sufficiently by the interstitial O^{2-} ions to attain superconductivity. The doping level is essentially identical to that of $(La_{0.925}Sr_{0.075})_2CuO_4$, thus giving a comparable T_c near 35 K.

Phase separation can be effectively suppressed by partial substitution of Bi^{3+} ions for La^{3+} ions, which may be due to the prevention of oxygen diffusion by lone pairs of Bi^{3+} ions.

3.3.3 Oxygen Ordering and Nonstoichiometry in $YBa_2Cu_3O_{6+z}$

The oxygen content of $YBa_2Cu_3O_{6+z}$ ("123" phase) varies continuously, depending on temperature and O_2 partial pressure [7]. The change in z and the corresponding one in the hole concentration are caused by incorporation and release of oxygen on the CuO_z sheet between two BaO sheets. The superconducting behavior of $YBa_2Cu_3O_{6+z}$ is controlled by charge transfer between the CuO_2 sheet and the charge reservoir consisting of three successive sheets, BaO, CuO_z, and BaO.

Figure 3.25 shows a calculated phase diagram of $YBa_2Cu_3O_{6+z}$ [10] together with patterns illustrating oxygen occupation on the CuO_z sheet. When the oxygen deficiency is complete ($z = 0$), the limiting formula of $YBa_2Cu_3O_6$ is reached, and Cu(1) with charge number +1 is linearly coordinated to two O atoms (Figure 3.26). On the other hand, Cu(1) is coordinated to four O atoms with a square-planar arrangement at $z = 1$. $YBa_2Cu_3O_6$ is tetragonal (P4/mmm, a = 0.38639 nm and c = 1.1849 nm) whereas $YBa_2Cu_3O_7$ is orthorhombic (Pmmm, Figure 3.4). The 123 phase exhibits a very wide range of oxygen nonstoichiometry between these two compositions.

In the z region stabilizing the tetragonal (T) form ($z < 0.5$ at 700 K), O atoms randomly occupy the equivalent positions at $(1/2, 0, 0)$ and $(0, 1/2, 0)$.

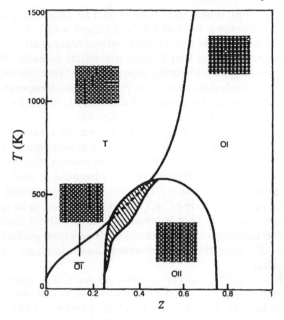

Figure 3.25. Calculated phase diagram of $YBa_2Cu_3O_{6+z}$. OI and T are the normal orthorhombic and tetragonal phases, respectively. OII is an ordered orthorhombic phase with a supercell doubled along the a-axis. For each phase, the pattern of oxygen occupation on the z = 0 plane is given. Filled and open circles denote O atoms and vacant O sites, respectively (adapted from [8] by Jorgensen [9]).

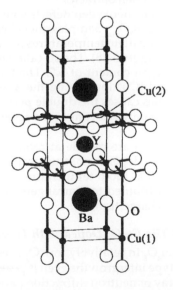

Figure 3.26. Crystal structure of $YBa_2Cu_3O_6$. Cu^+ ions in two-coordination are located at the corners of the unit cell.

The orthorhombic form, where the O site at $(0, 1/2, 0)$ is occupied selectively and partially, is stable for $z > 0.5$ at 700 K (Ortho-I $\equiv$ OI). Figure 3.27 shows the occupation factors of the two O sites versus temperature in O_2 at 0.101 MPa. The occupancies of the two O sites are equal to each other at higher temperature, increasing with lowering temperature. Then, the tetragonal form is converted to the orthorhombic form near 700°C. As the temperature is lowered below this temperature, the occupancy of the $(0, 1/2, 0)$ site increases rapidly whereas that of the $(1/2, 0, 0)$ site approaches zero.

Average structures of the tetragonal form seen by x-ray and neutron diffraction appear to support the completely random arrangement of the O defects among the two O sites on the a- and b-axes. However, this model is unreasonable from a crystal chemical point of view because it leads to the formation of a large amount of unstable Cu(1) atoms in three-coordination.

The real local structure on the CuO_z sheet is believed to be quite different from the average structure. That is, relatively short linear chains of $[CuO_4]$ square planes run parallel to a- and b-axes with an equal probability, as illustrated in Figure 3.25. Thus, three-coordinated Cu(1) atoms are situated at only the tails of the chains.

When z is less than 0.5, oxygen uptake on the CuO_z sheet increases the concentration of localized holes in the charge reservoir rather than the amount of delocalized holes transferred from the charge reservoir to the CuO_2 sheet. The T_c versus z curve in $YBa_2Cu_3O_{6+z}$ (Figure 3.28) is exceptionally gentle because oxygen introduction into the CuO_z sheet initially enhances the hole concentration of O atoms bonded to Cu(1) atoms. The amount of holes doped into the CuO_2 sheet, which is much more distant from the CuO_z sheet, increases with increasing oxygen content. This charge transfer converts $YBa_2Cu_3O_{6+z}$ from a semiconductor to a superconductor.

Order-disorder phenomena of oxygen defects somewhat affect the superconducting behavior of $YBa_2Cu_3O_{6+z}$. Samples quenched from, or equilibrated in low O_2 pressures at, intermediate temperatures show a broad 60 K plateau in the T_c versus 6+z curve (Figure 3.28). The so-called ortho-II (OII) phase has its stable field near an oxygen content of 6.5. The OII form has alternating locations of $^{IV}Cu^{2+}$ ions and $^{II}Cu^+$ ions along the a-axis (Figure 3.29). Such ordering of oxygen vacancies causes the double periodicity along the a-axis. The OII phase is almost certainly relevant to the appearance of the 60-K plateau.

O atoms in the CuO_z sheet are mobile at room temperature in samples with a z value near 0.5. Time-dependent structural phenomena in quenched $YBa_2Cu_3O_{6.41}$ were examined at room temperature (Figure 3.30). A rise in T_c with annealing time was attributed to rearrangement of O atoms on the $z = 0$ plane and the resulting charge transfer.

3.3.4 Stacking Faults in $YBa_2Cu_4O_8$ Doped with Foreign Metals

Preparation of $YBa_2Cu_4O_8$ in relatively low O_2 pressure results in the extensive inclusion of 123-type intergrowths even if peaks due to impurities are hardly observed in its x-ray or neutron diffraction pattern. Therefore, annealing at a high partial pressure of O_2 (typically 20 MPa at 1010°C) is required to obtain intergrowth-free samples of $YBa_2Cu_4O_8$. Relationships between struc-

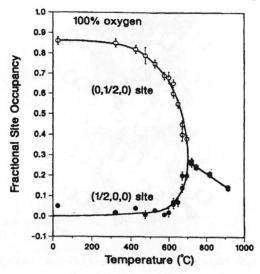

Figure 3.27. Occupation factors of the two O sites in $YBa_2Cu_3O_{6+z}$ as a function of temperature for a sample heated in an atmosphere of O_2 [11].

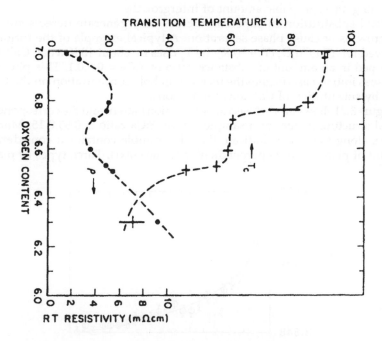

Figure 3.28. Dependence of T_c on the oxygen content, $6 + z$, in $YBa_2Cu_3O_{6+z}$ for $0.3 \leq z \leq 1$ [12]. Vertical bars indicate the 10%–90% resistive transition width, with the midpoint marked with a cross. The room-temperature resistivities are also shown.

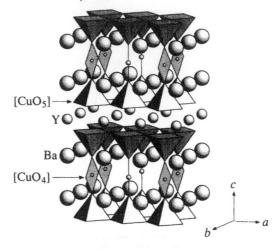

Figure 3.29. Idealized crystal structure of OII-type $YBa_2Cu_3O_{6.5}$.

tures and superconducting properties are not clear-cut with those compounds containing an appreciable amount of intergrowths.

Metal substitution in superconductors may generate defects such as intergrowths or cause phase separation. A typical example of the former is observed for $YBa_2Cu_4O_8$ ($T_c \approx 80$ K). Its T_c rises by about 10 K when Ca^{2+} ions are doped in an amount-of-substance ratio of Y:Ca = 0.9:0.1. This phenomenon was initially ascribed to the increase in hole concentration on the CuO_2 sheet by substitution of Ca^{2+} ions for Y^{3+} ions.

Figure 3.31 shows an a*-c* electron diffraction pattern and the corresponding crystal-structure image for a sample whose Y:Ca ratio is 0.85:0.15. Marked streaks along the c*-axis indicate that this sample contains a considerable amount of planar defects perpendicular to the c-axis. Three types of planar

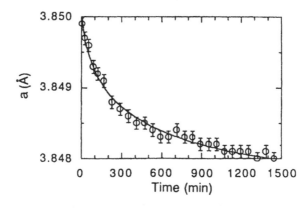

Figure 3.30. Lattice constant, a, vs annealing time at room temperature for $YBa_2Cu_3O_{6.41}$ [13].

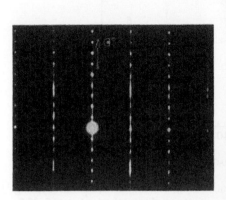 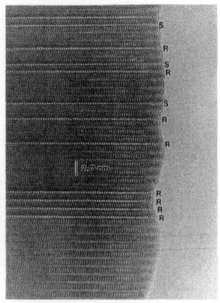

Figure 3.31. (a) Selected-area electron diffraction pattern (a^*-c^*) and
(b) corresponding HRTEM image for $YBa_2Cu_4O_8$ doped with Ca [14]. S and R in (b)
denote 123-like intergrowth layers and planar defects rotated by 90°, respectively.

defects were observed in its crystal-structure images: (1) replacement of the
double chains of $[CuO_4]$ square planes by single chains (123-type intergrowths),
(2) insertion of additional Ca(Y) sheets between two CuO sheets in the double
chain, and (3) 90° rotation of the double chains. Therefore, the rise in T_c on Ca
doping may result not only from the metal substitution but from formation
of the complex planar defects described above.

A similar situation also exists in the solid solution $Y(Ba_{1-x}Sr_x)_2Cu_4O_8$ ($0 \leq$
$x \leq 0.4$). T_c in this system remains virtually constant ($T_c \approx 75$ K) in the whole
x range. "Chemical pressure" applied by the substitution of small $^XSr^{2+}$ ions
(r = 0.136 nm) for large $^XBa^{2+}$ ions (r = 0.152 nm) seems to be ineffective in
changing T_c. An HRTEM study showed that a considerable amount of 123-type
intergrowths results from Sr doping. Thus, both the contraction of the crystal
lattice and generation of the planar defects must be, at least partially, responsible for the insensibility of T_c to the Sr content.

3.3.5 Oxygen Defects in $La_{2-x}Sr_yCa_{1+x-y}Cu_2O_{6\pm z}$

$La_{2-x}Sr_yCa_{1+x-y}Cu_2O_{6\pm z}$ is isomorphous with $(La_{1-x}Ca_x)_2(Ca_{1-y}La_y)Cu_2O_6$
($MM'Cu_2O_6$), but three kinds of cations, La^{3+}, Sr^{2+}, and Ca^{2+}, are distributed
at the two metal sites, M and M'. Interstitial O atoms are accommodated at
the normally vacant O(3) site between two CuO_2 sheets (Figure 3.17). The
inter-CuO_2-sheet spacing increases as Ca^{2+} at the M' site between these sheets

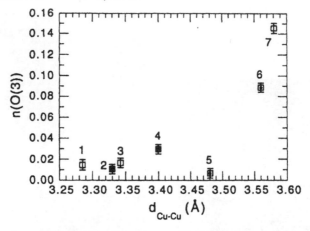

Figure 3.32. Occupation factor of the O(3) site, n(O(3)), vs the inter-CuO$_2$-sheet spacing d$_{Cu-Cu}$ [15]. Open, solid, and dotted squares represent samples synthesized at 40.5, 25.5, and 5.07 MPa of O$_2$ partial pressure, respectively.

is replaced by the larger Sr^{2+} and La^{3+} ions. The occupation factor of O(3) increases sharply when the Cu–Cu interatomic distance exceeds 0.35 nm (Figure 3.32).

Some evidence exists that Coulomb attraction involving these interstitial O atoms enhances the amount of La^{3+} on the M' site. T$_c$ goes down systematically with increasing occupancy of the O(3) site for samples that would otherwise be expected to be superconducting. This fact supports the idea that incorporation of the interstitial O(3) defects, that is, local formation of [CuO$_6$] octahedra, is very harmful for superconductivity in this solid solution.

3.3.6 Oxidation Behavior of Pb$_2$Sr$_2$(Y$_{1-x}$Ca$_x$)Cu$_3$O$_{8+z}$

Pb$_2$Sr$_2$(Y$_{1-x}$Ca$_x$)Cu$_3$O$_{8+z}$ (z = 0) is a typical example of a superconducting oxide where the desired oxygen defects are produced by synthesis under reducing conditions. Superconducting compounds, in which Cu atoms between two PbO sheets are in the oxidation state of +1, are formed only in a low O$_2$ partial pressure. A subsequent anneal in O$_2$ at lower temperature causes the incorporation of additional oxygen (z ≤ 1.8) to fill the oxygen vacancies (Figure 3.16), but the resulting compounds are no longer superconducting. Then, oxygen vacancies are the defects that lead to superconducting behavior in this system. Introduction of additional oxygen inevitably accompanies the oxidation of Pb^{2+} to Pb^{4+} and Cu$^+$ to Cu^{2+} in the charge reservoir. The total effect is to lower the hole concentration on the CuO$_2$ sheet to make this oxide nonsuperconducting.

Oxidation of Pb$_2$Sr$_2$(Y$_{1-x}$Ca$_x$)Cu$_3$O$_{8+z}$ at a high temperature of 520°C accompanies disordering of Pb and Cu in the charge reservoir, but they remain ordered at temperatures below 450°C. Similar metal disordering on oxidation occurs much more readily in the related compound PbBaSrYCu$_3$O$_{7+z}$ where each Cu sheet is sandwiched by PbO and (Ba,Sr)O sheets.

3.3.7 Lattice Defects in Bi- and Tl-based Superconductors

Considerable amounts of intergrowths are usually inserted in Bi-, Tl-, and Hg-based superconductors containing Ca^{2+} ions, with the result that their real (average) compositions deviate considerably from the ideal ones. In their powder diffraction patterns, 00l reflections tend to be broader than other reflections, and streaks along the c*-axis are often observed in their selected-area electron diffraction patterns. $Bi_2Sr_2Ca_2Cu_3O_{10+z}$ would contain only $(CuO_2)_3(Ca)_2$ type conduction sheets (Table 3.2) if no planar defects were formed during its preparation process. HRTEM, however, revealed the extensive inclusion of intergrowths of $(CuO_2)_2(Ca)$, $(CuO_2)_4(Ca)_3$, and other types.

The locations of the O atoms in the BiO and TlO sheets in Bi- and Tl-containing superconductors are poorly defined owing to unusually large displacements from their ideal positions. Such displacements occur because the natural dimension of the BiO or TlO sheet does not match that of the CuO_2 sheet. The dimension of the CuO_2 sheet containing very firm Cu–O bonds controls the overall lattice constants a and b. Consequently, the BiO/TlO sheets are "stretched", a situation that leads to the relaxation of O atoms in these

Figure 3.33. HRTEM image obtained by projection along the [100] direction for the superlattice ($a \approx 0.54$ nm, $b \approx 2.6$ nm, and $c \approx 3.06$ nm) of $Bi_2Sr_2CaCu_2O_{8+z}$ [16]. The two rows indicated by arrows correspond to BiO sheets. A unique incommensurate modulated structure can be seen.

sheets from their ideal positions to achieve desirable Bi–O/Tl–O bond lengths.

In $Bi_2Sr_2CaCu_2O_{8+z}$, which is basically isostructural with $Tl_2Ba_2CaCu_2O_8$, excess oxygen corresponding to z invades into the BiO sheet to expand it. The result is the appearance of an incommensurate modulated structure where atomic positions are waved in a complex manner. The most outstanding is the displacement of Bi atoms from average positions. HRTEM revealed that Bi-concentrated and Bi-diluted bands alternate with each other along the b-axis (Figure 3.33). The R-type sheet, BiO, owes its distortion to the invasion of excess oxygen corresponding to z. The interstitial O atoms are located in the Bi-diluted bands, in which the Bi–Bi distance is larger than that in the Bi-concentrated band. Unfortunately, the excess oxygen, which is a light element, can be hardly seen in Figure 3.33.

$Tl_2Ba_2CuO_{6+z}$ is overdoped with hole carriers, because of the presence of excess O atoms corresponding to z and, in addition, Cu atoms invading Tl sites partially. It crystallizes in the tetragonal and orthorhombic polymorphs. These two have different chemical compositions; a larger amount of Cu that substitutes for Tl and a smaller amount of excess oxygen favor formation of the tetragonal form. Both forms change from a superconductor with the maximum T_c of over 85 K to a metallic nonsuperconductor as its oxygen content is increased. The difference in z between the nonsuperconductor and the 85-K superconductor is about 0.1 per formula unit in the tetragonal form and 0.05 in the orthorhombic form. Excess O atoms occupy not TlO sheets but interstitial positions between a couple of TlO sheets. Each interstitial O atom has

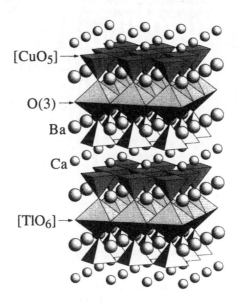

Figure 3.34. Crystal structure of $TlSr_2CaCu_2O_{7-z}$ represented with $[CuO_5]$ and $[TlO_6]$ coordination polyhedra.

as its nearest neighbors four Tl atoms in approximate tetrahedral coordination. The incorporation and release of the interstitial defects cause the changes in the overall oxygen content and the corresponding changes in hole concentration.

The interstitial O atoms become ordered when pressure is increased at room temperature but remain disordered on application of pressure at low temperature. Such pressure-induced oxygen ordering can account for the experimental fact that the T_c at high pressure for this oxide depends markedly on whether it is pressurized at room temperature or low temperature.

Overdoping of holes is also observed in $TlSr_2CaCu_2O_{7-z}$ (Figure 3.34) with single TlO sheets. The O site on the TlO sheet is appreciably deficient (occupancy = 0.88) even in an oxygen-rich nonsuperconducting sample. The occupancy of this site decreases to 0.80 in a superconducting sample with a T_c of 58 K. That is, the change in oxygen content and the corresponding one in hole concentration are caused by incorporation and release of O atoms on the TlO sheets.

3.3.8 T'-type Superconductors

In the n-type superconductor $(Nd_{1-x}Ce_x)_2CuO_{4-z}$, substitution of Ce^{4+} ions for Nd^{3+} ions leads to the doping of electrons on the CuO_2 sheet. The lack of apical oxygen is almost certainly essential for n-type superconductivity in this compound. In addition, oxygen deficiency plays an essential role to realize superconductivity in this system. An as-prepared sample does not show any superconductivity, but a reduced sample does. Alternatively, electron doping in Nd_2CuO_4 can be achieved by substituting F^- ions for a part of O^{2-} ions.

3.3.9 Infinite-Layer Compounds

In the infinite-layer compound $(Sr_{0.9}La_{0.1})CuO_2$ with a T_c of 42 K, both the metal and O sites are fully occupied, and there is no interstitial oxygen between two CuO_2 sheets. Thus, neither oxygen vacancies nor interstitial oxygen plays a role in electron doping, which is attained simply by partial replacement of Sr^{2+} ions with La^{3+} ions. By contrast, the local formation of lattice defects, that is, the vacancies of alkaline-earth cations and those O atoms on a part of CuO_2 sheets, was proposed to account for hole doping in $(Sr_{1-x}Ca_x)_{1-y}CuO_2$. This defect-structure model, which is unreasonable from a crystal-chemistry point of view, should be re-examined together with the appearance of superconductivity in this system.

3.4 Phase Equilibria and Phase Stabilities

3.4.1 Ln₂O₃-CuO Systems

Two types of stoichiometric compounds, Ln_2CuO_4 and $Ln_2Cu_2O_5$, exist in the pseudo-binary systems Ln_2O_3-CuO (Ln = rare-earth metals). The Ln_2CuO_4

phases crystallize in three different tetragonal (or pseudo-tetragonal) structures: T-type (K_2NiF_4-type), T'-type (Nd_2CuO_4-type), and T*-type. The T-type structure is obtained only for Ln = La, whereas the T'-type is stable for Ln = Pr, Nd, Sm, Eu, and Gd at ambient pressure. A T*-type superconductor was first isolated as $(Nd_{0.59}Sr_{0.41})(Nd_{0.73}Ce_{0.27})CuO_4$ ($\equiv MM'CuO_4$). This type of oxide can be synthesized with appropriate combinations of rare-earth and alkaline-earth metals at the M and M' sites, for example, La-Tb, La-Dy and La-Ln-Sr with Ln = Nd, Sm, Eu, Gd, Tb, and Dy. Superconducting transition occurs near 30 K in copper oxides with the three structures by doping a proper level of holes (T- and T*-type structures) or electrons (T'-type structure).

The Ln_2CuO_4-type phases are unstable for Ln = Y and Tb–Lu at ambient pressure, where $Ln_2Cu_2O_5$ ("blue phase") appears instead. The $Ln_2Cu_2O_5$ phases are the only stable compounds in the Ln_2O_3-CuO (Ln = Y and Tb–Lu) systems. In particular, $Y_2Cu_2O_5$ is well known as an impurity phase often formed in the process of preparing $YBa_2Cu_3O_7$. $Ho_2Cu_2O_5$ is orthorhombic ($P2_1nb$) with lattice constants of a = 1.2478 nm, b = 1.0813 nm, and c = 0.3495 nm. The rare-earth and Cu atoms in this structure occupy a highly distorted octahedral site and a square-planar site, respectively.

A definite correlation is seen between the ionic radii of the Ln ions and phase stabilities in the Ln_2O_3-CuO systems. The larger La^{3+} ion prefers the T-type structure whereas cations with intermediate sizes prefer the T'-type one. The T*-type structure is often stabilized when the system includes both larger and smaller A-site cations. Not Ln_2CuO_4 but $Ln_2Cu_2O_5$ appears for rare-earth cations smaller than Gd^{3+}. However, the T'-type compounds, Ln_2CuO_4, are formed under high pressure even for smaller rare-earth metals, Ln = Y and Tb–Tm.

A series of compounds, $La_{4+4n}Cu_{8+2n}O_{14+8n}$, exists in the La_2O_3-CuO system. The n = 2 and 3 members synthesized in bulk are stable only in very narrow temperature ranges near 1000°C and 1030°C, respectively. $La_2Cu_2O_5$ (n = 2) is not isomorphous with $Ho_2Cu_2O_5$ but crystallizes as monoclinic form (C2/c).

3.4.2 Ln_2CuO_4 Systems Doped with Foreign Cations or Excess Oxygen

The T-type superconductors, $(La_{1-x}A_x)_2CuO_4$ (A = Ca, Sr, and Ba), have been widely investigated mainly from the view point of basic research because their structures are relatively simple. $(La_{1-x}Sr_x)_2CuO_4$ has a wide solid-solution range, $0 \leq x \leq 0.67$. Its structure changes with increasing x, from the orthorhombic form ($0 \leq x < 0.05$), through the tetragonal form ($0.05 \leq x < 0.5$), to compounds with complicated superstructures ($0.5 \leq x \leq 0.67$). Oxygen vacancies are introduced into its crystal lattice in a higher x region. For instance, the chemical composition of this phase for x = 0.5 is $(La_{0.5}Sr_{0.5})_2CuO_{3.5}$. Superconductivity is observed in an x range from approximately 0.03 to approximately 0.16. Its T_c rises with increasing Sr content up to x = 0.075 but falls for a further increase in x.

The solid-solution ranges in $(La_{1-x}A_x)_2CuO_4$ (M = Ca and Ba) are much narrower than that in $(La_{1-x}Sr_x)_2CuO_4$: $x \approx 0.05$ for Ca and $x \approx 0.125$ for Ba. The Ca- and Ba-doped oxides show the HTT→LTO transition in the same manner as $(La_{1-x}Sr_x)_2CuO_4$. Only the LTO form of $(La_{1-x}Ba_x)_2CuO_4$ with $x \approx 1/16$ further transforms into the low-temperature tetragonal (LTT) form (Figure 3.13c): $P4_2/ncm$ (No. 138); $a = 0.53464$ nm and $c = 1.32319$ nm at 30 K for $x = 0.065$. Each $[CuO_6]$ octahedron tilts about the $[110]_t$ axis (t: tetragonal subcell, $a \approx 0.38$ nm) in the LTO form and about the $[100]_t$ axis in the LTT form.

The variation of T_c with x in $(La_{1-x}Ba_x)_2CuO_4$ is quite different from that in $(La_{1-x}Sr_x)_2CuO_4$. The T_c of the Ba system has two peaks near $x = 0.05$ and 0.08, going down at the intermediate range of x near $1/16$. This anomalous suppression of superconductivity has been discussed in connection with the low-temperature structural phase transitions. The LTT phase is stable in the intermediate x region below approximately 60 K, and no corresponding phase exists in the $(La_{1-x}Sr_x)_2CuO_4$ system. The correlation between the T_c depression near $x = 1/16$ and the appearance of the LTT form suggests that superconductivity is suppressed in the LTT form.

La_2CuO_4 can be made superconducting not only by doping A^{2+} cations but by introducing excess oxygen. Superconducting La_2CuO_{4+z} samples with z up to about 0.12 were prepared by high-pressure treatment with O_2, electrochemical oxidation, and oxidation in aqueous solutions of $KMnO_4$.

Ce-doped T'-type compounds, $(Ln_{1-x}Ce_x)_2CuO_4$ (Ln = Nd–Dy), are of particular importance because they show superconductivity with electrons as carriers. The solid-solution range of the system is $0 \leq x \leq 0.1$ for Ln = Nd.

3.4.3 AO-CuO Systems (A = Alkaline-Earth Metals)

The SrO-CuO system comprises three stable phases: Sr_2CuO_3, $SrCuO_2$, and "$Sr_{14}Cu_{24}O_{41}$". Three compounds, Ca_2CuO_3, $CaCuO_2$, and $CaCu_2O_3$, are also found in the CaO-CuO system. Ca_2CuO_3 is isostructural to Sr_2CuO_3. On the other hand, $CaCuO_2$ stable below approximately 740°C is not isomorphous with $SrCuO_2$. $CaCu_2O_3$ is stable only within a narrow temperature range near 1000°C. At least two stable compounds, Ba_2CuO_3 and $BaCuO_2$, exist in the BaO-CuO system. $BaCuO_2$ (cubic, Im3m; $a = 1.827$ nm) is isomorphous with neither $SrCuO_2$ nor $CaCuO_2$.

Figure 3.35 shows the subsolidus phase diagram of the SrO-CaO-CuO system. The 1:1 solid solution, $(Sr_{1-x}Ca_x)CuO_2$ ($x \leq 0.75$), is orthorhombic (Cmcm) with lattice constants of $a = 0.356$ nm, $b = 1.632$ nm, and $c = 0.392$ nm in $SrCuO_2$ ($x = 0$). Another 1:1 solid solution exists near the Ca-side end point of the solid solution with a very narrow homogeneity range: $x = 0.85 \pm 0.02$. This phase, well known as the infinite-layer compound, has a primitive tetragonal unit cell (P4/mmm; $a = 0.38611$ nm and $c = 0.31995$ nm for $x = 0.86$). A continuous series of solid solutions, $(Sr_{1-x}Ca_x)CuO_2$, with the infinite-layer structure is formed under high temperature and pressure. An Sr-rich part (typically $x = 0.3$) of the solution was claimed to exhibit p-type superconductivity below 110 K on introduction of Sr(Ca) vacancies. However, various experimental results have been throwing doubt upon superconductivity in this system. On the other hand, the partial substitution of La, Nd, Sm, and Gd for Sr in $SrCuO_2$ leads to superconducting behavior with electrons as carriers.

Sr_2CuO_3 crystallizes in orthorhombic form (Immm; $a = 1.268$ nm, $b = 0.391$ nm, and $c = 0.348$ nm). It contains not two-dimensional CuO_2 sheets but

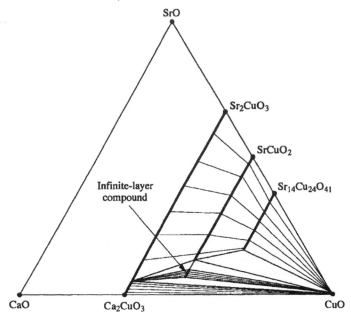

Figure 3.35. Phase diagram of the SrO-CaO-CuO system at 950°C. (From Phase Equilibria of the System SrO-CaO-CuO, R. S. Roth, C. J. Rawn, J. J. Ritter, and B. P. Burton, J. Am. Ceram. Soc. 72(8), 1545–1549 (1989). Copyright year 1989. Reprinted by permission of the American Ceramic Society).

one-dimensional chains of [CuO_4] square planes similar to those in the charge reservoir of $YBa_2Cu_3O_7$. A series of Ruddlesden-Popper-like compounds, $Sr_{n+1}Cu_nO2_{n+1+z}$ or $(Sr,Ca)_{n+1}Cu_nO2_{n+1+z}$ is formed in the SrO-CuO or SrO-CaO-CuO system at high presure under oxidizing conditions. Only a member with n = 1 (Sr_2CuO_{3+z}) has been isolated as a single phase. It is a nonsuperconductor with a complex superstructure which is a modification of the structure for Sr_2CuO_3. Other members with n = 2 and 3 are superconductors with T_cs between 80 K and 100 K.

3.4.4 Y_2O_3-BaO-CuO System

The subsolidus phase diagrams of the Y_2O_3-BaO-CuO system were reported by several groups. These results are not completely consistent with each other; there are some discrepancies in the Ba-rich part of the diagrams. Figure 3.36 shows a phase diagram for Y_2O_3-BaO-CuO at 875°C in air. Four quasi-ternary phases are included in their diagram: Y_2BaCuO_5 ("211" or "green phase," $Y_3Ba_8Cu_5O_z$ ("385"), $YBa_5Cu_3O_z$ ("152"), and the superconductor $YBa_2Cu_3O_z$ ("123"). Their crystal data except the 123 phase are (1) 211 phase: orthorhombic (Pbnm), a = 0.71276 nm, b = 1.21742 nm, and c = 0.56564 nm; (2) 385 phase: tetragonal (P4/mmm), a = 0.57875 nm and c = 0.80141 nm; (3) 152 phase:

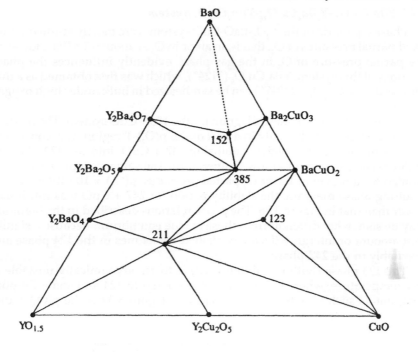

Figure 3.36. Phase diagram of the Y_2O_3-BaO-CuO system at 875°C [18].

orthorhombic (probably Immm), a = 0.4030 nm, b = 0.4090 nm, and c = 2.160 nm. An additional oxide, $YBa_4Cu_3O_z$ appears unless $BaCO_3$ is used as a starting reagent.

Not only subsolidus equilibria but solid-liquid ones are often of great importance for preparing good polycrystalline samples of superconductors or for growing their single crystals. Though the phase relationships of the whole Y_2O_3-BaO-CuO system have not been established as a function of temperature, many papers have dealt with its phase diagrams along some two-phase sections. Figure 3.37 gives a phase diagram along the 211-123 section. The 123 phase is decomposed peritectically into the 211 phase and a melt at about 1000°C.

On the basis of the diagram shown in Figure 3.37, a 123 sample with a large levitation force was produced by a melt-powder-melt-growth (MPMG) process. In this method, a Y-Ba-Cu-O sample is melted at above 1200°C and cooled down. The resulting mixture of Y_2O_3 and a solidified liquid phase is remelted between 1000°C and 1200°C and quenched to below 1000°C, where polycrystals of the 123 phase nucleate and grow. When the ratio of starting materials is selected in between the 211 and 123 compositions, fine particles of the 211 phase are dispersed in the matrix of the 123 phase in the final product. Such 211 particles are suggested to serve as effective pinning centers.

3.4.5 $YBa_2Cu_3O_7$-$Y_2Ba_4Cu_7O_{14}$-$YBa_2Cu_4O_8$ System

Phase equilibria of the Y_2O_3-BaO-CuO system were mostly studied under fixed partial pressures of O_2, that is, in air or in O_2 at about 0.1 MPa. However, the partial pressure of O_2 in the gas phase evidently influences the phase relations of this system. $YBa_2Cu_4O_8$ ("124"), which was first obtained as a thin film, and $Y_2Ba_4Cu_7O_{14+z}$ ("247") can be synthesized in bulk under high oxygen pressure.

Figure 3.38 shows a phase diagram for the 123 + CuO system. The mixture of 123 and CuO is stable only in a part of the $p(O_2)$-T region. It is converted, directly or through the region stabilizing 247 + CuO, into the 124 phase at high oxygen pressure or at low temperature. The same system was also examined for a region of lower O_2 partial pressures: $p(O_2) = 10^2$–10^6 Pa. In the resulting phase diagram, the stability region for 247 + CuO was much narrower than that in Figure 3.38. The phase relations described in the literature may be somewhat different from those at true equilibrium because a significant amount of intergrowths are contained sometimes in the 124 phase and inevitably in the 247 phase.

The 123 phase (without adding CuO) is thermodynamically unstable at low temperature, where it would be decomposed to 124 and some Cu-poor compounds such as Y_2BaCuO_5 and $BaCuO_2$. Figure 3.39 is the $p(O_2)$-T dia-

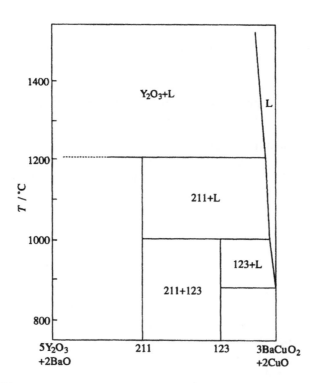

Figure 3.37. Phase diagram of the Y_2O_3-BaO-CuO system in O_2 along the 211-123 section [19].

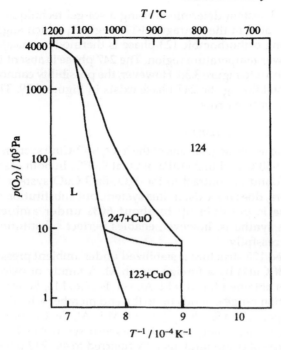

Figure 3.38. The p(O$_2$)-T diagram for the 123 + CuO system [20].

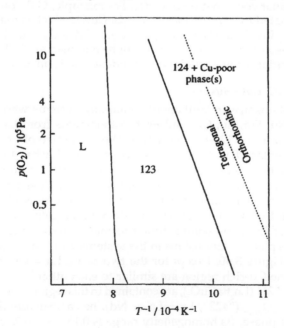

Figure 3.39. The p(O$_2$)-T diagram for the 123 system [21].

gram of the 123 system determined using a sol-gel technique. The 123/124 boundary is located in the "tetragonal-123 region", which suggests that the superconducting orthorhombic 123 phase is thermodynamically metastable at least in a lower temperature region. The 247 phase is absent in Figure 3.39 in marked contrast to Figure 3.38. However, the possibility cannot be excluded that a region stabilizing the 247 phase exists in Figure 3.39. The 247 region may be, if any, very narrow.

3.4.6 Y_2O_3-SrO-CuO Systems

The subsolidus phase relations of the Y_2O_3-SrO-CuO system were studied in air or O_2 at 950°C and in 20 MPa of O_2 at 980°C. In both cases, no ternary phases were found in contrast to the Y_2O_3-BaO-CuO system. The 123-type phase, therefore, does not exist in this system, but substitution of Sr for Ba in the 123 phase is possible up to about 60% under ambient pressure. High-pressure synthesis, however, enables perfect substitution, producing $YSr_2Cu_3O_z$ successfully.

The Sr-based 123 structure is stabilized under ambient pressure by partial replacement of Cu(1) by a foreign metal, M. A family of oxides, $YSr_2(Cu_{1-x}M_x)Cu_2O_z$, was obtained for M = Li, Al, Ga, Fe, Co, Mo, W, etc. In particular, the solid-solution range extends up to the end member with the stoichiometric metal composition $YSr_2MCu_2O_z$ when M is Al, Fe, Co, or Ga. A series of Ga-containing oxides, $LnSr_2GaCu_2O_7$, was also synthesized for Ln = Y and La–Yb. This type of compound, usually referred to as 1212 phases, has structures analogous to that of $YBa_2Cu_3O_7$ except that M does not necessarily adopt the square-planar coordination of Cu(1). For example, Ga^{3+} ions are tetrahedrally coordinated to four O^{2-} ions in $YSr_2GaCu_2O_7$, which is made superconducting by doping Ca at the Y site. $(Y_{1-x}Ca_x)Sr_2GaCu_2O_7$ with the continuous and full solid-solution range, $0 \leq x \leq 1$, can be obtained at 1200°C and 6 GPa. This solid solution shows a unique dependence of T_c on x.

3.4.7 Ln_2O_3-BaO-CuO System

Phase relationships in lanthanoid-containing systems were investigated for Ln = La, Nd, Sm, Eu, Gd, and Er. Most elements from La to Lu can be accommodated in the 123-type structure. When the size of Ln^{3+} is relatively large, it can replace Ba^{2+} to yield a solid solution, $Ln(Ba_{1-x}Ln_x)_2Cu_3O_z$. For instance, the solid solution extends from x = 0.25 to 0.4 for Ln = La and from 0.02 to 0.3 for Ln = Nd. The end member with x = 0 does not exist in both cases. The difference in ionic radii between Ba^{2+} and Ln^{3+} is a key factor governing the solid-solution range. That is, the x range gets narrower with decreasing size of Ln^{3+}, and solid-solution formation terminates at Ln = Gd.

Subsolidus phase relationships in the systems with Ln = Y and La–Er are compared in Figure 3.40. Except for the two cases, Ln = La and Nd, phase diagrams in the Cu-rich region are similar to each other. Two quasi-ternary phases, 123 and 211 (Ln_2BaCuO_5), are contained in this region. The solid solution, $Ln_{4-2x}Ba_{2+2x}Cu_{2-x}O_{10-2x}$ ("422", Ln = La or Nd), has a structure different from that of the 211 phase. Its homogeneity range is $0.15 \leq x \leq 0.25$ for La and $0.0 \leq x \leq 0.1$ for Nd.

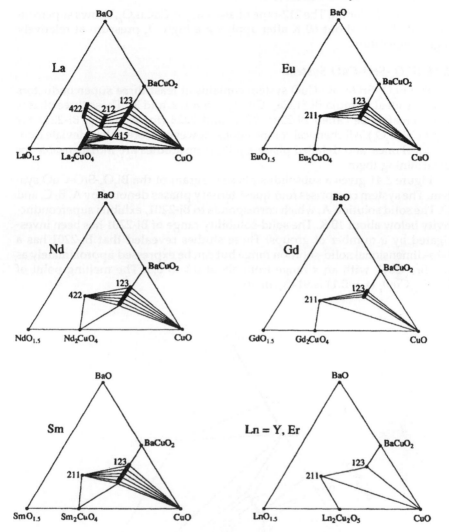

Figure 3.40. Phase relationships in the Ln$_2$O$_3$-BaO-CuO (Ln = La, Nd, Sm, Eu, Gd, Er, and Y) systems near the CuO-rich region at 950°C in air [22].

Only the La$_2$O$_3$-BaO-CuO system contain two additional phases: La$_4$BaCu$_5$O$_{13+z}$ ("415") and La$_{2-x}$Ba$_{1+x}$Cu$_2$O$_{6-z}$ ("212"). La$_4$BaCu$_5$O$_{13+z}$ has an oxygen-deficient perovskite lattice (tetragonal, P4/m; a = 0.86475 nm and c = 0.38594 nm). La$_{2-x}$Ba$_{1+x}$Cu$_2$O$_{6-z}$ is isomorphous with La$_{2-x}$A$_{1+x}$Cu$_2$O$_6$ (A = Ca and Sr). The 212 phase (tetragonal, I4/mmm) is closely related in structure to Sr$_3$Ti$_2$O$_7$, a member of the Ruddlesden-Popper series oxides. The solid-solution range extends with increasing size of A^{2+}: x ≈ 0.10 for Ca, 0 ≤ x ≤ 0.14 for Sr,

and $0 \leq x \leq 0.25$ for Ba. The 212-type phase, $La_{2-x}Sr_xCaCu_2O_6$, shows superconductivity below about 60 K after applying a high O_2 pressure at relatively
high temperature.

3.4.8 Bi_2O_3-SrO-CuO System

The Bi_2O_3-SrO-CaO-CuO system contains at least three superconductors
with the ideal formula $Bi_2Sr_2Ca_{n-1}Cu_nO_{4+2n}$ ($n = 1, 2,$ and 3). These three phases
will hereafter be called Bi-2201, 2212, and 2223 (for example, Bi-2212 for
$Bi_2Sr_2CaCu_2O_8$). All chemical compositions, however, more or less deviate from
the ideal ones, and detailed phase-equilibrium studies are indispensable for
determining them.

Figure 3.41 gives a subsolidus phase diagram of the Bi_2O_3-SrO-CuO system. The system comprises four quasi-ternary phases denoted by A, B, C, and
D. The solid solution A, which corresponds to Bi-2201, exhibits superconductivity below about 10 K. The solid-solubility range of Bi-2201 has been investigated by a number of groups. These studies revealed that Bi-2201 has a
two-dimensional solid-solution range but can be expressed approximately as
$Bi_{2+x}Sr_{2-x}CuO_z$ with an x range from about 0.1 to 0.5. The melting point of
$Bi_{2.1}Sr_{1.9}CuO_z$ ($x = 0.1$) is 914°C in air.

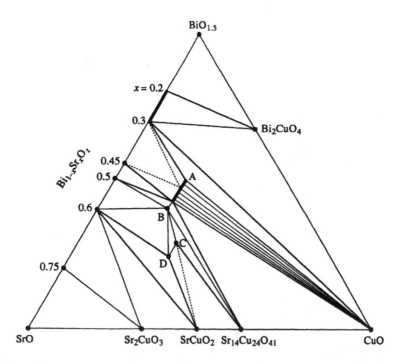

Figure 3.41. Phase diagram of the Bi_2O_3-SrO-CuO system at 840°C in air [23]. Dotted
lines indicate possible tie lines. The composition of phase B is different from that in [23].
A: $Bi_{2+x}Sr_{2-x}CuO_z$, B: $Bi_2Sr_2CuO_6$, C: $Bi_2Sr_3Cu_2O_z$, and D: $Bi_4Sr_8Cu_5O_z$.

The B phase exists near the superconducting Bi-2201 phase. It has a quasi-stoichiometric composition, $Bi_2Sr_2CuO_6$ and crystallizes in monoclinic form (C2/m or Cm; a = 2.4473 nm, b = 0.54223 nm, c = 2.1959 nm, and β = 105.40°). Its melting point is 904°C. The composition of $Bi_2Sr_2CuO_6$ is identical to the ideal one of the Bi-2201 superconductor, but it is not superconducting.

The C phase, $Bi_2Sr_3Cu_2O_z$, is isostructural to Bi-2212. Superconductivity appears below approximately 80 K by partial substitution of Ca for Sr in this phase. The D phase, $Bi_4Sr_8Cu_5O_z$, has an orthorhombic structure (Fmmm; a = 3.4035 nm, b = 2.405 nm, and c = 0.5389 nm).

3.4.9 Bi_2O_3-SrO-CaO-CuO System

Two superconducting phases, Bi-2212 (n = 2) and Bi-2223 (n = 3), appear in the Ca-containing system Bi_2O_3-SrO-CaO-CuO. Bi-2212 and 2223 show superconductivity below approximately 80 K and 110 K, respectively. Determination of their homogeneity ranges is difficult owing to a large number of compositional and processing parameters. In addition, mutual substitutions between the constituent cations are observed in these phases, for example, substitution of Ca for Sr and that of Bi for Sr and Ca, etc. Easy insertion of intergrowths of foreign phases with different n values further complicates the situation, in particular, in Bi-2223.

Many papers have been published on investigation of the single-phase

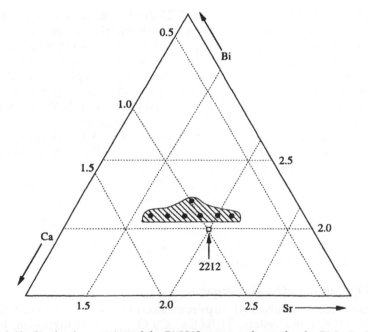

Figure 3.42. Single-phase region of the Bi-2212 superconductor for the $Bi_aSr_bCa_cCu_2O_z$ (a + b + c = 5) section [24]. Solid circles indicate experimental points where single-phase samples were prepared.

region for the Bi-2212 phase. Figure 3.42 shows the isothermal section of the system determined by fixing the amount-of-substance ratio of the metals at (Bi+Sr+Ca):Cu = 5:2. Pure Bi-2212 phases are obtained in a Bi-riche region, for example, $Bi_{2.1}(Sr_{1-x}Ca_x)_{2.9}Cu_2O_z$, rather than at the ideal 2212 composition. In addition, mutual substitution between Sr^{2+} and Ca^{2+} ions occurs in a wide range. Bi-2212 melts to form Bi-2201 and a liquid phase near 900°C in air.

$Bi_2Sr_2CaCu_2O_z$, which has the stoichiometric metal contents and a T_c as high as 96 K, could be successfully obtained by a heat treatment of a stoichiometric mixture with H_2 at 250°C and the final reaction in flowing (99.8% N_2 + 0.2% O_2) gas at 785°C. The pretreatment produced fine powders and promoted the subsequent reaction. These results suggest that the single-phase regions that have been determined by conventional solid-state reactions may not reflect the true equilibrium state.

The Bi-2223 superconductor with a T_c of about 110 K has not been isolated as a single phase but always mixed with Bi-2201, 2212, and/or other phases in the system. It takes an extremely long time to prepare Bi-2223 by solid-state reactions. The compositions of grains for Bi-2223 in multiphase samples were determined by electron microprobe analysis. The resulting compositions were distributed in a region that is slightly Bi-rich and Sr-poor with respect to the stoichiometric 2223 composition.

3.4.10 Bi_2O_3-PbO-SrO-CaO-CuO System

Bi atoms in the Bi-2201, 2212, and 2223 can be partially replaced by Pb, which accompanies various changes in their chemical and structural properties. Pure superconducting phases, in particular Bi-2223, can be obtained easily by Pb doping. Intergrowth formation is suppressed drastically in the Pb-doped 2223 phase. $(Bi_{1-x}Pb_x)_2Sr_2CuO_z$ is a superconductor isomorphous with $Bi_{2+x}Sr_{2-x}CuO_y$ while the monoclinic compound $Bi_2Sr_2CuO_6$ is nonsuperconducting.

Figure 3.43 shows the single-phase region of the slightly Bi(Pb)-rich section for $Bi_{1.84}Pb_{0.34}Sr_aCa_bCu_cO_z$ (a + b + c = 7). A single phase is obtained for this Bi/Pb ratio in a slightly Ca- and Cu-rich region. A low partial pressure of O_2, $p(O_2) \approx 8$ kPa, in addition to the substitution of Pb for Bi, is very effective for accelerating formation of the 2223 phase.

Compounds with the ideal 2212 and 2223 compositions lie on the line between Bi(Pb)-2201 and "CaCuO$_2$." Phase relationships were examined as a function of temperature along the line of $(Bi_{0.8}Pb_{0.2})_2Sr_2CuO_6$ –"$CaCuO_2$". $(Bi_{0.8}Pb_{0.2})_2Sr_2CuO_6$ and $(Bi_{0.8}Pb_{0.2})_2Sr_2CaCu_2O_z$ start to melt at 870°C and 876°C, respectively. On the other hand, the 2223 phase, $(Bi_{0.8}Pb_{0.2})_2Sr_2Ca_2Cu_3O_z$, is stable in a narrow temperature region between 835°C and 875°C.

3.4.11 Tl_2O_3-BaO-CaO-CuO System

A homologous series of superconductors was discovered in the Tl_2O_3-BaO-CaO-CuO system: $Tl_mBa_2Ca_{n-1}Cu_nO_{2(n+1)+m} \equiv Tl\text{-}m2(n-1)n$. In this family, the m = 2 members are essentially isostructural to $Bi_2Sr_2Ca_{n-1}Cu_nO_{4+2n}$. Five members (n = 1–5) in the m = 1 series and four members (n = 1–4) in the

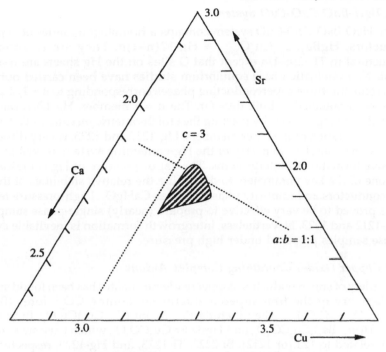

Figure 3.43. Single-phase region of the Bi(Pb)-2223 superconductor for the
$Bi_{1.84}Pb_{0.34}Sr_aCa_bCu_cO_z$ $(a + b + c = 7)$ section [25].

m = 2 series have been prepared so far, but very little is known about the phase equilibria of the system. Like the Bi-system, it is often difficult to prepare single-phase samples of the Tl-based superconductors. This is due to the formation of defects such as substitution of Tl for Ca, substitution of Cu for Tl, and formation of vacancies at the Tl or Ca site. Intergrowths of foreign phases are further inserted into the parent structure quite readily. Another serious problem is the vaporization of TlO. Reaction periods for synthesizing the Tl-containing superconductors must be limited to be very short to minimize the vaporization of TlO, which may result in nonequilibrium states in some cases.

Single-phase samples of Tl-2201 are slightly Cu-rich because Cu replace a part of Tl. Nearly pure samples of Tl-2212 and 2223 can be obtained by starting from the stoichiometric compositions. However, a better Tl-2223 sample was reported to be synthesized from a Tl-poor and Ca-rich starting mixture, "$Tl_{1.7}Ba_2Ca_{2.3}Cu_3O_{10}$".

Preparation of the single Tl-sheet oxides, Tl-12(n–1)n, is more difficult than that of the double Tl-sheet oxides, Tl-22(n–1)n. X-ray and neutron diffraction studies revealed significant replacement of Ca by Tl in Tl-1212 and $TlSr_2CaCu_2O_{7-z}$. No single-phase regions have been determined for this series of the oxides.

3.4.12 HgO-BaO-CaO-CuO System

The HgO-BaO-CaO-CuO system contains a homologous series of super-conductors, $HgBa_2Ca_{n-1}Cu_nO_{2(n+1)} \equiv Hg\text{-}12(n\text{--}1)n$. They are essentially isostructural to Tl-12(n–1)n except that O sites on the Hg sheets are nearly vacant. No systematic phase equilibrium studies have been carried out on this system, but three superconducting phases corresponding to n = 1, 2, and 3 have been obtained in bulk thus far. The n = 1 member, Hg-1201, can be obtained as a single phase by reacting the stoichiometric mixture of Ba_2CuO_3 and HgO. Isolation of the other members, Hg-1212 and 1223, is very difficult owing to the complex chemistry of the Hg-containing system as well as the extensive formation of intergrowths. The vapor pressure of Hg is indicated to be one of the key parameters to determine the relative stabilities of these superconductors and impurity phases such as $CaHgO_2$. High-pressure tech-niques proved to be very effective to prepare (nearly) single-phase samples of Hg-1212 and 1223. Nevertheless, intergrowth formation is inevitable even in those samples prepared under high pressure.

3.4.13 Copper Oxides Containing Complex Anions

A variety of superconducting copper oxide carbonates has been found since the discovery of the first superconductor containing CO_3^{2-} ions, $(Ba_{1-x}Sr_x)_2Cu_{1+y}O_{2+2y+z}(CO_3)_{1-y}$. They include $(Y_{1-x}Sr_x)Sr_2(C_{1-y}Cu_y)Cu_2O_z$, $Bi_2Sr_4Cu_2CO_3O_z$, $Tl(Sr_{1-x}Ba_x)_4Cu_2CO_3O_z$, and $HgBa_2Sr_2Cu_2CO_3O_z$, which have structures closely related to 123 (or 1212), Bi-2223, Tl-1223, and Hg-1223, respectively. Furthermore, two superconductors in the $(Cu_{0.5}C_{0.5})Ba_2Ca_{n-1}Cu_nO_{2n+3}$ series were prepared at 1200°C and 5 GPa under oxidizing conditions: n = 3 phase (T_c = 67 K) and n = 4 phase (T_c = 117 K). The n = 3 and 4 members of the series structurally resemble Tl-1223 and 1224, respectively; sites corresponding to Tl ones are occupied in order by Cu and C in a 1:1 amount-of-substance ratio.

BO_3^{3-} and NO_3^- ions proved to replace a part of CO_3^{2-} ions. New copper oxides containing SO_4^{2-} and PO_4^{3-} ions were also synthesized. Thus, a new and wide area of copper oxides containing complex anions has been exploited in no time.

3.5 References

1. F. Izumi, Physica C **190**, 35 (1991).
2. F. Santoro, M. Marezio, and R. J. Cava, Physica C 156, 693 (1988).
3. A. F. Wells, *Structural Inorganic Chemistry*, 4th ed., Clarendon Press, Ox-ford (1984), 239, 252, 584.
4. A. Santoro, F. Beech, M. Marezio, and R. J.Cava, Physica C **156**, 693 (1988).
5. K. Yvon and M. François, Z. Phys. B **76**, 413 (1989).
6. Y. Maeno, A. Odagawa, N. Kakehi, T. Suzuki, and T. Fujita, Physica C 173, 322 (1991).
7. M. Marezio, Acta Crystallogr., Sect. A **47**, 640 (1991).
8. D. de Fontaine, G. Ceder, and M. Asta, Nature (London) **343**, 544 (1990).
9. D. Jorgensen, Phys. Today 44, June, 34 (1991).
10. D. de Fontaine, G. Ceder, and M. Asta, Nature (London) **343**, 544 (1990).
11. J. D. Jorgensen, M. A. Beno, D. G. Hinks, L. Soderholm, K. J. Volin, C. U.

Segre, K. Zhang, and M. S. Kleefisch, Phys. Rev. B **36**, 3608 (1987).

12. R. J. Cava, B. Batlogg, C. H. Chen, E. A. Rietman, S. M. Zahurak, and D. J. Werder, Phys. Rev. B **36**, 5719 (1987).

13. J. D. Jorgensen, Shiyou Pei, P. Lightfoot, H. Shi, A. P. Paulikas, and B. W. Veal, Physica C **167**, 571 (1990).

14. Y. Matsui, F. Izumi, Y. Yamada, T. Matsumoto, Y. Kodama, S. Ikeda, K. Yanagisawa, and S. Horiuchi, J. Electron Microsc. **40**, 221 (1991).

15. H. Shaked, J.D. Jorgensen, B. A. Hunter, R. L. Hitterman, K. Kinoshita, F. Izumi, and T. Kamiyama, Phys. Rev. B **48**, 12941 (1993).

16. Y. Matsui, H. Maeda, Y. Tanaka, and S. Horiuchi, Jpn. J. Appl. Phys. **27**, L372 (1988).

17 R. S. Ross, C. J. Rawn, J. J. Ritter, and B. P. Burton, J. Am. Ceram. Soc. **72**, 1545 (1989).

18. G. Krabbes, W. Bieger, U. Wiesner, M. Ritschel, and A. Teresiak, Physica C 103, 420 (1993).

19. M. Murakami, M. Morita, K. Doi, K. Miyamoto, and H. Hamada, Jpn. J. Appl. Phys. **28**, L399 (1989).

20. J. Karpinski, S. Rusiecki, E. Kaldis, B. Bucher, and E. Jilek, Physica C **160**, 449 (1989).

21. H. Murakami, T. Suga, T. Noda, Y. Shiohara, and S. Tanaka, Jpn. J. Appl. Phys. **29**, 2720 (1990).

22. W. Wong-Ng, B. Paretzkin, and E. R. Fuller, Jr., J. Solid State Chem. **85**, 117 (1990).

23. Y. Ikeda, H. Ito, S. Shimomura, Y. Oue, K. Inaba, Z. Hiroi, and M. Takano, Physica C **159**, 93 (1989).

24. K. Knizek, E. Pollert, D. Sedmidubsky, J. Hejtmanek, and P. Pracharova, Physica C **216**, 211 (1993).

25. S. Koyama, U. Endo, and T. Kawai, Jpn. J. Appl. Phys. **27**, L1861 (1988).

3.6 Recommended Readings for Chapter 3

1. Hk. Müller-Buschbaum, Angew. Chem., Int. Ed. Engl. **28**, 1472 (1989).

2. B. Raveau, C. Michel, M. Hervieu, D. Groult, and J. Provost, J. Solid State Chem. **85**, 181 (1990).

3. J. D. Jorgensen, Neutron News 1, 24 (1990).

4. R. M. Hazen, Chap. 3, and C. H. Chen, Chap. 4, in *Physical Properties of High Temperature Superconductors* II, ed. D. M. Ginsberg, World Scientific, Singapore, 1990).

5. Special Issue: High-Temperature Superconductivity, Phys. Today **44**, June (1991).

6. A. W. Sleight, in *High Temperature Superconductivity: Materials, Mechanisms and Devices*, ed. D. P. Tunstall and W. Barford, Adam Hilger, Bristol, 1991, pp. 97–143

7. M. T. Anderson, J. T. Vaughey, and K. R. Poeppelmeier, Chem. Mater. **5**,

151 (1993).

8. *Phase Diagrams for High T$_c$ Superconductors*, J. D. Whitler and R. S. Roth, Am. Ceram. Soc., Westerville, 1991.

4

Statics and Dynamics of the Vortex State in Type II Superconductors

S. Sengupta and D. Shi

4.1 The Mixed State

When a magnetic field is applied to a type I superconductor ($\kappa < 1/\sqrt{2}$), the flux lines are completely expelled. This phenomenon of complete dia­magnetism is known as the *Meissner effect* [1]. The situation is different, how­ever, in a type II superconductor ($\kappa > 1/\sqrt{2}$). For a type II superconductor, above a certain critical field H_c (the lower critical field) flux lines start penetrating the superconductor, and at H_{c2} (the upper critical field) the superconducting property is lost [2]. This state where the flux lines penetrate the sample is known as the *mixed state*. [3].

The basic unit of the mixed state is a vortex, which essentially contains a quantum of flux line, Φ_0 given by

$$\Phi_0 = \frac{ch}{2e} = 2.07 \times 10^{-7} \text{ G - cm}^2. \tag{4.1}$$

The vortex consists of a central normal cylindrical core of radius ξ surrounded by a cylindrical superconducting region where a supercurrent circulates around the core to generate a single quantum of flux, Φ_0. This superconducting region is extended to a distance λ, the penetration depth.

The response of a type II superconductor to an applied magnetic field can be visualized as follows [4–6]. At low magnetic fields ($H < H_{c1}$), the flux lines are completely expelled and the superconductor is in the Meissner state. In this state, the energy of an isolated vortex is always greater than the reduction in the field energy that would occur if a flux line penetrated the superconductor. At the lower critical field, H_{c1} ($H = H_{c1}$), the condition of these two energies is just satisfied, and the flux lines start to enter the superconductor with the further increase of the applied field. The equilibrium vortex density at any applied field above H_{c1} is determined by the interaction forces that arise between the vortices as they come together. As the applied field is increased, the vortices approach each other; eventually their core overlaps, and the superconductivity disappears at the upper critical field, H_{c2}.

4.2 Properties of an Isolated Vortex

In this section we discuss the properties of an isolated vortex and then consider how two parallel vortex lines interact.

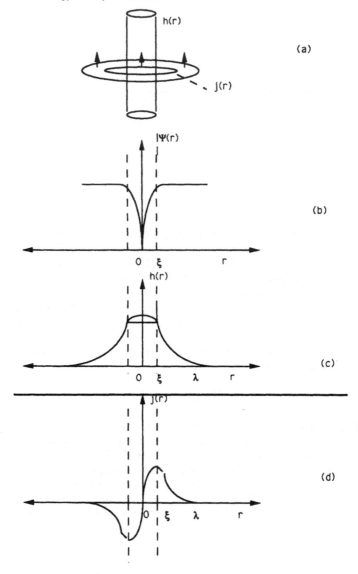

Fig. 4.1. Structure of an isolated vortex: (a) a single vortex, (b) the order parameter, (c) local field, and (d) the supercurrents.

4.2.1 *Magnetic Field and Current Density*

As mentioned earlier, a vortex can be visualized as a cylindrical normal core with radius ξ, surrounded by a cylindrical superconducting region extended up to a distance of λ. A supercurrent flows in this superconducting region to generate one quantum of flux, given by Equation (4.1). The structure of an isolated vortex line is shown schematically in Figure 4.1.

To understand the properties of an isolated vortex, we consider a type II superconductor in the London approximation [4,5,7]. This model is valid for extreme ($k \gg 1$) type II superconductors like high-T_c superconductors and is a good approximation at fields slightly higher than H_{c1} ($H_{c1} < H \ll H_{c2}$), where the interaction between vortices can be neglected.

In the high-κ approximation, the field outside the core of the vortex is given by the London equation

$$\lambda^2 \nabla \times \nabla \times h(r) + h(r) = \hat{z} \Phi_0 \delta_2(r). \qquad (4.2)$$

Here $h(r)$ is the local magnetic field of the vortex, $\hat{z}$ is an unit vector along the direction of the vortex, and $\delta_2(r)$ is a two-dimensional δ function at the location of the core. The right-hand side of the equation is required to satisfy the condition of the flux quantization. Integrating Equation (4.2) over an area S surrounding the vortex line, we obtain at a distance larger than the λ

$$\int_S h \cdot ds + \lambda^2 \oint_l (\nabla \times h) \cdot ds = \Phi_0. \qquad (4.3)$$

The line integral is taken along the circumference of the area S. At a distance much greater than λ, the current density is zero and hence

$$\nabla \times h = \frac{4\pi}{c} j = 0. \text{ Thus,}$$
$$\int_S h \cdot ds = \Phi_0. \qquad (4.4)$$

Using the Maxwell's equation $\nabla \cdot h = 0$, we can rewrite Equation (4.3) as

$$\nabla^2 h - \frac{h}{\lambda^2} = -\frac{\Phi_0}{\lambda^2} \hat{z} \delta_2(r), \qquad (4.5)$$

with the solution as

$$h(r) = \frac{\Phi_0}{2\pi\lambda^2} K_0\left(\frac{r}{\lambda}\right), \qquad (4.6a)$$

and supercurrent density, j, as

$$j(r) = \frac{\Phi_0 c}{8\pi^2\lambda^3} K_1\left(\frac{r}{\lambda}\right). \qquad (4.6b)$$

Here K_0 and K_1 are the zero- and the first-order Hankel function of imaginary argument. The function $K_0\left(\dfrac{r}{\lambda}\right)$ decreases as $\exp(-r/\lambda)$ at large distances and diverges logarithmically as $\ln(\lambda/r)$ as $r \to 0$, with a cutoff at $r \sim \xi$. Thus Equation (4.6a) can be written in the two-asymptotic approximation as

$$h(r) \approx \frac{\Phi_0}{2\pi\lambda^2} \ln\left(\frac{\lambda}{r}\right), \qquad\qquad \xi < r << \lambda, \qquad\qquad (4.7a)$$

$$h(r) \approx \frac{\Phi_0}{2\pi\lambda^2} \left(\frac{\pi\lambda}{2r}\right)^{1/2} \exp\left(-\frac{r}{\lambda}\right), \quad r >> \lambda. \qquad\qquad (4.7b)$$

4.2.2 Energy per Unit Length

The energy per unit length of a vortex can be written as

$$U_{vor} = U_{core} + U_{mag}. \qquad\qquad (4.8)$$

Here U_{core} is the energy associated with the core of the vortex, and U_{mag} is associated with the field energy and the kinetic energy of the supercurrents. Assuming the core to be a cylinder of radius ξ, we can approximate U_{core} as

$$U_{core} = \frac{H_c^2}{8\pi} \pi\xi^2. \qquad\qquad (4.9)$$

Similarly, we can estimate U_{mag} as [4,5]

$$U_{mag} = \int_S \left(\frac{h^2}{8\pi} + \lambda^2 \mid \nabla \times h \mid^2\right) dS \qquad\qquad (4.10a)$$

$$\approx \left(\frac{\Phi_0}{4\pi\lambda}\right)^2 \ln\kappa.$$

Using the relationship $\Phi_0 = 2\sqrt{2}\pi\lambda\xi H_c$, we obtain

$$U_{mag} = \frac{H_c^2}{8\pi} 4\pi\xi^2 \ln\kappa. \qquad\qquad (4.10b)$$

Comparing Equation (4.9) with (4.10b), one can easily see that the U_{mag} is larger than the U_{core} by a factor of the order of $4\ln\kappa$. This fact implies that for an extreme type II superconductor ($\kappa >> 1$), the error in the calculations for neglecting the details of the core is not important.

4.2.3 Interaction between Vortex Lines

We now consider the interaction between two vortex lines parallel to each other [4,5]. In the case of $\kappa >> 1$, the medium is linear, and one can use the principle of superposition to find the field resulting from the presence of two vortices. If the centers of the two vortices are specified by r_1 and r_2, the resultant magnetic field $h(r)$ will be given by

$$h(r) = h_1 \left(| r - r_1 | \right) + h_2 \left(| r - r_2 | \right).$$ (4.11)

The increase in the total free energy per unit length can be calculated as

$$\Delta U = 2U_{vor} + U_{12} (r_1 - r_2).$$ (4.12)

Here the first term is the sum of two individual energies per unit of the vortex. The second term is the interaction energy between the two vortex lines and is given by

$$U_{12} = \frac{\Phi_0 h_1(r_2)}{4\pi} = = \frac{\Phi_0}{2\pi\lambda^2} K_0 \left(\frac{| r_1 - r_2 |}{\lambda} \right).$$ (4.13)

Therefore, interaction energy falls off as $(| r_1 - r_2 |)^{-1/2} \exp (-| r_1 - r_2 | /\lambda)$ at large distances and varies logarithmically at small distances.

The force in the x direction on the vortex line at r_2 due to the presence of the vortex line at r_1 is given by

$$f_{2x} = - \frac{\partial U_{12}}{\partial x_2} = - \frac{\Phi_0}{4\pi} \frac{\partial h_1(r_2)}{\partial x_2} = \frac{\Phi_0 j_{1y}(r_2)}{c}.$$ (4.14)

Thus the force per unit length on the vortex at r_2 due to the supercurrents at r_1 is given by

$$f_2 = j_1(r_2) \times \frac{\Phi_0}{c},$$

where the direction Φ_0 is parallel to the flux density. In general, the forces acting on the core of a vortex line due to the supercurrents of all other vortices can be written as

$$f = j_s \times \frac{\Phi_0}{c}.$$ (4.15)

Any applied transport current should also be included in j_s. Thus a vortex array can be stationary only if the force acting on the vortices at any given position is zero, which can be accomplished by a symmetrical distribution of vortices forming a vortex lattice [3–5].

4.3 Critical State Models

As mentioned in the preceding section, because of the vortex-vortex interaction the vortices tends to arrange themselves into a vortex lattice. This lattice is also sometimes referred as the *Abrikosov vortex lattice* or the *flux line lattice* (FLL). The vortex lattice can withstand static magnetic fields much higher than the thermodynamic critical field, H_c, but it cannot carry an electric current. In the presence of an applied current, a Lorentz force given by Equation (4.15) acts on the vortices. This force tends to move the vortices, resulting in an electric field, E, parallel to the current:

$$E = \frac{1}{c}(\mathbf{B} \times \mathbf{v}),\qquad(4.16)$$

where **v** is the velocity of the vortex. This results into a resistive dissipation of energy. Thus a type II superconductor will not sustain any dissipationless current unless the motion of the vortices is prevented by some mechanism. Such a mechanism is known as *flux pinning* since it "pins" the vortices in preferred locations. Pinning results from any spatial inhomogeneity in the material, which may be caused by structural defects or compositional variation. Any inhomogeneity in a type II superconductor will give rise to variation of the superconducting parameters, resulting in the lowering of the vortex energy in preferred pinning sites.

When a field greater than H_{c1} is applied to a type II superconductor, the field starts penetrating in the form of vortices. Vortices are nucleated at the surface, and they move into the superconductor until the force from the density gradient is balanced by the pinning forces, f_p. The system is then said to be in the *critical state* [8]. Mathematically the condition can be written as

$$f_p = f_L.\qquad(4.17)$$

The flux distribution for any cycle or field is therefore defined uniquely by the pinning force provided that the field is always perpendicular to the current. Thus, the critical state equation can be written as

$$j_c = \frac{c}{4\pi}(\nabla \times \mathbf{B}).\qquad(4.18)$$

The flux density is usually assumed to take its equilibrium values at the surface, as defined by the reversible magnetization curves.

The concept of critical state was first considered by Bean [8], and hence the state is also known as the *Bean critical state*. In the most general case, the Bean model says that there exists a limiting microscopic superconducting current density, $j_c(H)$, that a hard superconductor can carry. Any electromotive force, however small, will induce this full current to flow locally. In this case only two states of current flow are possible with a given axis of magnetic field: zero current for those regions that never felt the magnetic field, and full current flow perpendicular to the field axis, the sense depending on the sense of the electromagnetic force that accompanied the last local change of field.

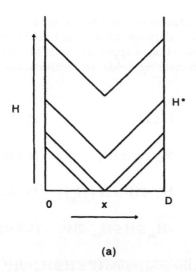

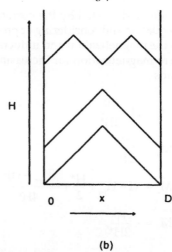

(a)   (b)

Fig. 4.2. The local field distribution in a slab at a constant temperature based on the Bean model (a) for increasing applied magnetic field and (b) for decreasing applied magnetic field.

To develop the magnetic hysteresis of a hard type II superconductor, Bean assumed that the current density is independent of magnetic field. To simplify the mathematics, he further considered an infinitely long sample with a uniform cross section of a slab of thickness D. The three-dimensional Maxwell equation then reduces to a one-dimensional problem.

Following Bean, let us now use practical units. From Ampere's law, $\frac{dB}{dt} = -\frac{4\pi j_c}{10}$ the field within the specimen decreases linearly with distance, as shown in Figure 4.2.

In the initial stages of magnetization, the current flows in the superficial layers whose thickness Δ is just enough to reduce the internal local field to zero (i.e., $\Delta = 5\,H/2j_c\pi$). This field-dependent penetration depth leads directly to the size-dependent magnetization curves. At fields of $H \geq j_c D\pi/5 = H^*$, currents flow through the entire volume of the specimen. The magnetization curves can then be calculated as

$$B = \frac{\int H dv}{\int dv} \tag{4.19a}$$

and

$$4\pi M = B - M \tag{4.19b}$$

In other words, B is the volume average of the local field, and $4\pi M$ is the average field created by the currents. Substituting the expression for H into this equation and considering appropriate boundary conditions, one can obtain both the magnetization and induction as a function of applied field. If M_+ and M_- are magnetization for increasing and decreasing fields, respectively, one obtains

$$-4\pi M_+ = H - \frac{H^2}{2H^*}, \qquad\qquad H \leq H^*, \qquad\qquad (4.20a)$$

$$-4\pi M_+ = \frac{H^2}{2H^*}, \qquad\qquad H_m \geq H \geq H^*, \qquad (4.20b)$$

$$-4\pi M_- = H - H_m + \frac{H^*}{2} + \frac{(H_m - H)^2}{4H^*}, \qquad H_m - 2H^* \leq H \leq H_m, \qquad (4.20c)$$

$$-4\pi M_- = -\frac{H^2}{2H^*}, \qquad\qquad -H_m \leq H \leq H_m - 2H^*, \qquad (4.20d)$$

where H_m is the maximum applied field. In the field region $H^* \leq H \leq H_m - 2H^*$, the hysteresis is connected to the full-penetration field by

$$\Delta M = M_- - M_+ = \frac{H^*}{4\pi}.$$

This leads to a direct relationship between the hysteresis width, ΔM, and the critical current density, j_c:

$$j_c = \frac{20\Delta M}{D}. \qquad\qquad (4.21)$$

Equation (4.21) is widely used to estimate j_c from magnetic hysteresis experiments. Since the critical current density is obtained magnetically, it is also sometimes mentioned as magnetic critical current density. For other types of geometric shapes, such as an infinite cylinder with diameter d or a rectangular parallelepiped with sides 2a x 2b, one gets

$$j_c = \frac{30\Delta M}{d}, \qquad\qquad \text{for a cylinder} \qquad\qquad (4.22a)$$

$$j_c = \frac{10\Delta M}{\left(1 - \dfrac{a}{3b}\right)} \qquad \text{for a rectangular parallelepiped.} \qquad (4.22b)$$

Equation (4.22b) reduces to Equation (4.21) for 2a = D and b →∞ and Equation (4.22a) for a = b. In general, the critical current density can be related to the hysteresis by the relation

$$j_c = \frac{A\Delta M}{d}, \qquad\qquad (4.23)$$

where A is a geometrical constant depending on the shape of the sample and d is the length scale of the current loops defined by the penetration of the field.

While developing the critical state model, Bean assumed that j_c is independent of the local field H. Further improvement in this model by considering the field dependence of j_c was reported later by various research groups [9–16]. In fact, the field dependence of j_c results from the complicated pinning mechanisms in hard superconductors.

The Bean model was first modified by Anderson and Kim [9]. They assumed that the critical current density is inversely proportional to the local field. Mathematically one can write

$$j_c(H, T) = \frac{j_c(0, T)}{1 + \dfrac{|H(x)|}{H_0(T)}}, \tag{4.24}$$

where H_0 is a materials parameter with the dimension of field. This model was found to agree well with some of the conventional hard superconductors.

Equation (4.24) indicates that the field dependence of j_c is associated with the term $\dfrac{|H(x)|}{H_0(T)}$, which may vary considerably among different systems. By considering more specific pinning mechanisms, Irie and Yamafuzi [10] developed a power-law field dependence of j_c which can be written as

$$j_c(H, T) = j_c(0, T)\left(\frac{|H(x)|}{H_0(T)}\right)^n, \tag{4.25}$$

where n is a materials parameter directly reflecting the pinning mechanisms.

Based on the magnetization data on cold-worked Nb-Zr wires, Feitz et al. [11] found that their experimental results were excellently fit to an empirical formula

$$j_c(H, T) = j_c(0, T)\exp\left(-\frac{|H(x)|}{H_0(T)}\right). \tag{4.26}$$

Equation (4.26) was obtained by several trial functions. The critical current density calculated by using this equation agreed well with the experimental transport j_c data in Nb-25 % Zr. Feitz et al. also pointed out that the Kim model was unable to fit the experimental data above 15 kG, while Equation (4.26) was valid up to 40 kG. Subsequently, the relationship (4.26) was applied to high-T_c superconductors and produced quite reasonable agreement between the theoretical and experimental results [13–15].

Recently, it has been pointed that all these critical state models are related and one can describe the magnetic hysteresis and j_c in terms of a generalized critical state model [16]. It has also been shown that all the critical state models described so far follow from the generalized model,

$$j_c(H, T) = \frac{j_c(0, T)}{\left(1 + \frac{|H(x)|}{H_0(T)}\right)^\beta},$$ (4.27)

where β is a dimensionless constant. Equation (4.27) reduces to the Bean model for $\beta = 0$ and to the Kim model (Equation 4.24) for $\beta = 1$. Equation (4.25) is obtained when the condition $\frac{|H(x)|}{H_0(T)} \gg 1$ is satisfied and $n = Q\beta$ in Equation (4.27).

Let us now consider the condition when $\frac{|H(x)|}{H_0(T)} = y \ll 1$ and $\frac{H_0(T)}{\beta} = H_0$. Equation (4.27) can then be written as

$$j_c(H, T) = \frac{j_c(0, T)}{(1 + y)^\beta}.$$

Taking the limit that $y \to 0$, we obtain

$$\lim_{x \to o} j_c(H, T) = \frac{j_c(0, T)}{\lim\limits_{x \to o}(1 + y)^\beta}$$

$$= \frac{j_c(0, T)}{\lim\limits_{x \to o}(1 + y)^{\left(\frac{1}{y}\right)\left(\frac{|H(x)|}{H_0}\right)}}$$

$$= j_c(0, T) \exp\left(-\frac{|H(x)|}{H_0}\right),$$

which is identical with Equation (4.26).

Equation (4.27) has three adjustable parameters: $j_c(0,T)$, H_0, and β. Here $j_c(0,T)$ is the macroscopic critical current density related to the flux pinning. The physical meaning of H_0 is not well understood. It is generally used in practice as an adjustable parameter that can be related to the critical fields (H_c, H_{c1}, H_{c2}) of the superconductor. (We point out that the power law and the exponential law are special cases with extreme conditions of $\frac{|H(x)|}{H_0(T)} \gg 1$ and $\frac{|H(x)|}{H_0(T)} \ll 1$, respectively.)

4.4 Flux Pinning in Type II Superconductors

This section focuses on various types of pinning interaction in type II superconductors.

4.4.1 Elementary Pinning Interactions

In the preceding section we pointed out that flux pinning results from the interaction of various inhomogeneities with the vortices. Here we consider the various types of elementary interactions of an isolated vortex line with inhomogeneities. Such interactions result in a change in vortex energy, giving rise to an effective flux pinning force, f_p.

4.4.1.1 Core Interaction

A vortex line can be pinned as a result of an inhomogeneity in the superconducting material by a variation in the core energy of the vortex. When a vortex line passes through a normal region, the energy saved is given by $U_{core}d$, where d is the dimension of the normal region along the length of the vortex line. As an example we consider the interaction between a vortex line with a spherical cavity of radius a (a > ξ) [17]. The energy recovered when the vortex line is in the center of the cavity (Figure 4.3) is given by

$$U = \frac{H_c^2}{8\pi} \pi \xi^2 2a. \tag{4.28}$$

To estimate the pinning force, f_p, acting on a vortex, we consider the center of the cavity to be the equilibrium position for the vortex. As the vortex line is displaced by a distance r (r < a) from the center of the cavity (as shown Figure 4.4), the change in energy is given by

$$\Delta U(r) = U(0) - U(r) = \frac{H_c^2 \xi^2}{4} [a - (a-1)]$$

$$= \frac{H_c^2 \xi^2 a}{4} \left[1 - \sqrt{1 - \left(\frac{r}{a}\right)^2}\right]. \tag{4.29}$$

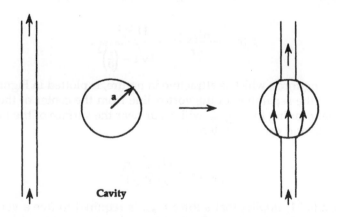

Flux line   **Flux line trapped in the cavity**

Fig. 4.3. Schematic of a flux line trapped in a spherical cavity.

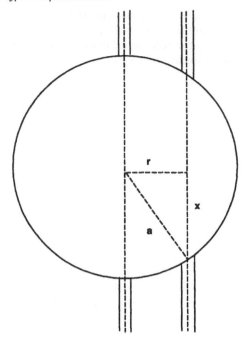

Fig. 4.4. Schematic diagram to estimate the elementary pinning force f_p due to core interaction between a vortex line and a spherical cavity.

The elementary pinning force f_p, acting on the vortex at the position r, is given by

$$f_p(r) = -\frac{\partial U(r)}{\partial r} = -\frac{H_c^2 \xi^2}{4\sqrt{1 - \left(\frac{r}{a}\right)^2}} . \qquad (4.30)$$

The pinning force, which is attractive in nature, is plotted in Figure 4.5 as a function of the separation of the vortex line from the center of the cavity. The maximal pinning force, $f_{p,max}$ will occur near the surface of the inclusion (i.e., $r = a - \xi$) and can be written as

$$f_{p,max} = -\frac{H_c^2 \xi^2}{4\sqrt{2}} \sqrt{\frac{a}{\xi}}. \qquad (4.31)$$

Equation (4.31) implies that a force $f_{p,max}$ is required to free a vortex line trapped by a cavity of radius a. Furthermore, the pinning force increases as $\sqrt{a}$ with the increasing size of the cavity.

4.4.1.2 Magnetic Interaction

When the interaction between a vortex line and an inhomogeneity results

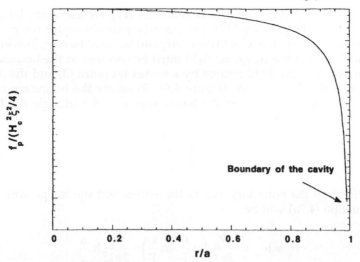

Fig. 4.5. The elementary pinning force $f_p(r)$ as a function of distance r from the center of the cavity. Note that the maximum pinning force, $f_{p,max}$, occurs near the surface of the cavity $(r = a - \xi)$.

in a variation in the magnetic energy U_{mag}, of a vortex line, the interaction is known as *magnetic interaction*. This interaction is purely electromagnetic and can be estimated using the London equations. The change in U_{mag} occurs because the inhomogeneities change the distribution of the magnetic fields and the supercurrents.

As an example we consider the interaction of a cylindrical cavity with an isolated vortex line. This problem was solved by Mkrtchyan and Shmidt [18]

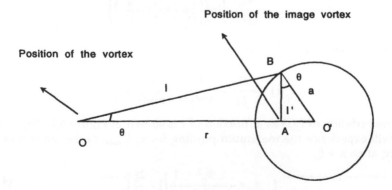

Fig. 4.6. Schematic diagram to estimate the elementary pinning force f_p due to the magnetic interaction between a vortex line and a cylindrical cavity.

and further developed by Timms and Walmsley [19]. We use a simpler image method to solve the problem [17]. The boundary condition for the problem is that the current perpendicular to the cavity surface must be zero. This implies (Equation 4.18) that the magnetic field must be constant at the boundary.

Let us consider the field created by a vortex (at point O) and the image vortex inside the cavity (at A) (Figure 4.6). To satisfy the boundary conditions, we choose the point A in the figure such that the triangle AO'B and OO'B are similar. Then

$$\frac{x}{a} = \frac{l}{l'} = \frac{a}{r},$$
(4.32)

and the field on the boundary due to the vortex and the image antivortex from Equation (4.7a) will be

$$h_v + h_{-v} = \frac{\Phi_0}{2\pi\lambda^2}\left(\ln\frac{\lambda}{l} - \ln\frac{\lambda}{l'}\right) = \frac{\Phi_0}{2\pi\lambda^2}\ln\frac{a}{r},$$
(4.33)

and should be constant. The general solution is the sum of the vortex-antivortex solution and the Meissner solution for the cylindrical cavity. If only a single quantum is trapped by the cavity, the solution is

$$h = \frac{\Phi_0}{2\pi\lambda^2}\left(\ln\frac{\lambda}{|l|} - \ln\frac{\lambda}{|l'|}\right) + \frac{\Phi_0}{2\pi\lambda^2}\ln\left(\frac{\lambda}{|1-r|}\right).$$
(4.34)

For a $<< \lambda$ we may neglect the contribution of the total flux from the field inside the cavity. The energy of the vortex interaction with the cylindrical cavity is given by

$$U = \frac{\Phi_0}{4\pi}[h_{-v}(O) + h_v(O)] + \frac{\Phi_0}{4\pi}h_v(O')$$

$$= \frac{\Phi_0^2}{4\pi^2\lambda^2}\left[\ln\left(\frac{\lambda}{r}\right) + \frac{1}{2}\ln\left(1 - \frac{a^2}{r^2}\right)\right].$$
(4.35)

The elementary pinning force is

$$f_p(r) = -\frac{\partial U}{\partial r} = -\frac{\Phi_0^2}{4\pi^2\lambda^2 r}\left(\frac{a^2}{r^2-a^2} - 1\right).$$

The variation of $f_p(r)$ as a function of r is shown in Figure 4.7. The vortex line will experience the maximum pinning force, $f_{p,max}$, at the surface of the cavity, at r= a + ξ,

$$f_{p,max} = -\frac{\Phi_0^2}{4\pi^2\lambda^2}\frac{1}{2\xi}\left[1 - \frac{2\xi}{a}\right],$$
(4.36)

which is same as that obtained by Mkrtychyan and Schmidt [18] and Timms

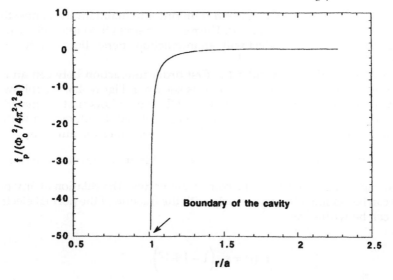

Fig. 4.7. The variation of $f_p(r)$ as function of r resulting from the magnetic interaction between a cylindrical cavity and a vortex line. The maximum force, $f_{p,max}$, is experienced by the vortex at the surface of the cavity.

and Walmsley [19].

For a spherical cavity, the maximum pinning force originating from the magnetic interaction was obtained by Shehata [20] to be

$$f_{p,max} = -2\pi \left(\frac{\Phi_0}{4\pi\lambda^2}\right)\sqrt{\frac{a}{\xi}} = -\frac{\pi\xi^2 H^2_c}{\lambda^2}\sqrt{\frac{a}{\xi}}. \qquad (4.37)$$

Comparing Equations (4.31) and (4.37), one can see that the pinning forces originating from the variation of the core energy is about λ^2 times larger than that of the pinning force resulting from the magnetic interaction. For an extreme type II ($\kappa \gg 1$) such as a high-T_c superconductor, the core pinning is thus generally stronger than that of magnetic pinning.

4.4.1.3 Interaction through Variation in the Elastic Energy

Interaction between vortex lines may originate from the small changes in the volume, ΔV, and the elastic constants, Δc, that occur when a material becomes superconducting. For an elastically isotropic medium, the fractional volume change, $\Delta V/V$, was found to be in the order of 10^{-7}. The change in the elastic constants, $\Delta c/c$, was estimated to be in the order of 10^{-4}. Structural defects such as dislocations and stacking faults, which generate a stress field, can interact with a vortex line by two mechanisms [6,21].

The first mechanism is known as the "volume effect" or the "first-order interaction." It was first estimated by Kramer and Bauer [22]. The interaction originates because the normal core of the vortex is contracted with respect to the superconducting region around it. In order to maintain the connectivity of the lattice, stresses are induced. The resulting stress field interacts with the dilation as a result of a defect with an interaction energy linear in the defect stress.

As an example we consider the first-order interaction between an edge dislocation parallel to the vortex line, as shown in Figure 4.8. We follow the derivation as reported by Kramer et al. [22]. The authors first estimated the stress field around a vortex line and then calculated its interaction with the stress field of the dislocation, using the Peach-Koehler formula [23].

The lattice dilation, $\varepsilon_v = \dfrac{\Delta V}{V_s}$, associated with the vortex line is a function of the radial distance r from the core of the vortex. The dilation at any point $\varepsilon_v(r)$ can be assumed to vary linearly with the fraction of the normal electrons and can be written as

$$\varepsilon_v(r) = \varepsilon_{v0}\left(1 - |\Psi|^2\right),$$

where Ψ is the order parameter. At low magnetic fields $|\Psi|^2 = 1 - \exp\left(-\dfrac{r^2}{\xi^2}\right)$ is a good approximation, and one obtains

$$\varepsilon_v(r) = \varepsilon_{v0}\exp\left(-\dfrac{r^2}{\xi^2}\right).$$

Since the analysis of the stress field is analogous to the stress arising from a cylindrically symmetric temperature gradient in an infinite cylinder, the principal stress associated with a vortex core can be written in cylindrical

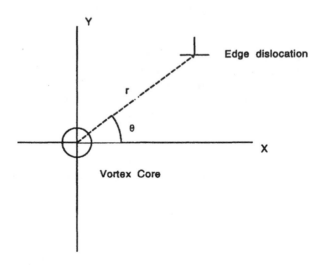

Fig. 4.8. Schematic diagram to estimate first-order interaction between an edge dislocation parallel to vortex line.

coordinates as

$$\sigma_{rr} = A \frac{\xi^2}{r^2} \left[1 - \exp\left(-\frac{r^2}{\xi^2} \right) \right],$$

$$\sigma_{\theta\theta} = A \left(2 + \frac{\xi^2}{r^2} \right) \left[\exp\left(-\frac{r^2}{\xi^2} \right) - \left(\frac{\xi^2}{r^2} \right) \right], \qquad (4.38)$$

$$\sigma_{zz} = 2A\exp\left(-\frac{r^2}{\xi^2} \right),$$

where, $A = -\dfrac{\varepsilon_v \mu(1+v)}{3(1-v)}$, μ is the shear modulus, and v is the Poisson ratio.

The force per unit line, f_p, acting on a dislocation characterized by the Burger vector, b, and a unit vector, t, can be estimated from the Peach-Koehler formula [23]:

$$f_p = -\sigma \times b.t. \qquad (4.39)$$

For an edge dislocation with $b = -b\hat{x}$ and $t = \hat{z}$ parallel to the vortex line situated along the z-axis, Equation (4.39) reduces to

$$f_p = 2g(r)b\sin\theta\cos\theta\hat{x} + [f(r) - 2g(r)\cos\theta]\hat{y}, \qquad (4.40)$$

where $f(r)$ and $g(r)$ are defined as

$$f(r) = -b\sigma_{\theta\theta}$$

$$g(r) = \frac{b}{2}(\sigma_{rr} - \sigma_{\theta\theta}).$$

Thus a parallel dislocation-vortex line interaction gives rise to a net attractive long-range force.

The second mechanism is quadratic in stress and is referred as "second-order interaction" [6,21]. It was first described by Fleischer [24], Webb [25], and Tott and Pratt [26]. Since the energy of a defect depends on the elastic constants of the surrounding medium, a vortex core (being stiffer than the superconducting matrix) may interact by modification of the defect energy. The energy is quadratic in defect stress.

As an example we consider the second-order interaction between a vortex line and a screw dislocation with a Burgers vector b perpendicular to the vortex line. We follow the derivation of Webb [25] and Campbell and Evetts [6]. We also assume the material to be cubic and isotropic. The stress field, τ, around a straight screw dislocation has a cylindrical symmetry and can be expressed as

$$\tau = \left(\frac{b}{2\pi r} \right) \frac{1}{S_{44}},$$

where S_{44} is the compliance. If the compliance S_{44} is changed locally by

dS_{44} because of the presence of a vortex line, the interaction energy between the vortex and the defect can be given by

$$U(r) = \frac{1}{2} \int \delta S_{44} \tau^2 dV.$$

In general, δS_{44} depends on the order parameter and hence is a function of r. If we assume that δS_{44} is constant within the core and is zero outside it, the interaction energy perpendicular to the vortex is

$$U(r) = \delta S_{44} \left(\frac{b}{2\pi r S_{44}} \right)^2 (\pi \xi^2 3r) = \delta S_{44} \left(\frac{b}{2\pi S_{44}} \right)^2 \frac{3\pi \xi^2}{r}. \qquad (4.41)$$

The pinning force can then be expressed as

$$f_p(r) = -\frac{\partial U}{\partial r} = \delta S_{44} \left(\frac{b}{2\pi S_{44}} \right)^2 \frac{3\pi \xi^2}{r^2}. \qquad (4.42)$$

Figure 4.9 shows the variation of f_p as a function of distance. The force is repulsive, and the vortices are expected to be pinned between the energy barriers provided by a forest of dislocations. The maximum pinning force, $f_{p,max}$, occurs when the outer edge of the core is just crossing the dislocation line (i.e., $r = \xi$). The first-order and the second-order interactions are usually are on same order of magnitude; however, they have different spatial variation, with the first-order interaction always dominating at large distances.

4.4.2 The Summation Problem

Let us turn now to the summation problem, in particular, the direct summation model and the elastic-lattice summation model.

4.4.2.1 Direct Summation Problem

In the preceding section we considered the interaction of an isolated vortex line with a defect, to estimate the elementary pinning force, f_p. To estimate the bulk pinning force, F_p, however, one has to consider the sum over all such elementary pinning forces [27–30]. Earlier theories used a *simple summation* like linear superposition to obtain an expression for F_p. The *direct summation model* implies that each vortex line is pinned individually by a single defect. If N_p is the density of the pinning sites, the volume pinning force, F_p, is given by

$$F_p = N_p f_p. \qquad (4.43)$$

In vortex lattice theory, the situation for direct summation can occur if the vortex lattice can be deformed plastically by the defects so that each vortex can be pinned at individual position of minimum energy. This situation can arise for strong pinning centers with large f_p or when the vortex lattice is very soft (i.e., at high magnetic fields) [30].

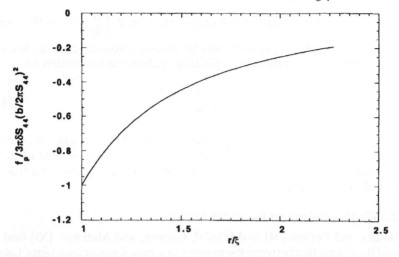

Fig. 4.9. The variation of $f_p(r)$ as a function of distance r resulting from the second-order interaction between a vortex line and a screw dislocation with a Burgers vector, b, perpendicular to the vortex line.

4.4.2.2 The Elastic Lattice Summation Model

Feitz and Webb [31] pointed out that a perfectly rigid vortex lattice cannot be pinned since the vortex lines cannot minimize their energy by bending because of the interaction with the pinning centers. A vortex lattice can be pinned only if the pinning centers can distort the lattice elastically or plastically, thereby enabling many of its vortices to become associated with the pinning centers. As pointed out earlier, the direct summation model can be applied when the pinning centers distort the vortex lattice plastically.

In the *elastic lattice summation model* the vortex lattice is distorted elastically as a result of interaction with the pinning centers [30,31]. The distortion is given by $\delta = f_p/c_{ij}a$, where c_{ij} is the vortex lattice elastic constant and a is the lattice spacing of the vortex lattice. The volume pinning force is then given by

$$F_p = \eta N_p f_p \qquad (4.44)$$

where η is an efficiency factor and N_p is the effective density of the pinning centers. If W_p is the width of the pinning potential, the efficiency factor is defined as

$$\eta = \frac{\delta}{W_p}.$$

The effective density of the pinning centers, N_p, is given by

$$N_p = n \left(\frac{B}{\Phi_0}\right) W_{p'}$$

where n is the density of the pinning centers, and $\left(\dfrac{B}{\Phi_0}\right)$ W_p is the vortex interaction cross section per unit volume. Thus the volume pinning force for a vortex lattice interaction with n pinning centers can be written as

$$F_p = n \left(\frac{B}{\Phi_0}\right)^{3/2} \frac{f_p^2}{c_{ij}}. \tag{4.45}$$

Because of the introduction of the elasticity of the vortex lattice, the volume pinning force, F_p, is found to be proportional to the square of the elementary pinning force, f_p. This result was obtained by Yamafuji and Irie [32], Feitz and Webb [31], and Labush [33].

4.4.2.3 Elastic Constants of the Vortex Lattice

Silicox and Rollins [34] and Friedal, Gennes, and Matricon [35] first described the interaction between the vortices in terms of elastic constants. Labush had estimated all the elastic constants by means of the local theory of elasticity [33,36]. Brandt had further shown that in the theory of elasticity of the vortex lattice, the nonlocal effects play an important role; he also estimated the dependence of the elastic constants on the wave vector k [37–44].

In Voigt's notation, the stresses, σ_i, are related to the strain, ε_j, as

$$\sigma_i = c_{ij}\varepsilon_j,$$

where i takes the value of 1 to 6 (which refers the suffices xx, yy, zz, yz, xz, and xy, respectively) and

$$\varepsilon_{xx} = \frac{\partial u_x}{\partial x} \text{ and } \varepsilon_{xy} = \left(\frac{\partial u_x}{\partial y} + \frac{\partial u_y}{\partial x}\right).$$

For a hexagonal lattice with a sixfold rotational symmetry, only five independent elastic constants are required to describe the stiffness matrix

$$[c_{ij}] = \begin{bmatrix} c_{11} & c_{12} & c_{13} & 0 & 0 & 0 \\ c_{12} & c_{11} & c_{13} & 0 & 0 & 0 \\ c_{13} & c_{13} & c_{33} & 0 & 0 & 0 \\ 0 & 0 & 0 & c_{44} & 0 & 0 \\ 0 & 0 & 0 & 0 & c_{44} & 0 \\ 0 & 0 & 0 & 0 & 0 & c_{66} \end{bmatrix}, \tag{4.46}$$

with $2c_{66} = c_{11} - c_{12}$.

If we assume that the vortex line is parallel to the z-axis and if the superconductor is isotropic, the properties in the x-y plane are isotropic. Since the line tension of a vortex line is independent of its length, no elastic constant is associated with the strain parallel to the z direction, ε_{zz}. Therefore, the third column and (by symmetry) the third row of the matrix $[c_{ij}]$ is zero. Thus the elasticity of a vortex lattice can be described as a function of three elastic constants: c_{11}, the bulk modulus; c_{44}, the tilt modulus representing the modulus for deformation that tilts a bundle of vortex lines from the z direction while leaving its cross section in the x-y plane; and c_{66}, the shear modulus in the x-y plane [6, 30].

Labush derived the expressions for c_{44} and $c_L = c_{11}$-c_{66} which controls the size but not the shape of the vortex lattice cell [33–36]. He also derived an expression for c_{66} for magnetic fields near H_{c2}, that is, in the region where the linearized Ginzburg-Landau equation can be applied. His results can be summarized as follows:

$$c_L = \frac{B^2}{4\pi} \frac{\partial H_{eq}(B)}{\partial B} \qquad (4.47a)$$

$$c_{44} = \frac{B}{4\pi} H_{eq}(B) \qquad (4.47b)$$

and $\qquad c_{66} = \frac{H_c^2}{8\pi} \frac{1}{(2\kappa^2 -1)\beta_A} 0.48 \left(1 - \frac{H}{H_{c2}}\right)^2,$ $\qquad (4.47c)$

where $\kappa = \lambda/\xi$ and β_A depends on the symmetry of the vortex array.

Brandt [37–40] derived an expression for c_{66} valid over the entire field range and recalculated other elastic moduli in both high and low magnetic fields. He pointed out the strong dispersive (nonlocal) nature of the bulk modulus c_{11} and the tilt modulus c_{44} [38]. The physical reason is that the elastic energy is determined by the variation of the magnetic energy. The magnetic field cannot vary, however,at distances less than the

effective penetration depth $\lambda' = \lambda \Big/ \left(1 - \frac{B}{H_{c2}}\right)^{1/2}$ (the factor $\lambda' = 1 \Big/ \left(1 - \frac{B}{H_{c2}}\right)^{1/2}$

is due to the overlapping of the vortex fields).

Therefore, the elastic energy caused by a local distortion of the vortex lattice is contained in a sphere with radius of the order of λ'. This implies that the elastic response of the vortex lattice is nonlocal and the elastic moduli cij depends on the wavelengths $2\pi/k$ of the Fourier components of the strain field [44]. Furthermore, under a long wavelength deformation with the wave vector, k, smaller than the $1/\lambda'$, the elastic energy does not increase with k, and the vortex lattice is soft against such deformations. Larkin and Ovchinnokov [45] later confirmed Brandt's result in the entire temperature range $0 < T < T_c$ from the microscopic BCS-Gor'kov theory. In case of an isotropic superconductor at magnetic fields $B/H_{c2} > 1/(2\kappa^2 + 3)$, the results of the nonlocal elasticity can be summarized as [41, 47]

$$c_{66} \sim \frac{H_{c2}^2}{4\pi}\left(1 - \frac{1}{2\kappa^2}\right)\left(\frac{B}{H_{c2}}\right)\left(1 - \frac{B}{H_{c2}}\right)^2\left(\frac{1 - 0.29\dfrac{B}{H_{c2}}}{8\kappa^2}\right),$$

$$c_{11}(k) \sim \frac{B^2}{4\pi}\left(1 - \frac{1}{2\kappa^2}\right)(1 + k^2\lambda'^2)^{-1}(1 + k^2\xi'^2)^{-1},$$

and

$$c_{44}(k) \sim \frac{B^2}{4\pi}\left[(1 + k^2\lambda'^2)^{-1} + \frac{1}{k_{BZ}^2\lambda'^2}\right],$$

where $\xi' = 2\xi / \left(1 - \dfrac{B}{H_{c2}}\right)^{1/2}$. Note that the shear modulus c_{66} is not dispersive.

The functions $c_{11}(k)$ and $c_{44}(k)$ are the nonlocal elastic moduli for compression and tilt, respectively. At low magnetic fields (i.e., $B/H_{c2} <$ minimum of 0.2 or $1/2\kappa^2$) where the vortex lines interacts only with their nearest neighbors, the vortex lattice behaves like a local elastic medium [41].

4.4.3 Larkin Approach to Flux Pinning

In this section we present Larkin's explanation for the *flux pinning* of the vortex lattice in type II superconductors. The concept of collective flux pinning is also discussed.

4.4.3.1 Destruction of the Long-Range Order

In a type II superconductor, the vortex lattice is arranged in a periodic structure. Larkin [46] showed that the *long-range* order of the vortex lattice is destroyed in presence of inhomogeneities. He identified a correlation volume, V_c, in which the vortices are arranged almost periodically like that of a homogeneous superconductor. The periodicity, however, is lost outside the correlation volume.

Two types of inhomogeneities are often encountered in a type II superconductor:

(i) Defects leading to the inhomogeneities in the mean free path of the electron. The zero-field thermodynamic properties, like T_c, are not affected by this type of inhomogeneity.

(ii) Defects causing an inhomogeneous effective interaction between electrons are magnetic in nature. This type of defect smears out the phase transition.

If the order parameter $\Psi(r)$ is small and varies slowly in space, the free energy of a type II superconductor with random defects can be described by the Ginzburg-Landau equation as

$$F = F_n + \int dV \left[\alpha \, | \Psi |^2 + \frac{\beta}{2} \, | \Psi |^4 + \frac{\gamma}{2m^*} \left| \left(\frac{\hbar}{i} \nabla - \frac{e^*}{c} A \right) \Psi \right|^2 + \frac{h^2}{8\pi} \right], \quad (4.48)$$

where F_n is the normal state free energy. The parameters α, β, and γ are the random function of space coordinates due to a random potential. The first type of inhomogeneity strongly influences γ. The second type of inhomogeneity influences the parameter α, which is proportional to T_c-T. The effect of inhomogeneities on β is very small and can be disregarded.

Assuming that α and γ deviate by a small amount, we can represent the free energy by two terms as

$$F = F_0 + F_1. \quad (4.49)$$

Here F_0 is the free energy corresponding to the defect-free superconductor, and

$$F_1 = \int dV \left[\alpha_1 \, | \Psi |^2 + \frac{\gamma_1}{2m^*} \left| \left(\frac{\hbar}{i} \nabla - \frac{e^*}{c} A \right) \Psi \right| \right].$$

The parameters α_1 and γ_1 are the deviation from their mean values α_0 and γ_0. To be specific,
$\alpha = \alpha_0 + \alpha_1$, and $\gamma = \gamma_0 + \gamma_1$ with $\alpha_1 < \alpha_0$ and $\gamma_1 < \gamma_0$.

The condition for the minimum of F_0 gives a periodic solution for $\Psi_0(r)$ and $h_0(r) = \nabla \times A(r)$ corresponding to the hexagonal Abrikosov lattice [3].

Larkin considered the first-order correction to the random functions α_1 and γ_1 to $\Psi_0(r)$ and $A_0(r)$ and calculated the deformation field $u(r)$ induced by the random potential on the lattice. Using the theory of elasticity, he obtained

$$\left\langle | u(R) - u(0) |^2 \right\rangle$$

$$= \int \frac{d^3k}{(2\pi)^3} (1 - \cos k.R) 2W(k) \left[\frac{1}{(c_{11}k_\perp^2 + c_{44}k_z^2)^2} + \frac{1}{(c_{66}k_\perp^2 + c_{44}k_z^2)^2} \right]. \quad (4.50)$$

Here the left-hand side represents the mean square displacement which characterizes the deviation of the position of the vortex lines from that of a periodic lattice. The parameter W is the pinning correlation function and characterizes the pinning strength. For distances large compared with the dimension of the lattice and the defects, the small k in Equation (4.50) is important. Disregarding the dependence of k of the parameters W and the elastic constants, we obtain

$$\langle | u(R) - u(0) |^2 \rangle = \frac{W}{4\pi} \left[\frac{(R_\perp^2 c_{44} + R_\parallel^2 c_{11})^{1/2}}{c_{44} c_{11}^{3/2}} + \frac{(R_\perp^2 c_{44} + R_\parallel^2 c_{66})^{1/2}}{c_{44} c_{66}^{3/2}} \right].$$

where $R_\perp$ and $R_\parallel$ are components of R perpendicular and parallel to the applied magnetic field.

Since $c_{11} >> c_{66}$, we obtain

$$\langle |\, \mathbf{u(R)} - \mathbf{u}(0)|^2 \rangle \approx \frac{W}{4\pi} \frac{(R_\perp^2 c_{44} + R_\parallel^2 c_{66})^{1/2}}{c_{44} c_{66}^{3/2}}. \tag{4.51}$$

The result signifies that even for a small pinning strength ($\sim W^{1/2}$), the mean square displacement increases linearly with increasing distance. The long-range order of the vortex lattice is destroyed when the mean square displacement

$\langle |\, \mathbf{u(R)} - \mathbf{u}(0)|^2 \rangle$ is on the order of a_0^2. Here a_0 is the vortex lattice spacing. If

r_f is the length scale of the random pinning potential, the linear dimension $\mathbf{R}_{cor}$ of the correlated region is defined by the relationship [45]

$$\langle |\mathbf{u(R}_{cor}) - \mathbf{u}(0)|^2 \rangle = r_f^2 \tag{4.52}$$

For small defects r_f is on the order of the coherence length, ξ. If L_c and R_c are the correlation lengths corresponding to the direction parallel and perpendicular to the direction of the applied field, one can define a correlated region with volume $V_c \approx R_c^2 L_c$ instead of a continuous displacement field. These correlated regions are weakly coupled to each other and can move independently over a distance on the order of r_f in presence of a driving force. The vortex lattice distortion within these correlated region can be neglected, and the lattice can be considered periodic. Recently, Chudnovsky [48,49] has shown that although the long-range positional order of an vortex lattice is destroyed in presence of a random potential, the orientational order still persists over a long range.

4.4.3.2 Collective Pinning

The concept of *collective pinning* can be applied for the case of the weak pinning when the local displacement of the vortex lines is very small ($<< a_0$) (45,47,50,51). In this case the vortices are pinned by an ensemble of defects rather than individual strong pinning centers. The situation for both cases is depicted in Figure 4.10. As mentioned in the preceding section, the pinning centers, although weak, will destroy the long-range order of the vortex lattice, and one can define a correlated region with volume, V_c, where the vortices are arranged almost periodically [46].

When a current with density, j, less than the critical current density, j_c, is passed through the superconductor, these correlated regions are displaced independently under the action of the driving force for a distance less than the scale of the random pinning potential, r_f. The resulting pinning force balances the driving force. Inside the correlated regions, the vortices are almost periodically arranged and the pinning centers are randomly distributed. The maximum pinning force acting on a correlated region with volume V_c is given

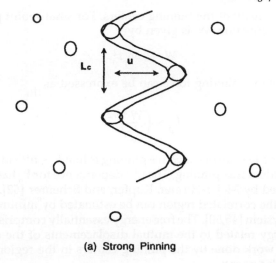

(a) Strong Pinning

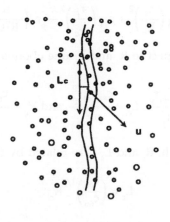

(b)   Weak Pinning

Fig. 4.10. Schematic diagram distinguishing a strong pinning and a weak pinning system.

by $f_p N^{1/2}$, where f_p is the force exerted by an individual pinning center on the vortex lattice and N is the number of pinning centers in the correlated region. At the critical state the driving forces are exactly balanced by the pinning forces. The volume pinning force is then

$$F_p = \frac{BJ_c}{c} = \frac{f_p N^{1/2}}{V_c} = f_p \left(\frac{n}{V_c}\right)^{1/2}, \qquad (4.53)$$

where n is the density of the pinning centers. For small-point pins, the pinning correlation function, W, is given by

$$W \approx 0.5 n_p f_p^2. \tag{4.54}$$

Thus the volume pinning force can be expressed as

$$F_p \approx \left(\frac{W}{V_c}\right)^{1/2}. \tag{4.55}$$

The characteristic feature of collective pinning is that $F_p \sim n^{1/2}$, in contrast with $F_p \sim n$ for individual flux pinning. The $n^{1/2}$ dependence for F_p has been experimentally verified by Meier-Hirmer, Kupfer, and Scheurer [52].

The size of the correlated region can be estimated by minimizing the free energy of the system [45,50]. The free energy essentially comprises two terms: the elastic energy related to the mutual displacements of the correlated regions, and the work done by the pinning centers in the regions in order to effect this displacement:

$$\delta F = c_{66}\left(\frac{r_f}{R_c}\right)^2 + c_{44}\left(\frac{r_f}{L_c}\right)^2 - f_p r_f \left(\frac{n}{R_c^2 L_c}\right)^{1/2}. \tag{4.56}$$

Minimizing with respect to R_c and L_c, one obtains

$$R_c \sim \frac{c_{66}^{3/2} c_{44}^{1/2} r_p^2}{n f_p^2}, \quad L_c \sim \frac{c_{66} c_{44} r_p^2}{n f_p^2}, \text{ and } V_c \sim \frac{c_{66}^4 c_{44}^2 r_p^6}{n^3 f_p^6}. \tag{4.57}$$

The correlation length, R_c and L_c, are related by the relationship

$$L_c = \left(\frac{c_{44}}{c_{66}}\right)^{1/2} R_c. \tag{4.58}$$

Equation (4.58) suggests that L_c is much larger than R_c.

4.5 Flux Motion

In this section we consider the dynamic properties of the vortex lattice. The motion of vortices can be categorized as three types:

(i) Flux motion that results when the driving force F_L is larger than the pinning forces. In this case the vortex lines are driven by F_L resulting in an energy dissipation. This type of motion is referred to in the literature as *flux flow* [4–6,30,53–60].

(ii) Flux motion that occurs near the critical state (i.e., $F_L \sim F_p$), where it is dominated by avalanches, with a power-law distribution in the avalanche sizes *and time*. The behavior of the flux motion can then be considered in the framework of *self-organized criticality* (SOC) [61–64].

(iii) Flux motion that results even if the driving force is much less than the pinning force. This type of motion is generally thermally activated and is often referred as *flux creep* [65,66].

4.5.1 Flux Flow

When an electric current with density j much larger than the critical current density, j_c, is applied to a superconductor, the vortices get depinned and start moving, resulting in a dissipation of energy. The forces acting on a moving vortex line satisfy the following equation [4]:

$$f_L + f_M + f_v + f_p = 0 \qquad (4.59)$$

The first term, f_L, is the Lorentz force (the driving force) experienced by the vortex line as a result of the passage of the current with density, j, is given by Equation (4.15):

$$f_L = \frac{(j \times \Phi_0)}{c}. \qquad (4.60)$$

The second term, f_M, as suggested by de Gennes [67] and de Gennes and Matricon [68], is the magnus force that compensates the Lorentz force. If we neglect the other two terms (f_p and f_v), the magnus force results in a vortex motion that is parallel to the current density, j (perpendicular to the driving force). This force is analogous to the lift force in hydrodynamics. It is expressed as

$$f_M = -fn_s e \frac{(v \times \Phi_0)}{c}. \qquad (4.61)$$

Here, f is a constant that indicates the effective fraction of the magnus force [69–71], n_s is the density of superelectrons, and v is the velocity of the vortex line.

The viscous drag, or the damping force, is represented by f_v and is opposite to the flux flow velocity. It can be expressed as

$$fv = -\eta v. \qquad (4.62)$$

Here, η is a parameter that represents the viscosity of the medium and contains all dissipative forces.

The fourth term, f_p, represents the frictional force acting on a moving vortex line due to the flux pinning centers. Directionally it is opposite that of the flux flow velocity, v. Yamafuji and Irie [72], Lowell [73], and Good and Kramer [74] have suggested that this frictional force originates from the elastic distortions introduced by the pinning centers. Furthermore, Lowell [73] pointed out that this frictional force is essentially independent of the velocity when the current density is appreciably higher than the critical current density, as was observed experimentally [74]. He also showed that the frictional force, f_p, is on the order of $\Phi_0 j_c$, as observed experimentally [75].

4.5.1.1 Flux Flow Resistivity, ρ_f

The flux motion induced by the driving force larger than the pinning forces leads to a electric field, E, given by Faraday's law (Equation 4.16). In the absence of flux pinning, and neglecting the magnus force for simplicity, one obtains

$$\frac{1}{c}(j \times \Phi_0) = \eta v.$$

The resistivity, ρ_f, is then given by [4,5]

$$\rho_f = \frac{E}{j} = \frac{B\Phi_0}{\eta c^2}. \tag{4.63}$$

The rate of energy dissipation per unit length of the vortex is

$$W = -f_v \cdot v = \eta v^2. \tag{4.64}$$

The coefficient of viscosity, h, can be determined by estimating the energy dissipated during the motion of a vortex. The energy dissipation has been calculated by many researchers using various approaches [76–89]. Bardeen and Stephen assumed that the dissipation during the vortex motion occurs primarily as a resistive dissipation at the vortex core [76]. In the local approximation and assuming the vortices with a core of radius ~ ξ, they obtained

$$\eta \approx \frac{\Phi_0^2}{4\pi\xi^2 c^2 \rho_n}.$$

Allowing dissipation outside the core, one obtains

$$\eta \approx \frac{\Phi_0^2}{2\pi\xi^2 c^2 \rho_n} \approx \frac{\Phi_0 H_{c2}}{\rho_n c^2}.$$

The flux flow resistivity, ρ_f, is

$$\frac{\rho_f}{\rho_n} \approx \frac{2\pi\xi^2 B}{\Phi_0} = \frac{B}{H_{c2}}. \tag{4.65}$$

The above result was also obtained empirically by flux-flow resistivity experiments in type II superconductors in the low-temperature limit. Deviation from the Bardeen-Stephen model, however, was observed experimentally on high-κ superconductors [90–95].

4.5.2 Flux Creep

When the driving force, F_L is smaller than the pinning force, F_p, the vortices are trapped at the pinning centers, and no further vortex motion is expected. At a finite temperature, however, the vortices can be thermally activated and hop from one pinning center to the other, driven by F_L and leading to dissipation of energy [65,66]. Flux creep will manifest itself in two ways, namely, in the decay of magnetization in magnetization measurements and in the appearance of a small resistive voltage in transport measurements. Flux creep measurement provides useful information about the dynamics of the pinned vortex state. We will consider more about the flux creep in the next section.

4.6 Magnetic Relaxation in a Type II Superconductor

Let us consider the effect of any external activation, like thermal fluctuation, on the critical state of the hard superconductor. In the presence of an external activation, the critical state becomes metastable, and the system relaxes to achieve an uniform distribution of the vortices. This situation results in the decay of current density, j. The magnetization, M, which is proportional to the current density therefore also decays with time. This phenomenon is known as *magnetic relaxation*.

Magnetic relaxation was first explained by Anderson in terms of a thermally activated flux creep model [65,66]. He assumed that the mechanism of the magnetic relaxation is a thermally activated motion of bundles of flux lines aided by the Lorentz force (or magnetic pressure) over the energy barriers of the flux pinning centers. Anderson argued that, because of the relatively long-range nature of the vortex-vortex interaction ($\sim e^{-r/\lambda}$), the local perturbation of the vortex density is energetically unfavorable. The arrangement of the vortices can be irregular only on a scale greater than λ. Within the length scale of λ the density must be uniform, and any variation must be a slight increase in the local density spread out over a region of radius $> \lambda$. He therefore concluded that flux lines will jump from one pinning center to another in "flux bundles" of radius $\sim \lambda$ rather than jumping individually (radius $\sim \xi$). He further pointed out that since the interaction function is a slowly varying function near $r \longrightarrow 0$, the arrangements of lines need to be crystallographically regular, and the bundles can slide past each other easily. If U is the effective activation energy required for a flux bundle to hop, and if v is the hopping frequency, then

$$v = v_0 \exp\left(-\frac{U}{kT}\right), \tag{4.66}$$

where v_0 is an attempt frequency and *is not* greater than $10^{10}/s$. Anderson obtained the expression for U from the free energy considerations. If d is the average bundle size, the structure-sensitive pinning energy, U_0, is given by

$$U_0 = \frac{pH_c^2}{8\pi d^3}, \tag{4.67}$$

where p is the average fractional amount of pinning. In the presence of a driving force, the free energy of a flux bundle as a function of the bundle position, x, can be written as

$$U_p(x) = U_0 - jBVx, \qquad (4.68)$$

where $V = d^3$ is the bundle volume or, in other words,

$$U_p(x) = \frac{pH_c^{\,2}}{8\pi d^3} - jBd^2lx. \qquad (4.69)$$

Here l is the effective length of the flux line over which the forces act, presumably the distance between the pinning centers. The free-energy hump or the activation energy, U, that the bundle must climb in order to get from one metastable minimum to another is given by

$$U = \frac{pH_c^{\,2}}{8\pi d^3} - jBd^2lx_m. \qquad (4.70)$$

Here x_m gives the extremum points for $U_p(x)$ and reflects the size of the pinning barrier. The diffusion of flux lines is given by the rate at which the flux enters and leaves a small volume element

$$\frac{d<B>}{dt} = -\nabla.n[\Phi_0 \nu_0 \exp(-\frac{U}{kT})],$$

where n is an unit vector directed in the direction of the gradient of the magnetic pressure (i.e., in the direction of the Lorentz force). Solving the above equation, Anderson predicted a logarithmic decay of magnetization after the initial transient; this phenomenon was later experimentally observed by Kim, Hempstead, and Strnad [96] and later by File and Mills [97]. Although the prediction agreed well with the observation of Kim, Anderson emphasized that the theory may be extremely rough, since several important factors such as the details of the interaction between flux lines, change of free energy with current or field, and shape factors of the flux bundles are completely neglected. He stressed that the theory may need major modifications before application to other systems of superconductors or/and other field and temperature regimes.

4.6.1 Nonlogarithmic Decay of Magnetization

As noted, the Anderson and Kim model predicted a logarithmic decay of magnetization with time, a prediction that was experimentally verified by Kim and his coworkers [96]. Beasley, Labush, and Webb [98] also observed a logarithmic decay of the magnetization in Pb-Tl alloys; however, they realized that in general the effective activation energy, U, is a nonlinear function of the current density (see Figure 4.11). The term U_0 as used by Anderson and

Kim is actually the intercept of the tangent to the U-j curve at a current density j_0 [98–102]. With the approximation of $U_0 << U$, Beasley and his colleagues found that

$$\frac{dU}{dj} = -\frac{U_0}{j_0}$$

(4.71)

Using the rate equation (4.66), they further obtained

$$j(t) = j_0[1 - \frac{kT}{U_0} \ln (\frac{t}{t_0})],$$

(4.72)

where t_0 is some arbitrary time with j ($t = t_0$) = j_0. If U varies linearly with j, however, U_0 becomes the pinning energy as used by Anderson and Kim [65,66].

Magnetic relaxation in high-T_c superconductors has been observed by various reasearch groups [99–141]. The higher T_c in these materials open the door to study magnetic relaxation at high temperatures. Two discrepancies were realized, and the validity of the Anderson-Kim model was seriously questioned. Earlier workers [102,108–119], using Equation (4.72) to find the pinning energy, U_0, of the newly discovered materials, found that the so-called pinning energy, U_0, increases with increasing temperature and diverges at the irreversibility line (see Figure 4.12). This result contrasts sharply with expectations. It was later realized that the discrepancy stems from the nonlogarithmic decay of magnetization [99–102] and U_0 is not the pinning energy but rather the intercept of the tangent to the U-j curve at a current density j_0 [98–102].

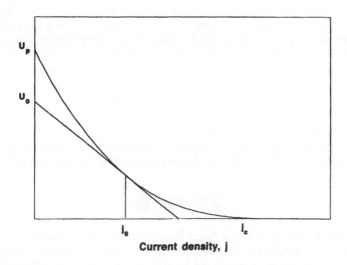

Fig. 4.11. Schematic of the effective activation energy, U, as a function of the current density, j. Note that U_0 is the intercept of the tangent of the U-j curve at the point j_0.

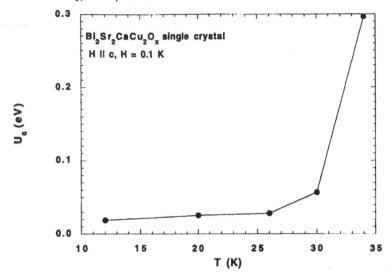

Fig. 4.12. Plot of U_0 (estimated from Equation 4.71) as a function of temperature T for a $Bi_2Sr_2CaCu_2O_x$ single crystal with H I I c at a field of 0.1 T. Note that U_0 is fairly constant at low temperature regimes and then starts rapidly increasing at higher temperatures and finally diverges near the irreversibility line.

Figure 4.13 shows the magnetic relaxation for a $Bi_2Sr_2CaCu_2O_x$ single crystal with a field parallel to the c-axis at 12 K, 20 K, 26 K, 30 K, 34 K, and 40 K [121]. As is evident from the figure, the nonlinearity increases with decreasing driving force. This effect becomes pronounced as the temperature is increased. We emphasize, however, that the decay of magnetization has only been observed experimentally. The simplified Anderson-Kim model predicts a logarithmic decay of the magnetization, and estimation of pinning energy U_0 based on Equation (4.72) may therefore lead to wrong conclusions. Various models have been proposed to explain the nonlogarithmic decay of the magnetization [121–127,130,136–168]. One approach will be to assume a nonlinear U-j relationship.

4.6.2 *Nonlinear U-j Relationships*

The nonlogarithmic decay of magnetization can be explained by assuming a nonlinear U-j relationship while keeping the basic Arrhenius relationship

$$\frac{dj}{dt} = A\exp\left(-\frac{U}{kT}\right). \tag{4.73}$$

This relationship follows from Equation (4.66), as we will show by considering the following simple case. In a one-dimensional form the flux conservation equation as proposed by Beasley, Labusch, and Webb [98] can be written as

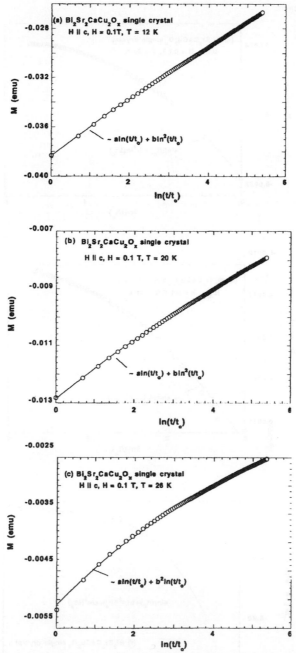

Fig. 4.13. Magnetic relaxation for a $Bi_2Sr_2CaCu_2O_x$ single crystal with a field of 0.1 T parallel to the c-axis at (a) 12 K, (b) 20 K, (c) 26 K, (d) 30 K, (e) 34 K and (f) 40 K. Note that the nonlinearity increases with decreasing driving force, with the effect getting more pronounced at higher temperatures.   *Figs. 4.13 continued over*

Figs. 4.13 continued

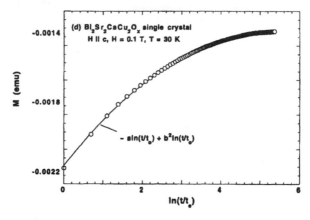

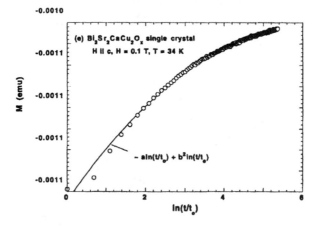

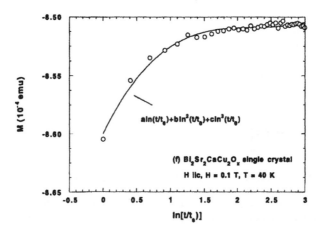

$$\frac{dB}{dt} = \nabla \cdot [\mathbf{n}B\omega v_0 \exp(-\frac{U}{kT})], \tag{4.74}$$

where ω is the average distance a flux bundle can hop and $\mathbf{n}$ is an unit vector directed in the direction of the gradient of the magnetic pressure (i.e. , in the direction of the Lorentz force).

Considering a slab of thickness d and integrating over a sample volume, one obtains

$$\frac{d<B>}{dt} = 4\pi \frac{dM}{dt} = \frac{H\omega v_0 \exp\left(-\frac{U}{kT}\right)}{d}, \tag{4.75}$$

which can be rewritten as

$$\frac{dM}{dt} = A \exp\left(\frac{-U}{kT}\right).$$

From Bean's model, we know that $M-M_{eq} \propto j$ and hence $dM/dt = dj/dt$. M_{eq} is the equilibrium magnetization, and this leads to Equation (4.73).

At a constant temperature and field, the effective activation energy can be expanded in the neighborhood of some current density j_0 at time t_0 to obtain [121,123,124]

$$U(j) = U(j_0) + \left[\frac{\partial U}{\partial j}\right]_0 (j - j_0) + \frac{1}{2}\left[\frac{\partial^2 U}{\partial j^2}\right]_0 (j - j_0)^2 + \dots$$
$$= U(j_0) + \alpha(j - j_0) + \frac{1}{2}\beta(j - j_0)^2 + \dots \tag{4.76}$$

where α and β are the slope ($\left[\frac{\partial U}{\partial j}\right]_0$) and the curvature ($\left[\frac{\partial^2 U}{\partial j^2}\right]_0$), respectively, of the U-j curve at $j = j_0$. Let us now consider only the first- order term (linear approximation). At a constant temperature and field the preexponential factor, A, can be assumed to be constant within the small variation of j considered. Integrating Equation (4.73) within the time interval t_0 and t, one obtains

$$j(t) = j_0 + \frac{kT}{\alpha} \ln\left(\frac{t}{t_0}\right). \tag{4.77}$$

Here t_0 is some arbitrary reference time with $j(t_0) = j_0$. Thus one obtains a logarithmic magnetic relaxation when one assumes a linear U-j relationship as considered by Anderson [65], Anderson and Kim [66], and Beasley, Labusch and Webb [98].

Considering the second-order term and once again integrating Equation (4.73) between the time interval t_0 and t, one obtains

$$j(t) = j_0 + \frac{kT}{\alpha} \ln\left(\frac{t}{t_0}\right) - \frac{k^2T^2\beta}{2\alpha^3} \ln^2\left(\frac{t}{t_0}\right) \tag{4.78}$$

$$= j_0 + a\ln\left(\frac{t}{t_0}\right) + b\ln^2\left(\frac{t}{t_0}\right),$$

where $a = \dfrac{kT}{\alpha}$ and $b = \dfrac{k^2T^2\beta}{2\alpha^3}$.

Thus one can see that the introduction of a second-order term in Equation (4.76) leads to an additional term in Equation (4.78) which is quadratic in lnt. Similarly considering higher-order terms, one obtains additional terms in Equation (4.77) that are proportional to higher powers of lnt. Thus a nonlogarithmic decay can be explained by assuming a nonlinear U-j relationship.

Nevertheless, the origin of the nonlinearity is not yet fully understood. In the following sections, we introduce various models that have been proposed to explain the phenomenon [122,136,137,142–170].

4.6.3 Models Based on Hopping of Flux Bundles

In the Anderson-Kim theory [65,66] of magnetic relaxation, each vortex or flux bundle is effectively modeled as a single, nearly independent, "zero-dimensional" particle in a pinning potential of barrier height U_0. As discussed earlier, this generally leads to a logarithmic magnetic relaxation. If we assume a shape of the pinning barrier [98,122,146–149,164,169,170] and/ or the probability of reverse hopping [5,141,143,150,164,166], however, we obtain a nonlinear U-j relationship, and the nonlogarithmic decay of magnetization can be explained.

4.6.3.1 Shape of the Pinning Potential

Nonlinearity in the U-j relationship as a result of the local pinning potential profile was realized by Beasley, Labusch, and Webb [98] in late 1969. To understand the effect, let us consider a pinning potential with the characteristic energy scaling as U_0 (which may be barrier height in some particular cases), g(x) as some sort of shape factor, and x as the displacement from the center of the pinning center. The pinning potential experienced by an isolated vortex or a flux bundle can then be written as a function of the distance from the pinning center

$$U(x) = U_0 g(x), \tag{4.79}$$

and the elementary pinning force, f_p, acting on a flux bundle is given by

$$f_p = -\frac{\partial U}{\partial x}. \tag{4.80}$$

In the presence of a driving force, the pinning potential will be modified as

$$U(x) = U_0 g(x) - jBVx. \tag{4.81}$$

The effective activation energy, $U(j)$, as defined by the Arrhenius relationship can then be written as the difference between two adjacent maxima and minima. The condition for an extrema for Equation (4.81) is

$$\frac{\partial U}{\partial x} = 0 = U_0 g'(x) - jBV$$

or

$$\frac{\partial g(x)}{\partial x} = \frac{jBV}{U_0}. \tag{4.82}$$

If x_{max} and x_{min} are two adjacent maxima and minima satisfying the above condition, then

$$U(x) = U(x_{max}) - U(x_{min}) = g(x_{max} - x_{min}) - jBV (x_{max} - x_{min}). \tag{4.83}$$

Since x_{max} and x_{min} generally become a function of j, U becomes a nonlinear function of j. Let us now consider three particular cases that results in three different U-j relationships.

(i) Saw-Tooth Potential

We first consider a saw-tooth potential as shown in Figure 4.14. Mathematically it can be represented as

$$g(x) = \frac{|x|}{w} \qquad -\frac{w}{2} \leq x \leq \frac{w}{2},$$

where w is the width of the pinning potential.

In the presence of a driving force, the pinning energy profile becomes

$$U(x) = U_0 \frac{|x|}{w} - jBVx,$$

with $x_{max} = -\frac{w}{2}$ and $x_{min} = -\frac{w}{2}$.

The effective activation energy becomes

$$U(j) = U_0 - jBVw$$

$$= U_0\left(1 - \frac{j}{j_c}\right), \tag{4.84}$$

which is the linear U-j law and will lead to logarithmic magnetic relaxation. The linear U-j law also results if one considers a square-wave potential given by

$$g(x) = 0 \qquad -\frac{w}{2} \le x \le \frac{w}{2}$$

$$= U_0 \qquad x = \pm\frac{w}{2}.$$

The resulting U-j relationship for both the saw-tooth and the square-wave potential is plotted in Figure 4.15.

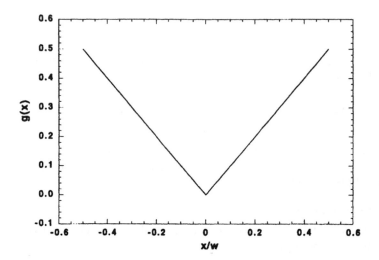

Fig. 4.14. g(x) for a saw-tooth local pinning potential.

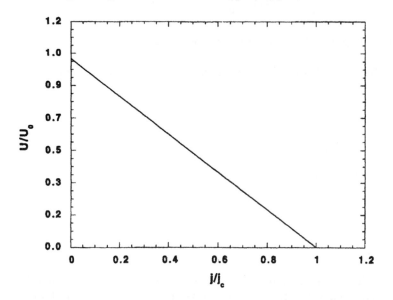

Fig. 4.15. U-j curve for a saw-tooth local pinning potential.

(ii) Sinusoidal Potential

A sinusoidal potential is shown in Figure 4.16. This type of potential is a natural choice for a random periodic potential and was first considered by Beasley, Labusch, and Webb [98]. The shape factor $g(x)$ is given by

$$g(x) = -\frac{1}{2}\cos\left(\frac{\pi x}{w}\right).$$

In the presence of a driving force, the pinning energy will be given by

$$U(x) = -\frac{U_0}{2}\cos\left(\frac{\pi x}{w}\right) - jBVx,$$

with
$$x_{max} = w\sin^{-1}\left(\frac{2jBVw}{\pi U_0}\right) \text{ and } x_{min} = \pi - w\sin^{-1}\left(\frac{2jBVw}{\pi U_0}\right).$$

The effective activation energy can be then written as

$$U(j) = U_0\sqrt{1 - \frac{4j^2B^2w^2V^2}{\pi^2U_0^2}} - 2jBVw\cos^{-1}\left(\frac{2jBVw}{\pi U_0}\right) \tag{4.85}$$

$$= U_0\left\{\sqrt{1 - \left(\frac{j}{j_c}\right)^2} - \left(\frac{j}{j_c}\right)\cos^{-1}\left(\frac{j}{j_c}\right)\right\},$$

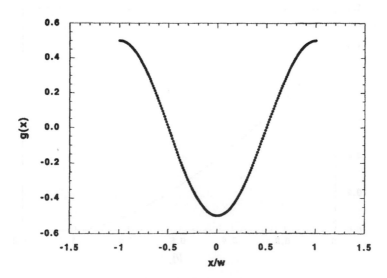

Fig. 4.16. $g(x)$ for a sinusoidal local pinning potential.

with $U_0 = \dfrac{2j_c BVw}{\pi}$.

The corresponding U-j relationship is shown in Figure 4.17. Equation (4.85) can be approximated as

$$U(j) = U_0 \left(1 - \frac{j}{j_c}\right)^{3/2}. \tag{4.86}$$

In fact, if g(x) is a smooth function x, the effective activation energy can be represented by Equation (4.87) provided that the flux creep occurs only in one dimension.

(iii) Logarithmic Potential

Based on the experimentally observed U-j relationship, a logarithmic potential (Figure 4.18) was proposed by Zeldov et al. [69,170]. The shape factor for an logarithmic potential can be written as

$$g(x) = 1 + \ln(\frac{x}{x_p}) \qquad x_p < x < w$$

$$= \frac{x}{x_p} \qquad x \leq x_p.$$

In the presence of a driving force, the effective pinning potential becomes

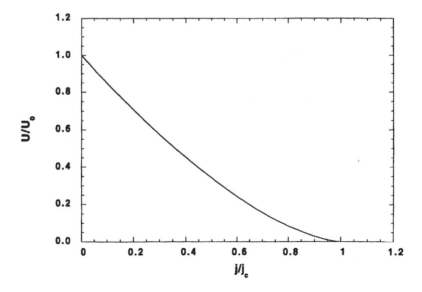

Fig. 4.17. U-j curve for a saw-tooth local pinning potential.

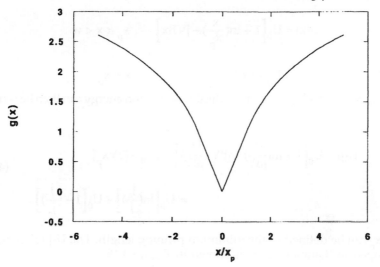

Fig. 4.18. g(x) for a logarithmic local pinning potential.

$$U(x) = U_0\left\{1 + \ln(\frac{x}{x_p}) - jVBx\right\} \qquad x_p < x < w$$

$$= U_0\left(\frac{x}{x_p}\right) - jBVx \qquad x \leq x_{p'}$$

with $x_{max} = \dfrac{U_0}{jBV}$ and $x_{min} = 0$. Therefore,

$$U(j) = U_0\left[1 + \ln\left(\frac{U_0}{jBVx_p}\right)\right] - \frac{jBVU_0}{jBV}$$

$$= U_0\left\{\ln(\frac{j_c}{j})\right\}, \qquad\qquad (4.87)$$

with $j_c BVx_p = U_0$.

Generally, x_p is on the order of the coherence length. The above pinning potential can also be slightly modified so that U_0 is better defined [122]. We take

$$g(x) = 1 + \ln(\frac{x}{x_p}) \qquad x_p < x < w$$

$$= U_0 \qquad x \leq x_p.$$

In the presence of a driving force, the effective pinning potential becomes

$$U(x) = U_0\left[1 + \ln(\frac{x}{x_p}) - jVBx\right] \qquad x_p < x < w$$

$$= U_0 - jBVx \qquad\qquad x < x_p$$

with $x_{max} = \dfrac{U_0}{jBV}$ and $x_{min} = x_p$. The effective activation energy can then be written as

$$U(j) = U_0\left[1 + \ln(\frac{U_0}{jBV}) - jBV\,(\frac{U_0}{jBV})\right] - [U_0 - jBVx_p] \tag{4.88}$$

$$= U_0\left[\ln(\frac{j_c}{j})\right] - U_0\left[1 - (\frac{j}{j_c})\right].$$

Here x_p can be defined as the minimum pinning length. The U-j relationship developed in Equation (4.88) is shown in Figure 4.19.

4.6.3.2 Reverse Hopping

Tinkham [5] first pointed out that a more realistic jumping rate can be obtained by considering the probability of reverse hopping, that is, jumping against the driving force (Figure 4.20). This process becomes important at higher temperatures and/or at low driving force. The concept of reverse hopping was expanded by Kes et al. in their thermally activated flux motion model [143]. To demonstrate this concept, we start with the linear Anderson-Kim

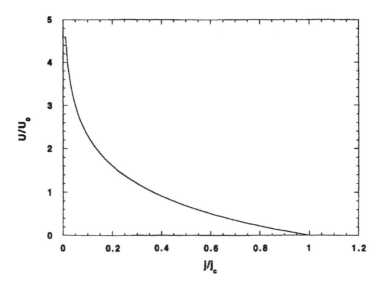

Fig. 4.19. U-j curve for a logarithmic local pinning potential.

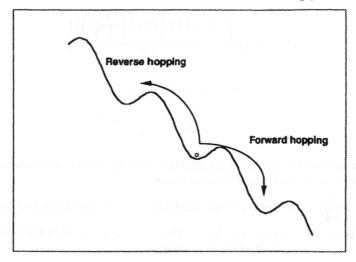

Fig. 4.20. Schematic of the hopping process.

model. If $\nu(+)$ and $\nu(-)$ are the jumping rate at the forward and reverse direction, respectively, we get

$$\nu(+) = \nu_0\exp[-\frac{U(+)}{kT}] \quad \text{and} \quad \nu(-) = \nu_0\exp[-\frac{U(-)}{kT}].$$

The resulting jumping rate is given by

$$\nu(j) = \nu(+) - \nu(-).$$

Using the Anderson-Kim model, we obtain

$$\nu(j) = \nu_0\exp\left[-\left(\frac{U_0 - jBVx}{kT}\right)\right] - \nu_0\exp\left[-\left(\frac{U_0 + jBVx}{kT}\right)\right]$$

$$= 2\nu_0\exp\left(-\frac{U_0}{kT}\right)\sinh\left(\frac{jBVx}{kT}\right) \tag{4.89}$$

$$= 2\nu_0\exp\left(-\frac{U_0}{kT}\right)\sinh\left(\frac{U_0 j}{kTj_c}\right).$$

Thus we get a non-Arrhenius type of equation with the preexponential factor given by $2\nu_0 \sinh\left(\frac{U_0 j}{kTj_c}\right)$,which is a function of current density.

Alternatively, Equation (4.89) can be written in an Arrhenius form with a nonlinear U-j relationship as

$$v(j) = v_0 \exp\left\{-\frac{U_0 - kT\ln\left[2\sinh\left(\frac{U_0 j}{kTj_c}\right)\right]}{kT}\right\}.$$

Using Equation (4.66), one therefore obtains

$$U(j) = U_0 - kT\ln\left[2\sinh\left(\frac{U_0 j}{kTj_c}\right)\right]. \tag{4.90}$$

Now the preexponential factor is a constant, and U varies nonlinearly with j.
   Let us now consider two limiting cases.

(i) $x = \dfrac{U_0 j}{kTj_c} \gg 1$. This corresponds to the low temperature and/or large driving force. In this case one obtains the linear U-j law as in the limit $x \gg 1$, $\sinh(x) \sim \exp(x)/2$, and hence we obtain

$$U(j) = U_0\left(1 - \frac{j}{j_c}\right).$$

One therefore gets a logarithmic decay of magnetization, as expected at low temperatures and/or large driving force.

(ii) $x \ll 1$. This corresponds to the high temperature and/or low driving force $(j \ll j_c)$. In this limit $\sinh(x) \sim x$, and therefore

$$U(j) = U_0 - kT\ln\left[2\left(\frac{U_0 j}{kTj_c}\right)\right] \tag{4.91}$$
$$= U_0 - kT\ln\left[2\left(\frac{U_0}{kT}\right)\right] - kT\ln\left[2\left(\frac{j}{j_c}\right)\right].$$

Thus one obtains the nonlinear U-j law, resulting in a nonlogarithmic decay.

4.6.4 Models Based on Nucleation of Vortex Loops

In models based on nucleation of vortex loops, the vortices are considered as an elastic object in a random potential [50,142,144,155–158,161,163, 167,172,173]. The vortex motion is shown schematically in Figure 4.21. A nucleus first develops because of any external fluctuations (e.g., thermal fluctuation). If the fluctuating segment is too small, the nucleus collapses. If the fluctuating segment is long enough, however, the nucleus expands, and the vortex moves to the next metastable state. The velocity of the vortex decreases as $\exp(-U(j)/kT)$ with decreasing current density. The effective activation energy, U, increases rapidly as $(j_c/j)^\mu$ and diverges at low current density level, thus predicting zero electrical resistance in the limit of zero current density. This physical property separates the low-temperature vortex glass phase as proposed by Fisher [150] from the high-temperature vortex liquid phase which is characterized by zero critical current density. In the vortex glass phase, the effective activation energy, U(j), for the thermally activated flux motion is the energy of a "critically sized" vortex loop. Similar results were obtained in the theory of collective creep by Feigel'man et al. [142] and independently by Nattermann [156]. These authors assumed that the vortices are pinned collectively by dense, randomly pinned weak pinning centers like oxygen vacancies in high-T_c superconductors. These defects induce elastic distortion of the flux line lattice. The free energy associated with the distortion is

$$F = \int dv [c_{11} - c_{66}] \frac{(\nabla \cdot \mathbf{u})^2}{2} + c_{66} \frac{(\nabla_\perp \mathbf{u})^2}{2} + c_{44} \frac{\left(\frac{\partial \mathbf{u}}{\partial z}\right)^2}{2} + U_p(\mathbf{u},\mathbf{r}).$$

Here, the two-dimensional vector $\mathbf{u}$ ($\mathbf{r}$) describes a local displacement of the flux line lattice; c_{11}, c_{44}, and c_{66} are, respectively, the bulk, tilt, and shear modulus; and $U_p(\mathbf{u},\mathbf{r})$ is the random pinning potential.

In the absence of external current, the flux line lattice is in some local metastable state. Under the action of external current, some other metastable states become more favorable. Transitions between different metastable states are due to the thermal activation through the energy barrier whose characteristic scale is U(j). The energy barrier [the effective activation energy U(j)] is on the order of the elastic energy of the hopping flux bundle. If $R_\parallel$ and $R_\perp$ are the bundle size in the direction parallel and perpendicular to the direction of hopping, and L is the length along the direction of the applied magnetic field, then the bundle volume V_B is given by

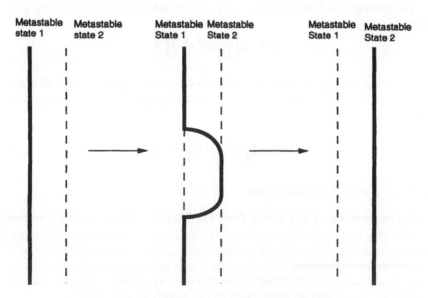

Fig. 4.21. Flux motion through nucleation of vortex loops.

$$V_B = R_{||} R_\perp L.$$

These flux bundles are made up of a large number of subbundles with volume V_c. The subbundles are formed independently from one another because of the competition between the shear (and tilt) deformation energy and the random pinning potential. The subbundles hop together, since the large value of c_{11} prohibits independent hopping of subbundles. The optimal hops are determined by the condition that the energy gain due to the Lorentz force is on the order of the pinning and elastic energies. Near the critical state or for the critical current density, this condition is fulfilled for neighboring states that differ a distance of $u_{hop}(j_c) \sim \xi$. At $j \ll j_c$, the hopping distance $u(j) \sim j^{-\alpha}$ should be much larger to compensate for the smaller driving force. This leads to a current dependence of the activation energy $U(j) \sim j^\mu$ with $\mu > 0$. In other words, the activation energy $U(j)$ diverges as the current density becomes zero. This phenomenon signifies that the flux creep vanishes in the lower current limit [50,142].

The value of μ depends on the dimensionality of the system, the field, and the current density. For the three-dimensional case, at low temperatures and low magnetic fields but relatively high currents, the exponent $\mu = 1/7$. At intermediate currents, in the regime of small-bundle hopping and strong intermediate spatial dispersion of the elasticity moduli, the power is given by $\mu = 3/2$. In the low-current regime where the flux bundle determines the creep, there is no dispersion of the elasticity moduli and $\mu = 7/9$. For a two-dimensional system, $\mu = 9/8$ for collective pinning in the large-current limit [50,142]. In the limit of larger flux bundles and smaller currents, Nattermann and Fisher found $\mu = 1/2$ [157].

This power-law behavior leads to a current relaxation law for $j \ll j_c$ given by [161]

$$j(t) = j_0 \left[\frac{kT}{U_0} \ln\left(\frac{t}{t_0}\right) \right]^{-\frac{1}{\mu}}$$

Considering the initial Anderson-Kim–like behavior near the critical state ($j \sim j_c$), Feigel'man, Geshkenbein, and Vinokur [161] predicted an interpolation law given by

$$j(t) = j_0 \left[1 + \frac{kT}{U_0} \ln\left(\frac{t}{t_0}\right) \right]^{-\frac{1}{\mu}}, \tag{4.92}$$

with $\mu = 1$ near the critical state.

4.6.5 Construction of U–j Curve from the Magnetic Relaxation Experiment

The effective activation energy, U, as a function of the current density, j, can be extracted from magnetic relaxation experiments (or the decay of the current density). Both methods described here rely on the Arrhenius relationship, Equation (4.73). The first method was developed by Maley et al. [99]. Taking the logarithm on both sides of the equaiton, one obtains

$$\ln\left|\frac{dj}{dt}\right| = \ln A - \frac{U}{kT}$$

or

$$U = kT\left(C - \ln\left|\frac{dj}{dt}\right|\right), \tag{4.93}$$

where $C = \ln A$ is assumed to be a constant in the temperature and current density regime considered. Experimentally $|dj/dt|$, can be determined at different temperatures; and by adjusting the constant C, a smooth U-j curve can be obtained. The choice of a single additive constant, however, leads to a detoriation of the quality of a fit at higher temperatures. Therefore, it was suggested [120] that the absolute energy scale for the flux motion must also reflect the scaling of fundamental pinning-related parameters on temperature and field. It was found that by using a scaling function g(t), a smooth curve was obtained even at higher temperatures. The scaling function g(t) was found to vary as $(1-t)^{3/2}$. The reduced temperature, t, was given by T/T_x where T_x is a characteristic temperature that depends on the system and the field and temperature regime studied. For a $YBa_2Cu_3O_x$ single crystal, T_x was found to coincide with the temperature of the irreversibility line (120). For a $Bi_2Sr_2CaCu_2O_x$ single crystal, however, one must use two characteristic temperatures, T_x, to obtain a smooth U-j curve. It was interpreted that the two T_xs corresponds to the three-dimensional and two-dimensional regimes, respectively, in the $Bi_2Sr_2CaCu_2O_x$ system [128]. In general, because of the temperature dependence of U, one can write

$$U(j,H,T) = g\left(\frac{T}{T_x}\right)U(j,H).$$

The function $g(T/T_x)$ is used to scale the implicit temperature dependence to $T = 0$ K. Figure 4.22 shows a typical U-j curve developed by this method for a $YBa_2Cu_3O_x$ single crystal in a magnetic field of 0.2 T applied parallel to the c-axis.

Another method that gives equivalent results was elaborated by Sengupta et al. [121,123]. They developed the U-j relationship from the nonlogarithmic decay of magnetization. Equation (4.78) was used to fit the magnetic relaxation data. The slope α and the curvature β of the U-j curve can be calculated from the fitting parameters a and b. The U-j curve can then be developed at each temperature from the isothermal magnetic relaxation. It should be pointed out that U obtained by this method is actually the change in activation energy when the current density decays from j_0 (current density at time t_0) to some current density j_f (current density at time t_f). To obtain a U-j curve with a significantly large range of j at a constant temperature and field, one has to study long-time relaxation, since the phenomenon of relaxation is approximately logarithmic with time. Therefore, it would take years to get a considerable portion of the U-j curve.

One can take advantage of the functional form of the U-j relationship, however, and can construct a considerable portion of a U-j curve by studying isothermal magnetic relaxation at constant field with temperature as a parameter. By changing the temperature, one changes the critical current density, j_c. In other words, a large portion of the U-j curve can be extracted by considering the temperature dependence of the effective activation energy. In the most generalized form, U can be written as

$$U\left(U_0, \frac{j}{j_c}\right) = U_0 f\left(\frac{j}{j_c}\right),\qquad(4.94)$$

where U_0 gives the characteristic energy scale and j_c is the critical current density. At a constant applied field, U_0, j, and j_c are functions of temperature, and j changes with time. The characteristic energy, U_0, is a weak function of temperature (at least at low temperatures). Then, assuming U_0 to be constant within two nearby temperatures, one can use the relationship of the form

$$U\left(\frac{j}{j_{c0}}\right) = U\left(\frac{j}{j_{c1}}\right) + C(j, j_{c0}, j_{c1})\qquad(4.95)$$

to develop a U-j curve at the temperature with a critical current density, j_{c0}. When U varies logarithmically with j/j_c, the function C becomes a constant and depends only on j_{c0} and j_{c1}. To demonstrate this point, we write $z_0 = j/j_{c0}$ and $z_1 = j/j_{c1}$. We now expand $U(z_0)$ with a Taylor's series in the neighborhood of z_1 as

$$U(z_0) = U(z_1) + (z_0 - z_1)\left[\frac{\partial U}{\partial z_0}\right]_{z1} + \ldots$$

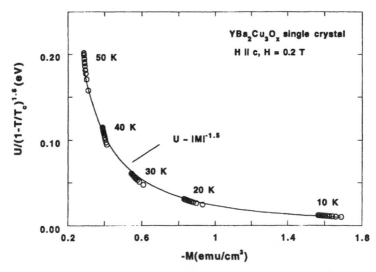

Fig. 4.22. U-j curve developed using the method in [99,120].

The function C would be a constant if and only if $(z_0 - z_1)\left[\dfrac{\partial U}{\partial z_0}\right]_{z1}$ is a constant, that is,

$$(z_0 - z_1)\left[\frac{\partial U}{\partial z_0}\right]_{z1} = K$$

or

$$\frac{\partial U}{\partial z_0} = \frac{K}{z_0\left(1 - \dfrac{z_1}{z_0}\right)}.$$

Now j_{c0} and j_{c1} are constants, and hence $(1 - z_1/z_0)$ is also a constant. Therefore,

$$\frac{\partial U}{\partial z_0} = \frac{K'}{z_0},$$

$$\text{where} \quad K' = \frac{K}{\left(1 - \dfrac{z_1}{z_0}\right)}.$$

Integrating, we get

$$U = K' \ln(z_0) = K' \ln\left(\frac{j}{j_{c0}}\right).$$

Therefore, a constant can simply be added to the U-j curve obtained at different temperatures to obtain a U-j curve for the lowest temperature if U varies logarithmically with j/j_{c0}. Experimentally, U(j) is generally found to vary logarithmically with j/j_{c0}. Even if U is not a logarithmic function of j/j_{c0}, however, Equation (4.95) can still be used to develop the U-j curve; but in that case, C also becomes a function of j. The constant C for various temperatures can be evaluated with the help of experimentally determined values of α and β and Equation (4.78). If the conditions $\beta \sim -\alpha/j$ and $|j/j_0 - 1| \ll 1$ are satisfied, Equation (4.76) is equivalent to $U \approx U_0 \ln(j/j_{c0})$. Experimentally, these conditions are generally satisfied. This method to construct a U-j curve cannot be used at very high temperature, however (i.e., near the irreversibility line where $U_0(T)$ becomes a strong function of temperature). In that case, the temperature dependence of $U_0(T)$ should also be considered. Figure 4.23 shows a typical U-j curve developed by this method for a $Bi_2Sr_2CaCu_2O_x$ single crystal in a magnetic field of 0.5 T applied parallel to the c-axis.

4.6.6 On the Physical Meaning of U-j Curves

As pointed out earlier, U gives the measure of the barrier to the flux motion. Thus, U does not reflects the true pinning force but rather the forces opposing the flux motion. The situation is quite analogous to that between static and dynamic friction in mechanics. In other words, the terminology used in stud-

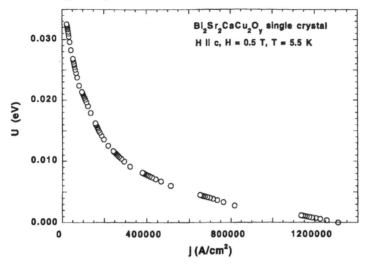

Fig. 4.23. U-j curve for $Bi_2Sr_2CaCuO_x$ single crystal with H ‖ c axis developed by using the method in [121–123].

ies of flux pinning have different physical implications. In transport measurements, where the flux lines are somehow static, the flux pinning energy, U_p, is more relevant, since the depinning process occurs as j increases from zero to j_c. On the other hand, in magnetic relaxation measurements, the effective activation energy becomes a more appropriate parameter, since the flux lines are already in motion. It is expected that U should approach U_p as j approaches zero, where the flux motion stops. Experimentally, however, U is always found to diverge at a low current density level. In a typical magnetic relaxation experiment, the relaxation rate decreases with decreasing current density level. Near the irreversibility line, where the current density is negligibly small, the decay process is extremely low (as the driving force is very small), resulting in a diverging U. Physically, this effect can be explained from various models. The glassy models like the collective creep theory or the vortex glass predict a diverging current density. This behavior is also expected from the reverse hopping, which is expected to have a large contribution at low current density. Near the irreversibility line, an equal number of forward and reverse jumps can be expected, resulting in a negligibly small decay rate which will imply a diverging U. The local pinning potential, which varies logarithmically with the distance from the pinning center, is also expected to have a similar effect on the effective activation energy, U.

Although it is difficult to obtain U_p by suppressing j to zero, a finite effective activation energy, U, at a given driving force would still be physically meaningful in studying the flux-pinning mechanisms. For instance, the U-j curves for materials with high pinning strengths will have large U at the same current density level j (or driving force). Physically, this means that at the neighborhood of some driving force, the current density of the strongly pinned material would decay more slowly (or equivalently, the flux line would encounter larger barrier to the flux motion).

To demonstrate this concept, we compare two superconducting samples with nominal composition $Bi_2Sr_2Ca_2Cu_3O_x$ (2223) and $Bi_2Sr_2Ca_4Cu_5O_x$ (2245), which are known to have different flux pinning strengths. These samples were made by the splat-quenching method. The quenched glass samples were annealed at 870°C in air and then slowly cooled to room temperature [174].

Figures 4.24a and 4.24b show the hysteresis and the critical current density as a function of field for both samples. The grains of the annealed samples appeared to be thin rectangles. The intergranular j_c was estimated by using Bean's model. As evident from the figure, a larger hysteresis and higher j_c are observed for the 2245 compound. (It should be noted that both 2223 and 2245 are nominal compositions.) After quenching and annealing, these samples crystallize and form mostly the superconducting 2212 phase, with various amounts of calcium- and copper-rich precipitates. The difference in the j_c in the 2223 and the 2245 samples can be attributed to the flux pinning by these precipitates directly or by the defects associated with these precipitates. Extensive transmission electron microscopy has revealed that only a few precipitates exist in the 2223 samples, whereas a large amount of fine precipitates was found in the 2245 samples [175]. The size of the precipitates varies widely, less than 0.01 μm to almost 0.1 μm. Many lattice defects such as stacking faults and severe lattice distortion also exist in the 2245 as a result of calcium and copper supersaturated in the sample during quenching experiments. It was found that the lattice varied about 8% between two local regions separated by approximately 50 nm. This severe lattice distortion did not result in any suppression of T_c. In contrast, a uniform lattice structure was observed in the 2223 sample, and practically no copper- and calcium-rich precipitates were observed.

Figure 4.25 shows U as a function of magnetization at 10 K for 2223 and 2245 samples. As can be seen from the figure, at a constant driving force (i.e., constant M), U is higher for the 2245 sample than for the 2223 sample. From Equation (4.73) one can say that at the neighborhood of any particular M, the magnetization for the 2223 sample is decaying faster than that of 2245 sample. To further illustrate this point, we have plotted the magnetization as a function of effective time for both samples at 10 K (Figure 4.26). This curve is equivalent to the U-j curve and is constructed by adjusting the characteristic time t_0.

Figure 4.27 shows the decay rate (lndM/dt) as a function of driving force. As expected, the 2245 sample has a lower decay rate than that of the 2223 sample in the neighborhood of any M at a constant temperature. This result implies that since the thermal energy ~kT is the same, the vortices are facing a higher barrier while moving in the 2245 sample than in the 2223 sample. This barrier can be attributed to the flux pinning by the calcium- and copper-rich precipitates and/or other lattice defects observed in 2245 sample [127].

4.6.7 Self-Organized Criticality (SOC)

The dynamical behavior of the flux lines has been traditionally interpreted by the Anderson-Kim model based on thermally activated flux motion [66]. The Anderson-Kim model assumes flux motion in the analogy of the diffusion process with the activation energy being a core physical parameter (i.e., a flux line may gain enough energy through thermal activity and overcome

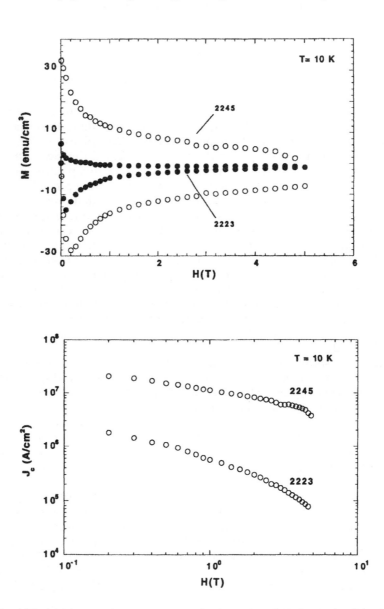

Fig. 4.24. (a) Magnetic hysteresis at 10 K for the 2223 and 2245 samples; (b) critical current density as a function of applied magnetic field at 10 K for the 2223 and 2245 samples.

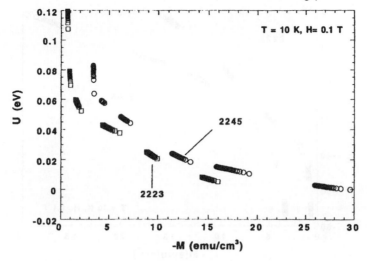

Fig. 4.25. U vs M curve for the 2223 and 2245 samples at T = 10 K.

a potential barrier). The flux motion considered by Anderson and Kim is characterized by a slow and individual event, thus often regarded as "flux creep." In the flux-creep process, two physical parameters are used to characterize flux motion: flux hopping distance, l, and flux hopping time, t_0. Generally, for a given system, l and t_0 are intrinsic to the system. For instance,

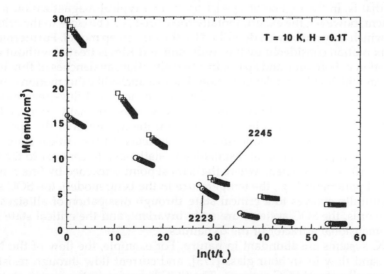

Fig. 4.26. Magnetization, M, as a function of ln(t/t0) for the 2223 and 2245 samples at T = 10 K. Note that ln (t/t0) is obtained by adjusting the characteristic time, t_0. This curve is equivalent to the U vs M curve for the 2223 and 2245 samples at 10 K (Figure 4.25).

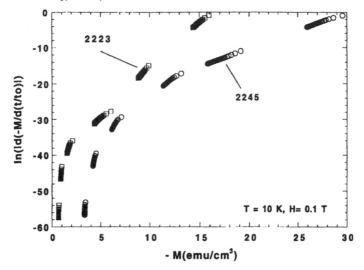

Fig. 4.27. The decay rate $\ln |dM/dt|$ as a function of M for the 2223 and 2245 samples at 10 K. This curve is calculated from the M vs $\ln (t/t_0)$ curve shown in Figure 4.26. Note that in the neighborhood of any M the 2245 sample has a lower decay rate than that of 2223 sample.

in type II high-T_c superconductors, $l \sim 100$ Å and $t_0 \sim 10^{-10}$ s. Therefore, l and t_0 may be regarded as intrinsic scales of the system. In the thermally activated flux-creep model, the current density regime is far away below the Bean critical state, that is, in the regime of $|j - j_c| \gg 1$. In a typical magnetization measurement, however, the experimentally measured j can be close to the critical state, which may not be well described by the flux-creep model. Furthermore, the flux motion considered by the Anderson and Kim is treated without any correlation in both time and space. In a real situation, avalanches of flux lines can occur with highly correlated motion. The avalanche-like flux motion cannot be well described in the framework of thermally activated flux-creep model.

The self-organized criticality (SOC) theory was originally proposed by Bak, Tang, and Wiesenfeld [61,62] to explain the dynamics of the spatially and temporally extended nonequilibrium systems. At the critical state, the system is marginally stable and statistically stationary. In contrast to the traditional critical phenomena where the critical point is reached by fine tuning a control parameter (e.g., the temperature in the Ising model), the SOC system naturally evolves to a critical state through dissipation of all sizes. In other words, the SOC systems are scale invariant, and the critical state is a low-dimensional attractor of the dynamics.

SOC systems are abundant in nature. For example, the flow of the Nile [177], sand flow in an hour glass [178], and current flow through resistors [179] are all natural SOC systems. The SOC behavior in the vortex state of type II superconductors has also been investigated [63,64,167,168,176,181]; indeed, the Bean critical state exhibits remarkable similarity to the sand flow.

Near the critical state, $F_p \sim F_L$, the flux motion takes place in form of avalanches and exhibit self-organized criticality. The term "self organized" implies that the system evolves into a critical state without detailed specification of the initial condition. To understand the basic idea of self-organized criticality in a transport system, let us consider the growth of a sand pile. A pile of sand can be constructed from scratch by randomly adding sand, a grain at a time. As the pile grows, the slope increases and eventually reaches a critical value such that further addition of a sand grain results in a avalanche. Alternatively, a pile with slope larger than the critical slope collapses until it reaches the critical slope where it is barely stable with respect to any perturbations. As the system evolves into a critical state, the characteristic size of the largest size grows, until at the critical point there are avalanches of all sizes up to the size of the system. The critical state can be viewed as an "attractor" of the dynamics. The avalanche size, s, and the duration, t, obey a power law distribution given by [61,62]

$$D(s) \sim s^{-\tau}, \text{ and } D(t) \sim t^{-\alpha}. \tag{4.96}$$

Here τ and α are the critical exponents. Researchers are uncertain, however, whether these spatio-temporal effects are due to the conservation laws in the nonlinear transport system or are the result of a true critical point in the system.

The concept of self-organized criticality has been applied to the study of vortex motion near the critical state of a type II superconductor. Tang [64] has proposed that the flux motion is originally thermally activated but that the subsequent hopping process is dominated by avalanches particularly at large driving force. He has also pointed out that the velocity of the vortex motion is determined by two factors, namely, the flux hopping rate and the flux avalanche size.

Experimentally from magnetic relaxation measurements, Wang and Shi [167,168,176] have determined that avalanches are dominant near the critical state and and that the contribution decreases as the system decays away from the critical state.

In a zero-field-cooled magnetization measurement, the flux profile can be seen as a process similar to that of an SOC sand pile. Consider a pile of sand initially covered by a box, forming a rectangular shape (top, Figure 4.28a). This corresponds to a zero-field-cooled experiment for a type II superconductor (Figure 4.28b). Below T_c, when the magnetic field is applied, the flux gradience dB/dx [15] can reach infinity at t = 0, producing a j much greater than j_c (top, Figure 4.28b). Once the covering box is removed, sand begins to flow with a decreasing slope through many avalanches of sand grains (middle, Figure 4.28a). Similarly, in the situation of flux motion, as a result of magnetic pressure, flux gradience is quickly reduced as t > 0 (middle, Figure 4.28b). For an SOC system, a feedback mechanism will "organize" itself into a critical state. Therefore, the slope of the sand flow will eventually reach the critical point (bottom, Figure 4.28a). In the vortex state, the magnetic pressure is to

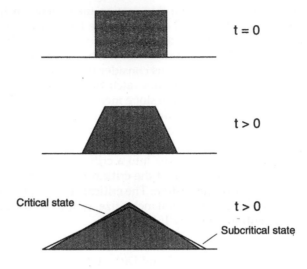

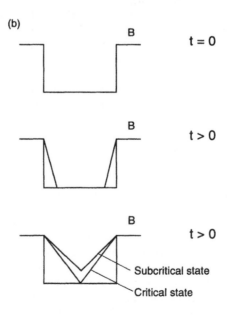

Fig. 4.28. Schematic of dynamics of (*a*) a sand pile, and (*b*) the vortex state.

be balanced by the pinning force, and the system will reach the Bean critical state where $j = j_c$ (bottom, Figure 4.28b). As pointed out above, the critical state is barely stable, and any external disturbances can trigger large avalanches, in turn decreasing the slope. In the case of a sand pile, the avalanches will occur if there is mechanical vibration. At finite temperatures, flux lines can be thermally activated, causing temporally and spatially correlated flux avalanches which will also reduce the slope dB/dx. Here, we denote any state smaller than the Bean critical state as the subcritical state where $|j_c-j| \ll j_c$.

In treating the current decay problem associated with flux creep, Anderson and Kim predicted a logarithmic dependence of j on time [66]. In a magnetic relaxation measurement of the alloy PbIn, Beasley, Labusch, and Webb [98] verified the Anderson-Kim prediction and found that the current decay had the form of $j/j_c = 1 - [kT/U_0 \ln(1 + t/\tau')]$, where j_c is the Bean critical current density and U_0 is the characteristic energy. Recently, however, it has been found that all type II superconductors actually exhibit nonlogarithmic current decay in an enlarged time window, which contradicts the Anderson-Kim model [116]. We will see that this discrepancy can be well resolved by the SOC theory.

Tang [64] has proposed that the time scale and the length scale of the velocity of flux lines are set by the thermal activation rate, ω, and avalanche size, s. He expressed ω by an Arrhenius expression:

$$\omega = \omega_0 \exp(-U/kT), \tag{4.97}$$

where ω_0 is the attempt frequency and $U = U_0(1 - j/j_c)^\beta$ [1], where β is a exponent constant. Tang defined s as the total flux line displacement involved in each avalanche. The avalanche size for an SOC-like system can be written as [61,62]

$$s = s_c (1 - j/j_c)^{-\infty}, \tag{4.98}$$

where s_c is the characteristic size and α is a critical exponent.

Combining Equations (4.97) and (4.98), we can write the flux conservation equation (4.98) in one-dimensional form as

$$\begin{aligned} dB/dt = -\nabla D &= -\nabla(Bv) = -\nabla(B\omega s) \\ &= -\nabla[Bs_c\omega_0(1 - j/j_c)]^{-\alpha}\exp[-U_0(1 - j/j_c)^\beta/kT], \end{aligned} \tag{4.99}$$

where D is the flux flow density, $v (= \omega s)$ is the velocity of flux line, and B is the magnetic field. Considering a slab of thickness d and integrating Equation (4.99) over a sample volume, we obtain the decay rate, dj/dt, by using Bean's model [8]:

$$dj/dt = -(4Bs_c v_0/\mu_0 d^2)(1 - j/j_c)^{-\alpha}\exp[-U_0(1 - j/j_c)^\beta/kT], \tag{4.100a}$$

or

$$dJ/d(t/\tau) = J^{-\alpha}\exp(-J^\beta), \tag{4.100b}$$

where $\varepsilon = U_0/kT$, $J = \varepsilon^{1/\beta}(1-j/j_c)$, $\tau = B_c d/2B\varepsilon s_c \omega_0$, and $B_c = \mu_0(d/2)j_c$, which is the lowest field at which complete flux penetration occurs. In the Anderson-Kim model, the decay rate can be written as $dJ/d(t/\tau') = \exp(-J)$, where $\tau' = B_c d/2B\varepsilon x_0 \omega_0$. Comparing Equation (4.100b) with the Anderson-Kim model, we note that the term $J^{-\alpha}$ is associated with pure flux avalanches, while the term $\exp(-J^\beta)$ originates from thermal effect. As was found out in previous studies [167,168,176], the avalanche effect is significantly reduced, and the flux lines exhibit independent creep behavior at small j (large time). In other words, the $J^{-\alpha}$ term is important at a large j (small time) and reaches unity as j has decayed to a significantly low level. This dynamical crossover implies that the $\exp(-J^\beta)$ term dominates at $\alpha \sim 0$ and $\beta \sim 1$. We therefore have

$$\alpha = \beta - 1. \qquad (4.101)$$

With Equation (4.101), we found good agreement between experimental data and Equation (4.100) in previous studies [167,168].

By substituting α with $\beta - 1$ in Equation (4.100) and integrating it, we can write the decay as

$$j = j_c - j_c[kT/U_0\ln(1 + \beta t/\tau)]^{1/\beta}, \qquad (4.102a)$$

or

$$j/j_c = 1 - [kT/U_0\ln(1 + \beta t/\tau)]^{1/\beta}, \qquad (4.102b)$$

where j/j_c is equal to $\Delta M/\Delta M_c$ where ΔM and ΔM_c are the hysteresis differences at a given time and at the critical state, respectively [123]. Equation (4.102) is an expression for all type II superconductors by considering the avalanche dynamics of flux motion. It is of great interest to compare Equation (4.102b) with the expression for current decay from the Anderson-Kim model [98]: $j/j_c = 1 - [kT/U_0\ln(1 + t/\tau')]$. Obviously, the Anderson-Kim model is a special case of Equation (4.102) with $\beta = 1$. Thus, $\alpha = \beta - 1$ is shown to be a natural consequence of the dynamical crossover observed previously [167,168,176]. In the Anderson-Kim model, τ' is a constant on the order of 10^{-4} s [149]. This would be a typical characteristic time scale for an individual hopping event. In the SOC model, however, the flux motion is characterized by the avalanche size which varies dynamically with j or t. The avalanche size can approach system size at large j (or small t) and decrease with decreasing j (or increasing t). As avalanches occur, the spatially and temporally correlated flux collisions result in a much larger characteristic time, on the order of 10–1000 s (as we will show later).

Wang and Shi [167,168,176] have investigated the magnetic relaxation over the long time period at a wide range of temperatures and fields in single crystals of high-Tc superconductors ($YBa_2Cu_3O_x$) and in conventional type II superconductors (multifilamentary wire of Nb_3Sn). They found that the logarithmic time dependence of (Anderson-Kim prediction) is valid only in the small j regime (i.e., as the system has decayed far away from the critical state).

Wang and Shi carried out magnetization measurements on a Quantum

Design superconducting quantum interference device (SQUID) magnetometer. The samples were first zero-field cooled to a desired temperature, T, below the transition temperature, T_c. Both increasing and decreasing magnetic fields were then applied parallel to the c axis for a single crystal of $YBa_2Cu_3O_x$ and to the multifilamentary wire of Nb_3Sn with the magnetic field normal to the length of the wire at 5.5 K. The magnetization, M, of the samples was measured as a function of time, t. The initial data point of the magnetization was taken after $t_0 = 100$ s. (Using standard procedures with a commercial SQUID, one cannot obtain the relaxation information earlier than 100 seconds after the field is settled.) The travel length of the sample in each scan was 3 cm for all measurements to avoid field inhomogeneity. The magnetization data for both increasing (M_+) and decreasing (M_-) was recorded and used to calculate the current density j by the Bean critical state model introduced earlier. Here the magnetic hysteresis diference $\Delta M = |M_+ - M_-|/2$ and $j = k\Delta M/d$, where k is a geometrical constant and d is the sample dimension [123].

In Figure 4.29a, the normalized critical current density, $j/j_c(H, T)$ is plotted against the time, t, for $YBa_2Cu_3O_x$ at H = 2 T and 3 T at 5.5 K [176]. If one

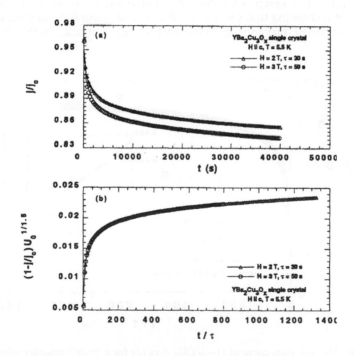

Fig. 4.29. (a) j/j_c vs t for a single crystal of Y123 at 5.5 K and field H = 2 T and 3 T perpendicular to the c-axis. The solid curves are fit to the Equation (4.102b) with $\beta = 1.5$; (b) the universal curve of $(1 - j/j_c)U_0^{1/\beta}$ vs t/τ for a single crystal of Y123 with t = 30 s and 50 s. The solid curves are fit to $Y = [kT\ln(1 + \beta X)]^{1/\beta}$ with $\beta = 1.5$, where $Y = (1 - j/j_c)U_0^{1/\beta}$ and $X = t/\tau$.

assumes that the shape of the pinning barrier has a functional form of $1/2U_p\cos(Z/Z_p)$, where Z is the position, then according to [98], $\beta = 1.5$. As shown in Figure 4.29a, the j/j_c vs t curves are fit well with Equation (4.102b) with U_0 being the only parameter. It has been found that the characteristic energy of $YBa_2Cu_3O_x$ is 0.06 eV, while the characteristic time changes from 30s to 50 s with increasing field. Further, by plotting $(1-j/j_c)U_0^{1/\beta}$ vs t/τ, a generic curve is obtained, as shown in Figure 4.29b.

Measurements on a Nb_3Sn wire showed similar behavior as evidenced in Figure 4.30 [176]. When the results were fit with Equation (4.102b), it was found that $U_0 = 0.07$ eV and $\tau = 50$ s for the fields indicated. Note that, in the small j regime as indicated by the arrow in Figure 4.30, the dynamical crossover [167,168] requires β to be unity, leading to the original Anderson-Kim model. Therefore, a logarithmic fit gives the best description at small j, as shown in the inset of Figure 4.30.

The change in τ indicates that the relaxation process slows down as the system decays away from the critical state. Near the Bean critical state where $j \sim j_c$, the system decays rapidly, and there is an observable change in τ. The decaying process slows down as j decreases, leading to a saturated τ. At small j where individual creep is a dominating behavior, the relaxation rate slows significantly, showing that τ is not sensitive to changing field or temperature.

To summarize, flux motion in the vortex state is similar to the sand pile,

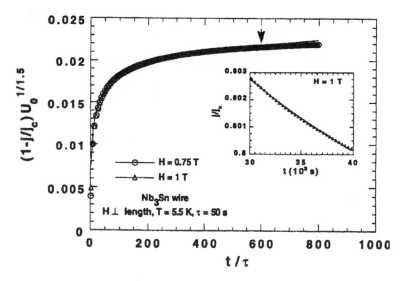

Fig. 4.30. The universal curve of $(1 - j/j_c)U_0^{1/\beta}$ vs t/τ for a multifilamentry wire Nb_3Sn with field H = 0.75 T and 1 T perpendicular to the length of the wire with $\tau = 50$ s. The solid curves are fit to $Y = [kT\ln(1 + \beta X)]^{1/\beta}$ with $\beta = 1.5$, where $Y = (1 - j/j_c)U_0^{1/\beta}$ and $X = t/\tau$. The inset shows the portion of the curve in the small j regime (starting from dynamical crossover as indicated by the arrow). The solid line is the fit to the Equation (4.102b) with $\beta = 1$.

which is a temporally and spatially extended SOC system. The expression for the time dependence of j based on SOC can describe the current decay in a wide range of j for type II superconductors. The relaxation measurements on both high-T_c and conventional superconductors have shown excellent agreement with this theoretical prediction. The variation in the characteristic time indicates that flux motion in the vortex state is a slowing-down dynamical process, with pronounced avalanches at the initial stage and independent creep far away from the Bean critical state.

Acknowledgement — This work was supported by the US Department of Energy (DOE), Basic Energy Sciences–Materials Science under Contract # W-31-109-Eng-38 (DS), and DOE Grant DE-FG02-90-ER45427 through the Midwest Superconductivity Consortium (SS).

4.7 References

1. W. Meissner and R. Ochsenfeld, Naturwiss. **21**, 787 (1933).
2. L. V. Shubnikov, V. I. Khotkevich, Yu. D. Shepelev, and Yu. N. Riabinin, Zh. Eksp. Teor. Fiz. **7**, 221 (1937).
3. A. A. Abrikosov, Zh. Eksp. Teor. Fiz. **32**, 1442 (1957). [Sov. Phys. JETP **5**, 1174 (1957)].
4. R. P. Huebener, *Magnetic Flux Structures in Superconductors*, Springer-Verlag, Berlin (1979).
5. M. Tinkham, *Introduction to Superconductivity*, McGraw-Hill, New York (1980).
6. A. M. Campbell and J. E. Evetts, Adv. Phys. **21**, 199, (1972).
7. F. London and H. London, Proc. Roy. Soc. (London) A **149**, 71 (1935).
8. C. P. Bean, Phys. Rev. Lett. **8**, 250 (1962); C. P. Bean, Rev. Mod. Phys. **36**, 31 (1964).
9. P. W. Anderson and Y. B. Kim, Rev. Mod. Phys. **36**, 39 (1964).
10. F. Irie and K. Yamafuji, J. Phys. Soc. Jpn. **23**, 255 (1967).
11. W. A. Feitz, M. R. Beasley, J. Silicox, and W. W. Webb, Phys. Rev. **136** A, 335 (1964).
12. J. H. P. Watson, J. Appl. Phys. **39**, 3406 (1968).
13. G. Ravi Kumar and P. Chaddah, Phys. Rev. B **39**, 4704 (1989).
14. D. -X. Chen and R. B. Goldfarb, J. Appl. Phys. **66**, 2489 (1989); D.-X. Chen, A. Sanchez, and J. S. Munoz, J. Appl. Phys. **67**, 3440 (1990); D.-X. Chen, A. Sanchez, J. Nogues,and J. S. Munoz, Phys. Rev. B **41**, 9510 (1990).
15. Donglu Shi, M. Xu, A. Umezawa, and R. F. Fox, Phys. Rev B **42** , 2062 (1990).
16. M. Xu, Donglu Shi, and R. F. Fox, Phys. Rev. B **42** , 10773 (1990).
17. S. Sengupta, Donglu Shi, J. S. LuO, A. Buzdin, V. Dorin, V. Todt, C. Varanasi, and P. J. McGinn (unpublished results).
18. G. S. Mkrtchyan and V. V. Shmidt, Sov. Phys. JETP **34**, 195 (1972).
19. W. E. Timms and D. G. Walmsley, Phys. Stat. Sol. (b) **71**, 741 (1975).
20. L. N. Shehata, Phys. Stat. Sol. (b) **105**, 77 (1981).

21. H. Kronmüller, *International Discussion Meeting on Flux Pinning in Superconductors*, Sonnenberg, Germany (1974), 1.
22. E. J. Kramer and C. L. Bauer, Phil. Mag. **15**, 1189 (1967).
23. M. Peach and J. S. Koehler, Phys. Rev. **80**, 436 (1950).
24. R. L. Fleishcher, Phys. Lett. **3**, 111 (1962).
25. W. W. Webb, Phys. Rev. Lett. **11**, 191 (1963).
26. L. E. Toth and I. P. Pratt, Appl. Phys. Lett. **4**, 75 (1964).
27. D. Dew-Hughes, Rep. Prog. Phys. **34**, 821 (1971).
28. D. Dew-Hughes, Phil. Mag. **30**, 293 (1974).
29. R. G. Hampshire and M. T. Taylor, J. Phys. F: Metals Phys. **2**, 89 (1972).
30. E. W. Collings, *Applied Superconductivity, Metallugy, and Physics of Titanium Alloys*, Plenum Press, New York (1986).
31. W. A. Feitz and W. W. Webb, Phys. Rev. **178**, 657 (1969). Erratum: Phys. Rev. **185**, 852 (1969).
32. K. Yamafuji and F. Irie, Phys. Lett. **25A**, 387 (1967).
33. R. Labush, Phys. Status Solidi **19**, 715 (1967).
34. J. Silicox and R. W. Rollins, Appl. Phys. Lett. **2**, 231 (1963).
35. J. Friedel, P. G. De Gennes, and J. Matricon, Appl. Phys. Lett. **2**, 119 (1963).
36. R. Labush, Phys. Stat. Solidi **32**, 439 (1969).
37. E. H. Brandt, Phys. Stat. Solidi (b) **77**, 551 (1976).
38. E. H. Brandt, J. Low Temp. Phys. **26**, 709 (1977).
39. E. H. Brandt, J. Low Temp. Phys. **26**, 735 (1977).
40. E. H. Brandt, Phys. Rev. B **34**, 6514 (1986).
41. E. H. Brandt and U. Essmann, Phys. Stat. Solidi (b) **144**, 13 (1987).
42. A. Sudbø and E. H. Brandt, Phys. Rev. B **43** , 10482 (1991).
43. A. Sudbø and E. H. Brandt, Phys. Rev. Lett. **66**, 1781 (1991).
44. E. H. Brandt, Super. Sci. Tech. **5**, S25 (1992).
45. A. I. Larkin and Y. N. Ovchinnikov, J. Low Temp. Phys. **34**, 409 (1979).
46. X. Ling, Ph.D. thesis,University of Connecticut (1992).
47. A. I. Larkin, Sov. Phys. JETP **31**, 784 (1970).
48. E. M. Chudnovsky, Phys. Rev. B **40**, 11355 (1989).
49. E. M. Chudnovsky, Phys. Rev. B **43**, 7831 (1991).
50. V. Vinokur, Lectures notes delivered at ANL.
51. P. Kes and J. van den Berg, *Studies of High Temperature Superconductors*, ed. by A. Narlikar, Nova Science Publishers, Vol. 5, (1990), 83.
52. R. Meier-Hirmer, H. Kupfer, and H. Scheurer, Phys. Rev B **31**, 183 (1985).
53. C. J. Gorter, Physica **23**, 45 (1957).
54. C. J. Gorter, Phys. Lett. **1**, 69 (1962).
55. C. J. Gorter, Phys. Lett. **2**, 26 (1962).
56. P. W. Anderson, Phys. Rev. Lett. **9**, 309 (1962).
57. Y. B. Kim, C. F. Hempstead, and A. R. Strnad, Phys. Rev. **129**, 528 (1963).
58. Y. B. Kim, C. F. Hempstead, and A. R. Strnad, Phys. Rev. **131**, 2486 (1963).
59. Y. B. Kim, C. F. Hempstead, and A. R. Strnad, Phys. Rev. **139**, A1163 (1965).
60. Y. B. Kim and M. J. Stephen, *Superconductivity*, ed. by R. D. Parks, Marcel Dekker, New York (1969), 1107.
61. P. Bak, C. Tang, and K. Wiesenfeld, Phys. Rev. A **38**, 364 (1988).
62. C. Tang and P. Bak, Phys. Rev. Lett. **60**, 2347 (1988).

63. O. Pla and F. Nori, Phys. Rev. Lett. **67**, 919 (1991).
64. C. Tang, Physica A **194**, 315 (1993).
65. P. W. Anderson, Phys. Rev. Lett. **9**, 310 (1962).
66. P. W. Anderson and Y. B. Kim, Rev. Mod. Phys. **36**, 39 (1964).
67. P. G. De Gennes, *Superconductivity of Metals and Alloys*, W. A. Benjamin, New York (1966).
68. P. G. De Gennes and J. Matricon, Rev. Mod. Phys. **36**, 45 (1964).
69. A. G. Van Vijfeijken and A. K. Niessen, Philips Res. Rep. **20**, 505 (1965).
70. A. G. Van Vijfeijken and A. K. Niessen, Phys. Lett. **16**, 23 (1965).
71. A. G. Van Vijfeijken, Philips Res. Rep. Suppl. **8**, 1 (1968).
72. K. Yamafuji and F. Irie, Phys. Lett **25A**, 387 (1967).
73. J. Lowell, J. Phys. C: Sol. State Phys. **3**, 712 (1970).
74. J. A. Good, and E. J. Kramer, Philos. Mag. **22**, 329 (1970).
75. A. R. Strnad, C. F. Hempstead, and Y. B. Kim, Phys. Rev. Lett. **13**, 794 (1964).
76. J. Bardeen and M. J. Stephen, Phys. Rev. **140**, A1197 (1965).
77. A. Schmid, Phys. Kondensierten Materie **5**, 302 (1966).
78. C. Caroli and K. Maki, Phys. Rev. **159**, 306 (1967).
79. C. Caroli and K. Maki, Phys. Rev. **159**, 316 (1967).
80. C. Caroli and K. Maki, Phys. Rev. **164**, 591 (1967).
81. C. Caroli and K. Maki, Phys. Rev. **169**, 381 (1968).
82. C. R. Hu and R. S. Thompson, Phys. Rev. B **6**, 110 (1968).
83. J. R. Clem, Phys. Rev. Lett. **20**, 735 (1968).
84. M. Tinkham, Phys. Rev. Lett **61**, 1658 (1988).
85. M. Tinkham and C. J. Lobb, *Solid State Physics* **42**, ed. H. Ehrenreich and D. Turnbull, Academic Press, New York (1989), 91.
86. H. A. Blackstead, Phys. Rev. B **44**, 6955 (1991).
87. K. H. Lee and D. Stroud, Phys. Rev B **46**, 5699 (1992).
88. J. C. Phillips, Phys. Rev B **46**, 8542 (1992).
89. H. A. Blackstead, Phys. Rev. B **47**, 11411 (1993).
90. J. Le G. Gilchrist and P. Monceau, J. Phys. C : Sol. State Phys. **3**, 1399 (1970).
91. J. Le G. Gilchrist and P. Monceau, J. Phys. Chem. Sol. **32**, 2101 (1971).
92. A. M. Clogston, Phys. Rev. Lett. **9**, 266 (1962).
93. B. S. Chandrashekhar, Appl. Phys. Lett. **1**, 7 (1962).
94. C. J. Axt and W. C. H. Joiner, Phys. Rev. Lett. **21**, 1168 (1968).
95. W. C. H. Joiner and J. Thompson, Solid State Commun. **11**, 1393 (1972).
96. Y. B. Kim, C. F. Hempstead, and A. R. Strnad, Phys. Rev. Lett. **9**, 306 (1962).
97. J. A. File and G. Mills, Phys. Rev. Lett. **10**, 93 (1963).
98. M. R. Beasley, R. Labusch, and W. W. Webb, Phys. Rev. **181**, 682 (1969).
99. M. P. Maley, J. O. Willis, H. Lessure, and M. E. McHenry, Phys. Rev. B **42**, 2639 (1990).
100. Donglu Shi and S. Salem-Sugui, Jr., Phys. Rev. B **44**, 7647 (1991).
101. M. E. Smith, Donglu Shi, S. Sengupta, and Zuning Wang, Appl. Superconductivity **1**, 151 (1993).
102. Y. Xu, M. Suenaga, A. R. Modenbaugh, and D. O. Welch, Phys. Rev. B **40**, 10882 (1989).

103. K. A. Müller, M. Takashige, and J. G. Bednorz, Phys. Rev. Lett. **58**, 1143 (1987).

104. A. C. Mota, A. Pollini, P. Visani, K. A. Müller, and J. G. Bednorz, Phys. Rev. B **36**, 401 (1987).

105. C. Giovannnella, G. Collin, P. Rault, and I. A. Campbell, Europhys. Lett. **4**, 109 (1987).

106. J. R. Carolan, W. N. Hardy, R. Krahn, J. H. Brewer, R. C. Thompson, and A. C. D. Chaklader, Solid State Commun. **64**, 717 (1987).

107. M. Touminen, A. M. Goldman, and M. L. Mecartney, Phys. Rev . B **37**, 548 (1988).

108. Y. Yeshurun and A. P. Malozemoff, Phys. Rev. Lett. **60**, 2202 (1988).

109. H. S Lessure, S. Simizu, and S. G. Sankar, Phys. Rev. B **40**, 5165 (1989).

110. Donglu Shi, M. Xu, A. Umezawa, and R. F. Fox, Phys. Rev. B **42**, 2062 (1990).

111. S. Sengupta, P. McGinn, N. Zhu, and W. Chen, Physica C **171**, 174 (1990).

112. M. Turchinskaya, L.H. Bennett, L. J. Swartzendruber, A. Roitburd, C. K. Chiang *Superconductors: Fundamental Properties and Novel Materials Processing*, Mat. Res. Soc. Symp. Proc. **169**, 931 (1990).

113. U. Atzmony, R. D. Shull, C. K. Chiang, L. J. Swartzendruber, L. H. Bennett, and R. E. Watson, J. Appl. Phys. **63**, 4179 (1988).

114. M. D. Lan, J. Z. Liu, and R. N. Shelton, Phys. Rev. B **44**, 2751 (1991).

115. J. Z. Liu, Lu Zhang, M. D. Lan, R. N. Shelton, and M. J. Fluss, Phys. Rev. B **46**, 9123 (1992).

116. Donglu Shi and M. Xu, Phys. Rev. B **44**, 4548 (1991).

117. T. Matsuhita, S. Funaba, Y. Nagamatsu, B. Ni, K. Funaki, and K. Yamafuji, Jp. J. Appl. Phys. **28**, L1754 (1989).

118. M. Murukami, M. Morita, and N. Koyama, Jpn. J. Appl. Phys. **28**, L1754 (1989).

119. C. Keller, H. Kupfer, R. Meir-Hirmer, U. Wiech, V. Selvamanikam, and K. Salama, Cryogenics **30**, 410 (1990).

120. M. E. McHenry, S. Simizu, H. Lessure, M. P. Maley, Y. C. Coulter, I. Tanaka, and H. Kojima, Phys. Rev. B **44**, 7614 (1991).

121. S. Sengupta, Donglu Shi, S. Salem-Sugui, Jr., Zuning Wang, P. J. McGinn, and K. DeMoranville, J. Appl. Phys., **72**, 592, (1992).

122. S. Sengupta, Donglu Shi, Sergei Sergeenkov, and P. J. McGinn, Phys. Rev. B **48**, 6736 (1993).

123. S. Sengupta, Donglu Shi, Zuning Wang, M. E. Smith, and P. J. McGinn, Phys. Rev. B **47**, 5165 (1993).

124. B. M. Lairson, J.Z.Sun, T. H. Geballe, M. R. Beasley, and J.C. Bravman, Phys. Rev. B **43**, 10405 (1991).

125. S. Sengupta, Donglu Shi, S. Salem-Sugui, Jr., Zuning Wang, M. E. Smith, and P. J. McGinn, Phys. Rev. B **47**, 5414 (1993).

126. Donglu Shi, S. Sengupta, M. Smith, and Zuning Wang, Cryogenics **32**, 527 (1992).

127. S. Sengupta, Donglu Shi, Zuning Wang, M. E. Smith, and P. J. McGinn, IEEE Trans. Appl. Superconductivity **3**, 1226 (1993).

128. S. Salem-Sugui, Jr., and Donglu Shi, Phys. Rev. B **46**, 6618 (1992).

129. Y. Ren and Peter A. J. de Groot, Physica C **196**, 111 (1992).
130. X. Ling, D. Shi, and J. I. Budnick, Physica C **185**, 2181 (1991).
131. X. X. Dai, P. H. Hor, L. Gao, and C. W. Chu, Phys. Lett. A **169**, 161 (1992).
132. A. Gurevich and H. Küpfer, Phys. Rev. B **48**, 6477 (1993).
133. Y. R. Sun, J. R. Thompson, Y. J. Chen, D. K. Cristen, and A. Goyal, Phys. Rev. B **47**, 14481(1993).
134. A. Spirgatis, R. Trox, J. Kötzler, and J. Bock, Cryogenics **33**, 138 (1993).
135. M. Jirsa, L. Pust, H. G. Schnack, and R. Griessen, Physica C **207**, 85 (1993).
136. C. J. van der Beek, P. Kes, M. P. Maley, M. J. V. Menken, and A. A. Menovsky, Physica C **195**, 307 (1992).
137. C. J. van der Beek, G. J. Neiuwenhuys, P. H. Kes, H. G. Schnack, and R. Griessen, Physica C **197**, 320 (1992).
138. Y. R. Sun, J. R. Thompson, D. K. Cristen, F. Holtzberg, A. D. Marwick, and J. G. Ossandon, Physica C **194**, 403 (1992).
139. J. R. Thompson, Y. R. Sun, L. Civale, A. P. Malozemoff, M. W. McElfresh, A. D. Marwick, and F. Holzberg, Phys. Rev B **47**, 14440 (1993).
140. D. Fiorani, A. M. Testa, and G. Calestani, J. Super. **5**,39 (1992).
141. Y. L. Ma, X. X. Dai, W. H. Zhong, H. F. Li, and P. H. Hor, Phys. Lett. A **183**,   425 (1993).
142. M. V. Feigel'man, V. B. Geshkenbein, A. I. Larkin, and V. M. Vinokur, Phys. Rev. B **63**, 2303 (1989).
143. P. Kes, J. Aarto, J. van den Berg, C. J. van der Beek, and J. A. Mydosh, Supercond. Sci. Technol. **1**, 242 (1989).
144. V. M. Vinokur, M. V. Fiegel'man and V. B. Geskenbein, Phys. Rev. Lett. **67**, 915 (1991).
145. C. W. Hagen and R. Griessen, Phys. Rev. Lett **62**, 2857 (1989).
146. P. Manuel, C. Aguillon, and S. Senoussi, Physica C **177**, 281 (1991).
147. D. O. Welch, IEEE Trans. Magnetics **27**, 1133 (1991).
148. D. O. Welch, M. Suenaga, Y. Xu, and A. Ghosh, *Advances in   Superconductivity II*, ed. T. Ishiguro and K. Kajimura, Springer-Verlag, Tokyo (1990).
149. R. Griessen, Physica C **172**, 441 (1991).
150. M. P. A. Fisher, Phys. Rev. Lett. **62**, 1415 (1989).
151. Z. Koziol, P. F. Châtel, J. J. M. Franse, Z. Tarnawski, and A. A. Menovsky, Phyica C **212**, 133 (1993).
152. L. Niel, Cryogenics **32**, 975 (1992).
153. V. B. Geshkenbein and A. I. Larkin, Sov. Phys. JETP **68**, 639 (1989).
154. P. H. Kes and C. J. Vanderbeek, Mat. Res. Soc. Symp. Proc **275**, 157 (1992).
155. R. S. Markiewicz, Physica C **171**, 479 (1990).
156. T. Nattermann, Phys. Rev. Lett. 64, 2454 (1990).
157. K. H. Fischer and T. Nattermann, Phys. Rev. B **43**, 10372 (1991).
158. M. V. Feigel'man and V. M. Vinokur, Phys. Rev. B **41**, 8986 (1990).
159. L. Burlachkov, Phys. Rev. B **47**, 8506 (1993).
160. B. I. Ivlev and N. B. Kopin, Phys. Rev. Lett. **64**, 1828 (1990).
161. M. V. Feigel'man, V. B. Geshkenbein, and V. M. Vinokur, Phys. Rev. Lett. **43**, 6263 (1991).
162. H. G. Schnack, R. Griessen, J. G. Lesink, C. J. van der Beek. and P. Kes, Physica C **197**, 337 (1992).

163. V. Geshkenbein, A. Larkin, M. Feigel'man, and V. Vinokur, Physica C **162-164,** 239 (1989).

164. C. W. Hagen, R. P. Griessen, and E. Salomons, Physica C **157**, 199 (1989).

165. C. J. van der Beek, Ph.D. thesis, Leiden University (1992).

166. D. Dew-Hughes, Cryogenics **28**, 674 (1988).

167. Z. Wang and Donglu Shi, Phys. Rev. B **48**, 4208 (1993).

168. Z. Wang and Donglu Shi, Phys. Rev. B, **48**, 9782 (1993).

169. E. Zeldov, N. M. Amer, G. Koren, A. Gupta, M. W. McElfresh, and R. J. Gambino, Appl. Phys. Lett. **56**, 680 (1990).

170. E. Zeldov, N. M. Amer, G. Koren, and A. Gupta, Appl. Phys. Lett. **56**, 1700   (1990).

171. D. Hu, Physica C **205**, 123 (1993).

172. C. J. van der Beek, V. M. Vinokur, and P. H. Kes, Physica C **165-166**, 1139 (1990).

173. C. J. van der Beek and P. H. Kes, Phys. Rev. B **43**, 13032 (1991).

174. Donglu Shi, M. Tang, M. S. Boley, M. Hash, K. Vandervoot, H. Claus, and Y. N. Lwin, Phys. Rev. B **40**, 2257 (1989).

175. Donglu Shi, J. G. Chen, U. Welp, M. S. Boley, and A. Zangvil, Appl. Phys. Lett. **55**, 1354 (1989).

176. Z. Wang and Donglu Shi, Phys. Rev. B **48**, 16176 (1993).

177. W. H. Press, Comm. Mod. Phys. **C7**, 103 (1978), P. Dutta and P. M. Horn, Mod. Phys. **53**, 497 (1981).

178. K. L. Schick and A. A. Verveen, Nature **251**, 599 (1974).

179. R. F. Voss and J. Clarke, Phys. Rev. B **13**, 556 (1976).

180. D. Shi, X. Ling and J. I. Budnick, *Physical Phenomena at High Magnetic Fields*, ed. E. Manousakis, P. Schlottmann, P. Kumar, K. Bedell, and F. M. Mueller, Addison-Wesley, New York (1991), 389.

4.8 Recommended Readings for Chapter 4

In general for type II superconductors:

1.   *Superconductivity of Metals and Alloys*, P. G. DeGennes, W. A. Benjamin, New York, 1966.

2.   *Superconductivity: Fundamentals and Applications*, W. Buckel, Weinheim, Federal Republic of Germany, New York, 1991.

3.   *Introduction to Superconductivity*, M. Tinkham, McGraw-Hill, New York, 1980.

For the properties of the mixed state:

4.   *Magnetic Flux Structures in Superconductors*, R. P. Heuebener, Springer-Verlag, Berlin, 1979.

5.   A. M. Campbell and J. E. Evetts, Adv. Phys. **21**, 199 (1972).

6.   *Applied Superconductivity, Metallurgy, and Physics of Titanium Alloys*, Vol. 1 & 2, E. W. Collings, Plenum Press, New York, 1986.

For high-T$_c$ superconductors:

7. *Physics of High-T$_c$ Superconductors*, J. C. Phillips, Academic Press, Boston, 1989.

8. *Introduction to Superconductivity and High-T$_c$ Materials*, M. Cryot and D. Pavuna, World Scientific Singapore, River Edge, N. J., 1992.

5

Structural Defects in $YBa_2Cu_3O_{7-\delta}$ Superconductors

Y. Zhu

The discovery of the high-temperature superconductor $YBa_2Cu_3O_{7-\delta}$ in 1987 [1] produced widespread euphoria among the physics and material science communities. Despite the subsequent worldwide effort in superconductivity research, however, the progress of commercialization of this highly touted material has been rather disappointing. Now it has become clear that the barrier to the practical applications of the perovskite-based oxides is their low critical current density, J_c, especially the low critical transport current density, J_{ct} [2,3]. Improving these properties depends heavily on our knowledge of the crystal structures.

For structure-sensitive properties, a knowledge of the average, or ideal, structure is not sufficient; often, it is structural defects that play an important role in determining the properties. We have learned that in superconducting materials, some defects destroy superconductivity, while others promote it [4,5]. For example, defects of an appropriate size can act as flux-pinning centers in the crystal. By preventing the motion of magnetic flux lines, such defects can enhance the ability of the material to carry current. On the other hand, large-angle grain boundaries, which often act as weak links in the cuprate oxides (i.e., they subdivide the material into regions of strong superconductivity separated by weak superconducting interfaces), are key elements limiting the use of these materials as bulk superconductors. Thus, an understanding of intragranular defects is critical to point out the way to enhance flux pinning, while an understanding of the interfacial structure of grain boundaries is essential in elucidating the superconducting characteristics of the boundaries and, hence, in seeking possible engineering processes to eliminate the "bad" grain boundaries in the high-temperature superconductors.

It is important to note that the range of the superconducting coherence length, ξ, in the cuperate oxide is less than 20 Å ($\xi_{ab} \sim 10–20$ Å, $\xi_c \sim 2–4$ Å) [6], and the pertinent geometrical scale of importance for local structure is on the same order; therefore, characterization by using a near-ξ-scale source and high-resolution techniques often is required to study the defects. The details of a local atomic structure and of the stoichiometric variations associated with structural defects are difficult to obtain from the averaged molecular science information of X-ray and neutron diffraction studies [7]. In contrast, transmission electron microscopy (TEM) with its abilities of better than 2-Å resolution in imaging and 20-Å resolution in chemical composition provides unique capabilities for characterizing such local structure and has proved to be one of the most powerful methods for studying structural defects.

This chapter concentrates specifically on the applications of TEM in elucidating various structural defects in superconducting oxides. (A brief review of such defects was published in the *MRS Bulletin* in 1991 [8].) Structural details will be emphasized, as well as the techniques used to characterize these defects, such as image analysis of diffraction contrast (two-beam imaging) and phase contrast (many-beam imaging), interpretation of electron diffraction and diffuse scattering, structural modeling and computer simulation, chemical analysis using nanoprobe energy dispersive X-ray spectroscopy (EDX), and electron energy-loss spectroscopy (EELS). These techniques are complementary, and, very often, several techniques have to be used to solve one structural problem. Our focus here is on crystal defects in $YBa_2Cu_3O_{7-\delta}$, the most extensively studied system. An account of structural defects in related high-temperature superconductors can be found in an earlier review paper by Beyers and Shaw [9]. Among the many varieties of microstructural defects, the types of defects discussed are the primary ones observed in bulk materials; but the defects in

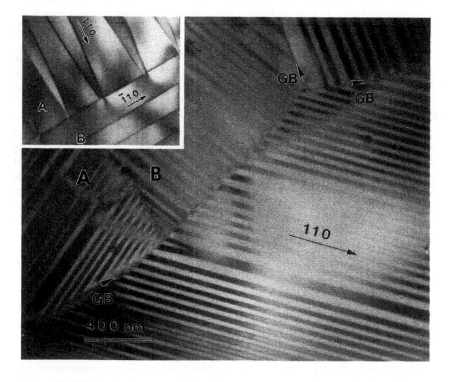

Fig. 5.1. A typical two-beam image of the twinning morphology in $YBa_2Cu_3O_{7-\delta}$ ($\delta \approx 0$) showing black-white twin lamellae. The interfaces between neighboring lamellae are twin boundaries. Grain boundaries are marked as arrowheads. Inset shows two orthogonally oriented twins. The impinging set of twin is denoted as twin A, and the orthogonally oriented set as twin B.

thin films and single or bicrystals are similar.

Flux pinning and physical properties that are associated with these defects are discussed in Chapter 4. Here we describe various techniques used to characterize structural defects and elucidate the essential microstructural characteristics that govern superconducting properties.

5.1 Intragranular Structural Defects

The first section of this chapter covers the structural features of major intragranular defects in $YBa_2Cu_3O_{7-\delta}$. Since oxygen distribution has a dramatic effect on microstructure and superconducting behavior, and $YBa_2Cu_3O_{7-\delta}$ has been the model system for studying such effects, the defects associated with oxygen ordering are addressed first. In particular, this section describes the structures of twin boundaries with different oxygen substoichiometry, δ, and twin termination; examines the superstructure of oxygen ordering in oxygen-deficient $YBa_2Cu_3O_{7-\delta}$; and reviews structural modulations in $YBa_2(Cu_{1-x}M_x)_3O_{7-\delta}$, which also involve oxygen ordering due to chemical substitution. Emphasis is placed on how to interpret the observed electron diffuse scattering and TEM image contrast. Also included is a brief discussion of conventional crystal defects, such as dislocations, stacking faults, and inclusions. Finally, the section concludes with a discussion of structural defects deliberately and selectively created by heavy-ion irradiation.

5.1.1 Twins

5.1.1.1 Twin Boundaries

The most prominent structural defects in $YBa_2Cu_3O_{7-\delta}$ ($\delta < 0.5$) are twins. Below about 750°C, $YBa_2Cu_3O_{7-\delta}$ undergoes a tetragonal-to-orthorhombic structural phase transition, resulting in twinning on the {110} planes. Figure 5.1 shows a typical twinning morphology in three adjacent grains. The platelet images are twin lamellae. The alternative black and white contrast corresponds to the pair sets of twins with a twinning rotation $\alpha = 90° - 2\arctan(a/b)$ [10].

The twinning rotation can be measured by the splitting of the Bragg spots, as marked by the arrow heads in Figure 5.2a. In bright-field imaging, when one set of twin lamellae fulfills a Bragg reflection (showing white contrast), the other set of twin lamellae is off the Bragg reflection (black contrast). In $YBa_2Cu_3O_{7-\delta}$, twins are formed to reduce strain energy due to the change in shape and volume resulting from the phase transformation. The spacing of the twin lamellae is determined by minimizing the total energy associated with the strain energy at the grain boundaries where the twins terminate and the interfacial energy of the twin boundary [11–14].

An important observation in a (001)* selected-area diffraction (SAD) is that sharp streaks perpendicular to the twin boundary are superimposed on the Bragg spots (Figure 5.2a, insets of Figure 5.2a and b). By using the streak to form a dark-field image, it was concluded that the streaks result from the boundary region, as shown in Figure 5.2b. Also, the intensity of the streak strongly depends on the oxygen content of the samples. Figure 5.3a is a

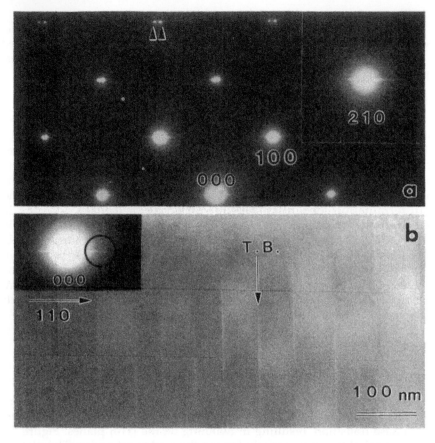

Fig. 5.2. (*a*) A selected area diffraction pattern of (001)* zone covering several twin boundaries from a YBa₂Cu₃O₇ sample. Note that sharp streaks superimpose on all the Bragg spots in a direction perpendicular to the twin boundary; (*b*) a dark-field image obtained by allowing half of a streak, marked by a circle on the inset, to pass through the objective aperture. The enhanced contrast of the twin boundaries unambiguously demonstrates that the streaks result from the boundaries.

high-resolution electron microscopy (HREM) image of a twin boundary in the (001) projection from a fully oxygenated YBa₂Cu₃O₇₋δ (δ ≈ 0) sample. We note that, in addition to the severe lattice distortion of the twin boundary region, there is a lattice displacement $\mathbf{R} = (1/2-1/3) \cdot 2d_{110}$ at the boundary. The displacement can be easily seen by looking at the (100) lattice row across the boundary. Detailed diffraction studies, including optical diffraction and digitized diffractograms from HREM twin boundary images and diffraction intensity calculations with various twin boundary models, suggest that the reciprocal streaks are mainly due to the displacement along the twin boundary [15]. The analysis of the fringe contrast of the twin boundary under

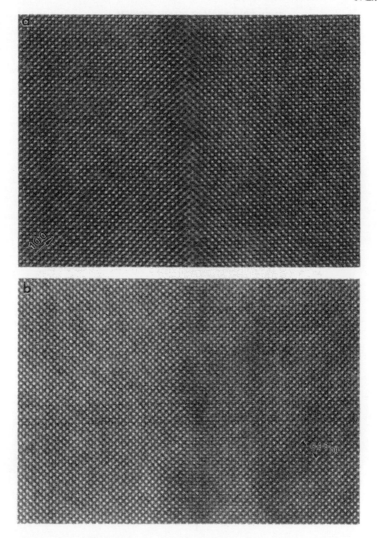

Fig. 5.3. High-resolution images of twin boundary of (001) projection of (*a*) $YBa_2Cu_3O_7$ and (*b*) $YBa_2Cu_3O_{6.6}$. The bright dots arranged in a near-square cell correspond to the projection of the fundamental perovskite unit cell. Image simulations suggest that the atom of the Ba(Y) columns appears in strong white contrast, and the Cu(O) columns in weak white contrast. The oxygen columns are shown between the Ba(Y) columns. The sharp, uniform atomic image implies that the columns of atoms are accurately aligned in the direction of the incident beam on both sides of the twin boundaries. By comparing the twin boundaries, we note that for the fully oxygenated twin boundary in (*a*) the boundary region is severely distorted, yet there is a lattice displacement at the boundary. The displacement can be easily seen by looking the (100) lattice across the boundary. In contrast, for the oxygen-deficient twin boundary in (*b*), there is no displacement and severe lattice distortion at the boundary.

TWIN BOUNDARY AT CHAIN OXYGEN

(a)

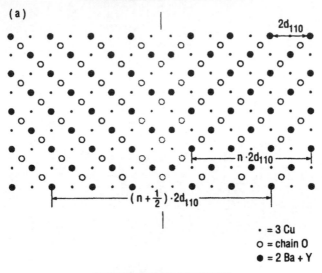

$2d_{110}$

$n \cdot 2d_{110}$

$(n + \frac{1}{2}) \cdot 2d_{110}$

• = 3 Cu
o = chain O
● = 2 Ba + Y

TWIN BOUNDARY AT CATIONS

(b)

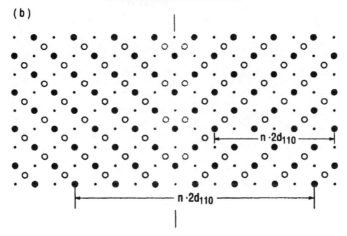

$n \cdot 2d_{110}$

$n \cdot 2d_{110}$

Fig. 5.4. Idealized drawing of the twin boundary region consistent with the observations of Figure 5.3a and Figure 5.3b, respectively. Most oxygen atoms are omitted, except for chain-oxygen atoms. (a) For twin boundary centered at chain oxygen, there is a displacement of cation sublattice along the boundary, while (b) for twin boundary centered at cations, there is no displacement of cation sublattice.

two-beam conditions also supports the existence of such displacement and suggests that the boundary is an α-δ interface [16]. The α component is due to lattice translation at the boundary, while the δ component is due to lattice rotation resulting from the twinning.

In contrast to the fully oxygenated twin boundaries, oxygen-deficient ones (seen mostly in samples with $0.1 < \delta < 0.5$) exhibit no lattice displacement at the boundary. The boundaries are usually broad and do not show strong contrast in HREM images (Figure 5.3b). Fringe-contrast analysis shows that the boundaries are pure δ-interfaces. These observations imply that there are two different types of twin boundaries: fully oxygenated ones and oxygen-deficient ones. For stoichiometric $YBa_2Cu_3O_{7-\delta}$ ($\delta \approx 0$), the twin boundaries appear to be centered on the (110) planes through the O atoms in the CuO layer accompanied by a displacement of the cation sublattice along the boundaries (Figure 5.4a); for oxygen-reduced $YBa_2Cu_3O_{7-\delta}$, the twin boundaries are centered through the cations, and the cation sublattice displacement is thus reduced or eliminated (Figure 5.4b). *In situ* experiments showed that the structure of these two types of twin boundaries can transform from one into the other. Such

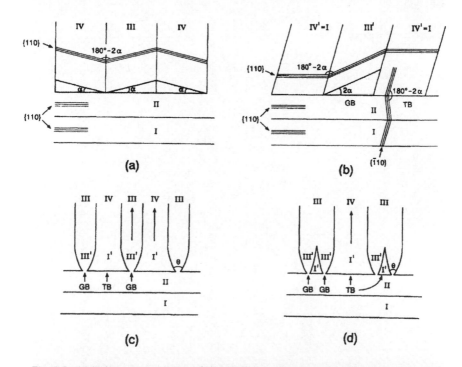

(a)

(b)

(c)

(d)

Fig. 5.5. (*a*) Relative orientation of the (110) invariant twinning planes for four twin variants. If there is no constraint, both impinging twins IV and III have a angular mismatch a with twin II. (*b*) If IV' and II become twinned by increasing the angular mismatch from a to 2a between III' and II, only III'-II has a mismatch. The symbol ' denotes the regions near the interface where two orthogonal sets of twins meet. (*c*) The gradual change of the orientation of the impinging twins is indicated by arrows. The interface consists of twin boundaries and grain boundaries. (*d*) Splitting of the tapered twins. Note that the additional twin variants (I') caused by the splitting have a twinning relationship with III' and II.

transition can be accomplished by the motion of the twinning dislocation and the associated twinning steps [17]. A review of the structural details of twin boundaries in YBa$_2$Cu$_3$O$_{7-\delta}$ can be found in [18].

5.1.1.2 Twin Tip

Twinning on the (110) and the equivalent (-110) planes in YBa$_2$Cu$_3$O$_{7-\delta}$ is frequently encountered within one crystal [19,20], as seen in the upper-left grain (denoted as twin A and twin B) in Figure 5.1. When two orthogonal twins meet, every second twin belonging to the set of impinging twins (twin

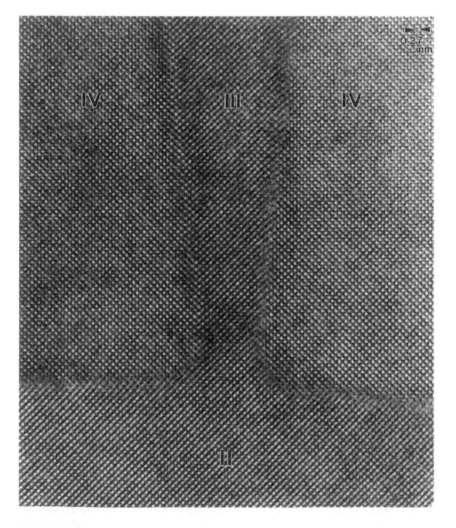

Fig. 5.6. A high-resolution image of a twin tip showing the termination of the twin. Note that the lattice in twin variant III does not align with that in II.

A) usually has a wedge shape but often does not taper down to zero width (inset of Figure 5.1).

In the region where the two sets of twin meet, there are four orientations (denoted as I, II, III, and IV in Figure 5.5a and Figure 5.5b), because the (110) and (-110) planes are the common invariant twinning plane for I, II and III, IV, respectively. If there is no constraint, a lattice mismatch will occur for both variants of the impinging twins at the interface of IV-II and III-II (Figure 5.5a).

Figure 5.6 shows an atomic image of such a wedge-shaped twin. There are five interfaces. Among them IV-III, III-IV, IV-II, and II-IV are twin boundaries; across these boundaries, the 1.87° twinning rotation can be clearly seen. The boundary between III and II is the end of the twin tip, approximately 5 nm wide. Based on the twinning geometry, the orientation of twin III and twin II should be equivalent. Nevertheless, when viewed along a prominent row of dots, at the region near the tip, the lattice of twin III does not align with that of twin II; there is about a 2° rotation.

Electron diffraction studies confirmed such misalignment [21]. By gradually moving an electron beam away from the twin tip to the region where the twin boundaries become parallel, a gradual change of rotation of the tapered twin (III′) from 2α to α can be seen. Further, twins IV′, which were not tapered, have the same crystallographic orientation as twin I (where ′ denotes the local orientation of the impinging twins near the interface where the orthogonal sets of twins meet) (Figure 5.5b–c). This configuration results in a zero angular mismatch for a twin boundary between twin IV′ and twin II, marked in Figure 5.5c as TB, rather than α as in Figure 5.5a. On the other hand, the angular mismatch increased from α to 2α for the other twins (III′-II) to retain the twinning rotation of 180°-2α between twin IV′ and twin III′. Thus, the interface of the orthogonally oriented twins comprises an alternation of twin boundaries and 2° small-angle grain boundaries. The description of twin-corner disclination can be used to explain the structureless grain boundary [22].

The shape of the tapered twin can be calculated by using a dislocation model based on the mechanical equilibrium distribution of the twinning dislocation in a double pile-up [23,24]. Such calculations show that the twin tip with a larger wedge angle is energetically unfavorable. If the wedge angle exceeds a certain limit (estimated about 15° for $YBa_2Cu_3O_7$) [21], the twin tip splits (Figure 5.5d). When the new twin variant grows away from the tip, it reduces the wedge angle of the tapered twin at the expense of increasing the length of the twin boundary. The equilibrium of the constraint at a large slope twin tip and the interfacial energy of the twin boundary and of the ~2° grain boundary determine the splitting and the final shape of the twin tip.

5.1.2 Oxygen/Vacancy Ordering

Oxygen ordering in the CuO basal plane (or chain plane), containing the -O-Cu-O- chains along the b-direction, is responsible for the tetragonal-to-orthorhombic phase transition [25–29] and hence for the formation of twins in $YBa_2Cu_3O_{7-\delta}$. Oxygen ordering also gives rise to ordered superstructures (oxygen/vacancy ordering) in oxygen-deficient samples [30–

33]. The characteristic feature of the oxygen/vacancy ordering is that the oxygen vacancies form long vacancy chains (-V-Cu-V-) along the b-direction (V-represents a vacant oxygen site in the -O-Cu-O- chain) rather than adopting a random distribution to form a superstructure. For example, alternating chains of filled and vacant oxygen sites give a superstructure with a wave vector $q = [1/2,0,0]$ at YBa$_2$Cu$_3$O$_{6.5}$, while the ordered removal of every fourth oxygen atom (or oxygen vacancy) from every other Cu-O chain gives a superstructure with a wave vector $q = [\pm1/4, \pm1/4, 0]$ for O$_{6.85}$ or O$_{6.15}$ [34]. In the latter case, the ordering scheme is the same in O$_{6.85}$ and O$_{6.15}$, but the roles of the oxygen and vacancies are reversed. Other ordered structures such as $q = [2/5,0,0]$ and $q = [1/3,0,0]$ are also observed [35,36]. Although oxygen itself is usually difficult to detect, oxygen ordering on a microscale can be seen in the diffraction pattern as well as in a dark-field image by TEM. Figure 5.7 shows an example of a 2a ordering. Figure 5.7a is a dark-field image using the superlattice reflection (1/2,0,0) marked by arrow I. The islands of white contrast in twin lamellae I, corresponding to the superlattice reflection, suggest the existence of a 2a (-O-V-O-V-) ordering along the [100] direction. The superlattice reflection marked by arrow II corresponds to the 2a ordering in twin lamellae II. In order to observe the ordering in twin lamella II, the superlattice reflection II must be used.

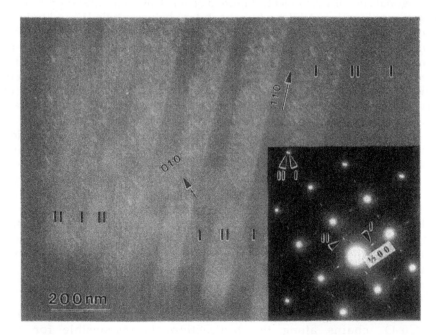

Fig. 5.7. 2a oxygen ordering in a nominal YBa$_2$Cu$_3$O$_{6.7}$ sample: (a) A dark-field image formed by the (1/2,0,0) superlattice spot (arrow I in (b)); (b) (001)* zone diffraction pattern. The (1/2,0,0) superlattice spot corresponding the 2a ordering in domain II is marked by arrow II.

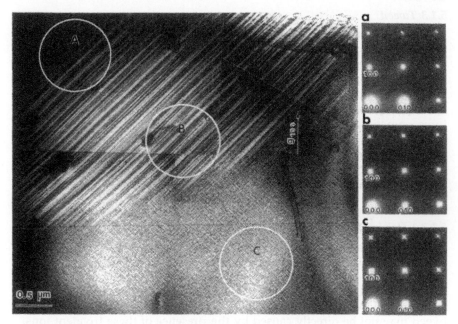

Fig. 5.8. Coexistence of twin and tweed resulting from the inhomogeneous concentration of dopant Fe in a nominal $YBa_2(Cu_{0.98}Fe_{0.024})_3O_{7+\delta}$ sample. (a), (b), and (c) show selected area diffraction of (001)* zone from the area of A, B, and C, respectively. Typical tweed image and the associated diffuse scattering are shown in region C and diffraction (c), respectively.

Analysis of electron diffuse scattering from the oxygen short-range ordering in $YBa_2Cu_3O_{7-\delta}$ ($0.1 < \delta < 0.8$) provides strong evidence that the material contains small domains (Magneli-like units [28]) with alternating planes of occupied and vacant chain sites along the [100] direction [37]. The extent of ordering in the (100) plane is > 5 nm, while the width of the ordering domain in the [100] direction is a function of δ, reaching the maximum (~10 nm) near $\delta \approx 0.4$. The dependence of the domain size on oxygen substoichiometry can be qualitatively explained by using a simple model involving only the local rearrangement of oxygen; results agree with the theoretical work [26–28] on this system.

For samples with a near-stoichiometric oxygen content ($\delta < 0.1$), oxygen or vacancy ordering does not usually generate a superstructure, although local ordering or clustering may still exist. Nanoscale oxygen fluctuation was observed in nominal $YBa_2Cu_3O_{7-\delta}$ ($\delta \approx 0$) by using high-resolution electron energy-loss spectroscopy (EELS) [38,39]. The observation confirmed the suggestion that, based on the observed anomalous double-peaked magnetization curve and intragrain weak-links, the possible presence of nanoscale oxygen-deficient regions can act as significant pinning centers in stoichiometric single crystals [40].

5.1.3 Tweed

Local oxygen ordering also occurs when a small fraction of the Cu atoms in the basal plane is replaced by certain trivalent cations. In $YBa_2(Cu_{1-x}M_x)_3O_{7-\delta}$ (M = Fe [41,42], Co [43,44], or Al [45,46]), when x < 0.02, the substitution causes the reduction of twin spacing (Figure 5.8, regions A and B). For x > 0.025, it generates a homogeneous structural modulation with an overall tetragonal symmetry of the Bravais lattice (Figure 5.8, region C). The modulated structures, sometimes called *tweed* because of their appearance in TEM images, have a roughly periodic domain contrast associated with streaks of diffuse scattering as seen in electron diffraction patterns (Figure 5.8c). Although substitution did not enhance flux pinning as originally expected, the oxygen-ordering–related structural modulation has attracted much attention [47–52].

5.1.3.1 Origin of the Tweed

Similar tweedy modulations have been encountered in some binary alloy systems that exhibit statistical fluctuations in their composition (or in an order parameter) or undergo a phase transformation with the reduction of the crystal symmetry. The cause of the tweed in those systems has been debated for many years [53–56]. In $YBa_2(Cu_{1-x}M_x)_3O_{7-\delta}$, several observations suggest that tweed may not be directly caused by the presence of these trivalent ions.

First, tweed was also observed in oxygen-reduced (M = Cu, δ = 0.6–0.8) but unsubstituted samples [57,58] (see Figures 5.9a, b). Second, the extra positive charge that results from replacing a divalent or monovalent Cu ion with trivalent ions, the dopant M, may attract an additional O^- ion to form a short Cu-O chain along the a-axis, thus forming a local "cross-links" around the dopant (Figure 5.10).Therefore, the overall oxygen content in $YBa_2(Cu_{1-x}M_x)_3O_{7+\delta}$ is more than seven; every two trivalent ions incorporate approximately one oxygen [41,59]. Third, oxygen reduction and low-temperature reoxidation experiments of substituted samples show that, in spite of a negligible net change in oxygen content and change in dopant distribution, tweed appearence can be changed substantially [60]. The disorder of the oxygen atoms occupying the "unpermitted *a* sites" nearest to the dopant changes the local symmetry of the lattice. It is the *oxygen configuration* in the basal plane that is responsible for the formation of the tweed. The "interstitial" oxygen atoms in the CuO planes cause the lattice to expand in the direction of the Cu-O chain and to contract perpendicular to the Cu-O chain, resulting in a displacive modulation along the <110> directions. Furthermore, the correlation of the oxygen ordering between the CuO planes generates a modulation along the [001] axis. The strain-induced, long-range oxygen-oxygen interactions and the nearest-neighbor dopant-oxygen attraction can account for the origin of tweed in the $YBa_2Cu_3O_{7-\delta}$ system [61,62].

5.1.3.2 <110> Structural Modulation

For x < 0.1, the modulation can be approximated as two dimensional along the [110] and the equivalent [-110] directions. The period of the modulation is inversely proportion to the dopant concentration x (Figure 5.9c-f), being

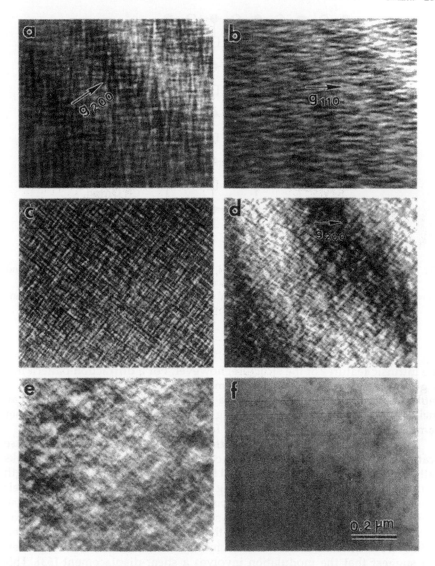

Fig. 5.9. (a)–(b): Tweed contrast from YBa$_2$Cu$_3$O$_{6.23}$ imaged with (a) **g** = 200, (b) **g** = 110. Note that in (b) only one set of tweed is visible, the other set perpendicular to the **g** is out of contrast; (c)–(f): tweed contrast for different dopant concentration x in YBa$_2$(Cu$_{1-x}$Fe$_x$)$_3$O$_{7+δ}$: (c) x = 0.03, (d) x = 0.05, (e) x = 0.10, (f) x = 0.15, imaged with **g** = 200.

approximately $p = 5.6 + x^{-1}$ (0.025 < x < 0.33). When x > 0.1, the contrast is too weak to be visible (Figure 5.9f); however, the structural modulations can be recognized by the associated diffuse scattering. Figure 5.8(c) (also Figure 5.14a) shows a typical tweed (001)* SAD pattern with cross-shaped diffuse scatter-

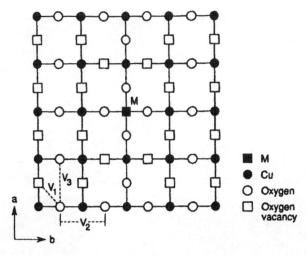

Fig. 5.10. Structure of basal plane of YBa$_2$(Cu$_{1-x}$M$_x$)$_3$O$_{7+\delta}$ with single M dopant. V$_1$, V$_2$, and V$_3$ are interaction parameters between nearest-neighbor oxygen atoms, next-nearest-neighbor oxygen atoms bridged by Cu, and next-nearest-neighbor oxygen atoms not bridged by Cu, respectively.

ing superimposed on every Bragg spot. Not all these reciprocal streaks reflect the intrinsic features of the structure, however. For a diffraction pattern in a high-symmetry Laue zone, we must consider the possibility of multiple-scattering effects, whereby a diffracted electron beam acts as an incident beam. Such multiple scattering can be minimized by using a thinner sample or by tilting the sample away from the zone axis.

After eliminating the multiple-scattering effect, the shape of the diffuse scattering corresponding to the tweed consists of two separate sets of streaks rather than a cross pattern, as sketched in Figures 5.11a and 5.11b. The length of the streaks increases with increasing distance from the origin, while their intensity falls off. For an ($h00$) reflection, two sets of equivalent streaks exist, while for an ($hh0$) spot, only one set of streaks exists. For an ($hk0$) ($h \neq k$) spot, two sets of asymmetric streaks exist; their relative lengths depend on their indices (see Figures 5.11 and 5.13). Such characteristics of the diffuse scattering suggest that the modulation involves a shear-displacement [63]. This observation is consistent with the image analysis, as the tweed contrast obeys the $\mathbf{g} \cdot \mathbf{R} = 0$ extinction rule under two-beam imaging conditions [64]. When the reflection of $\mathbf{g} = 110$ is used to form the image, the (-110) set of tweed image is out of contrast (Figure 5.9b). For a typical dopant concentration of $x = 0.03$, the average period (quasi-periodic) of the <110> modulation is about 40 Å, while for $x = 0.33$, the average period is 8–9 Å.

In order to shed light on the role of the dopants and oxygen ordering in structural modulation, the streaks of the diffuse scattering for $x = 0.03$ were simulated using a concentration wave/displacement wave approach. As discussed before, the "cross-links" and the local oxygen ordering in the CuO

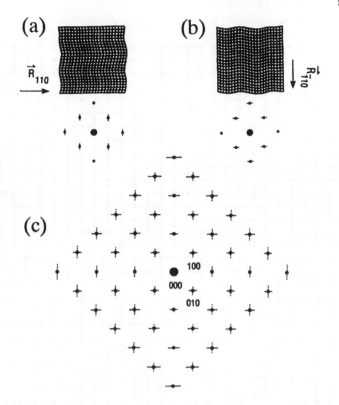

Fig. 5.11. A schematic drawing of the tweed domain and its (001)* diffraction pattern. The modulated lattices formed by (*a*) [110] shear and (*b*) [-110] shear are shown, together with the corresponding diffraction patterns. Note that there are no radial streaks; all the streaks are perpendicular to the invariant plane. Superimposition of the diffraction patterns of (*a*) (dashed line) and (*b*) (solid line) is depicted in (*c*). The tweed can be considered as two separate overlapping displacement waves, each having a set of streaks of diffuse scattering.

plane are essential in determining the lattice displacement of the crystal; therefore, low-energy configurations of the CuO plane were generated by using Monte Carlo simulations with a random distribution of dopant.

An example is given in Figure 5.12, which exhibits oxygen-ordered domains [65]. The detailed lattice distortion associated with that configuration then was derived using a linear elastic approximation (see [65] for detail). From a knowledge of the displacement, the intensity of the diffuse scattering (Huang scattering) finally was calculated with and without the addition of thermal diffuse scattering.

In Figure 5.13, the computed contour plots of diffuse scattering intensity around the (040), (240), and (440) reflections are compared with the observed diffuse scattering in $YBa_2(Cu_{0.97}Cu_{0.03})_3O_7$. Figure 5.13a–c shows the calculated

◆ Fe   ◇ Cu   ○ O

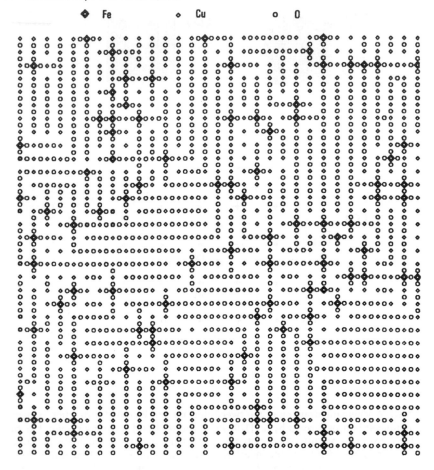

Fig. 5.12. A configuration of the basal plane for YBa$_2$(Cu$_{0.97}$Fe$_{0.03}$)$_3$O$_7$ obtained by a Monte Carlo simulation of 32 x 32 lattice with periodic boundary conditions at 300 K. Fe atoms, shown as large diamonds, are randomly distributed. The small diamonds and circles indicate the positions of Cu and O, respectively.

intensity caused by lattice distortion only, and Figure 5.13d–f shows the calculated total diffuse scattering intensity due to lattice distortion and thermal vibration. Figure 5.13(g)–(i) are the TEM observations at room temperature. Although displacement alone can count for streaking of the Bragg reflection, the good match of the calculated intensity profiles in Figure 5.13d–f with those in Figure 5.13g–i suggests that, at room temperature, thermal diffuse scattering also makes a significant contribution to the total intensity of diffuse scattering. The results clearly indicate that lattice distortion due to local oxygen ordering is responsible for the streaks of the diffuse scattering, as schematically shown in Figure 5.11.

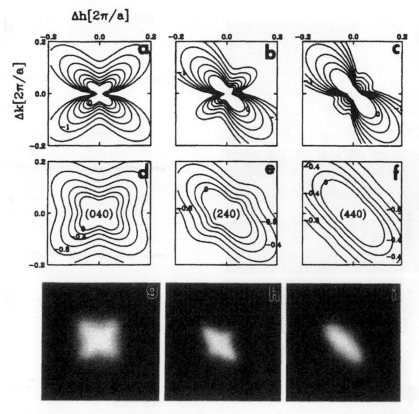

Δh[2π/a]

Δk[2π/a]

Fig. 5.13. Calculated and observed diffuse scattering around (040), (240), and (440) Bragg spots: (a)–(c) due to Huang scattering (lattice displacement); (d)–(f) the combined Huang scattering and thermal diffuse scattering; and (g)–(i) TEM observations.

5.1.3.3 3-D Structural Modulation

The feature of diffuse scattering is virtually three dimensional, although for $x < 0.1$ the [001] component is insignificant. The shape of the reciprocal spots can be determined by diffraction patterns of several different projections (Figure 5.14a–c). By analyzing the geometry of the intersection of the Ewald sphere and the reciprocal lattice (Figure 5.14d–f), it was found that the Bragg spot for $0.025 < x < 0.05$ consists of paddle-shape wings of diffuse scattering (Figure 5.14g). The edges of the wings do not point exactly in the [110]* direction but also have a small component along the c^*-axis. Measurement of the expansion of the wing along the c^*-axis suggests that the correlation length perpendicular to the a-b plane is about 400–500 Å. Studies of nanodiffraction using conventional TEM and scanning TEM with an annular dark-field detector [66] provide strong evidence of such quasi-periodic modulation, suggesting the existence of overlapped domains with a small change

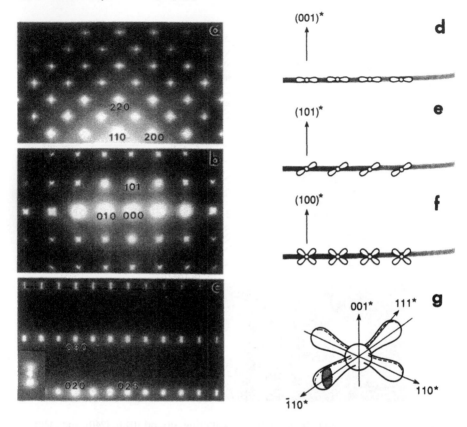

Figure 5.14. Selected area diffraction showing diffuse streaks superimposed on the fundamental reflections for x = 0.04: (*a*) (001)* zone. Note that the radial streaks are due to the double diffraction, and the nonradial streaks reflect the structural modulation. (*b*) (-101)* zone and (*c*) (100)* zone. Sketches of the intersection of the Ewald sphere (shaded curve) and the reciprocal lattice of (001)* zone (*d*), (101)* zone (*e*), (100)* zone (*f*) are included. (*g*) The shape of an (h00) reciprocal spot for x = 0.03–0.05.

in the crystal orientation, stacking sequence, or lattice parameters along the c-axis.

The structural modulation along the [001] direction is much more apparent in heavily doped YBa$_2$Cu$_3$O$_{7-\delta}$. When x reaches 0.10, the wings of diffuse scattering are replaced by satellite spots, or intensity maxima in (001)* projection, indicating an enhanced periodicity in the modulation. With a further increase in x, the satellite spots split along the c*-axis.

Figure 5.15a shows the diffraction pattern of the (001)* zone for x = 0.33; there are no distinct streaks or intensity maxima around the fundamental reflection in the (001)* zone. Both features are clearly visible when the crystal is tilted away from the zone axis, however, especially when the (101)* or

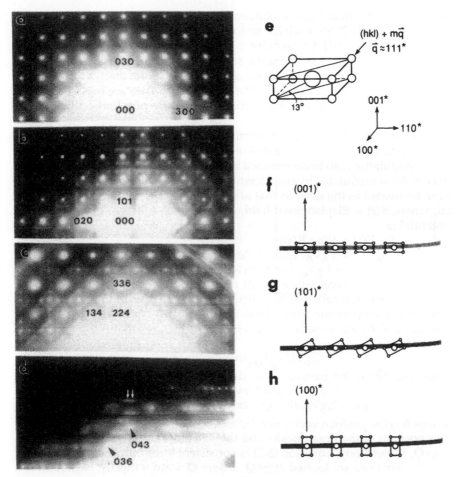

Fig. 5.15. Selected area diffraction showing satellite spots surrounding the fundamental reflections for x = 0.33: (a) (001)* zone, (b) (101)* zone, (c) (111)* zone, and (d) 10° away from the (100)* zone. A sketch of an (h00) reciprocal spot (e), where m is a fraction number (1/6 - 1/3) varying with the dopant concentration. (f)–(h): the intersection of the Ewald sphere (shaded curve) and the reciprocal lattice of (001)* zone (f), (101)* zone (g), and (100)* zone (h). Note that the satellite spots are invisible in the exact (001)* and (100)* orientations, but visible when the crystal is tilted away from these orientations.

(111)* orientation is parallel to the electron beam (Figures 5.15b and c). Figure 5.15d shows a diffraction pattern about 10° away from the (100)* zone. Figure 5.15f–h present sketches of the intersection of the Ewald sphere and the reciprocal lattice of (001)*, (101)*, and (100)*, respectively. The satellites are located above or below the Bragg spot (Figure 5.15e) and tend to point in the [11l]* directions (where l ≈ 1) or are located at hkl + mq, where mq is the superstructure diffraction around the fundamental reflection, ($mq \approx 1/6[11l]^*$ for $x =$ 0.15, $mq \approx 1/4[11l]^*$ for $x = 0.22$, and $mq \approx 1/3[11l]^*$ for $x = 0.33$). The periodicity along the [001] direction for $x = 0.33$ is about a c-lattice constant (by measuring the distance between the satellites along the c^*-axis, as indicated by arrow pairs in Figure 5.15d). This agrees with the fact that, at this concentration, almost all the Cu atoms in the CuO basal planes are replaced by the dopant, resulting in a periodic disorder on the scale of a unit-cell along the c-axis.

The satellites of the diffuse scattering also can be attributed to displacive modulation. The position of the satellites, which represents the reciprocal lattice of the modulation, can be determined by the kinematic diffraction theory [67]. If a crystal is modulated by displacement, the lattice potential at the position **r** can be treated as the same as that at position **r-R(r)** in a unmodulated crystal, where **R(r)** is displacement field, a continuous function of **r**. The lattice potential is

$$V(\mathbf{r}) = V(\mathbf{r} - \mathbf{R(r)})$$
$$= \Sigma g V g \exp(2\pi i g \cdot \mathbf{r}) \exp(-2\pi i g \cdot \mathbf{R(r)})$$
$$\approx \Sigma g V g \exp(2\pi i g \cdot \mathbf{r}) - 2\pi i \Sigma(g \cdot \mathbf{R(r)}) V g \exp(2\pi i g \cdot \mathbf{r}), \qquad (5.1)$$

where **g** is a reciprocal lattice vector of unmodulated crystal and **R(r)** is assumed to remain small, that is, $|\mathbf{R(r)}| << |\mathbf{g}|^{-1}$. Assuming that the displacements are periodic with a large periodicity, the product $\mathbf{g} \cdot \mathbf{R(r)}$ may be expanded as a Fourier series with coefficients $T_{Q,g}$, so that the final term of (1) becomes

$$\Sigma g \Sigma Q V_g T_{Q,g} \exp(2\pi i(g+Q) \cdot \mathbf{r}).$$

Then, the diffraction pattern of the displacively modulated crystal is given by a Fourier transform of (1), as

$$A(\mathbf{h}) = \Sigma g V g \delta(\mathbf{h} - \mathbf{g}) - 2\pi i \Sigma g \Sigma Q V g T_{Q,g} \delta(\mathbf{h} - \mathbf{g} - \mathbf{Q}), \qquad (5.2)$$

where **h** is the position vector in reciprocal space, and $\delta(x)$ is a delta function. The position of the Bragg peaks and the satellites are given by $\mathbf{h} = \mathbf{g}$ and $\mathbf{h} = \mathbf{g+Q}$, respectively. Equation (5.2) is consistent with our TEM observations that the satellites are located at **g+Q**, where $\mathbf{Q} = mq \approx m[11l]^*$ (where m is a variable of dopant concentration). The periodicity of the modulation is inversely proportional to the value of $|\mathbf{Q}|$. The intensity of the satellites at **g+Q** are proportional to $|V_g|^2 |T_{Q,g}|^2$.

The coefficients $T_{Q,g}$ have the form of $T_{Q,g} = Ae \cdot g$, where A and **e** describe the amplitude and the polarization of the displacement wave with wave vector **Q**. The satellites become extinct for those reflections for which the reciprocal vector is perpendicular to the polarization vector of the wave (i.e., $\mathbf{g} \cdot \mathbf{e} = 0$ (see Figure 5.16). In cases where the modulation is sinusoidal, two strong satellites per Bragg reflection will be visible. If the modulation is quasi-periodic, or near nonperiodic, the satellites broaden so that their intensity is smeared out over a larger area; if the intensity is redistributed throughout the entire Brillouin zone, the satellites disappear.

5.1.3.4 Simulation of Tweed Image Contrast

Tweedy domain-images are best seen for $x = 0.03$, viewing along the (001) projection. Note that there exists a quasi-periodic modulation along the [001] direction. Several domains stacking on top of each other contribute to the contrast when the incident electron beam is nearly parallel to the c-axis in two-beam imaging. The amplitude of the scattered wave, Φ_g, at the bottom surface of the crystal is [63]

$$\Phi_g = i\pi/(\xi_g)_0 \int^t e^{-2\pi i s z} \, dz, \tag{5.3}$$

where $\mathbf{R}$ is the displacement of the lattice planes relative to each other in the stacked domains, ξ_g is the extinction distance, t is the total thickness of the foil, $\mathbf{g}$ is the diffraction vector used for imaging, z is the distance from the center of the foil to the fault measured along the direction of the incident beam, and s is the deviation from the Bragg position. When s is sufficiently large, the so-called kinematical calculation gives the same result as two-beam dynamical calculations. The intensity of the reflected beam (i.e., the dark-field image) can be expressed by [64]

$$I = 1/[\xi_g s^2] \, [\sin^2(\pi t s + \pi \mathbf{g} \bullet \mathbf{R}) + \sin^2(\pi \mathbf{g} \bullet \mathbf{R})$$
$$-2\sin(\pi \mathbf{g} \bullet \mathbf{R}) \sin(\pi t s + \pi \mathbf{g} \bullet \mathbf{R}) \cos 2\pi s z]. \tag{5.4}$$

Figure 5.17 shows examples of the contrast calculations ($\mathbf{g} = 110$). For $\mathbf{R} = 0$ (i.e., a perfect crystal of constant thickness), there is no contrast (Figure 5.17a), while for $\mathbf{R} = $ constant (i.e., a stacking fault with displacement vector $\mathbf{R}$ running from top to bottom in a foil of constant thickness), we see the

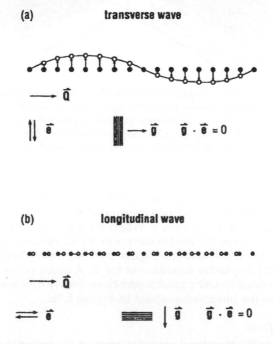

Fig. 5.16. Relationship of the displacement wave vector, **Q**, polarization vector **e**, and the corresponding reciprocal lattice planes, **g**, which cause the satellites (or diffuse scattering) to become extinct when **g.e** = 0: (a) transverse wave and (b) longitudinal wave.

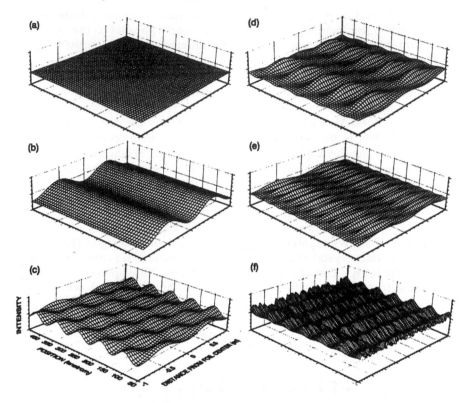

Fig. 5.17. Intensity profiles of two-beam dark field images from domain interface with displacement, R, when interface runs from top to bottom: *(a)* |**R**| = 0; *(b)* |**R**| = constant; *(c)*-*(e)* |**R**| = f(sinx), for different periods, p, and different deviations from Bragg position, s: *(c)* p = 8 nm, s = 3.25/t; *(d)* p = 8 nm, s = 3.5/t; *(e)* p = 4 nm, s = 3.25/t; and *(f)* |**R**| = f_1(sinx)+f_2(sinx), s = 1/t.

oscillation of contrast with depth of the fault (Figure 5.17b). When **R** varies as a function of position x, we obtain the modulated contrast shown in Figure 5.17c–d with different deviations, s. When the period of the modulation is reduced, the intensity of the modulation is reduced, consistent with the TEM observations. In these calculations (Figure 5.17c–e), a simple sinusoidal wave is used as a [110] displacive modulation for **R**. A quasi-periodic tweed contrast can be generated by using more sophisticated wave functions for **R** (Figure 5.17f), similar to the observed contrast in Figure 5.9b.

5.1.4 Other Defects

In addition to the structural modulations just discussed, a variety of conventional crystal defects can occur in YBa$_2$Cu$_3$O$_{7-\delta}$. These include dislocations, stack faults, and inclusions.

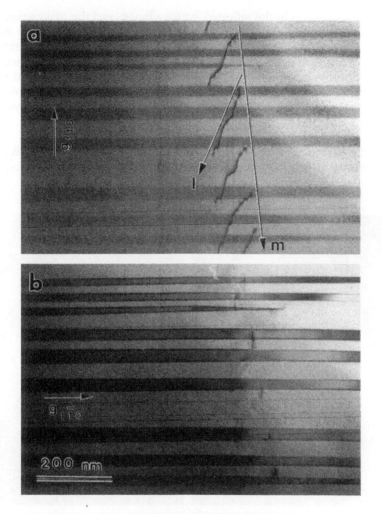

Fig. 5.18. [110] dislocations imaged with (*a*) **g** = 100, (*b*) **g** = 1-10. Note that when **g·b** = 0, the dislocations are out of contrast; however, the twinning contrast is still visible. The dislocation slip plane can be determined using the projected dislocation line direction **l** and the reference direction **m** under two different beam directions, **B**₁ and **B**₂.

5.1.4.1 *Dislocations*

The commonly observed dislocations in $YBa_2Cu_3O_{7-\delta}$ are perfect dislocations with a [100], [010], or [110] Burgers vector [68–70]. An example of [110] dislocations is shown in Figure 5.18a.

Although $YBa_2Cu_3O_{7-\delta}$ is highly anisotropic, the characteristics of the dislocations can be determined unambiguously by using the **g·b** criterion, especially when a high-order reflection is used. When **g·b** = 0, the Burgers vector

b of a dislocation is in the reflected lattice planes, and the dislocation is out of contrast (Figure 5.18b). By characterizing the line direction of the dislocations **ξ** (note, not the line projection **l**), we can learn whether the dislocation has a character of edge or screw, or a mixture of the two. For deformation-induced dislocations pileup, the slip plane normal **n** can be assessed by adding an additional reference direction **m**, and recording the dislocations under two different beam directions **B$_1$** and **B$_2$**. This process gives

$$\xi = (\mathbf{B}_1 \times \mathbf{I}_1) \times (\mathbf{B}_2 \times \mathbf{I}_2), \tag{5.5}$$

$$\mathbf{n} = [(\mathbf{B}_1 \times \mathbf{I}_1) \times (\mathbf{B}_2 \times \mathbf{I}_2)] \times [(\mathbf{B}_1 \times \mathbf{m}_1) \times (\mathbf{B}_2 \times \mathbf{m}_2)], \tag{5.6}$$

where the subscripts denote the corresponding beam directions, **B$_1$** or **B$_2$**. Using Equations (5.5) and (5.6), we can accurately determine the crystallographic

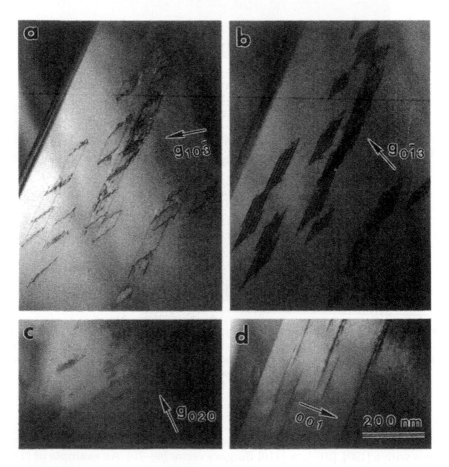

Fig. 5.19. Stacking faults with a displacement vector **R** = 1/6[301] and a pair of 1/2[100] partial dislocations: (a): **g·R** = 0; faults are invisible. (b): **g·b** = 0; dislocations are invisible. (c): **g·R** = **g·b** = 0; both dislocations and faults are invisible. (d): Fault planes are seen edge-on.

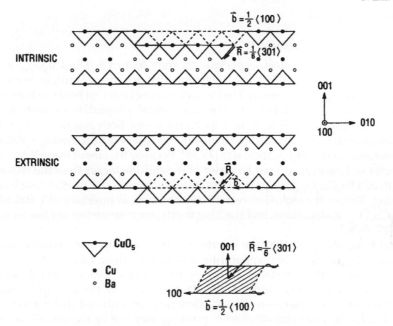

Fig. 5.20. Sketches of the intrinsic and extrinsic stacking faults formed through the shear arrangement of CuO_5. The triangles represent the CuO_5 octahedron, viewing along the [100]direction. Both chemical faults have a displacement vector $\mathbf{R} = 1/6[301]$, plane normal [001], and are bonded by two sessile dislocations with Burgers vector of $1/2[100]$.

orientations of a line and a plane. The slip plane in $YBa_2Cu_3O_{7-\delta}$ is always found to be (001), that is, the a-b plane.

5.1.4.2 Stacking Faults

Stacking faults lying in the (001) plane and bounded by a pair of partial dislocations are frequently encountered in $YBa_2Cu_3O_{7-\delta}$ [71–73]. These faults are usually wide, about a hundred nanometers, indicating low fault energy. Using $\mathbf{g \cdot b} = 0$ and $\mathbf{g \cdot R} = 0$ extinction rules, we can determine the Burgers vector $\mathbf{b}$ of the dislocations and the displacement vector $\mathbf{R}$ of the fault, respectively, as shown in Figure 5.19a and 5.19b. In $YBa_2Cu_3O_{7-\delta}$, most of the stacking faults are the $\mathbf{R}=1/6[301]$ and $\mathbf{b}=1/2<100>$ type, as sketched in Figure 5.20. The $1/6[301]$ displacement can result from an edge-shear arrangement of the Cu-O_5 truncated octahedron. Such faults are nonconservative, that is, cannot be generated by the motion of dislocations. They are chemical faults, mainly due to one or two extra CuO planes (extrinsic fault, $YBa_2Cu_4O_8$ or $YBa_2Cu_5O_9$; also see Figure 5.24), although faults due to the lack of a CuO plane (intrinsic fault) also were observed [74,75]. The intrinsic or extrinsic character of the faults can be determined by the characteristics of the fault fringe contrast [76].

5.1.4.3 Y₂BaCuO₅ Inclusions

Another type of defect, especially in melt-textured $YBa_2Cu_3O_{7-\delta}$, is the randomly distributed Y_2BaCuO_5 inclusion due to an incomplete peritectic reaction during crystal growth. The inclusions are of interest because they have been shown to be beneficial to the mechanical properties as well as the current-carrying capability of the superconductors. Both magnetic and transport measurements have indicated that the J_c increases with increasing Y_2BaCuO_5 content and decreasing particle size [77,78]. Because the size of these Y_2BaCuO_5 particles is, however, at least an order of magnitude larger than the coherent length of $YBa_2Cu_3O_{7-\delta}$, the particles themselves would not be effective in flux pinning. Rather, the inclusion-related defects, such as interface of $Y_2BaCuO_5/YBa_2Cu_3O_{7-\delta}$, dislocations, and stacking faults, are responsible for the increase in J_c [77,79,80].

We have observed that the interface acts as a source of dislocations, as shown in Figure 5.21 [38], apparently due to the lattice mismatch at the interface. Not all the interfaces are likely to generate dislocations, however, because of the high anisotropy of the elastic coefficient both of the inclusion and the matrix. Furthermore, like the deformation-induced dislocations and stacking faults, their contribution in pinning may not be significant because these defects are confined in the *a-b* planes, and they can be beneficial to pinning only when the magnetic field is parallel to the *a-b* planes.

5.1.5 Structural Defects Induced by Heavy-Ion Irradiation

Although numerous efforts have been made to create randomly distributed structural defects through synthesis processes, the effectiveness of such defects in pinning flux lines has been disappointing. Recently, as an alternative, the

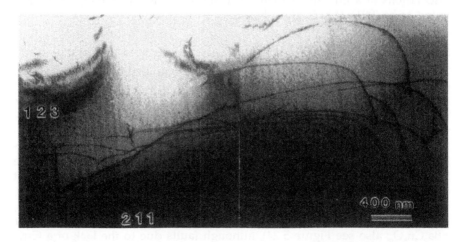

Fig. 5.21. [100] dislocations generated by a 211 particle. Here, 123 and 211 denote the $YBa_2Cu_3O_{7-\delta}$ matrix and Y_2BaCuO_5 inclusion, respectively.

creation of defects by heavy-ion (several hundred MeV) irradiation has become popular [81–83]. The resulting linear ion-tracks were found to provide strong flux pinning in temperature and field regimes where the other types of defects were ineffective [84–89]. The remainder of this section reviews recent findings on the structural features of radiation-induced defects. The investigations of such defects on flux pinning and critical current density were reported in [87–89].

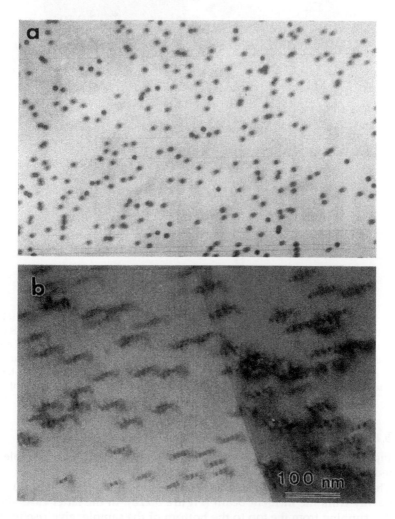

Fig. 5.22. Typical morphology of the Au^{+24} radiation-induced defects viewed (*a*) along the ion-track, and (*b*) about 18° away from the ion track.

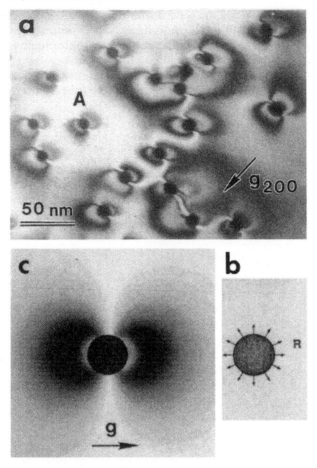

Fig. 5.23. (*a*) Strain contrast surrounding the columnar defects imaged with **g** = 200. Note that the contrast disappears in the direction perpendicular to **g**. (*b*) Sketches of the radial displacement field associated with the amorphous column. (*c*) Simulated strain contrast using dynamic diffraction theory. For E_{eff} = 2.0%, the calculated image matches the observed one with $\xi_g/\xi_0 = 0.01$, $\xi_g/\xi_g' = 0.01$, where ξ_g is the extinction distance corresponding to the reflection **g**. ξ_0' and ξ_g' are the mean absorption coefficient and anomalous absorption coefficient, respectively.

5.1.5.1 *General Features of Radiation-Induced Structural Defects*

Figure 5.22 shows typical morphology of a bulk $YBa_2Cu_3O_{7-\delta}$ sample irradiated with 300-MeV gold ions.

The view is along the direction of the incident ion-beam (Figure 5.22a) and away from the incident beam (Figure 5.22b). In the latter case, the ion tracks, running from the top to the bottom of the sample, give rise to thickness fringes as well as to contrast as a result of their intersection with the specimen surface and crystal matrix. Nanodiffraction and HREM of the damaged regions, viewed from the cross section of the ion track, shows that they are continuous columns of amorphous material. This feature is expected because the samples are extremely thin (< 0.5 μm) compared with the range of these ions in $YBa_2Cu_3O_{7-\delta}$ (> 14 μm) [90].

Careful observation shows that these columnar defects also generate lobes of contrast over an area two to three times larger than the diameter of the ion track (Figure 5.23a). Characterization with a weak-beam technique showed that the contrast surrounding the defects disappears only along the line perpendicular to the diffraction vector, **g**, and running across the center of the damaged area. The lobe contrast can be attributed to radial strain/displacement fields surrounding the amorphous columnar defect (Figure 5.23b). The strain and structural disorder of the amorphous region propagate into the crystal lattice in a direction perpendicular to the ion track.

The strain contrast around the ion tracks can be simulated by using the Howie Whelan equation [63,91]:

$$
\left(
\begin{aligned}
\frac{dA_0}{dz} &= -\frac{\pi}{\xi_0'}A_0 + \pi\left(\frac{i}{\xi_g} - \frac{1}{\xi_g'}\right)A_g \\
\frac{dA_g}{dz} &= \pi\left(\frac{i}{\xi_g} - \frac{i}{\xi_g'}\right)A_0 + \left(\frac{-\pi}{\xi_g'} + 2\pi i[s + \frac{d}{dz}(\mathbf{g}\cdot\mathbf{u})]\right)A_g
\end{aligned}
\right.
\tag{5.7}
$$

where A_0 and A_g are the transmitted and diffracted beams, respectively. ξ_g is the extinction distance corresponding to the reflection **g**. ξ_0' and ξ_g' are the mean absorption coefficient and anomalous absorption coefficient, respectively.

The strain field, **u**, can be modeled quantitatively by treating the amorphous region as a misfitting elastic cylinder in an elastic matrix. If one assumes that the matrix is approximately elastically isotropic in the *a-b* plane, and if one neglects the difference in elastic constants between the amorphous and crystalline regions, then, for an isolated amorphous track in an infinite matrix parallel to the c-axis, the linear elasticity theory [92] yields a radially symmetric displacement field u(r) in the matrix around the ion track, given by [90]

$$
u(r) = E_{eff}\frac{R_2}{r}
\tag{5.8}
$$

where R is the radius of the amorphous track and

$$
E_{eff} = \frac{(C_{11} + C_{12})E_{ab} + C_{13}E_c}{/2C_{11}},
$$

where the C_{ij} are elastic constants and E_{ab} and E_c are the fractional strains upon amorphization along the a (or b) and c directions, respectively. A detailed comparison between the calculated (Figure 5.23c) and observed con-

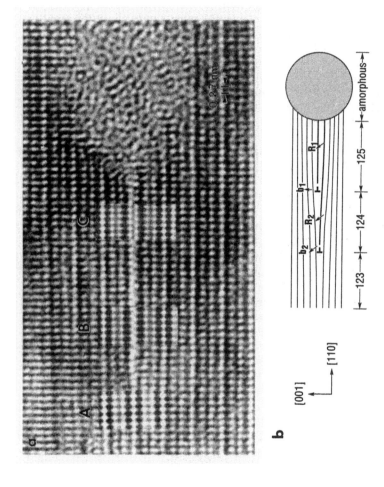

Fig. 5.24. (*a*) A mixture of YBa$_2$Cu$_4$O$_8$ (124) and YBa$_2$Cu$_5$O$_9$ (125) planar defects in a YBa$_2$Cu$_3$O$_7$ (123) matrix viewed along the [110] direction. Insets are simulated images. Inset A: the 123 matrix, inset B: 124 phase with $\mathbf{R}_2$ = 1/6[031]; and inset C: 125 phase with $\mathbf{R}_1$ = 1/6[032]. Defocus: 600 Å, specimen thickness: 20 Å. (*b*) Sketch of the observed stacking fault consisting of two partial dislocations with a Burgers vector of $\mathbf{b}_1$ = 1/6[001], and $\mathbf{b}_2$ = 1/6[031], respectively.

trast (Figure 5.23a, marked as A) for $YBa_2Cu_3O_7$ crystals yields $E_{eff} \approx 0.02$ [93].

Chemical disorder across the amorphous region of Au-irradiated samples also was analyzed by using EDX and EELS with a 20-Å probe. No distinct difference (< 3 %) was detected in the cation composition in and near the amorphous regions. For anion oxygen, however, a remarkable change occurred across the damaged area in the fine structure of the oxygen K-edge absorption spectrum, suggesting a marked change in electronic structure and, possibly, a loss of oxygen atoms from the amorphous region. More important, a drastic decrease of the hole density was observed in the crystal lattice surrounding the amorphous region compared with a region far away from the defect [90]. Such a reduction near the columnar defect enlarges the weak- or nonsuperconducting area to almost twice that of the amorphous region, about the extent of the observed strain field.

Planar defects associated with the amorphous columns often are observed in Au^{+24} and Ag^{+21} irradiated $YBa_2Cu_3O_{7-\delta}$ ($0.05 < \delta < 0.6$), especially when the electron beam is parallel to the a-b plane. The extension of the fault is about 3–5 times the radius of the amorphous column. The majority of defects are chemical faults comprising one or two extra CuO layers inserted between the double BaO layers of the original $YBa_2Cu_3O_{7-\delta}$ unit cell, similar to those observed in as-grown crystals. Figure 5.24 shows a Ag-irradiation–induced amorphous column and its associated stacking fault. HREM image simulations suggest that the planar fault starts with a local $YBa_2Cu_5O_9$ (125) structure with a displacement $R_1 = 1/6[032]$ (inset C), then changes to a $YBa_2Cu_4O_8$ (124) structure (inset B) with a fault displacement $R_2 = 1/6[031]$, and finally terminates in the $YBa_2Cu_3O_7$ (123) matrix (inset A). Thus, the planar defect consists of two stacking faults, each terminated by a sessile dislocation, $b_2 = 1/6[031]$ and $b_1 = 1/6[001]$, respectively.

5.1.5.2 Characteristics of the Radiation Damage

(i) **Ion species and its energy.** Zhu and his coworkers undertook a systematic study of the defects in thin films and bulk samples of $YBa_2Cu_3O_{7-\delta}$ created by irradiation with Au^{+24} (300 MeV), Ag^{+21} (276 MeV), Cu^{+18} (236 MeV), and Si^{+13} (182 MeV) ions. The degree of damage caused by the ions was found to vary with the rate of deposition of ion energy and the species of ion, the severity decreasing from Au, Ag, Cu, to Si. Irradiation with Au and Ag ions produces columns of amorphous material along the ion trajectories. For Au ions, the average density was 3.7×10^{10} ions/cm^2 for a dose of 3.86×10^{10} ions/cm^2. Thus, almost each Au ion produces a single columnar defect. For Ag ions, however, the measured density of the defects is only half of the original dose, showing that many Ag ions do not produce columnar defects. Amorphous columns are created only occasionally during irradiation with Cu ions, and are not induced with Si ions. The size of defects induced by Au, Ag, and Cu ions shows an approximate Gaussian distribution. The dominant diameter of the defects is 10.6 nm for Au-irradiated samples, 5.9 nm for Ag-, and 2.4 nm for Cu-irradiated samples.

(ii) **Crystallographic orientations.** At low magnification (Figure 5.22a), the defects appear as symmetrical circles along the ion trajectory in all orienta-

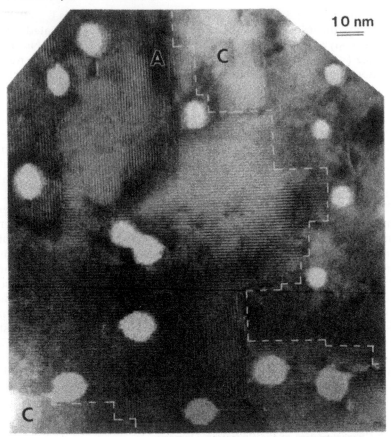

Fig. 5.25. High-resolution image of Au-irradiated YBa$_2$Cu$_3$O$_7$ thin film. The defects appear elliptical when the incident-ion is parallel to the [100] (or [010]) direction, but smaller and circular when parallel to the [001] direction. The areas denoted A and C have the a- (or b-) and c-axis parallel to the film normal, respectively.

tions. HREM, however, shows that the morphology of the defects depends on the direction of the incident ions in relation to the crystallographic axes of the material. When the ion track is parallel to the c-axis, the defects appear as circular disks. In contrast, when the ion track is parallel to the a or b axis, the defects appear somewhat larger and elliptical, and the edge of the amorphous region shows well-defined facets in the *a-b* planes.

The orientational dependence of the shape of the defect is clearly illustrated for an Au-irradiated thin film containing a mixture of grains with a and c orientations (Figure 5.25), where the areas denoted A and C have the a- (or b-) and c-axis parallel to the film normal, respectively. In region A, the amorphous column is large and elliptical, while in region C it is small and circular. Thus, an ion beam directed along the a- or b-axis causes more severe

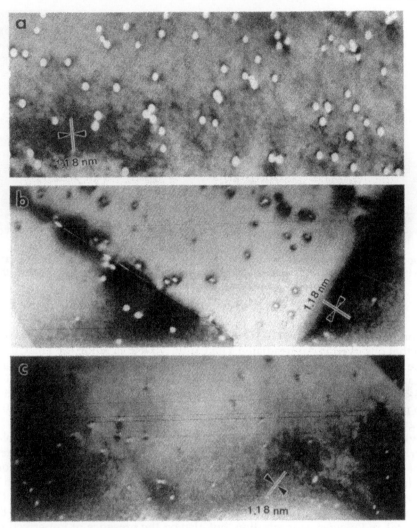

Fig. 5.26. Columnar defects induced by Ag ion radiation viewed along [100] direction: (a) $YBa_2Cu_3O_{6.3}$, (b) $YBa_2Cu_3O_{7-\delta}$, ($0 < \delta < 0.1$), and (c) $YBa_2Cu_3O_7$ (ozone-treated). In (b) and (c), the direction of the incident beam was slightly off from the [100] projection; therefore, only near the edge of the specimen are the defects end-on and appear as white contrast.

structural damage than a beam along the c-axis.

(iii) **Oxygen concentration**. The extent of radiation damage also depends remarkably on the oxygen stoichiometry of the sample, as demonstrated in Figure 5.26 for Ag-irradiated $YBa_2Cu_3O_{7-\delta}$ with (a) $\delta \approx 0.7$, (b) $\delta < 0.1$, and (c) $\delta \approx 0.0$ (oxygenated with ozone). All the micrographs show lattice images viewed

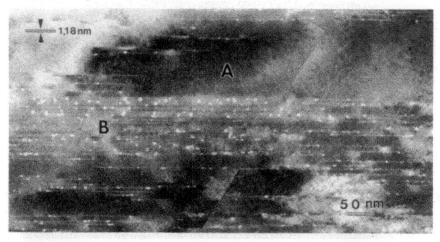

Fig. 5.27. (100) lattice image of an ozone-treated sample of $YBa_2Cu_3O_{7-\delta}$ ($\delta \approx 0$) irradiated with Cu ions. Note that the columnar defects were observed only at the location of the preexisting stacking faults.

along the [100] axis. For $YBa_2Cu_3O_{6.3}$, the average size of the defects is close to that observed in Au-irradiated $YBa_2Cu_3O_{7-\delta}$ ($\delta < 0.1$), while for $YBa_2Cu_3O_7$, the size is close to that observed in the Cu-irradiated material. The average diameter of the amorphous regions for $YBa_2Cu_3O_{6.3}$ was 7.8 nm, for $YBa_2Cu_3O_{7-\delta}$ ($\delta < 0.1$) it was 5.3 nm, and for $YBa_2Cu_3O_{7-\delta}$ ($\delta \approx 0.0$) it was 3.3 nm. It is important to note that a small change in oxygen substoichiometry (from $\delta = 0.1$ to 0.0) significantly reduced radiation damage.

(iv) Crystal imperfections. Preexisting imperfections in the crystal also play an important role in forming the columnar defects. Such an effect is not apparent for Au irradiation because each Au ion always produces a single amorphous column, regardless of the crystal. For a lighter and less energetic ion such as Cu, however, the formation of the amorphous zone is very sensitive to the characteristics of the crystal, depending not only on the orientation of the crystal and oxygen concentration, but also on the presence of imperfections in the crystal. Figure 5.27 is an image, viewing along the [100] orientation, from an area with preexisting planar defects. In contrast with preexisting defect-free areas of an ozone-treated sample irradiated with Cu, which show no amorphous columnar defects regardless of the orientation, here there is a high density of columnar defects, averaging ~2.2 nm. Furthermore, the distribution of these defects is not random. The radiation-induced defects seen in Figure 5.27 form only at the location (marked as B) of stacking faults (appearing as white lines, viewed here edge-on). In the area marked A, no stacking faults occurred, and consequently no columnar defects. We conclude that structural imperfections in the as-grown crystal enhance radiation damage, especially planar defects, such as stacking faults or grain boundaries when their plane normal is perpendicular to the incident ion beam.

5.1.5.3 Formation of the Ion Track

The difference in radiation damage can be attributed largely to differences in the thermal conductivity of the materials. A thermal-spike model was developed to provide a theoretical framework for the discussion of the variation of size and shape of the amorphous tracks under various conditions. In general, the radiation damage processes accompanying the passage of an energetic heavy ion are likely to encompass energy deposition into the electron gas of the target, the transfer of the energy from the electron gas to heat the lattice ions, the transport of heat in the lattice, and the phase changes and defect formation that accompany the rapid heating and quenching of the lattice. In the context of the modified thermal-spike model, we assume that in the zone along the ion track almost the entire region of the thermal spike where the temperature rises to or above the melting point initially becomes molten. On the other hand, the molten zone does not all remain amorphous, but some epitaxial regrowth occurs during the cooling-down period, albeit with some lattice defects remaining. Thus, the diameter of the amorphous region is smaller than that of the original molten zone. The observations of the predominant extrinsic stacking faults in heavy-ion irradiated samples [75] strongly support the likelihood of partial epitaxial regrowth of the molten zone after the passage of high-energy ions through the samples, agreeing well with calculations [90].

5.2 Intergranular Structural Defects—Grain Boundaries

The higher values of critical transport current density, J_{ct}, for single crystals of $YBa_2Cu_3O_{7-\delta}$ compared with those values for polycrystals suggest that improving superconducting characteristics of the boundaries is essential in achieving high transport current. Earlier, it was reported that the values of J_{ct} in thin films of $YBa_2Cu_3O_{7-\delta}$ are quite sensitive to the misorientation between adjacent grains separated by a grain boundary, and decrease drastically when the angle of the misorientation exceeds ~10° [3,94]. Later, however, it also was demonstrated that for polycrystalline thin films (a/b- and c-axis oriented) and bicrystals, some large-angle grain boundaries transport high currents [95–97]. Accumulated evidence shows that lattice mismatch at the boundaries contributes to variations in oxygen and cation concentrations at the boundaries [98,99], while the strain fields surrounding dislocation cores at the boundary suppress the superconducting order parameter [100,101]. These observations point out the importance of detailed structural characterization of the grain boundaries to elucidate the correlation between the structure and the properties and to overcome the major impediment to the practical applications of the cuprate superconductor.

This section first addresses the concept of the coincidence site lattice, which is widely used in grain boundary studies. Next, the discussion turns to the techniques used to accurately characterize the crystallography of arbitrary grain boundaries and to derive the boundary structure based on the O-lattice and Frank formula. To demonstrate the procedure, we provide an example. Recent studies on grain boundary oxygen are described, including the use of electron energy loss spectroscopy, the correlations between grain boundary oxygen and boundary misorientations, and the boundary phase transition during an oxygenation process. The chapter closes with a discussion of the

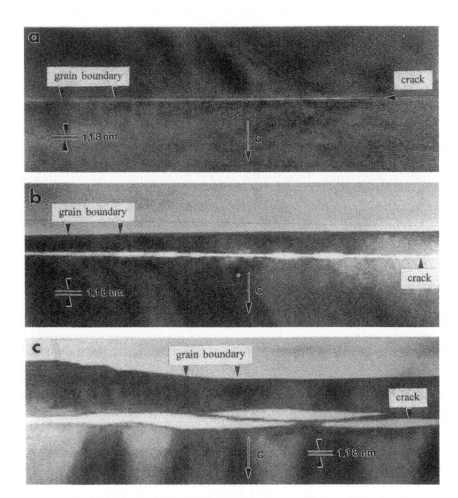

Fig. 5.28. Grain boundaries involving microcracks located at or near the boundary observed from a sample overexposed in ozone atmosphere: (*a*) a non-CSL boundary (measured orientation: 27.82° about [0.7534 0.5585 0.3471]) with a microcrack at the boundary. The crack does not extend through the entire grain boundary. The misorientation of the boundary was measured at the regions where the adjacent grains were structurally intact. (*b*)–(*c*) Microcracking near the grain boundary, but not at the grain boundary, suggests the boundaries are mechanically strong. The cracks originated from stacking fault sites during the oxygenation process. The measured boundary orientation is 60.0° [331] ($\Sigma 9$ boundary) for (*b*), while is 41.25° [0.220 0.051 0.974] (near the $\Sigma 5$ boundary, 36.87° [001]) for (*c*). All three grain boundaries have [001] boundary plane normal with the *a-b* planes of the bottom grain parallel to the beam.

effects of grain boundary dislocations on grain boundary chemistry.

5.2.1 The Concept of Coincidence Site Lattice

In general, the microstructure of a grain boundary is much more complicated than a local perturbation in an otherwise perfect crystal. Therefore, it is often appropriate to consider a boundary as a thin region with its own structure, such as periodicity and local chemistry. In order to understand the structures and properties of grain boundaries, two major theoretical approaches have been developed: (1) the coincidence site lattice (CSL) model [102], and (2) the structural unit model (SUM) [103]. Although SUM can provide detailed atomic positions of a grain boundary, it requires a precise knowledge of the interatomic potentials, thus limiting its application to simple crystals. The CSL model, based on geometric considerations between two crystals, has successfully explained the structural features observed at grain boundaries in many cubic systems with metallic, ionic, and covalent bonding [104–108]. An important reason for the popularity of the CSL stems from the fact that the description of a boundary structure can be verified by the experimentally determined grain boundary geometry and the configuration of grain boundary dislocations (GBDs) using TEM. A CSL can be produced by rotating two crystals relative to one another about a common crystallographic axis, using a lattice site as the origin. The ratio of the unit-cell volume of the CSL to that of the crystal is defined as Σ. The physical significance of the CSL is that if the CSL formed by two abutting grains is very dense (i.e., Σ is small), the two grains will share many lattice sites at the boundary, and the boundary should thus have low energy. Low Σ boundaries are, therefore, expected to form preferentially in polycrystalline materials compared with high Σ or non-CSL boundaries.

The CSL concept is fundamental for interfacial structures. However, for noncubic crystals, such as $YBa_2Cu_3O_{7-\delta}$, because of the irrational ratios of its lattice parameters, lattice coincidence does not generally exist. Nevertheless, a CSL can be formed if a small strain is applied to the crystal lattice [109–112]. Although it is uncertain to what extent the CSL, or the constrained CSL (CCSL), model is valid for noncubic systems, for $YBa_2Cu_3O_{7-\delta}$ evidence shows that the formation of grain boundaries with certain misorientations is more favorable than others in thin films [113–116], flux-grown bicrystals [117], melt-textured samples [118] and sintered materials [119]. The observed oxygen deficiency, cation segregation, and lattice distortion at the boundaries can be related to the structural features described by CSL or a modified CSL model [98,99]. Furthermore, low-Σ boundaries appear to possess better mechanical properties than noncoincidence boundaries. Figures 5.28a–c show three grain boundaries observed in an ozone overtreated sample. Microcracks, abundant in the sample as a result of the anisotropic thermal expansion of the crystal as the oxygen is taken into the sample, were observed at grain boundaries with non-CSL orientations (see, for example, Figure 5.28a). On the other hand, for a CSL ($\Sigma 9$, Figure 5.28b) or a near CSL (near $\Sigma 5$, Figure 5.28c) grain boundary, cracks were observed near the boundaries but not at the boundaries, suggesting that these boundaries are mechanical strong. A similar observation was made in Bi-based cuprate oxides [120].

5.2.2 Grain Boundary Crystallography

This section discusses methods for determining grain orientation, misorientation, and dislocation, as well as methods for constructing a grain boundary structure. As case study is presented to illustrate the methods.

5.2.2.1 Measurement of Grain Boundary Geometry

To determine the *orientation* of a grain, at least two crystallographic directions for each grain must be measured. One easy way is to choose an incident beam direction, **B**, and a direction along the shadow of the beam stop, **s**, as the two reference directions. These two directions can be recorded simultaneously with the beam stop pointing to the incident beam (000 center spot; see Figure 5.29a. High-quality Kikuchi patterns can be obtained by focusing the electron beam with a small convergent angle and tilting the crystal away from a low index zone axis. The orientations of the two reference directions then can be determined by matching the observed Kikuchi pattern with the calculated one (Figures 5.29a and 5.29b). Experimentally, the error of indexing from the measurement is much smaller than the local bending of the crystal itself [121]

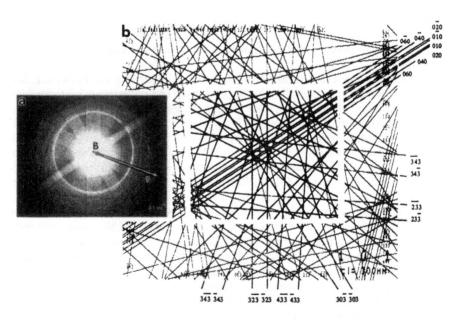

Fig. 5.29. (*a*) An experimentally observed Kikuchi pattern of the [101]* pole in YBa₂Cu₃O₇₋δ ($\delta \approx 0$). **B** and **s** represent the incident beam direction and the beam stopper direction, respectively. (*b*) A calculated Kikuchi pattern of the [101]* pole for YBa₂Cu₃O₇ crystal with $N = h^2+k^2+l^2/(c/a)^2 < 40$ showing good match with (*a*).

The *misorientation* of two adjacent grains, grain 1 and grain 2, can be described by a misorientation matrix **R**:

$$X_2 = R X_1.$$ (5.9)

Experimentally, **R** can be determined by using

$$R = [B_1, s_1, B_1 \times s_1] [B_2, s_2, B_2 \times s_2]^{-1}$$ (5.10)

where $[B_i, s_i, B_i \times s_i]$ ($i = 1$, or 2) is a 3 x 3 matrix formed by three column vectors to describe the orientation of the grain i. In order to determine the misorientation, only one pair of Kikuchi patterns from the adjacent grains is necessary. For a pure rotation, according to the properties of rotation matrix, the rotation angle α can be expressed by

$$\alpha = {}^1\!/_2 \arccos(r_{11} + r_{22} + r_{33} - 1),$$ (5.11)

where r_{ij} are the elements of the rotation matrix. The rotation axis $[c_1 c_2 c_3]$ is given by

$$c_1 : c_2 : c_3 = (r_{32} - r_{23}) : (r_{13} - r_{31}) : (r_{21} - r_{12}).$$ (5.12)

The misorientation of the grain boundary thus can be described by a rotation angle and a rotation axis, or Θ [*hkl*] parameters.

The geometry of *dislocation* arrays can be determined as described earlier. If two sets of nonparallel boundary dislocations are observed, then the grain boundary plane normal, **n**, can be accurately determined by

$$n = \xi_1 \times \xi_2.$$ (5.13)

The spacing of the dislocation arrays, D, at the grain boundary can be determined by measuring the projected spacing of dislocation D_p recorded along the beam direction **B**:

$$D = \frac{D_p}{\dfrac{(\xi \times n) \bullet B}{|\xi \times n| \ |B|}}$$ (5.14)

Experimental determination of the Burgers vector of a dislocation array at a grain boundary using the **g·b** = 0 criterion is more complicated than for one within the interior of a grain. For a grain boundary, three two-beam diffraction conditions are possible: (1) operating within only one grain, (2) operating simultaneously in both grains with the same diffraction vector, or (3) operating simultaneously in both grains but with different diffracting vectors in each grain. The **g·b** criterion is not valid for condition 3, is unreliable for condition 1, and is most suitable for condition 2 [122,123].

The Burgers vector of the dislocations was determined through three steps: (1) the possible Burger vectors were first derived by constructing the O-lattice and b-net (see below); (2) the **g·b** = 0 criterion was applied using several diffracting vectors from either crystal, or common to both crystals for all possible **b**s, including those that underwent dislocation reactions; and (3) the Burgers vector of the dislocations was finally determined by comparing the measured and calculated values (using the modified Frank formula, to be discussed shortly) of their line direction and line spacing.

5.2.2.2 Construction of the O-Lattice and B-Net

Based on the CSL model, a grain boundary has a localized structure of its own in which the atoms occupy sites that are displaced from normal lattice sites in two adjoining crystals. These lattice displacements accommodate the different orientations of the neighboring crystals (i.e., deviations from a perfect crystal or from a coincidence orientation), and they generate regions in the boundary where there is a good lattice fit, separated by localized regions of bad fit. The localized discordant areas are the cores of grain boundary dislocations. This concept can be applied to small- and to large-angle grain boundaries; the only difference between them is that, for the former, the deviation is accommodated by primary dislocations (i.e., dislocations with lattice vectors), while between the dislocations, the crystal structure is conserved, although slightly distorted elastically. For the latter, the density of the primary dislocations becomes so high that the deviation then is accommodated by secondary dislocations (with displacement-shift-complete (DSC) lattice vectors; see Figure 5.36) which reflect the perturbation of the primary dislocations.

Bollmann's O-lattice theory [102] is derived from the earlier concept of CSL. It is a geometrical approach to characterizing interfacial structure and describes the matching and mismatching of the misoriented lattices at an interface. If two misoriented crystal lattices are allowed to interpenetrate, there will be a periodic set of points in crystal space (not necessarily the lattice points of either lattice) where the two lattices shear the coincidence sites, known as the O-elements. These sites are the locations of the best match, or equivalently, sites of minimum strain. The O-elements can be separated from one another by the so-called O-cell wall: planes bisecting the connection between two O-elements. In this way, a cell structure representing the area of the maximum disregistry between the two lattices is introduced into the crystal space. Consequently, the intersection of a grain boundary with the cell walls is the dislocation network of the boundary.

The position of the O-elements (defined by vector $\mathbf{X}^{(O)}$) and the Burgers vector of the dislocations (defined by vector $\mathbf{b}^{(L)}$) can be determined by the O-lattice equation [102]:

$$\mathbf{X}^{(O)} = (\mathbf{I} - \mathbf{A}^{-1})^{-1}\, \mathbf{b}^{(L)}, \tag{5.15}$$

where $\mathbf{A}$ represents a homogeneous, linear transformation of lattice 1 into lattice 2, $\mathbf{I}$ is identity, and $\mathbf{b}^{(L)}$ are Burgers vectors of dislocation arrays. The O-elements may be a point, line, or plane lattice, depending on whether the rank of $(\mathbf{I} - \mathbf{A}^{-1})$ is 3, 2, or 1, respectively. When transformation $\mathbf{A}$ is a pure rotation, the O-element forms a line lattice parallel to the rotation axis. The projection of the lattice along the rotation axis is the O-lattice.

Equation (5.15) can be simplified by considering a pure rotation about the z-axis. The transformation matrix $\mathbf{A}$ is then given by

$$A = \begin{pmatrix} \cos\theta & \sin\theta & 0 \\ \sin\theta & \cos\theta & 0 \\ 0 & 0 & 1 \end{pmatrix}. \tag{5.16}$$

When θ is small, $\sin\theta \approx 0$, and $\cos\theta \approx 1$, Equation (5.15) becomes

$$
\begin{pmatrix} b_1^{(L)} \\ b_2^{(L)} \\ b_3^{(L)} \end{pmatrix} = \begin{pmatrix} 0 & \Theta & 0 \\ \Theta & 0 & 0 \\ 0 & 0 & 0 \end{pmatrix} = \begin{pmatrix} x_1^{(0)} \\ x_2^{(0)} \\ x_3^{(0)} \end{pmatrix}
\tag{5.17}
$$

Equation 5.17 suggests that the vectors $b^{(L)}$, which define a two-dimensional lattice of Burgers vector of the boundary dislocations along the rotation axis, known as the *b-net*, can be obtained directly from the O-lattice vectors $X^{(O)}$ clockwise 90° along the same axis, with the magnitude reduced by a factor of Q. With recent advances in computer graphics, one can draw a projected (two-dimensional) O-lattice accurately and, hence, the corresponding b-net. This approach is useful for predicting the possible Burgers vectors of dislocation arrays for a given boundary misorientation. Such an approach also can be applied to a large-angle grain boundary; however, Q then is a small angular deviation from a certain coincidence orientation, rather than the deviation from a perfect crystal, and the rotation is between two O-lattices rather than two crystal lattices. Thus, the rotation axis and rotation angle are named the O2-axis and O2-angle. Equation (5.15) becomes

$$
X^{(O2)} = (E^{-1} - A_c E^{-1} A^{-1})^{-1} b^{(DSC)},
\tag{5.18}
$$

where A_c is the rotation matrix of the corresponding coincidence boundary, and E is the constraint matrix for noncubic crystals. For an orthorhombic crystal

$$
E_{orthorhombic} = \begin{pmatrix} 1 & 0 & 0 \\ 0(b/a)_n/(b/a)_c & 0 & 0 \\ 0 & 0(c/a)_n/(c/a)_c \end{pmatrix},
\tag{5.19}
$$

where subscripts n and c refer to the natural material and material with the ideal c/a ratio to form a particular coincidence site lattice, respectively [124].

5.2.2.3 Frank Formula

Frank was the first to develop a theory that provided a dislocation description for the case of arbitrary boundary in which the axis of rotation **a** and the boundary plane have no special relative orientation [125]. The Frank formula gives the net Burgers vector **b** intersected by any vector **V** lying in the plane of the grain boundary:

$$
b = 2 \sin(\Theta/2) (V \times a).
\tag{5.20}
$$

For an arbitrary grain boundary, three sets of independent boundary dislocations are necessary [126], with no coplanar Burgers vectors b_i, b_j, and b_k, with line directions of ξ_i, ξ_j, and ξ_k, and with spacings of d_i, d_j, and d_k. ξ_i and ξ_i can be expressed as

$$
\xi_i = [a \times (bj \times b_k)] \times n,
\tag{5.21}
$$

$$
d_i = (2 \sin \frac{t}{2} \mid \frac{[a \times (b_i \times b_j)] \times n}{b_i \times (b_j \times b_k)} \mid)^{-1}
\tag{5.22}
$$

where **n** is the boundary plane normal, Θ is the deviation angle from the

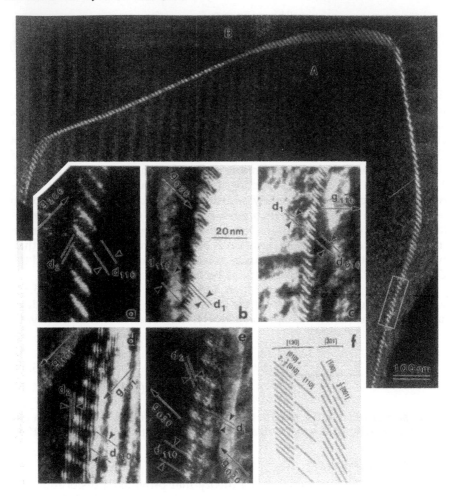

Fig. 5.30. A dark-field image of a dislocation network located at an entire grain boundary observed in bulk $YBa_2Cu_3O_{7-\delta}$: (*a*)–(*e*) the enlarged images of dislocation arrays from a segment of the boundary bounded by the rectangular box. All the images were recorded at the same magnification but under different two-beam conditions: (*a*) $g_A =$ 200, (*b*) $g_A = 020$, (*c*); $g_B = 1\text{-}10$, (*d*); $g_A = g_B = 10\text{-}1$, (*e*); $g_A = g_B = 0\text{-}20$. (*d*) was imaged away from the [1-11] zone axis, while others were imaged away from the [001] zone axis. d_{110} and d_{010} are denoted as spacing of the [110] and [010] dislocations, d_1 and d_2 are denoted as 1/2 [010] and [-100] (and/or 1/3 [001]) dislocations. A sketch of two groups of the dislocation arrays is shown in (*f*).

coincidence orientation, **a** is the rotation axis, and i, j, and k are the permutations of the *i*-th array of grain boundary dislocations [108]. For any given boundary misorientation and boundary planar normal, we can calculate the

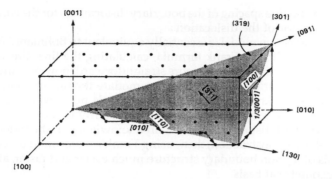

Fig. 5.31. The crystallography of the observed grain boundary (rotation angle 5.163°, rotation axis [0.6702 -0.2414 0.2398]) in orthorhombic coordinates. The rotation axis is very close to [3-11], and the rotation plane is very close to (3-19).

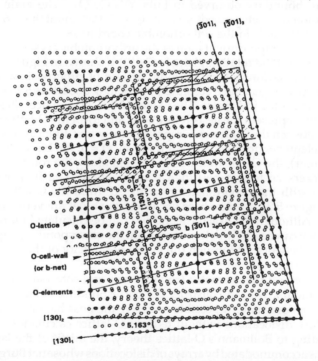

Fig. 5.32. A superimposed lattice pattern generated by rotating the (3-19) lattices 5.163° with respect one to another about the [3-11] axis. The O-elements (coincident-sites), denoted as black dots, represent the best-fit locations, while the positions between the O-elements represent the worst-fit locations. By bisecting the O-element, the O-cell-wall and b-net were generated. The Burgers vectors of the dislocations are assigned as [130] and [-301], with about 90° rotation from the O-lattice. In order to demonstrate the lattice mismatch in this figure, the large difference in scale between the O-lattice and the b-net was ignored, and for convenience, the lattice sites were chosen as O-element sites.

line direction and line spacing of the boundary dislocations for the corresponding Burgers vector of the dislocation.

Frank's dislocation model is essentially equivalent to Bollmann's O-lattice approach [127]. With the Frank formula, calculating the line directions and line spacings of the dislocations is more straightforward; however, the Burgers vectors of *all* dislocation arrays that accommodate the misorientation at the boundary must be known. Experimentally, this is not always possible. In contrast, the O-lattice approach allows us to determine the character of an individual array of dislocations. Based on a knowledge of the b-net derived from the O-lattice construction, then, using Frank's dislocation model makes the analysis of a grain boundary structure much easier and faster, at least on a purely geometrical basis.

5.2.2.4 Case Study

Figure 5.30 is a dark-field image showing the arrays of dislocation at an entire grain boundary observed in bulk YBa₂Cu₃O₇₋δ. The grain boundary misorientation was determined as a rotation of 5.163° about the [0.6702 -0.2414 0.2398] ($\approx$ [3-11]) axis in the orthorhombic coordinates.

The crystallographic orientation of the orthorhombic crystal, the observed rotation axis, and the rotation plane ($\approx$ (3-19)) is shown in Figure 5.31. Figure 5.32 shows a two-dimensional periodic pattern of superposition of the two (3-19) lattice planes generated accurately by rotating one lattice 5.163° in relation to another about the [3-11] axis. The O-elements (coincidence sites), denoted as black dots, represent the best-fit locations and form the O-lattice, while the positions between the O-elements represent the worst-fit locations. By bisecting the O-elements (for a very small rotation, $\Theta \approx 0$, the bisection should be perpendicular), the O-cell-wall was generated along the rotation axis. A lattice of Burgers vectors (b-net) then was assigned to each segment of the O-cell-wall with their coordinates rotating 90° from the O-lattice. As illustrated in Figure 5.32, the two base vectors of the b-net are very close to [130] and [-301]. Although the exact $X^{(O)}$ and $b^{(L)}$ can be calculated numerically, in practice, they can be easily determined from the corresponding regions of the best fit and the worst fit through the visual periodicity of the superimposed lattice pattern.

Thus, based on the geometry (rotation axis and rotation angle) of an arbitrary boundary, we can construct the corresponding O-lattice and b-net; such constructions make it possible to derive the possible Burgers vectors of the boundary dislocations without tedious O-lattice calculations.

According to Bollmann's O-lattice theory, the misfit of the lattice at the boundary is accommodated by arrays of dislocations whose net Burgers vectors must lie in a plane with its plane normal parallel to the rotation axis. Thus, two set of dislocations with a Burgers vector perpendicular to the rotation axis ([130] and [301] in Figure 5.31) would be sufficient to accommodate the boundary misorientation. Nevertheless, the dislocations predicated from the O-lattice construction are based on the geometry of the boundary, and they may not have low energy. Since the elastic strain energy of a dislocation is proportional to $|b|^2$, a dislocation b_1 would dissociate into b_2 and b_3, if $b_1^2 < b_2^2 + b_3^2$. Thus, in the present case, the following reactions will be energetically favorable:

$$a[130] \Rightarrow a[110] + a[010] + a[010],$$
$$a[010] \Rightarrow a/2[010] + a/2[010],$$
$$a/3\,[301] \Rightarrow a\,[100] + a/3\,[001].$$

With these dislocation reactions taken into account, these four types of dislocation arrays with Burgers vectors [110], [010], [-100], and 1/3[001] (as illustrated in Figure 5.30f) were then compared with experimentally observed ones.

Figure 5.30a–e shows the enlarged images of dislocation arrays from a segment of the boundary marked by the rectangular box in Figure 5.30. All the dislocations were imaged at the same magnification under two-beam conditions: $g_A = 200$ (a), $g_A = 020$ (b), $g_B = 1\text{-}10$ (c), $g_A = g_B = 10\text{-}1$ (d), $g_A = g_B = 0\text{-}20$, (e), where d_{110} and d_{010} are denoted as the spacings of the [110] and [010] type dislocations (with Burgers vectors [110] and [010]), respectively, and d_1 and d_2 are denoted, respectively, as 1/2[010] type, and [-100] type of dislocations (and/or 1/3[001] dislocations, because [-100] and 1/3[001] dislocations have a very close line direction and a line spacing). The configuration of the dislocations arrays is sketched in Figure 5.30f; the dislocations are 1/2[010] ([010]), [110], [100], and 1/3[001] with mixed edge and screw characters. Under nine different diffraction conditions, two groups of dislocation arrays were observed: 1/2[010], [010] and [110] in one group ([130] group), and [-100] and 1/3[001] in the other ([-301] group). The line directions of the different types of dislocation arrays in each group are nearly parallel. Under some diffraction conditions, the arrays of dislocation are not distinguishable; however, under others, they showed a different contrast, making it possible to define them. For example, Figure 5.30b, c, and e show 1/2[010] type dislocations denoted as d_1; however, in Figure 5.30c, [010] can be clearly differ-

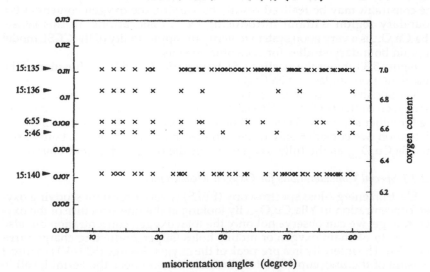

Fig. 5.33. Distribution of the ideal CCSL boundaries of $YBa_2Cu_3O_{7-\delta}$ ($\Sigma < 50$) as a function of oxygen content in misorientation-angle space.

entiated from 1/2[010], and both are out of contrast in Figure 5.30a and d. In Figure 5.30e [-100] dislocations show residual contrast, because the value of $\mathbf{g} \cdot (\mathbf{b} \times \xi)$, is very large. According to Hirsch et al., dislocations with mixed edge and screw characters will be out of contrast only when both grains are under the two-beam condition with $\mathbf{g} \cdot \mathbf{b} = 0$ and $\mathbf{g} \cdot (\mathbf{b} \times \xi) \leq 0.64$ [63]. Experiments using nine different diffractions showed the validity of the $\mathbf{g} \cdot \mathbf{b} = 0$ extinction criterion in YBa₂Cu₃O₇₋δ when the two-beam condition is fulfilled in either one grain or in both grains. The corresponding line directions and line spacings of these dislocations were compared with the calculated ones and showed good agreement (for details, see [128]).

In short, the grain boundary misorientation can be determined by the misorientation matrix. The Burgers vectors of grain boundary dislocations, which accommodate the misorientation at the boundary, can be visually derived by constructing the O-lattice and b-net ,then taking dislocation interactions into account. The determination of the Burgers vector can be further refined by comparing the calculated dislocation configurations with the experimentally determined Burgers, line directions, and lines spacings of the dislocations.

5.2.3 Grain Boundary Oxygen

As suggested by Bruggmann, Bishop, and Hartt [129], a three-dimensional CSL can be obtained only when the lattice parameter ratios $a^2 : b^2 : c^2$ are rational. Thus, for a crystal other than cubic, such as YBa₂Cu₃O₇₋δ, it is necessary to constrain the ratios of the lattice parameters at the boundary region to the closest rational values to form a constrained CSL, or CCSL. In a hexagonal system, such constraint was demonstrated to be accommodated by grain boundary dislocations [130,131]. In YBa₂Cu₃O₇₋δ, since the lattice parameters, particularly the c-axis, vary with the oxygen substoichiometry, δ [132,133], the constraint may be realized readily by varying the oxygen content at the boundary region without introducing extra dislocations. In this sense, YBa₂Cu₃O₇₋δ is a very good system to verify the applicability of the CCSL model in grain boundary studies for noncubic crystals.

Figure 5.33 shows the existing CSL boundaries ($\Sigma \leq 50$) as a function of the misorientation angles in YBa₂Cu₃O₇₋δ.

The possible $a^2 : c^2$ combinations with are 15:135, 15:136, 6:55, 5:46, and 15:140 for oxygen contents of about 7.0-6.3. Since the majority of the CSL boundaries are for either $a^2 : c^2 = 15:135$ ($\delta \approx 0.0$) or $a^2 : c^2 = 15:140$ ($\delta \approx 0.7$), in the following discussion, we consider only these two possibilities for the CSL boundaries in YBa₂Cu₃O₇₋δ, as the fully oxygenated or the oxygen-deficient boundary.

5.2.3.1 Special Considerations for Oxygen Measurements

Electron energy-loss spectroscopy (EELS) is very useful for studying oxygen concentration in YBa₂Cu₃O₇₋δ. By looking at the fine structure of the oxygen K-edge, we can observe not only the total oxygen concentration but also the density of chain oxygen or mobile holes (superconducting charge carriers) [134]. The intensity in a pre-peak of the oxygen K-edge (529 eV) is a direct measure of the unoccupied states (the hole density near the Fermi level) in the oxygen 2p band, which represents the transition of the electrons from the oxygen 1s into unoccupied states. X-ray absorption near-edge structure studies demonstrate that there is a one-to-one correspondence between the oxygen substoichiometry, δ, and the integrated intensity of the pre-peak [135].

Unfortunately, EELS measurement at a grain boundary is not a easy task. A small probe less than 20 Å is required, which results in a poor signal/noise ratios of a spectrum even with a low-temperature stage and a parallel EELS spectrometer. Thus, a field emission source is desirable. In order to avoid ambiguity in the measurements, "good" boundaries (clean, straight, structurally intact, and without grooving) are recommended. For samples with inhomogeneous oxygen concentrations in the interior of the grain, ozone

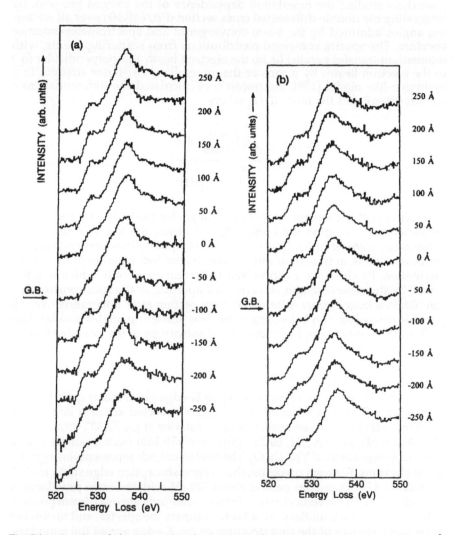

Fig. 5.34. A series of electron energy-loss spectra of the oxygen K-edge collected at 50 Å apart across (*a*) an oxygen-deficient grain boundary ($\Sigma = 31$; measured misorientation: 69.22°, [310]; $a^2:c^2 = 15:140$); and (*b*) a fully oxygenated grain boundary ($\Sigma = 3$, measured misorientation, 83.56°, [100]; $a^2:c^2 = 15:135$).

annealing can eliminate the problem [39].

Another important characteristic of EELS spectra of oxygen K-edge from the highly anisotropic crystals of YBa$_2$Cu$_3$O$_{7-\delta}$ is their orientational dependence. By changing the direction of a parallel incident beam, it was observed that the fine structure of the oxygen K-edge differed when the incident beam is parallel or perpendicular to the *a-b* plane [134,136,137], because the hole was mainly located at oxygen in two-dimensional Cu-O planes with O-2p$_{xy}$ symmetry. For grain boundary study, since the boundary has to be tilted edge-on, the freedom to change the crystal orientation was very limited. To minimize the crystallographic orientational anisotropy of the pre-edge feature, Zhu and coworkers studied the orientation dependence of the oxygen pre-peak by integrating the double-differential cross section d$^2\sigma$/(dEdΘ) over all scattering angles admitted by the beam convergence and spectrometer-entrance aperture. The spectra contained contributions from scattering events, with momentum transfer parallel (q$_\parallel$) to the electron beam and perpendicular (q$_\perp$) to the electron beam. By means of the generalized oscillator strength in a hydrogen-like model [138], the researchers calculated the differential cross section dσ/dE and the ratio $\bar{q}_\perp/\bar{q}_\parallel$ where

$$\bar{q} = \frac{\int_0^\beta q \left(\frac{d^2\sigma}{dEd\theta}\right) d\Omega}{\int_0^\beta \left(\frac{d^2\sigma}{dEd\theta}\right) d\Omega},$$

as a function of the integrated scattering angle for the O-1s core loss. At the characteristic collection semi-angle $\beta_E \approx 5.0$ mrad, the average amount of momentum transfer in the beam direction is equal to that perpendicular to the beam, that is, at this condition, the spectra are independent of the crystal orientation. By choosing an optimized collection angle of β, which is determined by the camera length and entrance aperture of the spectrometer, one can obtain a large energy-differential cross section after integrating over all possible directions of q belonging to various scattering angles smaller than β. Thus, the crystallographic orientational anisotropy of the pre-edge feature can be reduced.

5.2.3.2 Oxygen Absorption K-Edge

Figure 5.34 shows two types of oxygen K-edge observed at grain boundaries. Figure 5.34a is a series of EELS spectra collected across a large-angle grain boundary (measured misorientation: rotation angle 72.33°, rotation axis [0.95 0.31 0.02]; system Σ31, 69.22°, [310], a^2:c^2=15:140) from a sample with a nominal composition of YBa$_2$Cu$_3$O$_7$. The horizontal axis represents energy loss of the electrons. For each spectrum, the oxygen absorption edge starts at ~527 eV, followed by a pre-edge peak at about 529 eV, with the main peak located at about 537 eV. In the vertical axis, a distance scale indicates each beam position with respect to the boundary. In order to compare the spectra, and to observe the relative change of the fine structure of the K-edge across the boundary, the intensity of the oxygen main peak in each spectrum was normalized. The spectrum at the boundary has no visible oxygen pre-edge peak. Also, several additional spectra were collected along the grain boundary, with similar results. This drastic change of the oxygen pre-edge peak is likely to be due to a decrease in the hole content at the boundary, suggesting that the boundary is low in oxygen. A comparison of these results with the observations of Nücker et al. [134] suggests that the fine structure of K-edge at the boundary is similar to that observed in $YBa_2Cu_3O_{6.3}$. The oxygen-deficient area appears to be less than 100 Å wide. Many such boundaries have been observed, especially from samples with a small grain size.

Figure 5.34b shows another series of oxygen K-edge spectra from the same sample, but for a near $\Sigma3$ boundary (measured misorientation: 83.56°/ [100]; system: Σ 3, 90/[100], a^2:c^2=15:135). When all 11 spectra were compared, no major difference was detected in the intensity of the oxygen pre-edge peak, suggesting that no oxygen depletion occurred at the grain boundary. Similar measurements of the oxygen absorption edge were made for a number of grain boundaries; a comparison with the characteristics of the CCSL boundary structure can be found in [98]. In general, for a^2:c^2 = 15:135 coincidence boundaries, the EELS clearly shows the oxygen pre-peak, suggesting that the boundary is fully oxygenated, while for a 15:140 boundary, there is no oxygen pre-peak, indicating oxygen deficiency at the boundary. A recent study by Zhu et al. of the local changes of lattice parameters at the boundary region using convergent beam electron diffraction suggests that the oxygen/hole depletion can be related to the c-lattice expansion at the boundary [139].

5.2.3.3 Grain Boundary Structural Transition

At the typical sintering temperature (950°C), the crystal lattice has a larger c-lattice parameter ($c/a \approx 2c/(a+b) > 3.05$) than at lower temperatures, and the boundary formed at that c/a ratio will be oxygen deficient (a^2:c^2= 15:140). During slow cooling in an oxygen atmosphere, the specimen takes up oxygen and the c-lattice shrinks. However, the oxygenation state at the boundaries may differ from the overall oxygen content of the sample. In order to explore the possible mechanism for boundary structural transition during annealing and oxygenation process, two ideal tilt boundaries were compared: $\Sigma29$ (46.40° [100], c^2:a^2 = 15:135) corresponding to fully oxygenated grain boundaries and $\Sigma25$ (47.16° [100], c^2:a^2 = 15:140) corresponding to oxygen-deficient grain boundaries. Although the boundary energy comprises core energy, E_c, and elastic strain energy, E_{el} [140], because of the small difference in misorientation (< 1°) and the similar densities of coincidence sites Σ, the difference in their core energy, E_c, may be ignored. Thus, Zhu et al. [99] estimated the boundary energy by comparing the strain energy. They calculated E_{el} as a function of misorientation angle and the c/a ratio (or temperature, since the c/a ratio decreases with temperature). In each case, the strain energy tends to a minimum (i.e., the dislocation spacing tends to infinity), when the ideal c/a ratios and the exact coincidence misorientations are achieved. For one misorientation, say 46.5°, oxygen may easily diffuse into the boundary to lower the boundary energy, resulting in a fully oxygenated grain boundary; for another misorientation, say 47.5°, the boundary will remain oxygen deficient, because the change in the c/a ratio from 3.05 to 3.0 (through taking up oxygen) is energetically unfavorable. Thus, during oxygenation, some boundaries can reach a fully oxygenated state, while others remain oxygen deficient. It was estimated that among all possible CCSL boundaries ($\Sigma < 50$), only about 20% of the oxygen-deficient boundaries (a^2:$c^2 = 15$:140) are energetically favorable for transforming into the fully oxygenated boundaries (a^2:$c^2 = 15$:135).

5.2.4 The Effects of Grain Boundary Dislocations

We conclude this chapter with a brief discussion of the effects of the types of grain boundary dislocations: cation segregation and oxygen/hole depletion.

5.2.4.1 Cation Segregation at the Boundary

Zhu and his colleagues conducted energy dispersive X-ray spectroscopy (EDX) measurements with a 20-Å probe to detect local variations in cation concentration at arbitrary large-angle grain boundaries in bulk $YBa_2Cu_3O_{7-\delta}$ ($\delta \approx 0.0$) samples. Overall, the Y and Ba concentrations at the grain boundaries did not differ significantly from the crystal matrix. At some grain boundaries, however, there were Cu-rich regions, consistent with the observations from other groups [141–144]. In one case, Zhu et al. found a variation in Cu composition along the boundary (Figure 5.35a). To compare the relative change in local cation composition, the average of the peak intensities of the matrix area was normalized to Y = 7.7%, Ba = 15.4%, and Cu = 23.1% using C_A/C_B = $K_{AB}I_A/I_B$ [145], where C_A and C_B are concentrations of the elements A and B, I_A and I_B are the background subtracted peak intensities for A and B, and K_{AB} is the ratio of characteristic intensities measured on the grain boundaries and the crystal matrix. Seven locations were measured with a point-to-point separation about 80 Å along the boundary. The Cu concentration varied periodically at the boundary, but remained constant in the matrix (Figure 5.35b).

The misorientation for this grain boundary was 39.805° about [0.9999 0.0004 0.0005], deviating from the ideal $\Sigma5$ misorientation (rotation angle: 36.87°, rotation axis: [100]) by 2.94° (O2-angle) rotated about [0.9998 -0.0014 0.0070] (O2-axis). The boundary plane normal is [010], which is perpendicular to the rotation axis, implying that the boundary can be approximated as a tilt boundary. For a pure tilt boundary, only one set of edge dislocations is needed to ·accommodate the deviation from a coincidence orientation or from a perfect crystal. Based on the analysis of grain boundary dislocation, it was found that to accommodate the misorientation of 2.94° for this particular boundary geometry, an array of edge dislocations with a Burgers vector of [010] (DSC vector, see Figure 5.36) is required. The calculated spacing of the GBDs is about 75 Å, which agrees with the periodicity of the variations in Cu concentration observed at the boundary. Thus, the observed Cu segregation at the boundary can be associated with the single array of edge dislocations there. Cu solutes apparently can diffuse to the dilated region below the extra half plane and form clusters along the dislocation line.

Segregation of solute atoms associated with GBDs was observed by Michael, Linn, and Sass [146] in a small-angle [001] twist boundary in a Fe-Au bicrystal using EDX, and Cu-rich dislocation cores were observed in $YBa_2Cu_3O_{7-\delta}$ film

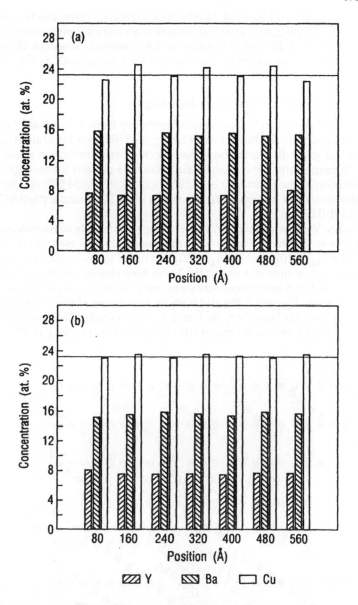

Fig. 5.35. (a) Measurements of Y, Ba, and Cu concentrations along the boundary (measured misorientation: 39.805° [0.9999 0.0004 0.0005], system: S5). Position denotes the distance from a reference starting point at the boundary. The Cu concentration varies periodically at the boundary (variation in the Y and Ba concentrations at the boundary was insignificant and within the uncertainty of the measurement). The average concentration of Cu in the matrix is indicated by the horizontal line. (b) The average cation concentration from two traces parallel to the boundary from neighboring grains.

by Gao et al. [147] and Jia et al. [148] using high-resolution electron micros-copy. Segregation of solute atoms at defects is an intrinsic phenomenon. Such segregation at the GBDs may not always be observed, however, depending on the spacing and line direction of the dislocations, as well as the extent of the segregation.

5.2.4.2 Oxygen/Hole Depletion at the Boundary

Figure 5.37 shows an V-shape grain boundary from a nominal YBa₂Cu₃O₇ sample imaged with the beam nearly along the [100] axis for grain A, and the [001] axis for grain B. Insets show the selected area diffraction pattern cov-ering both grains, and the corresponding Kikuchi-pattern for each grain. The misorientation of the boundary is 81.58° about [0.999 0.034 0.005]. The devia-tion of the measured misorientation from an ideal Σ3 boundary is 8.91° about [-0.949 0.111 0.099].

The entire V-shape grain boundary corresponds to only one misorientation; however, as is shown in Figure 5.37, the boundary has two different plane normals: $[001]_A/[010]_B$ and $[209]_A/[230]_B$. In order to understand whether the oxygen content or hole density at the boundary changes with the boundary plane normal, EELS measurements were made at the trace positions 1 and 2 (see Figure 5.37). Figures 5.38a and b show two series of the oxygen K-edge, collected across the boundary at 1 and 2, respectively. The spectrum from boundary plane 1 shows no major difference in the intensity of the oxygen

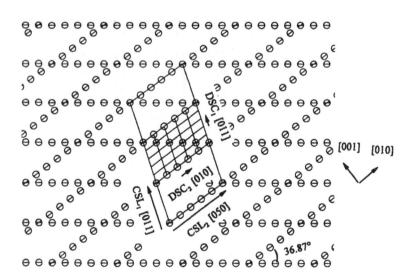

Fig. 5.36. Projection of an ideal (unrelaxed) Σ5 CCSL grain boundary (rotation angle: 36.87°, rotation axis: [100], c^2/a^2 = 135:15) viewed along the common rotation axis. ⊖ and ⊘ represent lattice 1 and 2, respectively. The unit cell of the coincidence-site-lattice is shown in the middle of the drawing. The unit vectors of the CSL are [011] and [050], while the unit vectors of the DSC are [011] and [010].

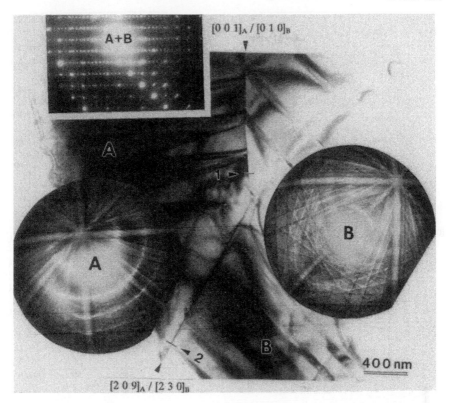

Fig. 5.37. The V-shape grain boundary with two different boundary planes. The boundary was imaged with the beam nearly parallel to the [100] axis for grain A, and the [001] axis for grain B. The selected area diffraction pattern, covering both grains and the corresponding Kikuchi-pattern for each grain are shown in insets. The measured grain boundary misorientation is 81.58° [0.999 0.034 0.005], which is close to a Σ3 boundary (90° [100]) with a deviation of 8.91° (O2-angle) about [-0.949 0.111 0.099] (O2-axis).

pre-peak compared with the interior of the grain (Figure 5.38a), while the spectrum from boundary plane 2 shows a degraded pre-peak (Figure 5.38b).

According to the CCSL model, an ideal Σ3 boundary (90°, [100], $a^2:c^2$=15:135) or a boundary close to the Σ3 orientation is expected to be fully oxygenated. We note that the observed grain boundary has a large deviation (8.91°) from the ideal misorientation, although the misorientation does not change with the boundary plane normal. The EELS results imply that the boundary plane normal, similar to boundary misorientation, plays an important role in determining the oxygen/hole content at the boundary. As discussed earlier, for a given boundary misorientation, the GBDs are defined by the intersection of the boundary plane with the planes with the worst lattice mismatch. Thus, a change in the boundary plane normal results in a change in the spacing and

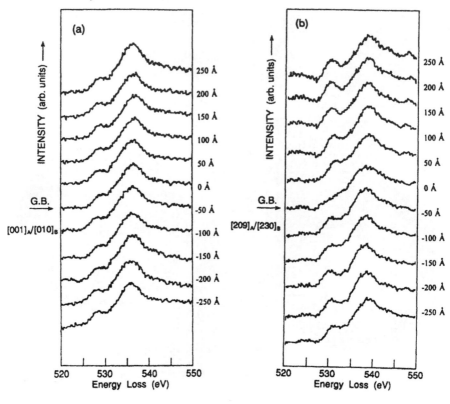

Fig. 5.38. Series of EELS spectra of oxygen K-edge collected 50 Å apart across the
boundary at the positions 1 and 2 in Figure 5.37 with a 20-Å probe. The distance
indicates each beam position respect to the boundary. (*a*) The spectrum acquired from
the boundary plane 1 shows no major difference in the intensity of the oxygen pre-peak
compared with those from interior of the grain, while (*b*) the spectrum acquired from
boundary plane 2 shows a degraded pre-peak at the boundary.

line direction of GBDs. It was calculated that to accommodate the 8.91° deviation about the [-0.949 0.111 0.099] axis from the ideal $\Sigma 3$ boundary, the
required Burgers vectors of the GBDs are [010] and 1/3[001] (DSC vectors).
For the [010] set, the density of GBDs is the same for both boundary planes.
In contrast, for the 1/3[001] set, the dislocation spacing for the $[209]_A/[230]_B$
boundary plane is 27 Å, but is about 100 times larger for the $[001]_A/[010]_B$
boundary plane (2860 Å). Thus, the hole depletion at the $[209]_A/[230]_B$ boundary
may be attributed to the closely spaced 1/3[001] dislocation array at the
boundary. Cai et al. suggested that to generate one hole in $YBa_2Cu_3O_{7-\delta}$, five
continuous undistorted oxygen chains are required [149]. Consequently, even
for a boundary with a high oxygen concentration, GBDs may cause the

reductionn or disappearance of the oxygen pre-peak at the boundary as a result of severe distortion of the lattice.

The structure of grain boundaries and the configuration of grain boundary dislocations also depend on the sample history. Wang et al. have observed that, statistically, prolonged oxygen annealing increases the numbers of low-Σ boundaries in bulk $YBa_2Cu_3O_{7-\delta}$ [119]. For standard sintered samples, however, especially those with small grains, the majority of large-angle grain boundaries may not be able to reach their equilibrium states. They usually exhibit a large deviation from coincidence orientations and possess closely spaced GBDs. Such GBDs may be detrimental to the properties of the grain boundaries. The strain field associated with a high density of GBDs can cause a depletion of hole density and suppress the superconducting order parameter and, hence, block the passage of superconducting current at the boundary [150,100]. Thus, far fewer large-angle grain boundaries are expected to have superconducting characteristics in "real" bulk $YBa_2Cu_3O_{7-\delta}$, which has undergone a standard sintering procedure.

5.3 References

1. M .K. Wu, J. R. Ashburn, C. J. Torng, P. H. Hor, R. L. Meng, L. Gao, Z. L. Huang, Y. Q. Wang, and C. W. Chu, Phys. Rev. Lett. **58**, 908 (1987).
2. D. Larbalestier, Physics Today **44**, 74 (1991).
3. D. Dimos, P. Chaudhari, and J. Mannhart, Phys. Rev. B **41**, 4038 (1990).
4. J. D. Jorgensen, Physics Today **44**, 34 (1991).
5. B. Raveau, Physics Today **45**, 53 (1992).
6. D. Clarke, T. M. Shaw, and D. Dimos, J. Am. Ceram Soc. **72**, 1103 (1989).
7. P. L. Gai, and J. M. Thomas, *Superconductivity Review*, C. P. Poole, Jr. ed., **1**, 1 (1992).
8. Y. Zhu, J. Tafto, and M. Suenaga, MRS Bulletin **16**, 54 (1991).
9. R. Beyers and T. M. Shaw, Solid State Physics, Vol. 42, ed. H. Ehrenreich and D. Turnbull, Academic Press, New York, (1989), 135.
10. J. C. Barry, J. Electron Microsc. Tech. **8**, 325 (1988).
11. A. L. Roitburd, Sov. Phys. Solid State **10**, 2870 (1969).
12. A. G. Khachaturyan, *Theory of Structure Transformations in Solids*, John Wiley, New York (1983).
13. C. J. Jou and J. Washburn, J. Mater. Res. **4** 795 (1989).
14. T. Roy and T. E. Mitchell, Philos. Mag. A **63**, 225 (1991).
15. Y. Zhu, M. Suenaga, J. Tafto and D.Welch, Phys. Rev. B **44**, 2871 (1991).
16. R. Gevers, J. van Landuyt, and S. Amelincx, Phys. Status Solidi **11**, 689 (1965).
17. Y. Zhu and M. Suenaga, Philo. Mag. A **66**, 457 (1992).
18. Y. Zhu and M. Suenaga, in *Interfaces in High-T_c Superconducting Systems*, ed. S. L. Shiné and D. A. Rudman, Springer-Verlag, New York (1993), pp. 140–175.
19. M. Sarikaya, B. L. Thiel, and I. A. Aksay, J. Mater. Res. **2**, 736 (1987).
20. T. M. Shaw, S. L. Shinde, D. Dimos, R. F. Cook, P. R. Duncombe, and C. Kroll, J. Mater. Res. **4**, 248 (1989).
21. Y. Zhu, M. Suenaga, and J. Tafto, Philo. Mag. A **67**, 1057 (1993).

22. A. H. King and Y. Zhu, Philos. Mag. A **67**, 1037 (1993).
23. F. C. Frank and J. M. van der Merwe, Proc. Roy. Soc. A **198**, 205 (1949).
24. J. P. Hirth and J. Lothe, *Theory of Dislocations*, McGraw-Hill, New York (1968).
25. J. D. Jorgensen, B. W. Veal, W. K. Kwok, G. W. Crabtree, A. Umezawa, L. J. Nowicki, and A. P. Paulikas, Phys. Rev. B **36**, 5731 (1987).
26. D. de Fontaine, G. Ceder, and M. Asta, Nature **343**, 544 (1990).
27. D. de Fontaine, M. E. Mann and G. Ceder, Phys. Rev. Lett. **63**, 1300 (1989).
28. A. G. Khachaturyan and J. W. Morris, Jr., Phys. Rev. Lett. **59**, 2776 (1987); A. G. Khachaturyan and J. W. Morris, Jr., Phys. Rev. Lett. **61**, 215 (1988).
29. D. Shi and D. W. Capone II, App. Phys. Lett. **53**, 159 (1988); D. Shi, Pys. Rev. B **39**, 4299 (1989); D. Shi, K. Zhang, and D. W. Capone II, J. App. Phys. **64**, 1995 (1988).
30. G. Van Tendeloo, H. W. Zandbergen, and S. Amelinckx, Solid State Comm. **63**, 603 (1987).
31. I. Reyes-Gasga, T. Krekels, G. Van Tendeloo, J. van Landuyt, S. Amelinckx, W. H. M. Bruggink, and H. Verweij, Physica C **159**, 831 (1989).
32. R. Beyers, B. T. Ahn, G. Gorman, V. Y. Lee, S. S. P. Parkin, M. L. Ramirez, K. P. Roche, J. E. Vazquez, T. M. Gür, and R. A. Huggins, Nature **340**, 619 (1989).
33. C. H. Chen, D. J. Werder, L. F. Schneemeyer, P. K. Gallagher, and J. V. Waszczak, Phys. Rev. B **38**, 2888 (1988).
34. M. A. Alario-Franco, C. Chaillout, J. J. Capponi, and J. Chenavas, Mat. Res. Bull **22**, 1685 (1987).
35. T. E. Mitchell, T. Roy, R. B. Schwarz, J. F. Smith, and D. Wohlleben, J. Electron Microsc. Tech. **8**, 317 (1988).
36. D. J. Werder, C. H. Chen, R. J. Cava, and B. Batlogg, Phys. Rev. B **37**, 2317 (1988).
37. Y. Zhu, A. R. Moodenbaugh, M. Suenaga, and J. Tafto, Physica C **167**, 363 (1990).
38. Y. Zhu, M. Suenaga, and A. R. Moodenbaugh, *Proc. International Workshop on Superconductivity*, sponsored by ISTEC and MRS, Honolulu, Hawaii (1992), 165.
39. Y. Zhu, R. L. Sabatini, Y. L. Wang, and M. Suenaga, J. Appl. Phys. **73**, 3407 (1993).
40. M. Daeumling, J. Seuntjens, and D.C. Larbalestier, Nature **346**, 332 (1990).
41. Y. Xu, M. Suenaga, J. Tafto, R. L. Sabatini, and A. R. Moodenbaugh, Phys. Rev. B **39**, 6667 (1989).
42. R. Wordenweber, G. V. S. Sastray, K. Hieneman, and H. C. Freyhardt, J. Appl. Phys. **65**, 1648 (1989).
43. Y. Maeno, T. Tomita, M. Kyogku, S. Awaji, Y. Aoki, K. Hoshino, A. Minami, and T. Fujita, Nature (London) **328**, 512 (1987).
44. W. W. Schmahl, A. Putnis, E. Salje, P. Freeman, A. Graeme-Baber, R. Jones, K. K. Singh, J. Blunt, P. P. Edwards, J. Loram, and K. Mirza, Philos. Mag. Lett. **60**, 241 (1990).
45. T. Siegrist, L. F. Schneemeyer, J. V. Waszczak, N. P. Singh, R. L. Opila, B. Batlogg, L. W. Rupp, and D. W. Murphy, Phy. Rev. B **36**, 8365 (1988).

46. J. M. Tarascon, P. Barboux, P. F. Miceli, L. H. Greene, G. W. Hull, M. Eibschutz, and S. A. Sunshine, Phys. Rev. B, **37**, 7458 (1988).
47. P. Bordet, J. H. Hodeau, P. Strobel, M. Marezio, and A. Santoro, Solid State Comm. **66**, 435 (1988).
48. G. Banngärtel and K. H. Bennemann, Phys. Rev. B **40**, 6711 (1989).
49. C. Y. Yang, S. M. Heald, J. M. Tranquada, Y. Xu, Y. L. Wang, A. R. Moodenbaugh, D. O. Welch, and M. Suenaga, Phys. Rev. B **39**, 6681 (1989).
50. B. D. Dunlap, J. D. Jorgensen, C. Segre, A. E. Dwight, J. L. Matykiewicz, H. Lee, W. Peng, and C. W. Kimball, Physica C **158**, 397 (1989).
51. X. Jiang, P. Wochner, S. C. Moss, and P. Zxchack, Phys. Rev. Lett. **67**, 2167 (1991).
52. T Kreckels, G. van Tendloo, D. Broddin, S. Amelinckx, L. Tanner, M. Menbod, E. Vanlathem, and R. Deltour, Physica C **173**, 361 (1991).
53. L. E. Tanner, Phil. Mag. A **14**, 111 (1966).
54. I. M. Robertson and C. M. Wayman, Phil. Mag. A **48**, 421, 443, 629 (1983).
55. S. M. Shapiro, J. Z. Larese, Y. Noda, S. C. Moss, and L. E. Tanner, Phys. Rev. Lett. **57**, 3199 (1986).
56. S. Muto, S. Takeda, R. Oshima, and F. E. Fujita, J. Phys. (Condens. Matter) **1**, 9971 (1989).
57. S. Iijima, T. Ichihashi, Y. Kubo, and T. Tabuchi, Jpn. J. Appl. Phys. **26**, L1790 (1987).
58. Y. Zhu, M. Suenaga, and A.R. Moodenbaugh, Philo. Mag. A, **62**, 51 (1990).
59. C. Y. Yang, A. R. Moodenbaugh, Y. L. Wang, Y. Xu, S. M. Heald, D. O. Welch, M. Suenaga, D. A. Fischer, and J. E. Penner-Hahn, Phys. Rev. B **42**, 2231 (1990).
60. Z.-X. Cai, Y. Zhu, and D. O. Welch, Philos. Mag. A, **65**, 931 (1992).
61. S. Semenovskaya and A. G. Khachaturyan, Phys. Rev. Lett. **67**, 2223 (1991); Phys. Rev. B **46**, 6511 (1992).
62. S. Semenovskaya, Y. Zhu, M. Suenaga, and A. G. Khachaturyan, Phys. Rev. B **47**, 12182 (1993).
63. P. B. Hirsch, A. Howie, P. B. Nicholson, D. W. Pashley, and M. J. Whelan, *Electron Microscopy of Thin Crystals*, Butterworths, London (1965).
64. Y. Zhu, M. Suenaga and J. Tafto, Philo. Mag. A **67**, 573 (1993); Y. Zhu and Z. X. Cai, Ultramicroscopy **52**, 539 (1993).
65. Z.-X. Cai, Y. Zhu, and D. O. Welch, Phys. Rev. B **46**, 11014 (1992).
66. Y. Zhu and J. Cowley, Philo. Mag. **69**, 397 (1994).
67. S. Amelinckx and D. Van Dyck, in *Electron Diffraction Techniques*, ed. J. M. Cowley, Oxford University Press, New York (1993).
68. S. Nakahara, S. Jin, R. C. Sherwood, and T. H. Tiefel, Appl. Phys. Lett. **54**, 1926 (1989).
69. J. Rabier and M. F. Denanot, Revue Phys. Appl. **25**, 55 (1990).
70. T. Yoshida, K. Kuroda, and H. Saka, Philo. Mag. A **62**, 573 (1990).
71. H. W. Zandbergen, R. Gronsky, K. Wang, and G. Thomas, Nature **331**, 596 (1988).
72. J. Tafto, M. Suenaga, and R. L. Sabatini, Appl. Phys. Lett. **52**, 667 (1988).
73. M. J. Kramer, L. S. Chumbley, R. W. McCallum, W. J. Nellis, S. Weir, and E. P. Kvam, Physica C **166**, 115 (1990).

74. Y. Zhu, H. Zhang, A. R. Moodenbaugh, and M. Suenaga, *Proc. 12th International Crogress for Electron Microscopy*, San Francisco Oress, Seattle (1990), 72.
75. Y. Zhu, H. Zhang, M. Suenaga, and D. O. Welch, Philo. Mag. A **68**, 1079 (1993).
76. S. Amelinckx and J. van Landuyt, in *Electron Microscopy in Mineralog*, ed. J. M. Christie, J. M. Colley, A. H. Hener, G. Thomas, and N. J. Tighe (1976), 68.
77. M. Murakami, S. Gotoh, H. Fujimoto, K. Yamaguchi, N. Koshizuka, and S. Tanaka, Supercond. Sci. Technol. **4**, S49 (1991).
78. D. F. Lee, X. Chaud, and K. Salama, Jpn. J. Appl. Phys. **31**, 2411 (1992).
79. M. Mironva, D. F. Lee, and K. Salama, Physica C **211**, 188 (1993).
80. Z. L. Wang, A. Goyal, and D. M. Kroeger, Phys. Rev. B **47**, 5373 (1993).
81. D. Bourgault, D. Groult, S. Bouffard, J. Provost, F. Studer, N. Nguyen, B. Raveau, and M. Toulemonde, Phys. Rev. B **39**, 6549 (1989).
82. B. Hensel, B. Roas, S. Henke, R. Hopfengrtner, M. Lippert, J. Ströbel, M. Vildic, and G. Saemann-Ischenko, Phys. Rev. B **42**, 4135 (1990).
83. L. Civale, A. D. Marwick, T. K. Worthington, M. A. Kirk, J. R. Thompson, L. Krusin-Elbaum, Y. Sun, J. R. Clem, and F. Holtzberg, Phys. Rev. Lett. **67**, 648 (1991).
84. M. Konczykowski, F. Rullier-Albenque, E. R. Yacoby, A. Shanlov, Y. Yeshurun, and P. Lejay, Phys. Rev. B **44**, 7167 (1991).
85. W. Gerhauser, G. Ries, H. W. Neumuller, W. Schmidt, O. Eibl, G. Saemann-Ischenko, and S. Klaumunzer, Phys. Rev. Lett. **68**, 879 (1992).
86. V. Hardy, D. Grout, M. Hervieu, J. Provost, and B. Ravean, Nucl. Instrum. Methods B **54** 472 (1991); Physica C **191**, 255 (1992).
87. R. C. Budhani, Y. Zhu, and M. Suenaga, Appl. Phys. Lett. **61**, 985 (1992).
88. R. C. Budhani and M. Suenaga, Solid State Commun. **84**, 8312 (1992).
89. R.C. Budhani, M. Suenaga, and S.H. Liou, Phys. Rev. Lett. **69**, 3816 (1992).
90. Y. Zhu, Z. X. Cai, R. C.Budhani, M. Suenaga, and D. O. Welch, Phys. Rev. B **48**, 6436 (1993).
91. M. F. Ashby and L. M. Brown, Philos. Mag. **8**, 1083 (1963).
92. A. E. H. Love, *The Mathematical Theory of Elasticity*, Sections 99 and 100, Dover, New York (1944).
93. Y. Zhu, Z-X. Cai, and D. O. Welch, unpublished.
94. D. Dimos, P. Chaudhari, J. Mannhart, and F. L. Legoues, Phys. Rev. Lett. **61**, 1653 (1988).
95. S. E. Babcock, X. Y. Cai, D. L. Kaiser, and D. C. Larbalestier, Nature **347**, 167 (1990).
96. D. M. Hwang, T. S. Ravi, R. Ramesh, S.-W. Chan, C. Y. Chen, L. Nazar, X. D. Wu, A. Inam, and T. Venkatesan, Appl. Phys. Lett. **57**, 1690 (1990).
97. C. B. Eom, A. F. Marshall, Y. Suzuki, B. Boyer. R. F. W. Pease, and T. H. Geballe, Nature **353**, 544 (1991).
98. Y. Zhu, Z. L. Wang, and M. Suenaga, Philo. Mag. A **67**, 11 (1993).
99. Y. Zhu, Y. L. Corcoran, and M. Suenaga, Interface Science 1, **359** (1994).
100. M. Kawasaki, P. Chaudhari, and A. Gupta, Phys. Rev. Lett. **68**, 1065 (1992).
101. K. Jagannadham, and J. Narayan, Mater. Sci. and Eng. B **8**, 201 (1991).

102. W. Bollmann, *Crystal Defects and Crystalline Interfaces,* Springer-Verlag, Berlin (1970);, *Crystal Lattices, Interfaces, Matrices* , published by the author (1982).
103. A. P. Sutton and V. Vitek, Phil. Trans. Roy. Soc. A **309**, 1, 37, 55 (1983).
104. D. A. Smith and R. C. Pond, Int. Metals Rev. **205**, 61 (1976).
105. R. W. Balluffi, ed., *Grain-Boundary Structure and Kinetics,* ASM, Metals Park, Ohio (1980).
106. C. P. Sun and R. W. Balluffi, Philos. Mag. A **46**, 49, 63 (1982).
107. R. W. Balluffi and P. D. Bristowe, Surf. Sci. **144**, 28 (1984).
108. C. T. Forwood and L. M. Clarebrough, *Electron Microscopy og Interfaces in Metals and Alloys,* IOP Publishing Ltd., Adam Hilger (1990).
109. H. Grimmer and D. H. Warrington, Acta Crystallogr. A **43**, 232 (1987).
110. R. Bonnet and F. Durand, Philos. Mag. **32**, 997 (1975).
111. R. Bonnet, E. Cousineau, and D. H. Warrington, Acta Crystallogr. A **37**, 184 (1981).
112. A. Singh, N. Chandrasekhar, and A. H. King, Acta Cryst. B **46**, 117 (1990).
113. S.-W. Chan, D. M. Hwang, R. Ramesh, S. M. Sampere, and L. Nazar, AIP Conf. Proc. No.200, *High T_c Superconducting Thin Films,* ed. R. Stockbauer, American Institute of Physics (1990), 172.
114. L. A. Tietz and C. B. Carter, Physica C **182**, 241 (1991).
115. A. F. Marshall and C. B. Eom, Physica C **207**, 239 (1993).
116. D. H. Shin, J. Silcox, S. E. Russek, D. K. Lathrop, B. Moeckly, and R. A. Buhrman, Appl. Phys. Lett. **57**, 508 (1990).
117. D. A. Smith, M. F. Chisholm, and J. Clabes, Appl. Phys. Lett. **53**, 2344 (1988).
118. Y. Zhu, H. Zhang, H. Wang, and M. Suenaga, J. Mater. Res. **6**, 2507 (1991).
119. J.-Y. Wang, A. H. King, Y. Zhu, and M. Suenaga, presented in MRS Fall Meeting, Boston (1992).
120. N. Tomita, Y. Takahashi, M. Mori, and Y. Ishida, preprint (1992).
121. Y. Zhu, H. Zhang, H. Wang, and M. Suenaga, J. Mater. Res. **6**, 2507 (1991).
122. P. Humble and C.T. Forwood, Phil. Mag. A **31**, 1011 (1975).
123. C. T. Forwood and P. Humble, Phil. Mag. A **31**, 1025 (1975).
124. K. Shin and A. H. King, Mater. Sci. & Eng. A **113**, 121 (1989).
125. F. C. Frank, *Proc. Symposium on Plastic Deformation of Crystalline Solids,* Mellon Institute, Pittsburgh, May (1950), 150.
126. S. Amelincx and W. Dekeyser, Solid State Phys. **8**, 325 (1959).
127. J. W. Christian, *New Aspects of Martensitic Transformations,* Kobe: Japan Institute of Metals (1976).
128. Y. Zhu, et al. Philo. Mag. A **69**, 717 (1994).
129. G. A. Bruggerman, B. H. Bishop, and W. H. Hartt, *The Nature and Behavior of Grain Boundaries,* ed. H. Hu, Plenum, New York (1972), 83.
130. F.-R. Chen and A. H. King, Acta Crystallogr. B **43**, 416 (1987).
131. F.-R. Chen and A. H. King, Phil. Mag. A **57**, 431 (1988).
132. J. D. Jorgensen, B. W. Veal, A. P. Paulikas, L. J. Nowicki, G. W. Crabtree, H. Claus, and W. K. Kwok, Phys. Rev. B **41** 1863 (1990).
133. R. J. Cava, A. W. Hewat, E. A. Hewat, B. Batlogg, M. Marezio, K. M. Rabe, J. J. Krajewski, W. F. Peck, and L. W. Rupp, Jr., Physica C **165**, 419 (1990).

134. N. Nücker, J. Fink, T. C. Fuggle, P. J. Durham, and W. M. Temmerman, Phys. Rev. B **37**, 5158 (1988); Phys. Rev. B **39**, 6619 (1989).
135. T. Takahashi, H. Matsuyama, T. Watanabe, H. Katayama-Yoshida, S. Sato, N. Kosugi, A. Yagishita, S. Shamoto, and M. Sato, *Proc. 3rd Intern. Symp. on Supercond.*, ed. K. Kajimura and H. Hayakawa, Springer-Verlag, Berlin (1990), 75.
136. P. E. Batson and M. F. Chisholm, J. Elec. Micro. Tech. **8**, 311 (1988).
137. O. Eibl, P. van Aken, and W. P. Müller, Phys. stst. sol. (a) **128**, 129 (1991).
138. R. F. Egerton, *Electron Energy-Loss Spectroscopy in the Electron Microscopy*, Plenum Press, New York (1986).
139. Y. Zhu, J. M. Zuo, A. R. Moodenbaugh, and M. Suenaga, Philo. Mag. A, in press (1994).
140. W. T. Read and W. Shockley, Phys. Rev. **78**, 275 (1950).
141. S. E. Babcock, T. F. Kelly, P. J. Lee, J. M. Seuntjens, L. A. Lavanier, and D. C. Larbalestier, Physica C **152**, 25 (1988).
142. S. E. Babcock and D. C. Larbalestier, Appl. Phys. Lett. **55**, 393 (1989).
143. D. M. Kroeger, A. Choudhury, J. Brynestad, R. K. Williams, R. A. Padgett, and W. A. Coghlan, J. Appl. Phys. **64**, 331 (1988).
144. K. B. Alexander, D. M. Kroeger, J. Bentley, and J. Brynestad, Physica C **180**, 337 (1991).
145. G. Cliff and G. W. Lorimer, J. Microsc. **103**, 203 (1975).
146. J. R. Michael, C.-H. Lin., and S. L. Sass, Scripta Metall. **22**, 1121 (1988).
147. Y. Gao, K. L. Merkel, G. Bai, H. L. M. Chang, and D. J. Lam, Physica C **174**, 1 (1991).
148. C. L. Jia, B. Kabius, K. Urban, K. Herrmann, J. Schubert, W. Zander, and A. I. Braginski, Physica C **196**, 211 (1992).
149. Z. X. Cai et al., unpublished.
150. M. F. Chisholm and S. J. Pennycook, Nature **351**, 47 (1991).

5.4 Recommended Readings for Chapter 5

1. *Electron Microscopy of Thin Crystals*, P. B. Hirsch, A. Howie, D. P. Nicholson, D. W. Pashley, and M. J. Whelan, London, Butterworth, 1965.

2. Practical Electron Microscopy in Materials Science, J. W. Edington, London, Van Nostrand Reinhold, 1976.

3. *Diffraction and Imaging Techniques in Materials Science*, S. Amelinckx, R. Gebers, and J. Van Landuy, eds., Amsterdam, North Holland, 1978.

4. *Crystal Defects and Crystalline Interfaces*, W. Bollmann, Berlin, Springer, 1970.

5. *Grain-Boundary Structure and Kinetics*, R. W. Ballufi, ed., Ohio, ASM, 1970.

6. *Electron Microscopy of Interfaces in Metals and Alloys*, C. T. Forwood and L. M. Clarebrough, New York, Hilger, 1991.

7. *Interfaces in High-T_c Superconducting Systems*, S. L. Shine and D. A. Rudman, eds., New York, Springer, 1993.

6

Transport Critical Currents

H. Jones and R. G. Jenkins

This chapter is about transport critical currents. Definitions will be given in the next section, but it is important to set the scope of the chapter. For high-T_c superconductors (HTS), transport current measurements are important as technological usage becomes nearer, particularly in power engineering applications such as superconducting magnets and machines. Therefore emphasis will be given to measurements on significantly sized conductors that incorporate bulk material, involving currents perhaps up to hundreds of amperes and certainly tens of amps. We will concentrate entirely on *direct* measurement using an injected transport current; thus, we will not address transport critical currents inferred from magnetization measurements. For magnets, high-T_c bulk materials are of great interest, as their upper critical fields (B_{c2}) are known to be very high at low temperatures, and insert coils have been built that have added incremental fields to backgrounds to give total fields in excess of 23 tesla [1]. Such high fields would not be possible for low-temperature superconductors (LTS), with the exception of Chevrel-phase materials (which have still to be developed). For this reason measurements at low temperatures (i.e., in the liquid helium regime) and in high magnetic fields will feature strongly. Much of the chapter will be general because since many of the principles discussed will refer equally to LTS as well as HTS. Where there are unique features, or features that are particularly important for HTS materials, these will be addressed specifically.

The chapter is divided into sections that consider different aspects of transport critical currents. We begin by defining some of the terms used. Then, we discuss the voltage-current (V-I) characteristic in general terms and also consider intrinsic phenomena, particular to HTS, that govern the nature of transport currents in these materials. Our focus then turns to the methods used to carry out transport critical current measurements, as well as to problems that can arise particularly with HTS. After enlarging on the effects of field, temperature, and sample orientation, we will then discuss the technological implications of transport critical current characterizations. Finally, we draw some conclusions about the state of the art in the area of transport critical currents.

6.1 Terminology

In this section we define and elaborate on some of the terms used within the context of this chapter. The definitions are not rigorous but are appropriate for the discussions that follow in the later sections. As stated in the introduction, we are restricting ourselves mainly to the consideration of transport currents in HTS conductors, which can sustain high currents—tens or hundreds of amps—in applied magnetic field and remain in the superconducting state. All the topics addressed in this section will be discussed in detail in the ensuing sections.

6.1.1 Superconductivity and the Critical Parameters

In Figure 6.1 we show the I,B,T surface that bounds the superconducting state. For practical purposes, below this surface one has a *superconductor*, above; a *normal conductor*. In other words, provided the temperature is below a certain value (the *critical temperature*, T_c) and provided the current flowing does not exceed the *critical current* (I_c), it is a superconductor in zero magnetic field. In a magnetic field, whether self-generated by the transport current or applied externally, the terminology becomes slightly more complicated.

The critical field for these purposes is known as the *upper critical field* (B_{c2}) to distinguish it from the *lower critical field* (B_{c1}), the point at which flux lines start to penetrate. Thus, the lower critical field delineates the transition from the pure diamagnetic state of a type I superconductor to the mixed state of a type II superconductor. The *mixed state* is thus the region between B_{c1} and B_{c2}. We will concern ourselves no further with type I superconductors as they are of little interest practically. For many years, the I,B,T surface shown was associated with LTS materials, of which the most commonly used are NbTi and Nb$_3$Sn and are routinely available commercially as practical conductors, usually consisting of fine filaments of superconductive material embedded in a copper or copper-bearing matrix.

T_c and B_{c2} are intrinsic parameters of any superconductor. I_c is extrinsic and is affected by many factors such as processing routes. The critical current is a meaningful parameter only when related to a particular conductor. There-

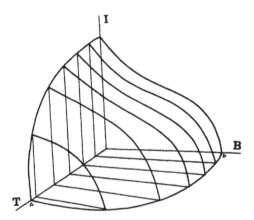

Fig. 6.1. The I,B,T surface, below which superconductivity occurs; above, normal conducting behavior. The vertical axis could equally be labeled J (current density). The triangular symbols on the horizontal axes mark T_c (critical temperature) and B_{c2} (upper critical field) at zero transport current. I_c and B_{c2} are intrinsic parameters. I_c (or J_c) is an extrinsic parameter.

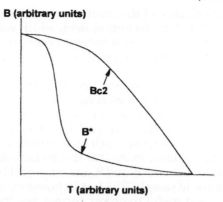

B (arbitrary units)

Bc2

B*

T (arbitrary units)

Fig. 6.2. B(T). This qualitative figure illustrates the difference between B_{c2} at zero transport current and B*, the critical field with a nonzero transport current.

fore, in general terms, we usually refer to the *critical current density* (J_c), which is I_c/a, where a is the cross-sectional area of the conductor.

A cautionary note needs to be sounded here. For LTS, the values of J_c given are usually calculated using the cross-sectional area of the superconductor composite; that is, they include any matrix. For HTS, however, there is a tendency to quote only the superconductor area, thus giving an optimistic value by traditional standards.

6.1.2 High-Temperature Superconductors

Until Bednorz and Muller's discovery of the superconducting ceramic LaBaCuO in 1987 [2], T_cs had not exceeded ~23 K (in Nb$_3$Ge, a material that did not develop technologically). The two "bread and butter" superconductors, NbTi and Nb$_3$Sn, have T_cs of 9.3 K and 18 K, respectively. Hence LaBaCuO, with a T_c of 32 K, was indeed a *high-temperature superconductor*. The rapid subsequent discovery of YBaCuO (T_c ~90 K) and BiSrCaCuO (T_c ~105 K for the 2223 phase) for the first time gave superconductors that would operate in liquid nitrogen. The highest T_c so far is that of TlBaCaCuO at 125 K.

All these "liquid nitrogen" superconductors are being developed for transport current carrying wires and tapes, but particular interest has been shown in the BSCCO material, both the 2223 phase (T_c = 106 K) and the 2212 phase (which has a T_c of only ~90 K but, generally speaking, can be processed more easily into practical conductors). Note the shorthand: Bi$_2$Sr$_2$Ca$_2$Cu$_3$O becomes BSCCO 2223, and Bi$_2$Sr$_2$Ca$_1$Cu$_2$O becomes BSCCO 2212, or simply 2223 and 2212. These two materials feature most in this chapter.

In Figure 6.2 we see how the B_{c2} at a given temperature, T, varies as a function of field. In classical superconductors this can be approximated by the parabolic relation

$$B_{c2}(T) = B_{c2}(0) [1-(T/T_c)^2],$$ (6.1)

where $B_{c2}(0)$ is B_{c2} at absolute zero of temperature.

This is true in the absence of transport current. Also shown in Figure 6.2 is the variation of B*(T). B* is the critical field, which we define as the field at which the critical transport current becomes vanishingly small at a given temperature [3]. (Note: later we will call it B_s to be consistent with the literature referred to.)

In other words, it is the *useful* critical field from the point of view of transport current applications. In LTS materials there is very little difference between the B_{c2} and B* lines; but, in the case of HTS, there can (as shown in the figure) be a big difference between B* and B_{c2}. From a fundamental standpoint the B* line has various names such as the *irreversibility line* or the *flux lattice melting line*. No evidence exists that it is an intrinsic property of HTS, and the size of the region between B* and B_{c2} can vary considerably from material to material and sample to sample. The fact is, however, that only below this line can an HTS support useful transport current density in the presence of a magnetic field. Hence, for the time being, it would seem that high current densities in significant fields at the higher temperatures are unlikely. (It should be stressed that Figure 6.2 is qualitative and purely for illustrative purposes. A quantitative version is shown in Figure 6.20.)

6.1.3 Transport Current

Transport current is simply the net current flowing from one end of a conductor to the other, as distinct from localized circulatory currents such as eddy currents or, in superconductors, screening currents.

6.1.4 Flux Pinning

Flux pinning is the mechanism whereby the dissipative movement of flux lines under the Lorentz force is prevented by *flux pinning centers*. These may be various defects such as dislocations, precipitates, and grain boundaries. Good flux pinning is essential for significant transport currents to be sustained in the superconducting state. Pins can be broken by thermal activation. Thus, critical currents (generally speaking) are low at temperatures approaching T_c in HTS materials in magnetic fields.

6.1.5 Criteria for Critical Current

Here we define the two criteria most commonly used to define critical currents.

Electric field: In the V-I characteristics (discussed later in this chapter) I_c is determined as the current that gives rise to a voltage appropriate to an arbitrarily determined electric field (E) along the superconductor. The most commonly used values of E are 10 and 100 µV m^{-1}.

Resistivity: The current at which the effective resistivity reaches a given level. Commonly used values are 10^{-12} and 10^{-14} Ωm.

Researchers still disagree about the best criterion to use for I_c determination. In our experience, however, for HTS, electric field is by far the most commonly stated criterion.

6.1.6 Index of Transition

The *index of transition* (n) is a measure of the "sharpness" of transition.

Scientists have empirically determined that, at transition,

$$V = KI^n,$$ (6.2)

where K is a constant of proportionality.

Thus, measurements at two electric field criteria separated by an order of magnitude, such as 10 and 100 $\mu V\ m^{-1}$, would give two values of I_c: I_{c1} and I_{c2}. Hence, the index would be given by

$$n = [\log(I_{c2}/I_{c1})]^{-1},$$ (6.3)

where $I_{c2} \geq I_{c1}$.

6.1.7 Practical (or Technical) Superconductor

One can easily define what is meant by a *practical superconductor* when one is talking about LTS. For HTS, on the other hand, such a definition is extremely difficult. We will define it here as a conductor of significant length that can be fashioned into one of the usual geometries for critical current measurements (discussed further below). Typically such a conductor would be a composite of HTS and Ag in wire or tape form.

6.2 E-J Characteristics

The transport critical current, I_c, of a superconductor is not a quantity that can be measured directly. Instead, we measure the V-I characteristic of the sample, using techniques described later. A value for I_c may then be obtained by applying arbitrary electric field or resistivity criteria. Critical currents determined in this way may be used to compare different samples, and their variation with applied fields and temperatures may tell us about the factors limiting the current-carrying capability of the material. We must be careful, however. By extracting a single "critical" value of current from the V-I curves, we are discarding an enormous amount of important information on the sample.

For example, consider the idealized voltage-current characteristics of two 1-cm-long samples shown in Figure 6.3. Sample 1 is completely resistive, and sample 2 is a superconductor. If we determine their "transport critical currents" by applying an electric field criterion of 1 μVcm^{-1}, we arrive at the same result for both. Thus, if one looks only at the transport critical current data, the samples appear identical; however, from the measured V-I characteristics, the experimenter knows differently. Although this is an extremely simple example, it does illustrate the fact that since I_c values are arbitrary, knowledge of the V-I curves from which they were extracted is necessary if they are to be useful. In addition, the V-I characteristic of a superconductor contains much more information than the "critical" current.

In this section we begin in quite a general way by discussing the V-I (or, equivalently, the electric field-current density, E-J) characteristics of homogeneous, isotropic, defect-free type II superconductors. Then we introduce the effects of inhomogeneities able to "pin" flux lines, as well as the effects of thermal fluctuations. Attempts to model these characteristics will be described, and finally we will look at the particular considerations for HTS materials.

6.2.1 The E-J Characteristic of a Homogeneous, Isotropic, Defect-Free Type II Superconductor

A type II superconductor exhibits two critical fields, H_{c1} and H_{c2}. Below

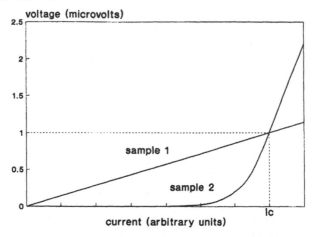

Fig. 6.3. Idealized V-I characteristics of two 1-cm-long samples. Sample 1 is resistive, and sample 2 is superconducting. The critical currents obtained by applying an electric field criterion of 1 μV/cm are the same for both.

H_{c1} the boundary energy of the normal-superconducting state interface is positive, and the full Meissner effect [4] is observed; flux is excluded from the bulk of the superconductor, just as in a type I material. The critical current, when defined as that which causes the material to undergo the transition to the normal state, is given by the Silsbee rule [5]; that is, it produces a mag-

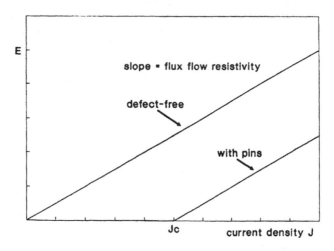

Fig. 6.4. The E-J characteristics of type II superconductors with and without microstructural defects able to pin flux lines. The slope of the lines is equal to the flux-flow resistivity, ρ_f.

netic field at the surface of the superconductor that, when added to any externally applied field, equals the lower critical field, H_{c1}. For a round wire of radius r, in zero applied field the critical current is then

$$I_c = H_{c1} \cdot 2\pi r. \tag{6.4}$$

In applied fields between H_{c1} and H_{c2} the normal-superconducting boundary energy is negative, and, as predicted by Abrikosov [6], flux enters the bulk of the superconductor in the form of quantized "fluxons." The superconductor is said to be in the *mixed state*. In the absence of transport current, these fluxons, by mutual repulsion, form a regular triangular lattice.

In all but the lowest fields above H_{c1}, to a good approximation the magnetic flux densities inside and outside the superconductor are equal; hence one commonly sees the lower and upper critical fields expressed as B_{c1} and B_{c2} in tesla. For a treatment of the errors involved in such an approximation, the reader is referred to [7].

Now, with the superconductor in the mixed state, if we try to pass a transport current, the flux lines experience a Lorentz force and will move unless prevented. Alternatively, from Maxwell's equations for a superconductor (i.e., a material of infinite conductivity),

$$\text{curl } \mathbf{B} = \mu_o \mathbf{J_s}, \tag{6.5}$$

the passage of transport current demands a nonuniform distribution of flux lines, which by mutual repulsion try to return to a regular triangular lattice. This motion leads to the dissipation of energy in essentially two mechanisms. First, as shown by Bardeen and Stephen [8,9], there is a constant electric field in the normal core of a moving fluxon in the direction of motion that drives a dissipative current through that normal region. Second, Tinkham [10] proposed a mechanism in which a fraction of the condensation energy was dissipated (to the production and/or heating of normal electrons) each time the order parameter at a particular point in the superconductor was forced to vary in time as a flux line passed through. Gorkov and Kopnin [11] extended this approach to derive an effective viscosity in the limit of low vortex velocity.

Thus, flux flow under the Lorentz force is lossy and equivalent to motion through a viscous medium. An electric field is developed along the superconductor in the direction of current flow, and the material appears to have a resistance even though there is a continuous superconducting path between the fluxons from one end to the other. The electric field is given by

$$\mathbf{E} = \mathbf{v} \times \mathbf{B} \tag{6.6}$$

or

$$\mathbf{E} = \rho_f \mathbf{J}, \tag{6.7}$$

where $\mathbf{v}$ is the velocity of the flux lines and ρ_f is the flux flow resistivity. The E-J characteristic of a defect-free type II superconductor in the mixed state is shown in Figure 6.4.

6.2.2 The E-J Characteristic of a Type II Superconductor with "Pins"

For a defect-free type II superconductor, all current flow in magnetic fields greater than B_{c1} is lossy. Clearly a type II superconductor will be able to carry a lossless current in the mixed state only if in some way the flux lines can be "pinned." Fortunately, pinning is provided by microstructural defects in the crystal such as dislocations, impurities, and precipitates. The subject is described in detail in Chapter 5; here it is sufficient to appreciate that defects that pin flux lines do so because they act as free-energy wells. The depth of the well, hence the strength of the interaction, is related to the magnitude of the difference between the superconducting properties of the defect and the perfect crystal.

The pin-fluxon interactions over the whole superconductor provide a pinning force per unit volume, which can oppose the Lorentz force and thus enable the material to carry a lossless current in the mixed state. The maximum value of this force, F_p, determines a "critical current" above B_{c1}, defined as that current at which flux first starts to flow.

If we define critical current density, J_c, as that J at which the Lorentz force equals the maximum pinning force,

$$F_p = J_c \times B, \tag{6.8}$$

then neglecting thermal effects, the E-J relationship becomes

$$E = \rho_f (J - J_c). \tag{6.9}$$

This is shown in Figure 6.4. Only when the Lorentz force exceeds F_p does flux begin to flow and generate an electric field along the superconductor.

We can increase J_c by increasing the pinning force per unit volume; in the LTS material NbTi, for example, this is achieved by cold working, which produces many dislocations in the microstructure. We cannot increase J_c indefinitely, however. Even if we could prevent flux motion entirely, there is an intrinsic limit to the supercurrent density. This limiting value is called the "depairing current," J_{cd}, and corresponds to the current density at which the kinetic energy of the superconducting electron pairs equals the condensation energy. From the phenomenological London theory [12] it can be shown that

$$J_{cd} = H_c / \lambda, \tag{6.10}$$

where λ is the penetration depth. Typical values for the depairing current are 10^8 and 10^9 Acm^{-2} for low- and high-temperature superconductors, respectively.

Now, in any real superconductor the defects that determine F_p will not be perfectly uniformly distributed; their type, size, orientation, and density will vary throughout the material. Hence it is appropriate to treat the superconductor as a collection of component regions, each with its own characteristic ρ_i, F_{pi}, and J_{ci}. The E-J characteristic of the material is given by a summation of terms like Equation 6.9. Thus, we can write (after Jones et al. [13])

$$E = \Sigma \, \rho_i (J - J_{ci}). \tag{6.11}$$

We can extend this approach further to allow for a continuous distribution of local pinning forces and hence critical current densities. We introduce the distribution function $f(J_{ci})$, where $f(J_{ci}).dJ_{ci}$ equals the fraction of the material with critical current densities between J_{ci} and $J_{ci}+dJ_{ci}$. After Baixeras and Fournet [14]:

$$E = \rho_f \int_0^J f(J_{ci})(J-J_{ci})dJ_{ci}, \tag{6.12}$$

where

$$f(J_{ci}) = (1/\rho_f)d^2E/dJ^2, \tag{6.13}$$

where we have taken the flux flow resistivity of each region to be the same (equal to ρ_f). An E-J characteristic for an arbitrary distribution function is shown in Figure 6.5.

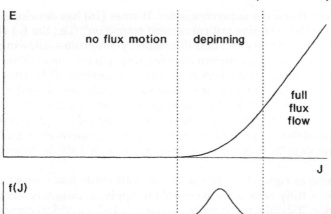

Fig. 6.5. The E-J characteristic of a type II superconductor with a distribution of pinning strengths, hence a distribution of local critical currents, throughout its microstructure. The distribution function f(J) is shown below.

At very low current densities, below the low J tail of the distribution, the Lorentz force is everywhere smaller than the maximum local pinning force; hence all flux is pinned, and E = 0. As J increases further and we enter the distribution of local critical currents, an increasing number of local J_{ci}'s are exceeded, and an increasing fraction of the flux line lattice is in motion. This part of the E-J characteristic is nonlinear. Eventually, at sufficiently high J, all J_{ci}'s have been exceeded, all flux is in motion, and we see a linear portion, with slope corresponding to the full flux-flow resistivity.

It should be noted that Equation 6.12 is completely general; no assumptions have been made as to the nature of the distribution. By inserting an appropriate distribution function, however, this equation can form the basis for an attempt to model E-J characteristics. We will return to this point later.

Clearly, from Equation 6.13 the distribution function for a real sample can be extracted from the E-J characteristic by calculating the second derivative. Warnes and Larbalestier [15] have done this for low-temperature superconducting composites incorporating NbTi and Nb$_3$Sn. They found that the distribution was essentially Gaussian.

At this point we briefly discuss the effect on the E-J characteristic of combining a superconductor with a normal metal, in a composite conductor. Until now, we have considered the characteristics of the superconductor alone. In most high-current, high-field applications, however, the superconductor, in the form of a single filament or a number of filaments, is embedded in a normal metal matrix, for reasons involving the stability of the superconducting state to thermal fluctuations. The presence of this normal metal, of course, modifies the E-J curve, since, at any finite electric field, some current sharing occurs between it and the superconductor. Warnes [16] has developed models explicitly for the "resistive critical current transition" (i.e., the E-J characteristic) in monofilamentary and multifilamentary composites, allowing for axial as well as axial and radial current transfer, respectively, through the matrices around "weak links" in the superconducting filaments. This work is applicable to both low- and high-temperature superconductors if used in composite form; the reader is referred to the original paper for a complete discussion of results. Here we briefly quote the result that the measured critical current distribution of the composite is narrower than the underlying distribution of critical currents in the superconducting component, and also is shifted to higher currents.

Returning to Figure 6.5, we see that at sufficiently low current densities, dissipation is truly zero. If we were able to apply an infinitely small electric field criterion, it should, in theory, be possible to define a critical current density in an unarbitrary way as that current density at which flux motion first began. When we introduce the effects of thermal fluctuations, however, this is no longer possible, as is discussed in the next subsection.

6.2.3 Effects of Thermal Fluctuations

The microstructural defects that pin flux lines do so because they are free-energy wells. In the absence of an applied force, a fluxon will sit at the bottom of the well. As an increasing force is applied, the fluxon will move further and further from its equilibrium position (of lowest energy) until, at sufficiently high force, it will break free. This was the situation considered in the preceding subsection. There is, however, another mechanism by which the fluxon can escape the well, namely, *thermal activation*. At any finite temperature, a finite probability exists that a fluxon will "hop" out of the potential well (i.e., over a potential barrier) owing to its thermal energy, even in the absence of an applied force.

The theory of the thermal activation of flux lines over free energy barriers coming from the pinning effects of inhomogeneities was first introduced by Anderson in 1962 [17], who called it "flux creep." His motivation was the desire to explain the appreciable temperature dependence of critical currents and critical fields in conventional superconductors at temperatures as low as $0.1\ T_c$, where none of the properties of the bulk superconducting state varied noticeably. Its particular relevance to HTS materials operating at elevated temperatures was first recognized by Dew-Hughes in 1988, who in an excellent paper [18] developed a flux-creep model starting from the general formalism of stress-assisted thermal activation. His model is described below.

First, we assume that some entity sits at the bottom of a potential well of arbitrary shape and depth, U. In the absence of any imposed stress, the entity can, by thermal activation, hop out of the well at a rate R, in both the forward and backward directions, given by

$$R = R_0 \exp\ (-U/kT), \tag{6.14}$$

where R_0 is an attempt frequency. As the forward and backward hopping rates are the same, no net motion is observed. Now, when a stress is imposed, favoring hopping in what we will call the forward direction, the rates of forward (R_f)

and backward (R_b) hopping become

$$R_f = R_0 \exp[-(U - W) / kT],$$ (6.15)

$$R_b = R_0 \exp[-(U + W) / kT],$$ (6.16)

where W is the work done by the imposed stress in moving the entity from its equilibrium position at the bottom of the well to the next equilibrium position. This leads to a net forward rate of hopping $R_{net} = R_f - R_b$ given by

$$R_{net} = 2 R_0 \exp(-U/kT) \cdot \sinh (W/kT).$$ (6.17)

Now considering the particular case of flux creep in superconductors, we referred above, somewhat vaguely, to a pinned "entity." As Anderson pointed out in his original paper, this will not necessarily be a single fluxon. In general, it will be a "bundle" of flux lines, bound together to some extent by the interaction of their magnetic fields and wave functions. He estimated that the volume of such a bundle would be d^3, where the dimension d is similar to the London penetration depth. In his model, Dew-Hughes assumes that the bundle volume is the entire volume of flux line lattice associated with one pinning center. If the number density of pins is n_p, then at sufficiently high fields (where each pin is intersected by a flux line) this volume $v_c = 1/n_p$. Within each bundle, the flux lines are in a regular triangular lattice. When a flux line is unpinned, this volume moves to the next equilibrium position, which is when the following flux line now intersects the pin. The distance moved each time is therefore equal to the interflux-line spacing a_0 (= $1.07(\Phi_0 /B)^{0.5}$).

The potential well depth, U, can be equated to the difference between the free energy of the flux line when it is in and out of the pin. U will, in general, be a function of B and T.

The pinning energy per unit volume is simply U x n_p; and as the distance moved by the flux line during the unpinning process is a_0, the pinning force per unit volume, F_p, is given by

$$F_p = J_0 B = U n_p / a_0,$$ (6.18)

where J_0 is the critical current density in the absence of flux creep.

The imposed stress is due to the Lorentz force, and the work done in moving the flux bundle from one equilibrium position to the next is

$$W = J B v_c a_0 = JU/J_0$$ (6.19)

(note that when $J = J_0$, W = U). Finally, recalling Equation 6.6, we have $E = v \times B$, where v is the velocity of the flux lines. This is simply the hopping rate multiplied by the displacement each time. Thus,

$$E = R_{net} a_0 B = 2 a_0 B R_0 e^{-U/kT} \cdot \sinh(W/kT)$$ (6.20)

$$E = 2 a_0 B R_0 e^{-U/kT} \cdot \sinh(JBv_c a_0/kT).$$ (6.21)

Immediately we see that at any finite temperature, dissipation is nonzero; a critical current cannot be uniquely defined since it must be extracted from E-J characteristics by applying arbitrary criteria.

It is instructive to consider two limiting cases, in which the expression above can be simplified. First, if W >> kT (which also implies that U >> kT), as is the case for very high stresses or very low temperatures, then

$$E = a_0 B R_0 \exp [(W - U)/kT]$$ (6.22)

$$E = a_0 B R_0 \exp [U(J - J_0)/J_0 kT].$$ (6.23)

Second, if W << kT (implying U < kT), corresponding to high-temperature operation, then

$$E = [(2 a_0 B R_0 J U) /kT] . e-U/kT. \qquad (6.24)$$

Thus, flux creep at high temperatures results in a resistivity that is linear with B.

Let us return to Figure 6.5, which shows the E-J curve for a type II superconductor with a nonuniform distribution of pins in its microstructure. What happens when we incorporate the effects of thermal activation described above?

We find that at low temperatures, such as those encountered in the operation of conventional low-T_c superconductors (typically 2–4 K), the shape of the curve is essentially unaltered. To a very good approximation, the second derivative is still proportional to the distribution function. The slight difference is that at very low current densities (and hence at very low electric fields), the curve is exponential. On this portion of the characteristic, however, the electric fields, which are due to flux creep, may be too small to observe with conventional four-terminal techniques.

At high temperatures, on the other hand, even when using four-terminal techniques, flux creep may result in no "zero" dissipation region being observed. Instead we see a resistive characteristic at low J whose slope is proportional to the applied field, B. The nonlinear portion of the characteristic (which in the absence of thermal effects corresponded to an increasing number of component critical currents being exceeded) may also be significantly affected, to the extent that the second derivative can no longer be considered accurately to represent $f(J_{ci})$.

In the following section we consider attempts to model E-J characteristics without taking into account thermal fluctuations. These attempts were originally developed for application to low-temperature superconductors, where the errors involved in neglecting thermal activation are small. For HTS materials operating at elevated temperatures, however, thermal effects are much more important; particular HTS considerations are dealt with later in the chapter.

6.2.4 Models for E-J Characteristics

Having described the various factors affecting the E-J characteristics of isotropic type II superconductors, namely, dissipative flux flow, distribution of pinning strengths and thermal activation, we will briefly discuss two models describing these characteristics. They were developed for conventional, LTS materials and therefore neglect thermal effects (which for these materials, as described in the preceding section, have negligible influence on E-J curves measured by four-terminal techniques).

6.2.4.1 The Empirical Two-Parameter (2-P) Law

Perhaps the most widely used expression to describe the E-J characteristic is
$$E = \alpha J^n, \qquad (6.25)$$
where α is the so-called alpha parameter (a constant) and n is the "index of transition." This expression is attractive for its simplicity but cannot accurately describe real E-J curves over the whole range of the transition (i.e., from zero dissipation to full flux flow). For example, it is common practice when analyzing E-J curves to extract critical current values using two electric field criteria (typically 10 and 100 μVm^{-1}) since, from these values, we can calculate "n." We find that if different criteria are used, n changes (see [19]). Equation 6.25 has been shown, however, to hold over several orders of magnitude

of E for the low-current portion of the J_c distribution by Volker [20] and Plummer and Evetts [21]. And, despite its limited validity, by calculating n values we have a useful indication of the sharpness of the transition. Also, by plotting n versus applied field, one can determine whether, in the case of LTS materials, critical currents are being limited by intrinsic or extrinsic factors [15]. If they are intrinsically limited, n will tend to increase linearly with decreasing field, whereas if the limitations are extrinsic, such as sausaging of filaments, then as B decreases, n will initially increase linearly but then reach a plateau. Finally, it should be noted that this model is purely empirical and involves no speculation as to the underlying mechanisms determining the shape of the E-J curve.

6.2.4.2 The Three-Parameter (3-P) Fit

As we stated earlier, Equation 6.12 is general; no assumptions having been made as to the nature of the distribution. From measurements on real materials Warnes and Larbalestier [15] showed it to be essentially Gaussian, and this led Hampshire and Jones [22,23] to develop a model, described below, for the E-J characteristic based on a normal distribution of critical currents.

They write:

$$E = \rho_1 \int_0^J f(J_i) \, (J - J_i) \, dJ_i,$$ (6.26)

where

$$f(J_i) = (1/(2\pi)^{1/2}) \, (\beta/J_c) \cdot \exp\{-1/2 \, [\beta \, ((J_i - J_c)/J_c) \,]^2\}.$$ (6.27)

These two equations contain three adjustable parameters: β, the synchronization constant, is a measure of the spread of the distribution of critical currents; J_c is the mean critical current density; and ρ_1 is the interaction resistivity. Hampshire and Jones derive limiting forms of the E-J characteristic for three different current density regimes:

$$\text{i) } J \rightarrow \infty \qquad E = \rho_1 (J - J_c)$$ (6.28)

$$\text{ii) } J \sim J_c \qquad E = \rho_1 J_c/\{\beta(2\pi)^{1/2}\}$$ (6.29)

$$\text{iii) } |\beta(J_i - J_c)/J_c|^2 \gg 1$$

$$E = \rho_1\{J_c^3/[(J - J_c)^3\beta^3]\} \cdot (2\pi)^{1/2} \cdot \exp\{-1/2[\beta(J-J_c)/Jc]^2\},$$ (6.30)

where the condition for Equation 6.30 is equivalent to the low-voltage limit, and expressions relating their three physical parameters to the empirical parameters of the 2-P law:

$$n = (\pi/2)^{1/2}\beta,$$

$$\text{and } \alpha=(1/2\pi)^{1/2}\rho_1/(\beta[J_c^{(\beta\sqrt{(\pi/2)}-1)}]) \Big\} \text{ for } J \sim J_c.$$ (6.31)

An important feature of Hampshire and Jones's model is the use of ρ_1, the interaction resistivity, many orders of magnitude smaller than the full flux-flow resistivity, ρ_F. The reason for this is that at currents above the distribution function, only defects in the flux-line-lattice are assumed to be in motion, and these represent only a tiny fraction of the total number of fluxons. A justification of this approach is as follows. In crystals, properties such as mechanical strength are largely determined by defects and not the properties of the lattice itself. For example, plastic flow takes place by the motion of crystal dislocations. Similar features are found in the FLL, and therefore it seems reasonable that these FLL defects should dominate its dynamics. Kramer [24] demonstrated that dislocation dynamics, in particular the activation of Frank-Read sources of single dislocations [25], could be applied to flux motion. Thus it was this mechanism that formed the basis of Hampshire and Jones's model. The validity of flux flow in high current density materials being characterized by defect motion in the FLL was demonstrated by applying the 3-P fit to the analysis of data obtained in the characterization of NbTi, V_3Ga, and Nb_3Sn conductors. Typical results yielded a $\rho_l < 10^{-5} \rho_F$. Thus, at currents above the normal distribution of critical currents, the V-I characteristic was linear, with a resistivity many orders of magnitude smaller than the bulk flux-flow resistivity.

6.2.5 Specific HTS Considerations

The preceding part of this section has dealt with the E-J characteristics of isotropic type II superconductors. We described how flux motion (or flow) was dissipative and showed that the E-J curve was determined by depinning under the Lorentz force or thermal activation. It was stated that for conventional, LTS materials operating at low temperatures (typically 4.2 K), thermal effects were negligible in terms of the E-J curves obtained using four-terminal techniques and that the nonlinear part of the characteristic was due to a distribution of pinning strengths. For HTS materials a number of other factors make the situation more complicated, and these are discussed below in terms of their effects on the E-J curves.

Distribution of Pinning Strengths: Just as in LTS materials, defects in the microstructures of high-T_c superconductors act as flux-pinning sites. Clearly, in any real material, there will be a distribution of pinning strengths (or pinning potentials) which contributes to the nonlinear portion of the E-J characteristic.

Operation at Elevated Temperatures: When HTS materials operate at high temperatures such that thermal energies exceed pinning potentials (i.e., kT > U), then according to Dew-Hughes, the E-J characteristic in the flux- creep regime should become linear (from Equation 6.24) with a slope proportional to the applied field. Such curves have been seen from many samples, including for example YBCO tested at 77 K [26]. At higher currents where, in the absence of thermal effects, depinning would begin, thermal activation now affects this nonlinear portion such that the second derivative is no longer accurately proportional to the distribution function; this nonlinear part is a complicated result of thermal effects and the distribution of pinning energies.

Anisotropy: Unlike the most commonly used conventional LTS materials, NbTi and Nb_3Sn, the three main families of high-T_c superconductors under investigation for technological applications are highly anisotropic. In order of increasing anisotropy they are $YBa_2Cu_3O_x$ (YBCO), $Tl_2Ba_2Ca_{n-1}Cu_nO_{2n+4}$ (TBCCO), and $Bi_2Sr_2Ca_{n-1}Cu_nO_{2n+4}$ (BSCCO). All have a perovskite layer structure, with superconducting Cu-O_2 layers in the *a-b* plane separated by insulating planes in the c direction.

This anisotropy has important implications for pinning and the structure of the flux line lattice (hence for the E-J characteristic). Palstra et al. [27] demonstrated that materials with large electronic anisotropy have intrinsically smaller pinning energies, and such materials exhibit thermally assisted dissipation down to temperatures far below T_c. The structure of the flux line lattice is affected because supercurrents are confined to flow predominantly in the Cu-O_2 planes, with only weak Josephson coupling between layers. Thus, where a flux line passes through one of these planes, it generates a 2D "pancake" vortex of supercurrent. In the mixed state, therefore, the array of flux lines penetrating the superconductor generates a stack of 2D triangular lattices of 2D pancake vortices [28]. Although coupling between pancake vortices in adjacent layers is weak, at low temperatures and fields it may be strong enough to bind them together in a 3D lattice (i.e., with distinct flux *lines*). At higher T and B, however, the system will break down into weakly coupled 2D pancake vortex lattices [29]. This situation has two important effects. First, to a first approximation, pinning potential can be written as the product of condensation energy and correlation volume:

$$(1/2\mu_o)B_c^2 \cdot v_c = U. \tag{6.32}$$

When, at high T and B, the 3D lattice breaks up, this correlation volume is greatly reduced; hence, pinning potentials that were already small are reduced further. Second, in a lattice of weakly coupled pancake vortices the shear modulus is very low; and so, even though some pancakes may be pinned, those not directly pinned are able to shear past unhindered. This phenomenon leads to appreciable flux flow and a linear E-J relation.

Thus, high intrinsic anisotropy results in small pinning energies and a weakly coupled lattice of 2D pancake vortices. At high T and B, coupling is broken, resulting in a further reduction in pinning energies and a weak flux line lattice, E-J, characteristic being affected accordingly. It is important, however, to mention a special case where the anisotropy leads to enhanced pinning. The insulating layers between Cu-O planes act as free-energy wells for vortices aligned in this plane. Thus, for this particular orientation, pinning is strong, evidence of which will manifest itself in the E-J curves. This type of pinning is known as *intrinsic pinning* [30].

Finally, it is appropriate to mention an alternative approach to the question of thermal effects on the flux line lattice in a highly anisotropic superconductor. Fisher [31] studied the effect of thermal motion on vortices in thin films and found that above a certain temperature, T_m, vortices no longer form a regular lattice but instead become a fluid. Also, Kosterlitz and Thouless had earlier shown [32] that any two-dimensional lattice becomes unstable to the formation of dislocations above a temperature T_m, thus melting by a second-order transition, with the shear modulus dropping discontinuously to zero. Although we might expect pinning to be enhanced in a vortex liquid because of the ability of flux lines to distort to take up positions of strongest pinning, the disappearance of the lattice shear modulus (i.e., it loses all stiffness) means that no hinderance remains to affect those vortices not directly pinned shearing past their pinned neighbors. Thus, a melting of the flux line lattice leads to unhindered flux motion under the Lorentz force and a linear E-J relationship. Now, strictly, 2D superconductivity can never be realized in

HTS materials because of finite interaction between layers, but a 2D-like state may be realized under the influence of sufficiently large transport current densities [33].

Granularity: Early work on bulk samples of HTS materials showed that they tended to behave like collections of superconducting grains connected by weak links, that is, Josephson junctions [34,35]. Much effort has been directed to describing the B and T dependence of the J_cs of bulk samples, modeled by using arrays of junctions with a distribution of lengths (see below), but not much information is available on the effects of granularity directly on the E-J characteristics. We state, however, that for highly granular materials, the properties of the weak links dominate the current-carrying capability of the sample, and in general a highly granular material will have a very broad E-J curve (i.e., n is very low). This result can be explained qualitatively in the following way. In the limit of a bulk sample comprising completely randomly aligned grains, then for any direction of applied field there will be a wide distribution of effective pinning strengths and tunneling currents.

Also, and fortunately, fabrication processes have been and are being developed that are able to produce well-textured high-T_c–based superconducting composites with good intergranular connection (for example, the powder-in-tube, or PIT, process [36]). For materials produced in this way, granularity is not a major limiting factor; with very few weak links, the E-J curve (and hence the current-carrying capability) is instead dominated by the distribution of pins, thermal activation, and intrinsic anisotropy.

Incorporation in a Low-Resistivity Matrix: In common with conventional low-T_c superconductors, for practical, high-current-density, high-field applications, HTS materials are incorporated into composite conductors with a normal metal. Typical geometries are tapes with a single or multilayer HTS core, or tapes externally coated with superconductor. The normal metal used is silver (or alloys of silver with small quantities of nickel and magnesium for strength), because it is chemically compatible with the HTS compounds, which are very reactive and easily contaminated to the extent that superconductivity is destroyed, and permeable to oxygen, enabling the correct oxygen content (so crucial to superconductivity in HTS) to be obtained during reaction.

Annealed silver has a very low resistivity at low temperatures, and some evidence suggests that it may dominate the E-J curves of composite conductors at high current densities [37]. At sufficiently high J, where the superconductor alone would normally be in a state of full flux-flow, for a composite, the superconducting component might remain in a flux-creep state, carrying only part of the total transport current, with the remainder being carried by the silver. For example, Gurevich et al. [37] for silver-clad BSCCO tapes estimated that ρ_{Ag} was approximately 10^{-3}–$10^{-4} \times \rho_n$, the normal state resistivity, where full flux-flow resistivity is given approximately by

$$\rho_f = \rho_n (B/B_{c2}). \tag{6.33}$$

Thus, for many materials the E-J characteristic may be linear at high J, with slope equal to the resistivity of the matrix, not the full flux-flow resistivity. It is important also to remember that the resistance of silver is field and temperature dependent; as a rule of thumb, magnetoresistance effects lead to an increase in silver's resistivity by one order of magnitude as the field is

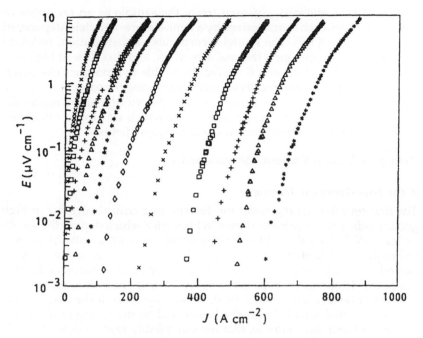

Fig. 6.6. Electric field vs current density characteristics at 80 K in fields applied parallel to the *a-b* plane for a textured bulk sample of BSCCO 2223. Applied fields from right to left are 0.1 T, 0.14 T, 0.18 T, 0.25T, 0.4T, 0.6T, 0.8 T, 0.9 T, 1 T, 1.2 T, and 1.4 T (after Kottman et al. [39]).

increased from zero to 10 tesla [38].

Above, we have discussed some of the factors that determine the E-J characteristics of HTS materials from which transport critical currents may be extracted. Some of these factors will be mentioned again when we cover explicitly the field and temperature dependence of critical currents.

It should be apparent that with so many factors influencing the E-J curve, its form for any particular sample is extremely complicated; there is no simple expression or model to describe it. Experimentally determined E-J characteristics from Kottman et al. [39] for a bulk, textured sample of BSCCO 2223 at 80 K are shown in Figure 6.6. In agreement with the above statements, these data cannot be described unambiguously by a power law or exponential relation over the whole range of electric field. Above $E = 10\ \mu V/m$, a power law seems to apply; below $10\ \mu V/m$, better agreement is obtained with an exponential dependence, corresponding to a classic flux-creep model. J_c values extracted from these curves by using an electric field criterion of $10\ \mu V/m$ are plotted in Figure 6.18 and discussed below.

We end this section by referring to one attractive model for HTS E-J characteristics, which includes the effects of some of the most important factors. It is Griessen's parallel resistor model [40,41] for well-connected (i.e., not weak-linked) HTS material. Griessen treats the sample as an ensemble of domains in parallel, each characterized by a temperature- and field-dependent activation energy, U(T,B). The model incorporates a distribution, m(U), of activation or pinning energies, thermal activation, and flux flow at high current densities. With it, Griessen finds that he is able to reproduce characteristic features of experimental results without introducing the concept of a phase transition or flux line lattice melting. For a full description of the model the reader is referred to Giessen's original papers, in which are plotted calculated E-J curves for a range of temperatures and magnetic fields.

6.3 Transport Critical Current Measurements

6.3.1 The Four-Terminal Technique

The four-terminal arrangement for determining critical currents in high magnetic fields is schematically shown in Figure 6.7 which depicts the qualitative features that are desirable for high-quality measurements. As we discussed earlier, the definition of I_c is arbitrary. It depends on measuring a voltage and therefore requires that the sample not actually be fully superconducting at that point. To get as close as possible to the true transition, we need to resolve very low levels of electric field. Obviously, then, if the distance between the potential taps is large, the larger will be the voltage for a given field. If we assume, for example, that we can reliably resolve down to 1 μV, then for the 10 μV m⁻¹ criterion we require 10 cm between the taps. For reasons we will go into later, we also need a significant separation between the current and voltage contacts, ideally a few centimeters. In other words, a minimum length of ~15 cm is required. We require a homogeneous magnetic

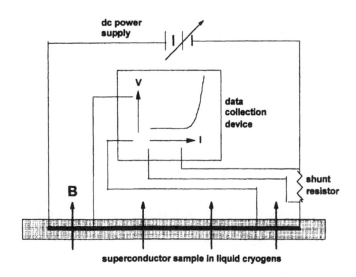

Fig. 6.7. Four-terminal measurement schematic.

field orthogonal to the conductor axis to give the most rigorous Lorentz force conditions for our test, and we would like these fields to be as high as possible, certainly on the order 10 (or more) tesla. The reality is that it would be almost impossible to achieve all these features in a straight sample such as depicted in the figure for real measurements, if for no other reason than it would be difficult, if not impossible, to achieve high fields homogeneously over such a sample length truly orthogonal to the sample axis. As in everything, compromises have to be made.

First, the measurements have to be representative of (it is hoped) a long length of the conductor in question. In even a modest-sized superconducting magnet, many kilometers of wire may be required, and it is not practicable to test the whole length but the magnet designer has to know what $I_c(B)$ is, so a "short sample" is measured. In the well-established LTS conductors, freely available commercially as wires, a typical sample is on the order of 1 m in length. In the case of HTS, however, at the time of writing, long lengths are not (yet) freely available, and other difficulties (dealt with in the next section) may mean that one is forced to use lengths only on the order of 1 cm.

Whatever the sample geometry, it is incorporated in a circuit as depicted schematically in Figure 6.7. The sample current is increased gradually, usually by using some form of ramp generator (sweep unit), and the current is monitored using a series resistor. The voltage across the central portion of the sample is monitored through a high-gain amplifier, and both current and voltage signals are fed into the data collection device. This could be a conventional X-Y recorder; a two-channel, digital storage oscilloscope; or, increasingly these days, some form of computer data-acquisition package which incorporates analogue/ digital conversion. The advantage of this last form of data collection is that the data can be collected, stored, and processed in one device. For instance, values of I_c at one or more E levels can be instantly extracted. Also, dV/dI and higher derivatives, important in analyzing I_c measurements, can be easily obtained. The advantage of the conventional X-Y recorder, on the other hand, is that one can learn a lot from the real-time observation of the V-I characteristic as it is generated (although most of the data-acquisition packages do incorporate a real time display on screen).

6.3.2 Sample Geometries

Figure 6.8 has been shown in various guises in a number of previous publications; it is essentially a catalogue of suitable sample geometries that are used in critical transport current determination. Table 6.1 gives a summary of the advantages and disadvantages of each format.

By far the most useful of the geometries is the coil, which allows a long length between the potential taps and adequate contact length and separation of current contacts from the taps. It allows access to the highest test fields, which are most easily obtained in small-bore solenoid magnets. Depending on the pitch chosen, the axis of the conductor is not quite orthogonal to the field, but usually the angle is so small that this can be ignored or a relatively easy correction can be made for very precise measurements.

Another factor that comes into play with coils is that of self-field of the

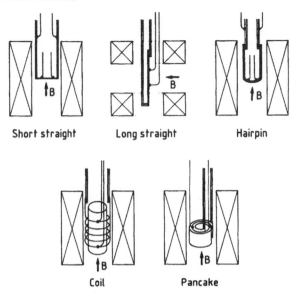

Fig. 6.8. Sample geometries.

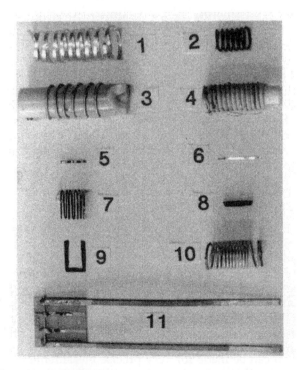

Fig. 6.9. Samples used in Oxford. Note the preponderance of the coil geometry.

Table 6.1. Sample geometry comparison.

Sample Geometry	Advantage	Disadvantage
Short straight	simple geometry solenoid magnet -> high field	high contact heating high current transfer high electric field criteria
Long straight	low contact heating medium electric field criteria simple geometry	some current transfer* split-pair magnet -> low field
Hairpin	low contact heating medium electric field criteria solenoidal magnet	some current transfer* moderate geometry
Coil	low contact heating low current transfer* low electric field criteria complicated geometry solenoid magnet	complicated geometry
Pancake	low contact heating suitable for tapes low current transfer* low electric field criteria solenoid magnet	moderate geometry ΔB radially

*Considerations for composites only.

coil, which can cause large errors, particularly at low fields, and therefore high critical currents, but this can be allowed for. For HTS all the sample geometries shown in Figure 6.8 have been used at one time or another [42]. Figure 6.9, which was first shown in Jones et al. [42], shows some of the samples used in Oxford. The suitability of the geometry depends first of all on whether one is making measurements on a bulk sample or a composite, such as powder-in-tube wire or tape. For instance, in the early days of HTS, many transport current measurements were made on small bars cut from other shapes, typically discs, which had been formed by pressing powder often with a binder of some description. This meant that one was immediately forced to use the short, straight geometry with all its attendant drawbacks, with the additional complications unique to HTS (see next section).

A useful geometry for measuring samples cut from the bulk is the hairpin [43], which proves to be a good compromise in some instances, although for highly textured samples it is not suitable. Transport current measurements are becoming increasingly common for the testing of practical conductors in composite tape and wire form that are now emerging. The lengths available are also increasing, and quantities in excess of several meters are not uncommon. It is becoming easier, therefore, to use traditional coil geometries with all their advantages.

6.3.3 Coil Geometry

Because the coil geometry is the best and most commonly used, we will consider it in more detail. For conventional superconductors it has been used for many years; and because it is now accepted by the industry as the usual means of characterizing conductors for commercial applications, it is becoming the subject of standardization exercises (see, for example, Tachikawa et al. [44]). A typical arrangement is shown in Figure 6.10. This shows the basic components of a sample and holder. There has always been a debate about the material that should be used for the sample holder and the method of fixing the sample to the holder. It is desirable that the sample be fixed to the holder because the Lorentz force, F_L, given by

$$F_L = B \times J \tag{6.34}$$

(where B is the magnetic field and J is the current density), can cause the sample to strain mechanically. For the brittle HTS materials, this mechanical strain can degrade the critical current in a similar way to the LTS A15 material Nb_3Sn.

Again, in analogy to Nb_3Sn, most small coil samples are prepared by the "wind and react" method. The unreacted conductor is first wound on a former and then heat treated. A ready reacted conductor would be severely strained in the process of winding a small-diameter ($\leq$ 50-mm) coil unless it was in very thin tape form.

For A15 superconductors there has been a long debate about whether the coil should be reacted on its test mandrel and subsequently fixed down or whether the fragile coil should be transferred from a reaction mandrel to a

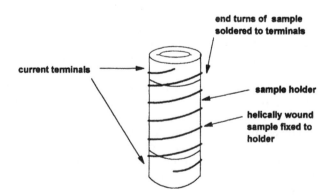

Fig. 6.10. Coil sample and holder arrangement. The current terminals are always copper. The sample holder can be metallic (e.g., stainless steel) or an insulator such as grp or a ceramic.

sample holder, after reaction, with all the inherent risks of handling damage. The advantage of the first method is that, apart from securing the reacted coil, no disturbance is necessary, but it means that the former must be of a refractory material—a metal or ceramic.

The second method gives greater freedom of choice of sample holder material and allows, for instance, the use of glass-reinforced plastic (grp) such as G.10. As stated earlier, the disadvantage is the handling risk. The debate is gradually being resolved in various standardization and pre-standardization efforts, with the balance of favor tipping towards the second method.

With HTS conductors the situation is aggravated by the fact that the reactions are at rather higher temperatures ($> 850^\circ$C) than those used for Nb_3Sn ($< 800^\circ$C), and the reaction atmosphere is invariably air or oxygen compared with the inert (e.g., argon) or vacuum reactions used for Nb_3Sn. These factors affect the choice of material for the reaction mandrel; it must withstand the high temperature and be resistant to oxidation. Another difficulty stems from the highly reactive nature of the new materials. For all practical purposes they coexist only with silver—hence its widespread use as a matrix material—and while this ostensibly acts as a barrier, "leaks" can occur, and the localized reaction that results spreads to a significant volume of the ceramic in the composites and the sample is degraded. (HTS coils will be discussed further below.)

6.3.4 Pancake Geometry

Figure 6.8 is based on Fickett, Goodrich, and Clark [45], who devised it to discuss sample arrangements for $I_c(B)$ measurements on LTS conductors. In fact all we have added to modify it for HTS is the pancake. This modification was done because many of the early attempts at practical HTS conductors were, and often still are, in the form of wide tapes ~1 cm or greater. With these, it is not possible to wind helical coils with diameters that will fit into typical high-field magnet bores (~50 mm or less). Thus, one is constrained to the short and straight or (if low field measurements only are acceptable) long and straight formats. In order to make measurements on longer lengths, the only alternative is the pancake; therefore, we include it as a legitimate sample format. In practice, the pancake is mostly used for prototype magnet construction, so a lot of our work at least is concerned with building small pancakes and characterizing these in self- and applied fields. These pancakes generally contain longer lengths (i.e., 2 to 5 m) than the usual short sample, so there is a grey area in which the distinction between a true short sample and a prototype magnet coil is blurred. Usually in this geometry more than one set of voltage taps is employed, often many more, and it is a useful geometry for homogeneity measurements. Jenkins et al. [46] describe this type of work, and the pancake format will feature later when we discuss the technological implications of transport critical currents in high-temperature superconductors.

6.3.5 Temperature

For I_c measurements as a function of B and T, a way must be found to vary the sample temperature. Nowadays a number of devices are available commercially that can achieve this. The most appropriate are continuous-flow cryostats (CFCs) or variable-temperature inserts (VTIs) for superconducting magnet cryostats, which are a specialized variant of a CFC. These work on the principle of pumping a stream of liquid He (or, for higher-temperature work, liquid N_2) through a heat exchanger with which the sample under investigation is in thermal contact either by mechanical linkage or via an exchange gas medium. An electrical heater is also incorporated and thermometry installed. By using temperature feedback, the parameters of cryogen flow rate and heater power can be controlled so that fixed temperatures, in the range 2–300 K or more, can be obtained, or, with more sophisticated control devices, a temperature sweep at a given fixed rate can be set up. These devices are useful for low-dissipation measurements such as those for superconducting electronics research. When it comes to making transport current measurements on high current-carrying samples, however, they tend to lack the refrigeration power necessary to maintain a given temperature.

We favor the "gas" probe approach originally developed by Hudson, Yin, and Jones in 1981 [47] following work by R. Hampshire et al. in 1969 [48] and further developed by D. P. Hampshire and Jones in 1987 [49] and then adapted for HTS work by Frost, Jones, and Belenli in 1993 [50]. The technique is based on the samples being contained in an isolated volume, which can contain exchange gas at pressures from zero to atmospheric, in contact with a heater. The probe is inserted in the bore of a magnet in liquid He. By adjusting the exchange gas pressure and heater power, a temperature can be set anywhere between 4.2 K and T_c or higher. A thermometer, typically a field-independent capacitance element, is arranged in closed loop with the heater control. As the transition takes place, dissipation occurs and the temperature rises, thus causing the heater power to reduce so that a large dynamic range can be achieved (that is, much of the resistive transition can be observed at the set temperature before control is lost). Another advantage is that the heavy current leads, necessary for high-current measurements, are thermally anchored in the liquid He bath before they enter the isolated volume.

Although the devices described above are ideal for a continuously variable range of temperatures, a much simpler method is to immerse the sample directly in a liquid cryogen and change its temperature by changing pressure.

Liquid Nitrogen (LN$_2$): LN$_2$ can span the range 77 K down to 63 K by pumping on the cryostat containing the liquid. Furthermore, if an accurate pressure measuring device such as a manometer is incorporated, the temperature of the bath can be accurately determined by the vapor pressure curves. The method has many advantages. Heat transfer to the liquid is good and the temperature is unambiguous; vapor pressure temperature measurement is unaffected by magnetic fields, for example. It is possible to go below 63 K, and temperatures of ~55 K can be achieved by hard pumping. Below 63 K, however, the N_2 is a solid; thus, heat transfer is not so good, and the sample can sustain mechanical damage. Temperatures above 77 K can be attained by using overpressure. The critical point of liquid N_2 is 126 K, and the corresponding pressure is 34 bar.

Liquid Helium (LHe): Other cryogenic liquids can be used. At the bottom end of the temperature range is liquid He. By pumping on liquid He one can access the superfluid He regime at T ≤2.2 K (the lambda point). Superfluid helium has remarkable properties. For example, its thermal conductivity is effectively infinite, so heat transfer to it is excellent. This sublambda temperature range is useful for checking whether dissipative disruptions to measurements, such as bad contacts, are present. Overpressure is less useful: the critical point of He is 5.3 K, so it is hardly worth the effort.

Liquid Hydrogen (LH$_2$): Liquid hydrogen has a boiling point of 20.3 K. Pumping can lower the temperature to ~15 K. Pressurizing up to 13 bar can give 33 K. This is a very useful temperature range for HTS. The problem is that liquid H$_2$ is dangerous. In the UK, for example, the safety regulations are such that its use for HTS research is virtually impossible except in a highly specialized and very expensive facility. Nonetheless, liquid H$_2$ has been used successfully elsewhere [51]. Frustratingly, one of the most useful combinations for HTS I$_c$ measurements would be liquid H$_2$ and pulsed magnetic fields > 20 tesla. It seems unlikely, however, that this combination of a highly combustible substance and kilovolt-driven, mechanically stressed magnets situated only millimeters away would be lightly sanctioned.

Liquid Neon (LNe): Liquid neon has a boiling point of 27 K and a melting point of 25 K. Applying pressure up to 140 bar will give 43 K. The fact that it spans a useful range either side of 30 K makes it a very interesting medium for HTS work, as the B* line (see above and Figure 6.2) takes a sharp upward turn in this region for most materials investigated so far. It is also inert, and therefore safe, and has a latent heat of vaporization of 86 kJ kg^{-1}, which means it lasts a long time. The problem is that it is very expensive. One can, however, construct a small liquefier to produce *in situ* small, but adequate, quantities of LNe for I$_c$ measurements. At Oxford, for example, a liquefier has been constructed to a design by J. Cosier and has already been used to produce LNe for I$_c$ measurements (not yet published). Work in LNe is also taking place elsewhere; see, for example, Iwasa and Bellis [52].

6.3.6 Magnetic Fields

For these measurements the application of magnetic fields is important. As already stated, the most common magnets in use are superconducting magnets, currently available up to ~20 T. Indeed, 18-T magnets, with bores of ~40 mm, are relatively economic and are becoming a routine tool for work on superconductor characterizations.

A point that is frequently overlooked is that apart from being *high-temperature* superconductors, these materials are also *high-field* superconductors. The evidence for the high B$_{c2}$ has been estimated of the order of 100 T. Remember that we are more concerned with B*; below temperatures in the range 20 K to 40 K, B* rises dramatically, and fields in excess of 20 T are required to make direct measurements of I$_c$(B) in this temperature range. The highest continuous fields in the world today are to be found in the large national magnet laboratories of which there are a handful in Europe, the United States, and Japan.

Hybrid magnets are combinations of multimegawatt, water-cooled copper solenoids surrounded by large superconducting magnets. These are the means whereby fields of ~30 T can be generated. They are found at only four centers worldwide. Hybrids to give fields of 40 T and 45 T are being built are NRIM in Tsukuba, Japan, and NHMFL in Tallahassee, Florida, respectively. The bore sizes of all these magnets is 30–50 mm (room temperature). Access to these facilities for HTS work is possible, but for routine work the only economic possibility is pulsed magnets.

A number of centers now are equipped with pulsed-field facilities that can generate fields of up to ~70 T over pulsed lengths of from 10 to 100 ms. Several hundred tesla is possible with much shorter pulses. $I_c(B)$ work on HTS is difficult with pulsed fields, but not impossible. At Oxford for instance, researchers are beginning such work in 20-mm-bore pulsed magnets up to 50 T.

It is not appropriate to go into further detail on magnet technology here. Several excellent works (given in the recommended readings at the end of the chapter) review the state of the art and list all the world facilities.

6.4 I_c Measurements on HTS

In an excellent paper, Goodrich and Bray [53] discussed in some detail the unique problems associated with I_c measurements on HTS materials. One clause worth reproducing from their conclusions is that "a number of problems developed when the I_c measurement, along with its traditional implications, was casually extended to a new class of materials." They also noted that any value of I_c reported without qualification is useless. In other words, unless all the measurement conditions (such as the electric field criterion used) is stated, any declared measurements on HTS can be regarded as dubious.

One major problem was the inconsistency of the properties of the material, both within a given batch, even within the same piece, and between batches; this problem had been largely eliminated in the traditional metallic superconductors. A second difficulty was the highly vulnerable nature of the ceramics. Initially at least they were prone to degradation by contamination with moisture in particular; and repeated measurements, which involved thermal cycling, particularly in air, proved to be very difficult to reproduce. Also, thermal cycling of bulk specimens often resulted in macrocracks and microcracks. Thus, in samples of any significant size, homogeneous superconducting behavior was often lacking simply because of mechanical discontinuities. Most of these remarks apply to measurements on bulk materials uncomposited with any matrix. As the materials science progresses and HTS superconductors become available in first-generation technological forms, such as wires and tapes, many of the problems are diminishing, but they still dog the subject, and it is therefore worth addressing the main problem areas in a little more detail.

6.4.1 Contacts

It is well known that making good electrical contacts to bulk high-T_c materials is difficult. After all, a primary requirement for any electrical contact, such as a soldered joint, is to eliminate any intervening oxides; these materials *are* oxides! For critical current measurements two types of contact need to be effected: one for the voltage and one for the current.

The voltage contacts need not necessarily be particularly good since voltages are usually measured by high input impedance instruments, and high contact resistance is not in itself a disaster; however, some care still needs to be exercised since obscure "battery" effects can interfere with the measurement, and the contact must be mechanically robust. Nothing is more frustrating than the voltage taps going open circuit during a measurement, because warming up and recooling may well have adverse effects, as discussed earlier. Products based on silver-loaded paint or epoxy resin can be used for connecting voltage taps to bulk samples.

The current connections are another matter. High-resistance contacts or unbalanced contacts can have annoying consequences. High-resistance can ruin a measurement or render it impossible. For practical conductors, currents on the order of 100 amperes are, or should be, routine. Sample currents are usually furnished by power supplies with maximum output voltages of ≤ 10 V. Thus, a contact resistance of just 1 ohm, resulting in a voltage drop of 100 V at 100 amps, would be too much for a typical power supply. Clearly the contact resistance needs to be at least an order of magnitude, ideally two or three, lower for this simple reason alone. There are, however, more serious, subtle effects that stem from high-resistance contacts, so a lot of work has been done on achieving good contacts to bulk HTS. Some techniques are remarkably successful; see, for example, the work of Ekin et al. [54]. Unfortunately, these techniques are usually suitable only for small currents (~10 mA) and are often complex and time consuming or require special preparation of the sample during synthesis.

For technological characterization, a quick method is desirable, particularly if any element of "screening" of a large number of samples is involved. Also, it is preferable that the sample itself need no elaborate preparation. An excellent compromise is ultrasonic soldering using indium as the contact medium [43]. A good investment is an ultrasonic soldering iron, but these are surprisingly difficult to locate and quite expensive. Even when thus equipped, a fair degree of skill is required, but contacts of typical area (~10 mm^2) can be achieved with resistances less than 10^{-5} ohm.

6.4.2 The Effects of Current Contact Problems

Ninety percent of contact problems result from resistance. Even if the resistance is not so high that the power supply voltage is exceeded, the dissipation can be such that excessive heating takes place. As well as contact resistance, contact area is important as dissipation per unit area, P, given by

$$P = (I^2 R_c)/a, \qquad (6.35)$$

(where I = current, R_c = contact resistance, and a = contact area) is particularly important in liquid cryogens because transition from the nucleate to the film-boiling regime [55] is sudden beyond a critical heat flux, which, in the case of liquid helium, is ~1 W/cm^2 and in the case of liquid nitrogen is ~10 W/cm^2. Film boiling results in a gas blanket, which insulates the surface. A large temperature increase, the "superheat," is the consequence. For LHe the superheat is ~10 K and for LN$_2$ ~100 K at the critical heat flux [56]. Therefore,

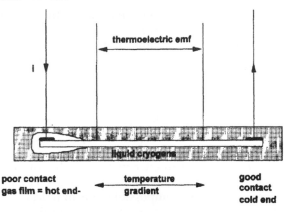

Fig. 6.11. Effect of unbalanced contacts.

for a 10 mm^2 contact area, a resistance of better than 10^{-5} ohm is necessary to stay under the critical flux at 100 A in LHe.

An excessive temperature rise at one, or both, of the current contacts can have a number of undesirable effects. It can simply drive the sample normal and thus give an I$_c$ that is inaccurately low. More subtly, it can exacerbate a number of what can be conveniently, if cumbersomely, referred to as magneto-thermal-galvanic effects. In [43] these are discussed in some detail; briefly, they arise mainly as a result of temperature *gradients* along the sample. These are particularly complex in the event of unbalanced contacts, that is, where one has a much higher resistance than the other. Figure 6.11 shows the situation. A temperature gradient, which includes the potential taps, will give rise to thermoelectric voltages as dissimilar materials form junctions at two different temperatures.

Figure 6.12, adapted from [43], shows some V-I characteristics in which the transition is distorted as a result of some of these various effects (named in the figure caption). It is worth noting that these become obvious only when the transport current is reversed; consequently, it is advisable to do just that during a set of measurements from time to time. A good transition should be symmetrical when current is reversed; that is, one should be a mirror image of the other (see Figure 6.12).

In fact, progress is such that for technological power-engineering applications it is unlikely that it will be necessary to contact directly to the bulk material since the conductors will be composites of HTS and, most likely, Ag. There is one notable exception: current leads for superconducting magnets where the very low thermal conductivity of the ceramic means that low boiloff from LHe cryostats can be achieved as a result of reduced heat leak. At the same time, no joule heating occurs when carrying current in the superconducting state. Low critical current densities seem to be no obstacle because cross-sectional areas of the leads are increased accordingly; yet the reduction in heat leak is significant when compared with even fully optimized metallic

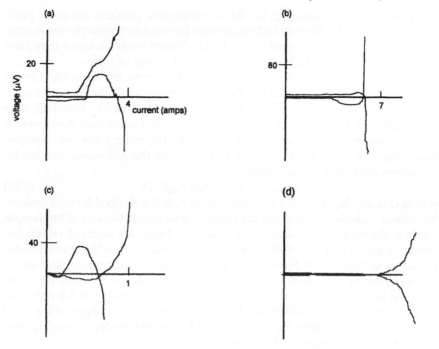

Fig. 6.12. Distortions and asymmetries of samples with resistive and unbalanced current contacts: (a) Thomson effect in YBCO; (b) Peltier effect in BSCCO; (c) both effects in YBCO; and (d) ideal, symmetrical transition.

leads of traditional construction. In practice, the leads are metal (brass or copper) tubes with a short element of bulk HTS (usually a tubular cylinder) in series, giving a thermal break in the lead assembly. Thus, joints *do* have to be made between the elements, and therefore contacts capable of carrying 100 amps or so with very little loss have to be made to the bulk. The dimensions of these assemblies are typically such that large contact areas can be achieved, but it should be noted that there is an upper limit to the degree to which contact areas can contribute to a good electrical joint. For a discussion of this topic, we refer to the treatment given by Wilson [57]. Later in this chapter we will describe a pair of experimental current leads under development.

6.4.3 High-T_c Composites

For most other purposes, composites of HTS and normal metal, invariably silver, are becoming commonplace as wires and tapes. Tape formats are most common, since they permit the introduction of texturing in the rolling and pressing processes, with all the attendant benefits for high current densities. In the next section, we discuss the anisotropy that results in textured samples. Generally speaking, for these composites contacts are not a problem, since normal soldering techniques can be used to contact to the matrix, although some care still needs to be taken. These wires or tapes may have single filaments or multifilaments (layers in the case of tapes).

Composites of this type have one problem, namely, that of current transfer from the matrix to the filament(s). This applies to all filamentary superconductors. The problem has been addressed and analyzed by Ekin [58], who has quantified the length of composite necessary to ensure that the current is fully transferred to the superconductor from the matrix for conventional materials such as NbTi and Nb₃Sn. Ekin derives the following relation for minimum current transfer length

$$X_{min} = (0.1/n)^{1/2}(\rho_m/\rho_c)^{1/2}D, \qquad (6.36)$$

where D is the diameter of a wire specimen, ρ_c is the desired level of resistivity criteria, ρ_m is the resistivity of the matrix transverse to the axis of the sample, and n is the index of transition. If the potential taps are situated within this current transfer length, a voltage will be superimposed on the characteristic, as shown in Figure 6.13. For typical LTS composites, a current transfer length on the order of some tens of wire diameters is required to make measurements at low electric field or resistivity criteria. To generalize, this length may be ~50 mm in a typical situation. This, of course, militates against the short straight sample format; hence the entry "high current transfer criteria" is listed under the disadvantages column in Table 6.1.

The situation with the emergent high-T_c composites may be somewhat worse (see Equation (6.36)), since n tends to be lower than for LTS (< 10 cf of order 100).

Another potential difficulty is the localized mechanical damage in the ceramic filament(s) with differential thermal contraction resulting from heating during the soldering process. Again, this hardly matters if a long current transfer length is available, but it is another objection to short straight geom-

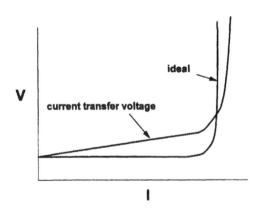

Fig. 6.13. Effect of insufficient separation of current and voltage contacts.

etries for high-resolution measurements.

6.4.4 High-T_c Coils

The preceding section went to some lengths to advocate coil geometry for transport critical current measurements. Since the beginning of HTS materials researchers have tried to use this format where possible, although in simple bulk form this is very difficult. Early attempts to form coils out of bulk material, however, were partially successful. The process usually involved extruding a filament of the ceramic which was held together by some organic binder. This was in the "green" or unfired state. After the coil was formed, the heat treatment would be applied, including debonding, that is, driving off the binder. Such coils are shown (items 2 and 7) in Figure 6.9.

Another novel approach was to machine a flat spiral in a layer of melt processed BiSSCO (approximately 300 μm thick), which was laid down on a MgO substrate. This approached the pancake format of Figure 6.8. A detailed description is given by Davies et al. [59]. Attempts to reproduce the technique on a cylinder of MgO, so that helices could be machined, were largely unsuccessful, since the BiSSCO would not adhere evenly to the cylinder during the melt processing. These bulk coils and spirals proved useful for applied research investigations because they represented more closely the dimensions of conductors that will ultimately be used in power applications such as magnets. In contrast, in thin films, considerable progress has been made in producing complicated track geometries by using techniques such as laser ablation and various masking methods, but, because of the low currents they support, such geometries are of little interest here. The drawback of these coils formed from bulk material is that they are difficult to texture; therefore, usually only low current densities ($\sim 10^3$ A cm^{-2}) are achieved.

Coil, or pancake, formats became much more practicable as methods of producing wires and tapes improved. Figure 6.9 (items 1, 3, 4, and 10) shows some coils wound from HTS/Ag composites, both wire and tape. As discussed earlier, the usual process route is "wind and react." With thin tapes, however, one can instead wind, react, and tighten slightly; in this case, transfer from a reaction mandrel to a sample holder is quite easy (with care).

6.5 Temperature and Field Dependence of Transport Critical Currents in High-Temperature Superconducting Materials

Earlier, we explained how transport critical currents were obtained from measured E-J characteristics by applying arbitrary electric field or resistivity criteria. We discussed the general factors affecting the E-J curves of type II superconductors and then described the additional factors relevant to HTS, namely, high-temperature operation, anisotropy and its effect on pinning energies and the properties of the flux line lattice, and granularity. As we discuss in this section the field and temperature dependence of critical currents explicitly, some of these HTS factors will be mentioned again, but with changed emphasis.

We begin by describing the temperature dependence of HTS J_cs, then deal with their dependence on both the magnitude and direction of applied field.

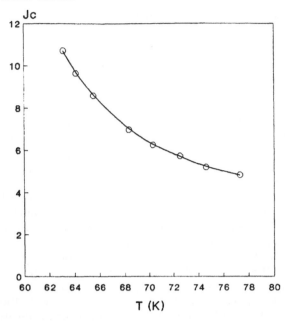

Fig. 6.14. Critical current density as a function of temperature in zero applied field for a short sample of BSCCO 2223/Ag PIT tape. The units of J_c are $10^3 A/cm^2$.

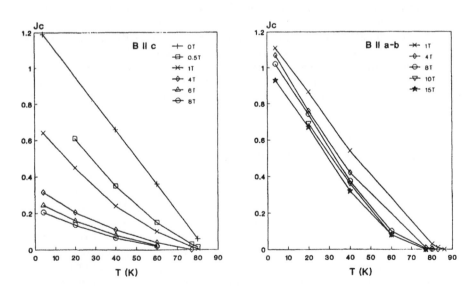

Fig. 6.15. Temperature dependence of J_c (units $10^7 A/cm^2$) of a YBCO thin film in a range of magnetic fields applied in two different orientations with respect to the *a-b* plane (after Schalk et al. [60]).

Finally, we discuss the so-called irreversibility line which effectively defines the region of B-T space in which an HTS is useful for high-current technology.

6.5.1 Temperature Dependence of J_c

Data on the temperature dependence of J_cs in HTS materials are shown in Figures 6.14 and 6.15. Figure 6.14 shows the critical current density (calculated from I_c values obtained from V-I curves by applying an electric field criterion of 10 $\mu V/m$) of a short sample of BSCCO 2223/Ag composite tape, produced by the powder-in-tube technique, in zero applied field as a function of temperature from 77.3 K to 63.1 K, the temperature range accessible by pumping on liquid nitrogen. J_c increases with decreasing temperature.

Figure 6.15 shows the data of Schalk et al. [60] on the J_c (defined at 10 $\mu V/$ cm) of a YBCO thin film as a function of temperature in a range of magnetic fields applied parallel to and perpendicular to the a-b plane. The anisotropy of J_c with respect to applied field orientation is discussed shortly. The shape of the J_c vs T curve can provide information about the mechanism most important in determining the J_c of the sample. In general, critical current will be limited by a combination of flux flow, thermal activation, and intrinsic anisotropy or by weak links between grains.

A flux-flow model has been proposed by Tinkham [61] that predicts a J_c dependence at temperatures well below T_c of the form

$$J_c(B,t) = J_c(B,0) \ (1 - \alpha(B)t - \beta t^2), \tag{6.37}$$

where t is the reduced temperature, T/T_c. The constant β comes from the temperature dependence of the free energy difference between pinned and unpinned flux quanta:

$$U(B,t) = U(B,0)(1-\beta t^2), \tag{6.38}$$

and the coefficient $\alpha(B)$ is given by

$$\alpha(B) = (kT_c/U(B,0)). \ \ln(aB\Omega/E_{min}), \tag{6.39}$$

where a is the average hopping distance of the flux quanta, Ω is the attempt frequency for escape, and E_{min} is the electric field criterion that defines J_c. Thus, flux-creep effects are represented in the second (linear) term of (6.37), which dominates at sufficiently low temperatures. The model predicts that $J_c(B,0)$ should vary as $B^{-0.5}$ and has been demonstrated to fit data from numerous samples [62]. Although some deviation from J_c proportional to $B^{-0.5}$ may be observed at small fields, this can be attributed to the self-field of the sample, the greatest discrepancy occurring at low temperatures where J_c is highest.

6.5.2 Field Dependence of J_c

We first discuss the field dependence of J_c in highly granular, weak-linked materials and then deal with well connected, well-textured samples where weak links have effectively been eliminated. Finally, the anisotropic nature of J_c with respect to the orientation of applied field will be discussed.

6.5.2.1 J_c-B Dependence of Granular HTS Materials

As stated earlier, early bulk samples of HTS behaved like isolated grains joined by Josephson tunnel junctions [34,35], and their properties were dominated by these weak links. Transport critical current densities were found to

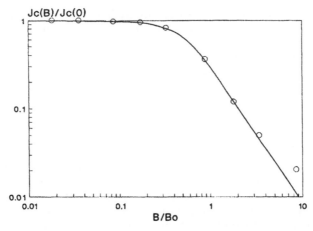

Fig. 6.16. Transport critical current density versus applied field for a bulk YBCO sample at 77 K. Solid line is the theoretical curve obtained from a model of a granular material incorporating a distribution of junction lengths and orientations. $B_0 = 0.57$ mT (after Peterson and Ekin [34]).

drop rapidly with the application of only small fields. For example, numerous authors reported that the transport J_c of bulk-sintered YBCO dropped by two orders of magnitude with the application of 10 mT, whereas magnetization J_cs were much higher and not rapidly changing functions of B. These results were consistent with a granular, weak-linked material, since the tunneling current is strongly reduced when a magnetic field penetrates a Josephson junction [63]. The tunneling current as a function of B is given by

$$J_c(B) = J_c(0) \, |\, [\sin(\pi B/B_0)]/(\pi B/B_0) \,|, \qquad (6.40)$$

where the characteristic field $B_0 = \Phi_0/d.L$, where d is the junction thickness, L the junction length, and B is the component of field normal to plane in which d and L lie. Experimental data from Peterson and Ekin [34] on a bulk-sintered YBCO sample are shown in Figure 6.16. These results were explained by modeling the sample as a collection of Josephson junctions with a distribution of lengths and angles to the applied field. The prediction of this model with $B_0 = 0.57$ mT is the solid line in the figure. From the fitting procedure Peterson and Ekin found that the average junction length was comparable to the average grain size in the sample, and hence they showed that the weak links were indeed at the grain boundaries (i.e., intergranular rather than intragranular). The model also predicts that at fields close to B_0, J_c should vary as $B^{-1.5}$.

Agreement with experiment is good; but as can be seen from Figure 6.16, at higher fields the data points turn and lie above the averaged Josephson weak-link curve. This is attributable to the fact that at high fields, current flow is predominantly along a percolative path of strong links. The weakest links (e.g., those for which L is small or those aligned unfavorably with respect to applied field) contribute to the bulk J_c at low fields, but in higher fields carry effectively zero current. Thus this model is applicable only in very low fields, and we do not expect much correlation between the high and low field dependence of J_c.

6.5.2.2 J_c-B Dependence of Well-Connected HTS Materials

The materials science of HTS has progressed enormously since the first granular bulk samples were produced. For example, by the use of appropriate fabrication techniques, weak links have effectively been eliminated in bulk BSCCO. For the 2212 phase, melt processing is employed, whereas for the 2223 phase, rolling and pressing produces the desired results. The samples produced have highly textured microstructures, consisting of platelike grains aligned with their c-axes parallel. Current flow is predominantly along the $Cu-O_2$ planes within grains, and transfer between grains occurs across large-area c-axis boundaries, hence at low current density. This picture of current flow has been called the "brick wall" model.

Examples of the field dependence of J_c in a number of well-textured materials are shown in Figures 6.17 and 6.18. Figure 6.17 shows the J_c-B characteristic at 4.2 K of a short sample of BSCCO 2223/Ag composite tape produced by the powder-in-tube technique. The electric field criterion used to define J_c was 100 μV/m. Initially J_c falls rapidly with field applied parallel to the c-axis, but then remains fairly constant. The shape of this curve is very characteristic of HTS at low temperatures. The initial rapid fall can be attributed to the suppression of tunneling currents in weak links, even though the sample was

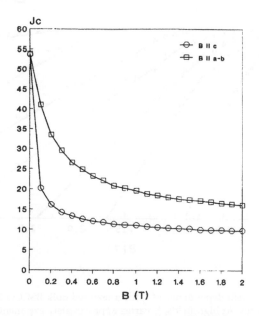

Fig. 6.17. Field dependence of J_c for a well-connected BSCCO 2223/Ag PIT tape at 4.2 K, illustrating anisotropy with respect to field orientation. Units of J_c are 10^3 A/cm^2.

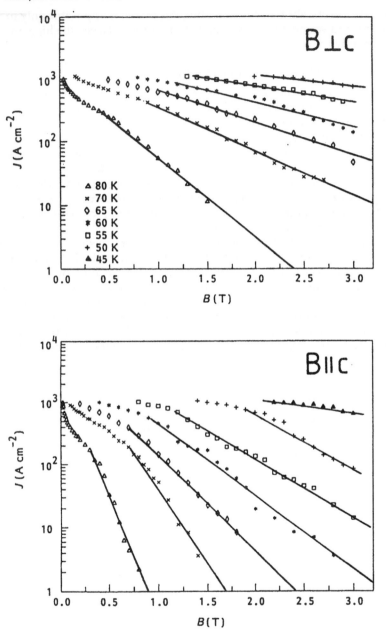

Fig. 6.18. Magnetic field dependence of J_c for a textured bulk BSCCO 2223 sample at a range of temperatures. At high fields, J_c varies approximately exponentially with B (see text and Equation (6.41)). Solid lines show the extrapolation by which values for the characteristic field, B_s, are obtained (after Kottman et al. [39]).

high quality with excellent grain alignment (micrographs can be found in [38]). The reason for this is that even in a very good sample some residual weak links can contribute to the overall current-carrying capability of the material at low fields. As the applied field is increased, however, they quickly drop out, and the current is now carried by the well-aligned, well-connected grains. Also shown in Figure 6.17 is the anisotropy in J_c with respect to the orientation of applied field (addressed later).

The data from Kottman et al. [39] plotted in Figure 6.18 show the J_c-B dependence of a textured bulk BSCCO 2223 sample, with J_c defined at 10 μV/m. At high fields, critical current density varies approximately exponentially with B:

$$J_c(B,T) = J_0(T) \exp [-\alpha(T)B/B_s(T)], \tag{6.41}$$

where B_s is some characteristic field at which J_c extrapolates to some near-zero value, J_{def} (Kottman uses a value of 1 A/cm^2) and α is given by

$$\alpha(T) = \ln J_0(T) + \ln J_{def} . \tag{6.42}$$

This exponential dependence has also been observed in silver-clad BSCCO 2223, in bulk melt-processed BSCCO 2212 [64], and in YBCO thin films [60] and is consistent with a J_c determined by flux pinning and thermal activation.

At low temperatures, the field dependence of J_c (especially for B parallel to the a-b plane) is very weak, corresponding to a high B_s. Unfortunately, Kottman was unable to collect data at temperatures below 45 K because of contact heating problems at high currents, but it is clear from results in numerous other publications (e.g., Figure 6 in [60]) that the 4.2-K data would lie on a line almost parallel to the field axis over the range 0–3 T. At high temperatures, however, J_c falls rapidly with field, and B_s is small. A plot of B_s vs temperature effectively defines the range in which the HTS material is useful and is intimately linked with the irreversibility line.

6.5.2.3 Anisotropy of J_c with Respect to the Orientation of Applied Field

We have already discussed the intrinsic anisotropy of HTS materials in terms of their microstructures. This in turn leads to anisotropy in J_c values with respect to the direction of applied field. Proof that the J_c anisotropy is indeed intrinsic comes from its observation in single crystals. This effect is explained as follows.

The order parameter is suppressed in the insulating layers between the Cu-O$_2$ planes; hence, for the special case of fields aligned parallel to the a-b plane they can act as free-energy wells for flux lines. The pinning they provide is very strong, because the flux lines are pinned over long lengths. This strong pinning is called "intrinsic pinning" and leads to relatively high J_cs for this particular orientation of field with respect to the microstructure. For fields aligned parallel to the c-axis, however, pinning is much weaker and is limited by the type and density of defects in the microstructure, and importantly by the intrinsic anisotropy which leads to a flux line lattice comprising weakly coupled 2D pancake vortices. Thus, for this orientation of B, J_c values are much lower.

Anisotropy of J_c with respect to applied field direction is thus strongly correlated to the intrinsic anisotropy of the microstructure, and in the HTS materials YBCO, TBCCO, and BSCCO is found to increase in that order.

Referring once more to Figures 6.15, 6.17, and 6.18, we see that for all values of B and T, the J_cs for fields applied parallel to the *a-b* plane are higher than those for fields parallel to the c-axis. For a well-textured sample at low temperatures, in fields large enough for any residual weak-link effects to have disappeared, the anisotropy between J_cs for B parallel and B perpendicular to the nominal *a-b* plane is a factor of approximately 2, as illustrated in Figure 6.17 for a BSCCO 2223 sample. Anisotropy in fact increases with improved texturing/grain alignment in bulk samples; clearly a microstructure of completely randomly aligned grains will lead to a bulk J_c showing no dependence on field orientation. Also, anisotropy increases with increasing temperature; at high temperatures intrinsic pinning remains strong, but for fields aligned parallel to the c-axis the coupling between 2D pancake vortices is broken, leading to reduced pinning energies and correlation volumes, flux motion, dissipation, and low J_cs.

So far we have considered and presented data on the J_c dependence on fields applied either parallel to or perpendicular to the c-axis. What about the detailed dependence of J_c on θ, the angle between B and the *a-b* plane? Many groups have measured this θ dependence and obtained results qualitatively the same as those from Ekin [65] for an YBCO sample at 76 K, plotted in Figure 6.19. A narrow peak (full width at half relative maximum ~16°) is observed at θ = 0, outside which J_c, although lower, is essentially independent of further increases in θ. These results correspond to an applied field magnitude of 0.5 tesla. Ekin found that as the magnitude of B increases, the central peak becomes narrower but the J_c on either side remains unchanged. At sufficiently high fields, however, the characteristic takes on a "head and shoulders" appearance, with J_c declining rapidly (after the flat region outside the narrow

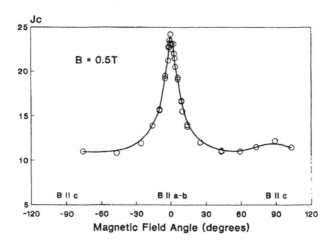

Fig. 6.19. Transport critical current density dependence on the angle between B and the *a-b* plane in oriented-grained YBCO. A sharp intrinsic pinning peak occurs for B parallel to the *a-b* plane. Units of J_c are $10^3 A/cm^2$ (after Ekin [58]).

peak for small angles) above some critical angle, θ_c. This led Ekin to suggest that as it was impractical to try to design a superconducting magnet to take advantage of the very narrow intrinsic pinning peak; one should instead allow an adequate angular margin to avoid the pinning transition at θ_c.

6.5.3 The Irreversibility Line

When the critical current density of an HTS material is plotted as a function of both field and temperature, we find that there is an extended region of the B-T plane over which, although the material is still in the superconducting state, J_c is effectively zero. The boundary of this region is known as the "irreversibility line," or IL, and lies under the B_{c2}-T line. We can think of the IL as separating the region of B-T space in which the HTS is technologically useful (with finite J_c) from the region where it is not. Its name comes from magnetization measurements on superconductors, where, at fields and temperatures below the IL, the properties are irreversible (showing hysteresis) but above the IL are reversible, the area of the magnetization loop collapsing to zero. The IL corresponds, therefore, to the effective loss of all flux pinning. It was first observed in HTS by Muller, Jakashige, and Bednorz [66] from field-cooled and zero-field-cooled magnetization measurements and found to have the form

$$B_{irr}(T) = B_{irr}(0) \left(1 - T/T_c \right)^n \qquad (6.43)$$

with the exponent $n \sim 1.5$. This expression (strictly valid only close to T_c) has

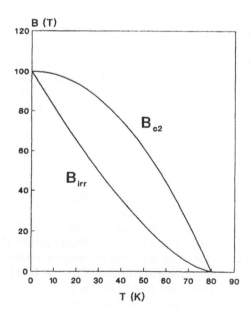

Fig. 6.20. Temperature dependence of irreversibility field, B_{irr}, and upper critical field, B_{c2}, as predicted by Equations (6.43) and (6.44) for a hypothetical HTS with $B_{c2}(0) = B_{irr}(0) = 100$ T and $T_c = 80$ K.

been used by several authors [67,68]; the form of IL it predicts is shown in Figure 6.20 together with the B_{c2}-T line given by

$$B_{c2}(T) = B_{c2}(0)(1-(T/T_c)^2).$$ (6.44)

Muller et al. suggested that the IL was due to a spin glass state arising from granularity, but also it has been seen in single crystals, thereby supporting an explanation in terms of depinning and/or a "melting" of the flux line lattice. Whether we think of the IL in terms of the line where thermal effects decouple the pancake vortices and reduce pinning potentials to the point where there is massive thermally assisted flux motion, or where the flux line lattice melts in a phase transition, the effect is the same; pinning effectively disappears, leading to dissipative flow and extremely small J_cs.

Equation (6.43) is by no means the only expression proposed to describe the shape of the IL; many variations can be found in the literature (see, e.g., [69–72]), some predicting an exponential dependence, and each derived from different models of the factors governing flux motion.

As with the J_c anisotropy with respect to applied field direction, the position of the IL is strongly correlated to the intrinsic anisotropy of the HTS. For example, in YBCO, the least anisotropic of the three main HTS families, the IL is displaced below B_{c2} by approximately 10–15 K over the field range 1–10 T [68,73], whereas in BSCCO B_{irr} extends from T_c to less than 30 K for fields in the range 0–1 T [73,27]. Attempts have been made to shift the IL to higher fields and temperatures by decreasing the insulator spacing between Cu-O planes in BSCCO and TBCCO, thereby reducing the intrinsic anisotropy, and have shown some success. In addition, the position of the IL is sensitive to the orientation of the applied field, although the special case for B parallel to the *a-b* plane is practically unimportant.

Increases in flux-pinning strength from defect creation by neutron irradiation have resulted in favorable shifts in the IL [74], but it is possible that intrinsic anisotropy may place an upper limit on its position. In contrast, although irreversibility lines do exist for LTS materials, it has been possible, by increasing the pinning force density, to shift them almost up to B_{c2}.

A variety of methods may be used to determine the IL, including field-cooled, zero-field-cooled, and ac magnetization techniques, the irreversibility temperature defined in the latter as that value at which the imaginary part of the ac susceptibility is a maximum. In terms of high-current, high-field technology, however, the most relevant data are those from transport current measurements. Using this method, one can define the IL as the position at which J_c has fallen to an arbitrary, near-zero value, J_{crit}. The irreversibility field, in this case referred to as the characteristic field, $B_s(T)$, is obtained by extrapolation to J_{crit} of $J_c(B,T)$ curves as described earlier. The exact position of the IL determined in this way is clearly dependent on the value chosen for J_{crit} but is more technologically meaningful. Plots of $B_s(T)$ corresponding to $J_{crit} = 1$ A/cm^2 obtained from the data in Figure 6.18 are shown in Figure 6.21, illustrating the dependence on field orientation.

6.6 Technological Implications

At the present time, encouraging values of J_c are being achieved in prac-

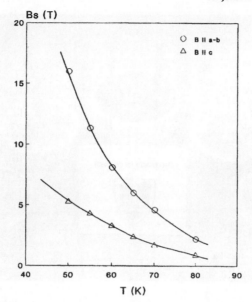

Fig. 6.21. Temperature dependence of the characteristic field, B_s, obtained by extrapolation from Figure 6.18, for a sample of textured BSCCO 2223. Note the dependence on field orientation (after Kottman et al. [39]).

tical HTS conductors. Typical values in short lengths of conductor are $>10^5$ Acm^{-2} at 4.2 K (B = 0); $>10^4$ Acm^{-2} at 4.2 K (B = 20 T); $>10^4$ Acm^{-2} at 77 K (B = 0) and $\sim10^3$ Acm^{-2} at 77 K (B = 1 T). These figures are taken from an excellent review article by Maley in 1991 [75], who goes into quite some detail regarding the characteristic features, good and bad, of the different types of HTS materials and therefore their advantages and disadvantages for various applications having regard for operating temperature, ambient field, and processing route.

To generalize Maley's conclusions, two major problems plague practical conductor development in long lengths: weak links and weak flux pinning. BSCCO tape conductors show promise because they lend themselves to mechanical processing which overcomes the weak-link problem, but in magnetic fields they still exhibit poor flux pinning at moderate temperatures and anisotropic behavior. YBCO, on the other hand, exhibits much stronger pinning at higher temperatures with a smaller sensitivity to magnetic field orientation but remains severely limited by weak links. TlBCCO is still in the early stages of development but also seems hampered by intrinsic anisotropy, although its very high T_c, 125 K, gives it a superior operating margin in liquid N_2.

In spite of all the problems, there is no denying that progress in the past few years has been immense. To end on an optimistic note, we will present two illustrative examples of how practical conductors are beginning to be

Fig. 6.22. Five HTS pancake coils. They are impregnated with wax and are destined for several demonstrator projects.

applied. We choose two projects involving magnet applications which, we believe, will be the first to exploit high transport current HTS conductors but, paradoxically, at *low* temperatures.

6.6.1 HTS Prototype Magnet Coils

The favorable behavior of HTS, particularly BSCCO, in very high fields at low temperatures means that they are a promising candidate for high-field inserts for magnets that are wound from LTS wires. One could envisage a magnet with an outer, low-field (10-T) winding of NbTi, then a middle section of Nb_3Sn up to 20 T, and in the center an HTS insert coil. This way magnets, operating in LHe for fields well in excess of 20 T could be achieved. In Figure 6.22 we show five of a set of six pancakes recently constructed. They are wound from BSCCO 2212/Ag tape produced by the dip-coating method in the Department of Materials of Oxford University, They have a working bore of 10 mm and an outer diameter up to 47 mm. The typical performance is a peak field of 0.15 T at 55 amps at 4.2 K. The size of the coils prevented scientists from applying a background field of more than 2 T during tests, but the coils typically added 0.1 T to this.

Another objective of these coils is to see whether ~0.5 T can be generated in the 5-mm air gap, between conical poles, of an iron yoke on which the 6 coils will be mounted refrigerated to 64 K. Individual performances suggest that this will be possible. A final aim is to use the coils, again with an iron core

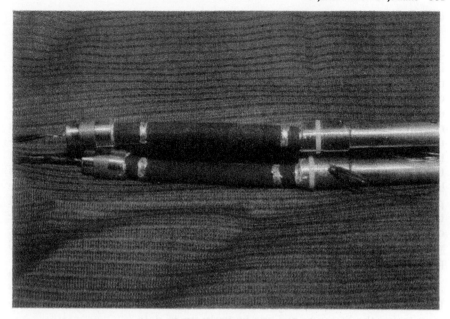

Fig. 6.23. A pair of current leads containing HTS sections (courtesy of Oxford
Instruments and ICI Superconductors).

and at 64 K, in a demonstrator for attraction-levitated transportation.

At this stage of development, this prototype coil technology is not particularly remarkable, but it exemplifies the type of work going on at many
centers throughout the world and points to the sort of serious applications
that will surely become increasingly possible as J_cs improve and homogeneous lengths of conductor become economically available.

6.6.2 Current Leads

The second, and final, example we cite is that of current leads which we
referred to earlier. Figure 6.23 shows a pair of leads that comprise brass tubes
with thermal breaks of YBCO rods 8 mm in diameter and 10 cm long. The
leads are being developed by the Research Instruments Division of Oxford
Instruments Ltd. and ICI Superconductors, who made the YBCO rods. Tests
on the leads show that with a transport current of 120 amps, there was effectively *no increase* in liquid helium boiloff above the zero amp value [76]. Although
the leads are not yet available commercially, this impressive result points to
still another application of these materials using transport current. Work on
current leads is ongoing at many other establishments; see, for example, Wesche
and Fuchs [77].

6.7 Summary and Concluding Remarks

In the foregoing sections we have explained the importance of transport critical currents in HTS. The parameter *critical current* (or critical current density) is possibly the most important piece of data for applications in power engineering, such as superconducting magnets. Its determination using transport currents is necessary since this method reflects most closely the conditions encountered in real applications. We have discussed the four-terminal technique at some length and explained the purpose of all the sample geometries used with particular emphasis on coils. The application of field and temperature to these measurements was outlined. We have, in some detail, discussed the factors that limit transport critical currents as well as those that affect its behavior with respect to field, temperature, orientation, and so on. Where possible, we have identified any problems that are unique to HTS in critical transport current measurements while, at the same time, pointing out the considerable common ground between HTS and LTS in this context. Finally, we have given illustrative examples of early attempts at applications of HTS practical conductors using transport currents.

In conclusion, we urge the reader to bear in mind that this is still a rapidly moving field of development. Much of what is written about any aspect of HTS could be out of date by the time it is published. Only by constantly monitoring the literature can one be truly up to date.

6.8 References

1. K. Sato et al., Cryogenics **33**, 243 (1993).
2. J. G. Bednorz and K. A. Muller, Z. Phys. B - Condensed Matter **64**, 189 (1986).
3. P. Kottman, H. Jones, A. J. Frost, and C. R. M. Grovenor, Supercond. Sci. Technol. **5**, 381 (1992).
4. W. Meissner and R. Ochsenfeld, Naturwissenshaften **21**, 787 (1933).
5. F. B. Silsbee, J. Wash. Acad. Sci. **6**, 597 (1916).
6. A. A. Abrikosov, Sov. Phys. JETP **5**, 1174 (1957).
7. J. Friedel, P. G. de Gennes, and J. Matricon, Appl. Phys. Lett. **2**, 119 (1963).
8. J. Bardeen and M. J. Stephen, Phys. Rev. **140**, A1197 (1965).
9. M. J. Stephen and J. Bardeen, Phys. Rev. Lett. **14(4)**, 112 (1965).
10. M. Tinkham, Phys. Rev. Lett. **13**, 804 (1964).
11. L. P. Gorkov and N. B. Kopnin, JETP **37**, 183 (1973).
12. F. London and H. London, Proc. R. Soc. London Ser. A**149**, 71 (1935); Physica **2**, 341 (1935).
13. R. G. Jones, E. H. Rhoderick, and A. C. Rose-Innes, Phys. Lett. **24A**, 318 (1967).
14. J. Baixeras and G. Fournet, J. Phys. Chem. Solids **28**, 1541 (1967).
15. W. H. Warnes and D. C. Larbalestier, Cryogenics, 26 (1986) p643.
16. W. H. Warnes, J. Appl. Phys. **63(5)**, 1651 (1988).
17. P. W. Anderson, Phys. Rev. Lett. **9(7)**, 309 (1962).
18. D. Dew-Hughes, Cryogenics **28**, 674 (1988).
19. R. G. Jenkins, H. Jones, N. Killoran, and W. Timms, IEEE Trans. Magn.

27(2), 1182 (1991).
20. F. Volker, Part. Accel. **1**, 205 (1970).
21. C. J. Plummer and J. E. Evetts, Paper MF-6, ASC'86, Baltimore (1986).
22. D. P. Hampshire and H. Jones, Cryogenics **27**, 608 (1987).
23. D. P. Hampshire and H. Jones, J. Phys. C **20**, 3533 (1987).
24. E. J. Kramer, J. Appl. Phys. **41**(2), 621 (1970).
25. R. Labusch, Phys. Lett. **22**, 9 (1966).
26. J. W. Ekin et al., J. Appl. Phys. **62**, 4821 (1987).
27. T. T. M. Palstra, B. Batlogg, L. F. Schneemeyer, and J. V. Waszczak, Phys. Rev. B **43**(4), 3756 (1991).
28. J. R. Clem, Phys. Rev. B **43**, 7837 (1991).
29. V. M. Vinokur, P. H. Kes, and A. E. Koshelev, Physica C **168**, 29 (1990).
30. M. Tachiki and S. Takahashi, Solid State Commun. **70**, 291 (1989).
31. D. S. Fisher, Phys. Rev. B **22**(3), 1190 (1980).
32. J. M. Kosterlitz and D. J. Thouless, in *Progress in Low Temperature Physics*, Vol. 7B, ed. D. F. Brewer, North Holland, Amsterdam (1978).
33. P. N. Mikheenko and I. S. Abaliosheva, Physica C **214**, 393 (1993).
34. R. L. Peterson and J. W. Ekin, Physica C **157**, 325 (1989).
35. J. Clem, Physica C **50**, 153 (1988).
36. K. Heine, J. Tenebrink, and M. Thoner, Appl. Phys. Lett. **55**, 2441 (1989.
37. A. Gurevich, A. E. Pashitski, H. S. Edelman, and D. C. Larbalestier, Appl. Phys. Lett. **62**(14), 1688 (1993).
38. R. G. Jenkins, H. Jones, I. Belenli, M. Yang, M. J. Goringe, and C. R. M. Grovenor, Cryogenics **33**(1), 81 (1993).
39. P. Kottman, H. Jones, V. Plechacek, and M. Polak, Cryogenics, **32**(11), 1005 (1992).
40. R. Griessen, Phys. Rev. Lett. **64**(14), 1674 (1990).
41. R. Griessen, Physica C **175**, 315 (1991).
42. H. Jones et al., Physica B **177**, 97 (1992).
43. H. Jones, L. Cowey, and D. Dew-Hughes, Cryogenics **29**, 795 (1989).
44. K. Tachikawa et al., IEEE Trans Mag. **25**, 2368 (1989).
45. F. R. Fickett, L. F. Goodrich, and A. F. Clark, NBS Int. Report IR80, 1642.
46. R. G. Jenkins et al., Paper 259, Proc. MT13, IEEE Trans. Mag., in press (1994).
47. P. A. Hudson, F. C. Yin, and H. Jones, IEEE Trans Mag. **17**, 1649 (1981)
48. R. Hampshire, J. Sutton, and M. T. Taylor, Proc. Conf. Low Temp. and Electric Power, International Institute of Refrigeration (1969).
49. D. P. Hampshire and H. Jones, J. Phys. E. **20**, 519 (1987).
50. A. J. Frost, H. Jones, and I. Belenli, Cryogenics **32**, 1014 (1992).
51. Y. Iwasa, Cryogenics **31**, 174 (1991).
52. Y. Iwasa and R. H. Bellis, Cryogenics **33**, 920 (1993).
53. L. F. Goodrich and S. L. Bray, Cryogenics **30**, 667 (1990).
54. J. W. Ekin et al., Appl. Phys. Lett. **52**, 1819 (1988).
55. G. Krafft, in *Cryogenic Engineering*, ed. B. A. Hands, Academic Press, London (1986),181.
56. J. M. Robertson, Phys. Rev. Lett. **61**, 158 (1988).
57. M. N. Wilson, *Superconducting Magnets*, Oxford University Press, Oxford (1983), 234.

58. W. Ekin, J. Appl. Phys. **49**, 3406 (1978).
59. K. Davies, R. G. Jenkins, C. Danjoy, C. R. M. Grovenor, and H. Jones, Applied Superconductivity **2**, 61 (1994).
60. R. M. Schalk et al., Cryogenics **33**(3), 369 (1993).
61. M. Tinkham, Helv. Phys. Acta **61**, 443 (1988).
62. J. Mannhart et al., Phys. Rev. Lett. **61**, 2476 (1988).
63. T. Van Duzer and C. W. Turner, *Principles of Superconductive Devices and Circuits* , Elsevier, New York (1981), Chap. 4.
64. H. W. Neumuller, G. Ries, J. Bock, and E. Preissler, Cryogenics **30**, 639 (1990).
65. J. W. Ekin, Cryogenics **32**(11), 1089 (1992).
66. K. A. Muller, M. Takashige, and J. G. Bednorz, Phys. Rev. Lett. **58**, 1143 (1987).
67. H. Kupfer et al., IEEE Trans. Mag. **27**, 1369 (1991).
68. Y. Yeshurin and A. P. Malozemoff, Phys. Rev. Lett. **60**, 2202 (1988).
69. T. Matsushita, T. Fujiyoshi, K. Toko, and K. Yamafuji, Appl. Phys. Lett. **56**, 2039 (1990).
70. L. Miu, Cryogenics **32**(11), 991 (1992).
71. A. Houghton et al., Phys. Rev. B **40**, 6763 (1989).
72. H. Fisher et al., Phys. Rev. B **43**, 130 (1991).
73. Y. Xu, M. Suenaga, Y. Gao, J. E. Crow, and N. D. Spencer, Phys. Rev. B **42**, 8756 (1990).
74. W. Kritscha et al., Europhys. Lett. **12**(2), 179 (1990).
75. M. P. Maley, J. Appl. Phys. **70**, 6189 (1991).
76. N. Kerley, Private communication, Oxford Instruments Ltd. , 1994.
77. R. Wesche and A. M. Fuchs, Cryogenics **34**, 145 (1994).

6.9 Recommended Readings for Chapter 6

1. *Introduction to Superconductivity and High T_c Materials*, M. Cyrot and D. Pavuna, World Scientific, Singapore, 1992.

2. *Engineer's Guide to High-Temperature Superconductivity*, J. D. Doss, Wiley Interscience, New York, 1989.

3. *Materials at Low Temperatures*, Chap. 13, J. W. Ekin, American Society for Metals, 1983.

4. "Magnets," F. Herlach and H. Jones, in *Encyclopedia of Applied Physics*, ed. G. L. Trigg, VCH Publishers, 1994.

5. Biennial series of Applied Superconductivity Conferences Proceedings, published in IEEE Trans. Mag.

6. Biennial series of International Conferences on Magnet Technology Proceedings, usually published in IEEE Trans. Mag.

7. Biennial series of Critical Currents Meetings Proceedings, published in Cryogenics.

7

High-Temperature Superconducting Thin Films

J. M. Phillips

As discussed elsewhere in this book, a large number of materials have been discovered with superconducting transition temperatures (T_c) above 77 K. These materials are all characterized by a modified perovskite crystal structure with Cu-O planes that are generally believed to be central to the phenomenon of high-temperature superconductivity (see, e.g., [1]). The two-dimensional nature of these planes also causes the materials to be highly anisotropic, a situation not encountered in "traditional" superconductors.

Because of their distinctive properties, the process of producing these materials in thin films differs from that for low-temperature superconductors or other heavily studied films such as semiconductors. The number of elements is relatively large (at least four), which complicates film growth by any technique, especially since the compounds do not melt or evaporate congruently. A large amount of oxygen, which is inherently incompatible with many traditional thin-film growth techniques, is necessary to form the superconducting phase. The crystal structure is complex compared with most traditional thin-film materials. Finally, thin-film growth of these materials has been proceeding for relatively few years; hence, no extensive literature exists concerning the growth of such materials.

In spite of the large number of different compounds, they may be classified into a rather small number of families. Only three of these have been studied extensively in thin-film form: $YBa_2Cu_3O_7$ (YBCO), $Bi_vSr_wCa_xCu_yO_z$ (BSCCO), and $Tl_vBa_wCa_xCu_yO_z$ (TBCCO). A number of important differences between these classes of material give each its own set of advantages and disadvantages for thin-film growth and electronic applications.

YBCO is by far the most studied of the high-temperature superconducting (HTS) compounds in thin-film form (see, e.g., [2]). It consists of four elements, three of which are metals (the least of any superconductor with a T_c above 77 K). The T_c is over 90 K. Critical current density (J_c) in high-quality thin films is high, above 10^6 A/cm² at 77 K and over 10^7 A/cm² at 4.2 K. The flux pinning is relatively strong in most films, so that the magnetic field dependence of J_c is not too great, even at 77 K. For applications requiring weak flux pinning, methods have been developed to attain such properties while preserving a large J_c in low magnetic field [3]. The superconducting properties of YBCO are particularly sensitive to oxygen loss, making it difficult to obtain high-quality surfaces and interfaces. As in all high-temperature superconductors, the coherence length is short, ~0.1–0.2 nm in the <001> direction and

~1.5 nm in the other directions. High-quality films have been grown by many different techniques.

BSCCO films are much less studied than YBCO layers (see, e.g., [4]). The T_c of the 2223 (v = 2; w = 2; x = 2; y = 3) phase is 110 K, which is attractive relative to the T_c of YBCO. Unfortunately, single-phase films with this phase have not been grown successfully, and most work has been carried out on the 2212 phase which has a T_c of 84 K. The material contains a total of five elements, four of which are metals. The total becomes six if Pb is added, as is sometimes done. The anisotropy of BSCCO is even greater and the coherence lengths even shorter than those of YBCO. An advantage of BSCCO over YBCO is the smaller sensitivity of its superconducting properties to high-angle grain boundaries and oxygen loss. BSCCO is plagued by weak flux pinning above 30 K, however, though the pinning at lower temperatures is sufficiently strong for there to be considerable promise for bulk applications in the lower temperature range. The highly layered structure, coupled with the characteristics of the elements involved, makes BSCCO amenable to growth by layer-by-layer techniques such as molecular beam epitaxy and atomic layer epitaxy. Artificially layered forms of BSCCO (that have not been fabricated in bulk form) have been made in this way [5]. It is important to note that such approaches have not been fruitful for YBCO.

TBCCO is particularly attractive because the 2223 phase has a T_c of 125 K (see, e.g., [6]). Its main disadvantage is the presence of Tl, which is toxic and, in the form of TlO, highly volatile. This makes it extremely difficult to grow superconducting thin films without an anneal in a Tl overpressure after growth. The first reports of such films have appeared only recently [7,8]. Like BSCCO, TBCCO contains five elements and is highly anisotropic, with a short coherence length. Its superconducting properties are also less sensitive to oxygen loss than those of YBCO. Thin films with $J_c \sim 10^6$ A/cm^2 at 77 K have been fabricated. This property coupled with weak flux pinning above 30 K—except perhaps for the 2223 phase—makes these films attractive for applications such as the superconducting flux flow transistor.

Another material that deserves mention in the context of thin film growth is the "moderate-temperature" superconductor $Ba_{1-x}K_xBiO_3$ (BKBO) (see, e.g., [9]). While the T_c is at most only 31 K, this material has other properties that have justified its study in thin-film form. Its potential advantages lie in its cubic (hence isotropic) crystal structure and in its coherence length, which is considerably longer than that of the higher-temperature superconductors. Thus, interlayers in superconductor-normal metal-superconductor (SNS) or superconductor-insulator-superconductor (SIS) junctions need not be as thin, and the detailed properties of the interfaces are not as important. BKBO is the only superconductor with such a high T_c that has demonstrated unequivocally BCS-like, tunnel-junction behavior. A final advantage of this material is its low growth temperature, which makes it potentially more amenable (than the other materials discussed here) to integration with semiconductors.

7.1 Thin-Film Fabrication

Thin-film fabrication of HTS materials was attempted very soon after the materials were discovered [10]. Nearly every growth method that had been used for the fabrication of other types of thin films was tried by one group or another [11]. Most of the early results were disappointing, with poor superconducting properties (except, possibly, for T_c), crystallinity, and morphology. In the years since the first report of successful HTS film growth, however, considerable progress has been made. The plethora of techniques explored has now been narrowed to a relatively small number that can produce films with state-of-the-art properties. While serious scientific and technological issues remain to be addressed, the progress to date gives confidence that the remaining problems can be solved. Because the body of work that has been done on $YBa_2Cu_3O_{7-\delta}$ (YBCO) thin films is far larger than the work on all other HTS films combined, references in this discussion concern YBCO unless otherwise specified.

7.1.1 General Methods of Thin-Film Growth

Two major categories of growth technique can be used when considering the growth of HTS thin films: *in situ* and *ex situ* film growth. *In situ* HTS films superconduct when they are removed from the growth chamber. *Ex situ* films do not superconduct after growth; they must be annealed in an oxygen-rich environment to become superconducting (this is frequently called a "post anneal"). *In situ* and *ex situ* films have different growth modes. *In situ* films grow in a more or less layer-by-layer manner, as illustrated schematically in Figure 7.1. As such, surface diffusion is particularly important, as the depos-

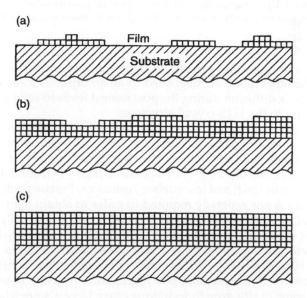

Fig. 7.1. Illustration of the *in situ* growth process. In (a), growth has nucleated in two-dimensional epitaxial islands which gradually merge. In (b), growth proceeds in a more or less layer-by-layer fashion until a uniform film of the desired thickness is achieved in (c).

(a)

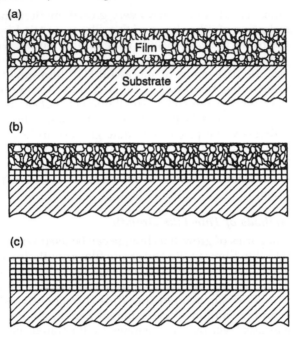

(b)

(c)

Fig. 7.2. Illustration of the *ex situ* growth process. An amorphous film is first deposited, as shown in (a). Upon annealing, solid-phase epitaxial growth initiates at the interface (b). The film eventually crystallizes throughout its thickness, as shown in (c).

ited atoms and molecules migrate to their equilibrium lattice sites. *Ex situ* films, on the other hand, are deposited in an amorphous state. As shown in Figure 7.2, bulk diffusion during the post anneal leads to solid phase epitaxy that gives rise to the HTS crystal structure.

In situ films can generally be grown at lower growth temperature, because of the dominance of surface diffusion for achieving epitaxy. Multilayers are readily grown by such techniques. The best films have smooth, almost featureless, surface morphology. They also have high J_c, with limited degradation in a magnetic field, and low surface resistance. Precise control of the film stoichiometry is not generally required in order to obtain high quality films [12]; control of the cation composition to within 10% is usually sufficient.

While it is possible to grow *in situ* HTS films of extremely high quality, these growth techniques do have a number of disadvantages. The growth apparatus is generally complex, given the requirements of substrate heating during growth and the presence of a high partial pressure of oxygen. Films grown by some *in situ* growth techniques may have a somewhat depressed T_c, even though their other properties are excellent [13]. In general, it is difficult to grow uniform, high-quality, large-area films using *in situ* techniques.

Ex situ films require a simpler growth apparatus than *in situ* films. Low oxygen pressure during film growth is generally acceptable. The best *ex situ* films have higher T_c than do some of the *in situ* films. Double-sided growth, which is important for some applications that require a ground plane in addition to a device film, is easily accomplished, and the growth of large-area uniform films is rather straightforward. The nonsuperconducting precursor films may, however, be unstable. Compared with *in situ* processes, higher processing temperature is required to render the films superconducting, since bulk diffusion processes must be activated. Multilayers are difficult or perhaps impossible to grow because of the necessity of an anneal after growth. The films generally do not have morphologies that are as good as the best *in situ* films. The best-quality films are less than 200 nm thick, and even these have somewhat higher surface resistance than the best *in situ* films. Finally, the best *ex situ* films are stoichiometric to within 1% [14].

HTS film-growth techniques can be divided between those using multiple sources and those requiring only a single source. *Multiple-source* evaporation offers the advantage of flexibility. Small changes in film composition are readily achieved in such a system. With good deposition monitoring and control, accurate control of composition is also possible. Atomic layering is also possible in principle, which allows the fabrication of artificially layered compounds that may not be stable in bulk form. Some people also believe that the control that is theoretically possible using multiple-source growth is necessary to achieve the highest quality films. Needless to say, multiple-source deposition systems tend to be extremely complex, especially if the issues of deposition monitoring and control are addressed properly. Accurate flux monitoring of each source is required. In addition, each source must operate stably so that the relative deposition rate of each material remains constant. Many param-

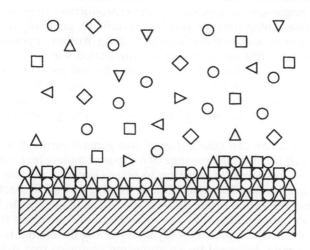

Fig. 7.3. A schematic illustration of film growth by physical deposition. Only the film constituents are in the vapor phase in the vicinity of the substrate. These collect on the substrate as shown.

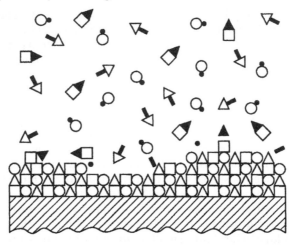

Fig. 7.4. A schematic illustration of film growth by chemical vapor deposition. The film constituents are contained in larger molecules in the vapor phase. These must be dissociated on or near the substrate in order for proper film growth to occur.

eters must be controlled. This requirement frequently leads to a problem of reproducibility in multiple-source systems.

Single-source growth is much simpler, both in theory and in practice. The apparatus can obviously be much less complex and less expensive. Accurate monitoring of the deposition source is less critical, and source stability is less crucial than in multiple-source growth. The existence of fewer critical parameters makes single-source apparatus more reproducible, in general. On the other hand, fine-tuning film composition is difficult, generally involving either the use of a new source material or the tuning of other deposition parameters that affect the film composition in complicated ways.

HTS thin films have been grown by both physical and chemical deposition techniques. *Physical deposition* (see Figure 7.3) involves the production of a vapor that includes only the species to be deposited (some combination of atoms, molecules, and radicals), possibly mixed with an inert ambient (e.g., Ar).

Chemical deposition can take either of two general forms. *Chemical vapor deposition* (CVD) involves the passage of a vapor of molecules containing the elements to be deposited (in addition to other elements, notably hydrogen and carbon) in a reactor. The conditions in the reactor lead to cracking of the molecules and deposition of the desired species, as illustrated in Figure 7.4. *Condensed-state chemical deposition* shown in Figure 7.5 involves the application onto a substrate of a precursor layer containing the elements of the film in the desired ratio, followed by an annealing sequence that drives off the undesired elements and leaves only the film constituents.

Physical deposition offers the advantage of having no extraneous elements present during film growth. Such additional species could lead to impurity

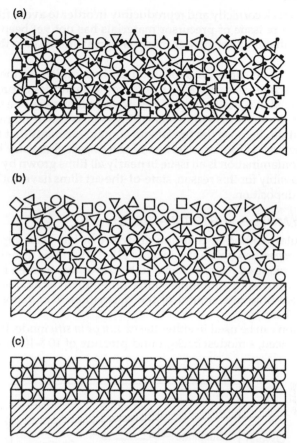

Fig. 7.5. A schematic illustration of film growth by condensed state chemical deposition. The precursor film contains the constituents in the desired film in larger molecules. These molecules must be dissociated in order to obtain the correct film composition.

incorporation in the film, which inevitably would have a bad effect on film properties. In addition, a variety of physical deposition techniques have been demonstrated to give high-quality HTS films. Without exception, however, physical deposition takes place in a vacuum system, which necessarily adds complexity to the deposition apparatus. Many physical deposition techniques are incompatible with high pressures of oxygen. This property must be circumvented in order to obtain HTS films.

CVD is, on the other hand, compatible with hostile gases, including oxygen. Furthermore, it is a familiar technique in the thin-film industry. CVD of high-temperature superconducting films, though, must make use of precursor molecules that have heavy elements in them (such as barium). This requirement makes it difficult to obtain volatile precursors. Moreover, the precursors must crack correctly and reproducibly in order to avoid film contamination. The entire issue of precursor materials has not yet been adequately addressed, so that the film grower must pay great attention to this aspect of CVD.

Condensed-state chemical deposition is the simplest of the deposition techniques described here. Because macroscopic quantities of precursor species can be mixed, it is quite straightforward to achieve accurate stoichiometric control of the precursor film, something that is much more difficult in vapor deposition. Because the precursors contain elements that must not be present in the final HTS film, selection of the correct chemistry is extremely important. Contamination is an issue in nearly all films grown by this general technique. Possibly for this reason, state-of-the-art films have not been grown by chemical deposition.

7.1.2 Specific Methods of HTS Thin-Film Growth

One popular method of thin-film growth that has been applied to HTS materials is *evaporation* (examples include [15–19]). In this technique, a separate source is required for each metal to be deposited, as seen in Figure 7.6. Because this is inherently a multiple-source film-growth technique, each source must have real-time rate control of the deposition. The deposition rate is typically 0.1–1 nm/s.

Evaporation can be used in either the *ex situ* or *in situ* mode. If *ex situ* films are to be produced, a modest background pressure of 10^{-5}–10^{-6} torr of oxygen

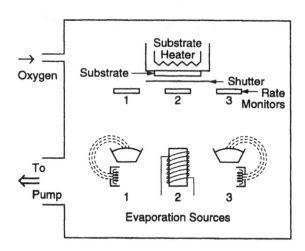

Fig. 7.6. An apparatus used for film growth by evaporation. There are three sources (labeled 1, 2, and 3). Sources 1 and 3 are electron gun evaporators in this example, such as might be used to evaporate Y and Cu in YBCO. Source 2 is a thermal effusion cell such as might be used to deposited Ba. Each source has its own rate monitor to control the film composition (labeled 1, 2, and 3). The substrate can be heated and can be shuttered from the incoming flux from the evaporation sources. Oxygen of one form or another can be bled into the chamber as discussed in the text.

is required. The most successful *ex situ* deposition technique is the BaF_2 process, which uses BaF_2 as the Ba metal source in $YBa_2Cu_3O_{7-\delta}$ (YBCO) [15,16]. These precursor films are very stable in air, whereas precursor films that contain Ba metal tend to react to form carbonates that are difficult to dissociate during the post anneal. *Ex situ* films are generally deposited onto an unheated substrate.

The BaF_2 process requires a two-stage anneal after growth to render the HTS films superconducting. The first stage of the anneal is performed in a wet-oxygen ambient in order to hydrolyze the BaF_2. During this annealing stage, which is at elevated temperature (~850°C), solid phase epitaxy occurs, giving the tetragonal 123 YBCO phase. Subsequently, a lower-temperature annealing stage in dry oxygen allows the full oxygenation of YBCO.

In situ films are deposited onto a heated substrate [17–20]. The substrate temperature is usually about 700°C for c-axis–oriented YBCO. The most difficult requirement for the deposition of *in situ* films is the necessary presence of oxygen during film growth. If molecular oxygen is used, the pressure must be in the 0.1–1 mtorr range in the vicinity of the substrate [17]. Such pressures are incompatible with evaporation sources, so that differential pumping must be used to protect them. Substrate heating is also difficult in so much oxygen. Various approaches to this problem include the use of Pt as a heating element, tungsten-halogen heating lamps, and the use of sealed heaters.

To avoid the necessity of differential pumping, some investigators use a reactive oxygen-containing molecule in place of O_2. The most common examples are ozone and NO_2. In the case of ozone, the use of pure O_3 allows the pressure in the vicinity of the substrate to be lowered by approximately one order of magnitude in comparison with that for O_2. Illumination of the substrate by ultraviolet light allows the pressure to be reduced by one more order of magnitude [18]. Because of its potential for explosion, extreme care must be exercised to generate and use O_3 safely.

Other investigators have used activated oxygen to lower the pressure requirement [19]. For example, a remote microwave oxygen plasma source has been developed that supplies about 8% activated oxygen to the substrate region. This allows the substrate region to operate in the 10^{-3} torr range, while the pressure in the main part of the chamber remains in the 10^{-5} torr range.

In its most sophisticated state, evaporation becomes molecular beam epitaxy [20,21]. Sources can be shuttered sequentially to enhance layer-by-layer growth, and artificially layered films of compositions not achievable in bulk form can be fabricated. This approach has been more successful for BSCCO than for YBCO.

Another method of HTS film growth that has been used with considerable success is *sputtering* [13,22–25]. Sputtering can be performed using either multiple-metal (or oxide) targets [22] or a single-oxide target [13]. The general principles are the same in either case, but far better results have been obtained using a single target; hence, this discussion will focus on that variant of the technique. Basically, sputtering involves the use of energetic ions (such as Ar+) to erode a target. The target species then deposit onto a heated substrate that is located near the target.

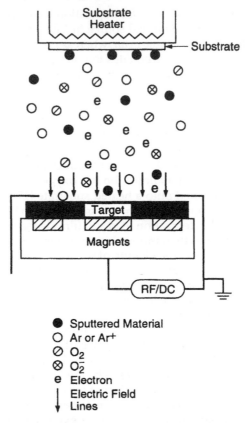

Fig. 7.7. The on-axis sputtering geometry. Note the possibility of bombardment of the substrate and growing film by O_2 ions, causing resputtering of the film.

In *on-axis sputtering* (see Figure 7.7), the substrate faces the target [23–25]. This configuration gives reasonable deposition rates (≥ 0.1 nm/s), but the geometry places the target within the plasma region, so that the deposited film is bombarded by ions. On-axis sputtering may result in film damage and, more important, in the case of HTS films, preferential resputtering of some atomic species. The films deposited also tend to be quite nonuniform.

Because of these problems, *off-axis sputtering* (illustrated in Figure 7.8) has become quite popular for HTS film deposition [13]. In this configuration, the substrate is oriented at 90°C with respect to the target, so that it is not in the plasma. This results in uniform and stoichiometric films with state-of-the-art properties. The major drawback of the off-axis geometry is the relatively low deposition rate, ~100 nm/h.

Pulsed laser deposition (PLD, also known as laser ablation) has also been used to grow films of high quality [26]. In this technique, an excimer laser, focused to an energy density of ≥ 1 J/cm^2, impinges on a single target of the

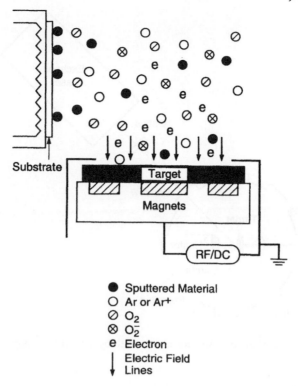

- ● Sputtered Material
- ○ Ar or Ar+
- ⊘ O$_2$
- ⊗ O$_2^-$
- e Electron
- | Electric Field
- ↓ Lines

Fig. 7.8. The 90° off-axis sputtering geometry. Because the substrate is out of the plasma, bombardment of the growing film by ions does not occur.

material to be deposited, as shown in Figure 7.9. The heated substrate faces the target, and ablated species (including atoms, molecules, and radicals) are transferred to the substrate. For HTS film deposition, growth typically takes place in ~100 mtorr of molecular oxygen. PLD has been touted as allowing the stoichiometric transfer of atomic species from the target to the substrate. In reality, the stoichiometry of the deposited film depends on a number of parameters besides the target composition [27]. These parameters include background gas pressure, laser energy density, and substrate temperature. Reasonable deposition rates of 0.1–0.5 nm/s are possible with PLD, but the area of uniform deposition is typically quite small. Another problem unique to this deposition technique is the frequent presence of particulates on the film. This is believed to arise from the use of a target that has nonuniform density, so that chunks of it may be blown off during the ablation process. The problem has largely been alleviated by the use of uniform high-density targets.

Chemical vapor deposition (CVD) and *metal-organic chemical vapor deposition* (MOCVD) have been receiving increased attention for the growth of HTS films

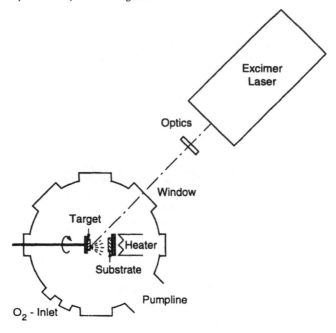

Fig. 7.9. A typical apparatus for film growth by pulsed laser deposition.

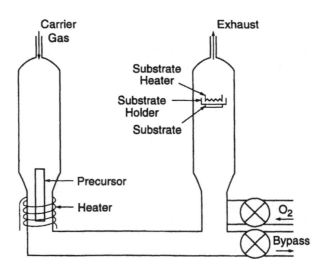

Fig. 7.10. An apparatus for film growth by solid source metallorganic chemical vapor deposition.

Chemical Deposition

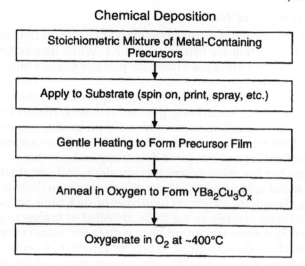

Fig. 7.11. A flowchart of the steps used in YBCO film growth by condensed state chemical deposition.

for the reasons mentioned above, in addition to the fact that it is quite straight-forward to deposit large-area films by these techniques [28]. Metal-containing precursor molecules are volatilized and passed through the reactor in the correct proportion in an inert carrying gas (see Figure 7.10). Oxygen is introduced separately. As mentioned above, the issues concerning precursors and possible contamination from them are just now being addressed. While films with high transition temperature and high critical currents have been produced, they tend to have worse morphology than state-of-the-art films and to have worse high-frequency characteristics, such as surface resistance. It is important to note, however, that progress in the quality of films that can be produced by this technique is occurring particularly rapidly.

A flow chart for the preparation of HTS films by chemical deposition is shown in Figure 7.11. Of the condensed-state chemical deposition techniques that have been attempted, by far the most successful is the *metal-organic deposition* (MOD) method [29]. In this technique, a solution of metal trifluoroacetates is spin-coated onto the substrate. During an anneal at modest temperature, the precursor film is decomposed to form a metal oxyfluoride film. This is then heat treated in moist O_2 to remove the fluorine and enable the epitaxial growth of YBCO. Finally, the film is oxygenated in pure O_2. After decomposition of the precursor film, the procedure is much like the BaF_2 process described above. The results are also similar. This is the only condensed-state, chemical-deposition technique that has yielded films that are nearly as good as state-of-the-art films. Because of the inherent simplicity of this approach, there is considerable incentive to determine the limits to the film quality that can be produced.

7.1.3 Major Unresolved Issues

The principal attributes of each growth technique for HTS films discussed in the preceding section are listed in Table 7.1. While tremendous progress has been made in HTS film growth over the past five years, a number of issues remain unresolved.

In the case of *ex situ* film growth, the BaF$_2$ process is the only physical deposition process that is still being pursued seriously. While films grown by this technique have excellent superconducting properties generally, the best films tend to have weak flux pinning, so that their current-carrying capability in a magnetic field is poor. For some device applications this can actually be an advantage, but for many others it is a serious problem. Performing the post anneal in reduced oxygen partial pressure results in films with stronger flux pinning, though the structural properties of such films are not generally as similar to single crystals as are those annealed at higher oxygen partial pressure [30–33]. A more serious problem with *ex situ* films is that multilayer growth has not been demonstrated, except for growth of a single film on each side of a substrate. Unless this difficulty is overcome, *ex situ* growth may be limited to certain applications niches.

In situ evaporation has two major problems. The first of these is the necessity to supply large amounts of oxygen during growth. The second, perhaps more serious, problem is that of growing large-area uniform films. The current goal is to grow high-quality films on 10-cm–diameter substrates. To accomplish this goal will require significant engineering advances.

On-axis sputtering has two major problems. First, as mentioned above, the composition of the HTS films is frequently not very stoichiometric or uniform because of the sputtering geometry. This has led to the second problem, namely, reduced film quality compared with what can be produced by other techniques.

Off-axis sputtering is capable of growing extremely high quality, smooth films. Its only major disadvantage at present is the low growth rate. This has been alleviated to some extent by going to a cylindrical sputtering geometry, but this approach has its own limitations. The targets for this geometry are very expensive, and the sputtering parameters do not scale exactly with those used in off-axis sputtering, contrary to predictions.

PLD is also capable of growing very good films. The issue of particulates in the films has been of concern, but it seems to be soluble by using targets of sufficiently high quality. Growth of large area, uniform films is an issue that has not yet been solved by this growth technique, although progress has been reported by several groups [34].

Films grown by MOCVD have excellent superconducting properties, except for rougher film morphology and larger surface resistance than the best films grown by other techniques. Because of the advantages of the technique, such as the capability of large-area growth and its familiarity in industry, these problems are receiving considerable attention.

Similarly, MOD is very attractive because of its simplicity. The major issue that must be resolved is the production of films of quality equal to that achievable by other techniques. Most notably, the film morphology must be improved.

Table 7.1. Major attributes of HTS thin-film growth techniques.

Growth Technique	Attributes
In situ Evaporation	Oxygen is incompatible with filaments Large amt. O_2 or activated or O_2 required Control Many parameters Multiple source Large area growth difficult
Ex situ Evaporation	No substrate heating during growth Little oxygen required during growth Anneal required after growth Only BaF_2 process is still studied Multiple source Multilayer growth difficult or impossible Large area and 2-sided growth
On-axis Sputtering	Single or multiple targets Compatible with oxygen Usually *in situ* Resputtering of film possible
Off-axis Sputtering	Usually single target Compatible with oxygen *In situ* No resputtering of film Slow growth rate
Pulsed Laser Deposition	Usually single target Compatible with oxygen Simple apparatus Particulates on film Usually *in situ* Usually small area of deposition
Chemical Vapor Deposition	Large area films straightforward Technique familiar in industry Precursor molecules not well known Film quality still improving Substrate heating straightforward
Metallorganic Deposition	Large area films straightforward Simple technique Films somewhat inferior

7.2 Substrates

One of the most important materials issues in HTS film growth concerns the choice of substrate. The best films that can currently be produced are grown on materials that have serious limitations. The search for viable substrate materials is an active area of research. Significant breakthroughs are required to identify materials that satisfy all of the requirements that the HTS films pose.

The quest for substrate materials that are capable of supporting excellent films of HTS materials has been in progress for nearly as long as HTS thin films have been prepared (see, e.g., [35]). The list of desirable substrate properties in Table 7.2 contains a number of entries that are common to good substrates for essentially all classes of thin films.

Table 7.2. Important qualities of a substrate for HTS thin films.

Desirable Substrate Properties
+ Good thermal match with film
+ Epitaxial template
+ Small lattice mismatch (usually)
+ Chemical compatability
+ Flat surface
+ Appropriate dielectric properties
+ Appropriate insulating/conducting properties
+ No phase transitions within processing window
+ No second phases or inclusions
+ Isotropic
+ Integrable with semiconductor technology

A good thermal expansion match is necessary, whether or not one is dealing with an epitaxial system. In the case of YBCO, this requirement is particularly important because of the brittleness of the superconductor. Thus, YBCO films on bulk Si substrates, even with a buffer layer of, say, yttria-stabilized zirconia (YSZ), develop cracks if they are over 50 nm thick, because of the large difference between the coefficients of thermal expansion of Si and YBCO [36]. Films on YSZ/Si layers on bulk sapphire do not have this problem as severely.

The best HTS films grown to date, as determined by a multitude of metrics including critical current density, morphology, and stability over time, are epitaxial on their substrates. This property most likely dictates that the lattice mismatch between the film and substrate should be as small as possible, although high-quality films have been grown on MgO, which has a mismatch of > 9% [37,38].

Of particular concern for the growth of high-quality YBCO films, whether or not they are epitaxial, is the chemical compatibility of the film with the substrate material. The constituents of YBCO are reactive with many substrates that might otherwise be good candidates (such as unbuffered Si) [39,40]. The relatively high temperatures required for growing even *in situ* films (≥ 700°C) [11] make the compatibility requirement more severe than it would be if high-quality films could be grown at lower temperature. In the case of *ex situ* films, the problem is even worse, since the maximum temperature that the film/substrate couple must withstand is usually 850°C [15]. The issue of chemical compatibility has generally meant that the substrates that support reasonably high quality YBCO films are themselves oxides.

An ideal substrate would have a flat surface and be free of twins and other structural inhomogeneities, although a number of materials in current use as YBCO substrates do have such problems [41]. It would be desirable, at the very least, to grow films on a substrate that has no phase transitions within the temperature regime required for film processing. In the case of microwave applications, where the dielectric properties of the substrate have an important effect on device performance, the existence of a twinning transition in the processing range is entirely unacceptable, since it precludes device modeling [42].

Device applications impose a number of other property requirements on the HTS substrate material. Microwave applications are not generally very sensitive to the dielectric constant of the substrate (as long as it is uniform and, preferably, isotropic), but they do depend on having a low value of the loss tangent. This requirement precludes the use of materials that contain magnetic ions, such as most of the rare earths. On the other hand, high-frequency applications (e.g., interconnects) require a low dielectric constant [43]. The search for substrates that can support the growth of high-quality epitaxial YBCO films has centered on materials having the perovskite crystal structure, usually oxides [10,44–46].

The properties of some of the more common HTS substrate materials are listed in Table 7.3. $SrTiO_3$ saw early success as a substrate material [10], which is not surprising in view of its rather small lattice mismatch with YBCO and its ready availability. The prohibitively large dielectric constant of this material ($\varepsilon = 277$ at room temperature), however, coupled with its unavailability in reasonable sizes, has spurred the search for alternatives.

MgO has received a good deal of interest in light of its ready availability and its modest dielectric constant ($\varepsilon = 9.65$) [37,38]. As mentioned above, it has a 9% lattice mismatch with YBCO, but with proper substrate preparation, high-quality epitaxial films can be grown. Even the best films, however, tend to have some high-angle grain boundaries, which is probably responsible for the fact that such films have poorer high frequency characteristics than do the best films grown on other substrates. MgO also reacts with water vapor, thus precluding its use as a substrate for films grown by the BaF_2 process or by MOD.

Sapphire (Al_2O_3) is of considerable interest as a substrate in view of its modest dielectric constant ($\varepsilon = 9.34$) and its commercial availability in large-diameter wafers. Sapphire reacts with YBCO, however, so that a buffer layer (such as MgO or $LaAlO_3$) must be used in order to obtain high quality films [17,20]. Because the crystal structure is not cubic, the dielectric properties of sapphire are anisotropic, making it difficult to model the performance of microwave devices.

In spite of serious problems with chemical reactivity, Si has received considerable attention as a substrate material because of the tantalizing possibilities offered by the integration of semiconductors with superconductors. The best results achieved to date have involved the use of YSZ as a buffer layer [36]. Even with this layer, however, the thermal mismatch of Si and YBCO is a serious problem limiting the ultimate YBCO thickness that can be grown.

Growth of HTS films on GaAs has proven to be even more difficult than similar depositions on Si because of the lower temperatures tolerated by the compound semiconductor. Recent progress has included the growth of reasonable-quality YBCO films ($T_c = 87$ K; J_c (77 K) = 1.5 10^5A/cm^2) on GaAs using a MgO buffer layer, possibly coupled with a $BaTiO_3$ layer on top of that [47].

Table 7.3. Properties of some of the common substrate materials.

	$\Delta a/a$ (%)	$\Delta b/b$ (%)	$\Delta c/c$ (%)	CTE (ppm)	ε	phase transf.	m.p. (K)
$SrTiO_3$	+2.0	+0.7	+0.1	10.4	277^	110*	2353
MgO	+9.0	+6.7	+7.4	10.5	9.65	---	3100
Al_2O_3	+9.4	+7.0	+11.2	8.4	9.34	---	2300
Si	+0.4	-2.4	-1.5	3.8	11.7	---	1683
YSZ	-3.6	-6.3	-5.8	8.8	25	---	3000
$LaGaO_3$	+1.5	+0.7	-0.5	10.3	25	420**	2023
$LaAlO_3$	-0.9	-2.2	-3.0	11.0	23	800*	2453
$NdGaO_3$	+0.3	+0.3	-1.3	8.0	20	>1300	1873
Sr_2AlTaO_6	+1.7	-0.1	0.0	9.0	23–30	---	---
$SrLaAlO_4$	-2.1	-4.8	7.4	7.4	27	---	1923

^ Room-temperature value.
* Second-order phase transformation.
** First-order phase transformation.

$LaGaO_3$ was identified as a potential substrate material rather early [44]. Its lattice matches and thermal expansion match with YBCO are quite good, and its dielectric constant at room temperature is smaller than that of $SrTiO_3$ by one order of magnitude ($\varepsilon = 25$). There were early reports of the growth of high-quality films on this substrate. $LaGaO_3$ has one serious drawback, however, namely, its first-order phase transition at 420 K [41], well within the processing region for HTS films grown by any technique. This transition gives rise to steps on the surface of the substrate, which can be particularly detrimental in the case of very thin or patterned films.

$LaAlO_3$ ($\varepsilon = 23$) also offers small lattice and thermal mismatches with YBCO

[45]. It, too, has a phase transition within the film processing regime (at 800 K), but the transition is second order [41]. While this leads to substrate twinning, there is no discontinuous volume change, and surface steps are not a problem. The twinning has not prevented the growth of high-quality films on this substrate, but it does make the fabrication of complex microwave devices (such as highly accurate multiple filters) impossible, since the dielectric properties of the substrate vary from point to point in a manner that cannot be controlled or predicted.

NdGaO$_3$ ($\varepsilon = 20$) was introduced as a possible alternative to the La-containing perovskites [45]. It has smaller lattice mismatches than either LaGaO$_3$ or LaAlO$_3$. It also has the advantage of having no phase transitions between its melting point and room temperature, so twin-free substrates are available for YBCO growth. Nd^{3+} is a magnetic ion, however, precluding its use as a substrate for microwave devices. There have been conflicting reports regarding the quality of film that can be grown on this substrate.

Sr$_2$RuO$_4$ is an oxide that has metallic conduction in two dimensions (and is semiconducting in the third). The oxide has been demonstrated to be a promising substrate material for HTS compounds [4]. It is noteworthy because of the small number of metallic substrates that have been demonstrated to support epitaxial YBCO growth.

7.3 Multilayers

Some of the potential electronic applications of high-temperature superconductors require multiple layers of superconductor interspersed with another material, such as a metal or insulator. In some cases, the nonsuperconducting layer is an integral part of an HTS device. In other cases the non-HTS film is not part of a device structure. The most obvious example of this application is the use of buffer layers between an HTS film and a substrate that is incompatible with it in some way. The most studied example is YBCO on Si, as already discussed.

When contemplating the fabrication of multilayers, one must ask what properties the nonsuperconducting layers must have. Many of the requirements are the same as those for substrate materials discussed in the preceding section. The most important requirement, of course, is that they not disrupt the superconducting properties of the HTS layers. This means that all of the materials in the multilayer must be chemically compatible, with no chemical reaction or interdiffusion at the interfaces. Experience has shown that, except for certain noble metals, this most likely requires that the individual layers themselves be oxides. Since unintentional grain boundaries are detrimental to the behavior of the HTS film, it is (at least at present) necessary to have essentially single-crystal HTS layers. This generally implies that the entire multilayer structure must be epitaxial. In turn, this means that the individual layers must be crystallographically compatible, with good lattice matches and similar symmetries. Since the HTS compounds are particularly brittle, a good thermal match is also necessary to avoid film cracking [49], especially if the layers are thick or numerous.

The issue of surface energy has not been discussed extensively in the context of HTS multilayers. In heteroepitaxy in general, the various layers must have comparable surface energy in order for the layers to be grown on each other with comparable ease (see, e.g., [50]). If this condition is not satisfied, the growth probably will not be symmetric; that is, it may be possible to grow a uniform layer of material A on material B, but when the materials are reversed, layer B will be highly nonuniform and possibly discontinuous. Since smooth surfaces are necessary for desirable and reproducible multilayer properties, such a siutation obviously is to be avoided. Finally, the layers must each have the electrical properties required for the successful fabrication of the device at hand. These properties include dielectric properties and insulating or conducting ability.

Most of the interlayers that have been grown successfully are insulators. $SrTiO_3$ was discussed earlier as a substrate material. Its advantages as an interlayer include its chemical and crystallographic compatibility with the HTS compounds and the rather similar growth conditions for the different materials [51]. The high dielectric constant of $SrTiO_3$ is not as severe a problem when the $SrTiO_3$ is an interlayer, since much less material is present than when it is used as a substrate. There are also reports that the dielectric constant of at least some $SrTiO_3$ films may be substantially lower than that of bulk material (though still high) [52]. $LaAlO_3$ has also been used successfully as an interlayer material [51] with similar results to those for $SrTiO_3$.

Yttria-stabilized zirconia (YSZ), with the cubic fluorite crystal structure, has been studied as a substrate material but has fallen out of favor because of the general superiority of various perovskite materials such as $LaAlO_3$ for

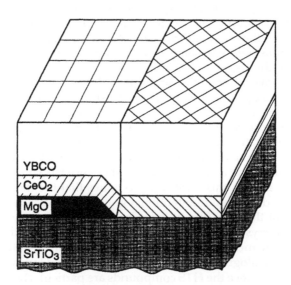

Fig. 7.12. An illustration of a technique used to grow reproducible grain boundaries in YBCO films. CeO_2 grows with different orientations on MgO and $SrTiO_3$. The orientation of the YBCO on top is determined by that of the CeO_2.

this purpose. It has, however, seen quite a bit of success as an interlayer in HTS structures. Its most common use in this context is as a buffer layer between an HTS film and an incompatible substrate [36]. Particular success has been demonstrated using YSZ as an interlayer between Si and YBCO in YBCO/YSZ/Si and YBCO/YSZ/Si/sapphire structures. One of the most notable features of YSZ is its permeability to oxygen, which results in oxidation of the YSZ/Si interface to form SiO_2 after the epitaxial relationship between the YSZ and Si is established. This property may be useful in some structures involving HTS layers, since the oxidation of these films is so critical to their performance.

CeO_2 is another fluorite structure insulator that has been used successfully as an interlayer [51]. Its properties are generally similar to those of YSZ.

The discovery of high-angle grain-boundary junctions with reproducible characteristics motivated the search for convenient ways to grow such boundaries reliably. The discovery that different classes of substrate support different orientations of YBCO under the same growth conditions [51] offered such an opportunity. By starting with a substrate of, say MgO, it is possible to grow a buffer layer of another material such as $SrTiO_3$ or CeO_2 on part of the substrate. The different orientations of the YBCO on the buffered and unbuffered regions gives a reproducible grain boundary at the edge of the buffered film, as illustrated schematically in Figure 7.12 [51].

Alkaline earth fluorides have been studied to some extent as substrates and interlayers, but they have more tendency to react with HTS compounds than do the above-mentioned oxides [54]. One possible use of these materials, however, lies in their ability to completely relieve strain at a highly mismatched interface, because of the relatively weak bonding across the interface [55]. This may be useful to relieve thermal mismatch between an HTS layer and a substrate.

Not nearly as many metals have been used successfully as part of an HTS multilayer structure. Ag, Au, and AgAu alloys have been used for metallization and in planar SNS structures because they do not react strongly with the HTS compounds [56]. They cannot be used as an interlayer, however, because epitaxial growth between these metals and the HTS materials is not possible.

The work on interlayer materials continues. The search for lower dielectric constants in oxides with perovskite-related crystal structures has led to examination of Sr_2AlTaO_6 [57]. While this material is capable of being grown on and of supporting high-quality YBCO, its dielectric constant of 23–30 is no better than that of other more established materials.

The need for a metallic material that can support epitaxial HTS materials has led to the identification of a number of metallic oxides, the most studied of which is $Sr_{1-x}Ca_xRuO_3$ [58]. This material has been shown to be compatible with YBCO and to support epitaxial growth of the HTS compound. $CaRuO_3$ has had some success as a barrier layer in junctions.

Another interlayer that has been used with considerable success is $PrBa_2Cu_3O_7$, (see, e.g., [59,60]). This structural cognate of YBCO is a semiconductor that is essentially insulating in the regime where YBCO superconducts. Because of its structural and chemical similarity to YBCO, very high quality superlattices have been made using these two materials. In fact, it is possible to fabricate superlattices with periods of one unit cell in this system.

Not all of the growth techniques discussed in the section on film growth are amenable to the fabrication of multilayer structures. In particular, *ex situ* techniques are not appropriate, since such methods would require the structure to be exposed to the atmosphere between the growth of different layers, resulting in interface contamination. All of the *in situ* techniques can be applied to multilayer growth, though it may be desirable to use single-target methods whenever possible because of the lower level of complexity of such approaches.

7.4 Characterization of HTS Thin Films

In beginning a discussion of characterization of high-temperature superconducting thin films, it is useful first to remember the two properties that distinguish a superconductor from other materials. The first is, of course, the existence of a critical temperature (T_c) below which the resistance of the material is zero. The second distinguishing feature is the perfect diamagnetism exhibited by a superconductor, such that in a magnetic field, screening currents are set up in the surface layer of the superconductor so that all flux is excluded from the interior. These two unique features form the basis for much of the characterization that is necessary to determine the superconducting properties of a sample.

Resistivity is typically measured by a four-point technique (illustrated in Figure 7.13).

A small current is passed between two of the points. The current must remain small in order to avoid sample heating and high current density. The voltage is measured across the other two points. Near the superconducting transition, voltages in the nV range need to be measured. With Ohm's law, the resistance, R, of the sample can be extracted as a function of temperature.

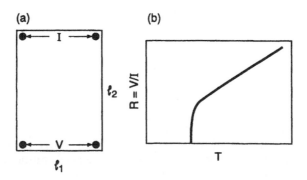

Fig. 7.13. (a) The orientation of contacts for measurement of the resistance of a thin film. Current (~10 µA) flows between the top two contacts, while voltage (often as small as nV) is measured between the bottom leads. A typical dependence of resistance on temperature for an HTS material is shown in (b).

The resistivity, ρ, can be obtained from the relation

$$\rho = H E R, \tag{7.1}$$

where H is a geometrical factor depending on the sample geometry. Representative values of H for various aspect ratios of rectangular samples are given in Table 7.4. E is the effective sample thickness, which is the actual thickness l_3 if

$$l_3 << (l_1 l_2)^{1/2} \tag{7.2}$$

where l_1 and l_2 are the lengths of the sides of a rectangular sample [61].

Another way of characterizing a superconducting sample is to make use of the Meissner effect (as illustrated in Figure 7.14). Perfect diamagnetism dictates that a superconductor subjected to a magnetic field will exclude flux from its interior by setting up screening currents on its surface. The Meissner effect, on the other hand, refers to the fact that a superconducting sample above T_c will present no barrier to magnetic flux penetration, but that this flux will be excluded from the bulk of the sample as it is cooled below its critical temperature. Measuring the magnetization during sample cooling in a field, then, allows one to determine the amount of the sample that is actually superconducting. Magnetization curves of a hypothetical sample are shown in Figure 7.15.

A noninvasive way to measure T_c of a superconducting film is to use two-coil mutual inductance [62] (as shown in Figure 7.16). Coils are placed on either side of a film. The drive coil is driven at a frequency of a few kilohertz, and the receive coil is used to detect the in-phase and quadrature components of the mutual inductance phase sensitively. As the film becomes superconducting, the in-phase component of the mutual inductance drops because of the screening provided by the film, and the out-of-phase component of the inductance dips.

After the T_c has been determined, it is usually of interest to measure other superconducting properties of the film as well. Perhaps the most important of these properties is the critical current density, or J_c. One of the most popular ways to do this is to measure the magnetization of the film as the applied magnetic field is varied. This technique necessitates the application of a model to describe the the way in which currents are flowing in the sample. The most usual model is that of Bean [63], which has more recently been modified for the case of thin films by Gyorgy et al. [64]. The model assumes that currents

Table 7.4. Representative values of H for various aspect ratios of rectangular samples necessary for calculating resistivity in Equation (7.1).

l_2/l_1	H
0.25	0.3207
0.5	0.8911
0.67	1.562
1.00	4.531
1.50	21.86
2.00	105.1
4.00	56,300

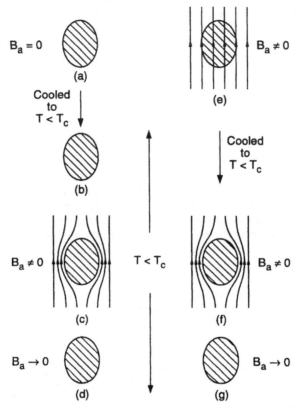

Fig. 7.14. The Meissner effect. In (a) the temperature is above T_c, and no external magnetic field (B_a) is applied. The sample is cooled in zero field in (b). In (c), a field is applied at $T < T_c$, resulting in flux exclusion from the superconducting sample by the setting up of currents on the sample. When the field is removed in (d), there is still no flux within the sample. If a warm sample is placed in a magnetic field, flux penetrates the sample (e). As the specimen is cooled below T_c as in (f), once again, persistent currents arise on the surface to cancel the field in the interior of the sample. As in the previous case, the currents vanish when the field is removed in (g).

flow within a grain only, which implies that the grain size must be known in order to determine J_c accurately. In addition, $\Delta M / \Delta H$ must be small, where M is the magnetization of the sample and H is the applied magnetic field. This means that this technique is not useful to extract J_c in low magnetic field. The J_c is ascertained by measuring the difference in magnetization at a given field as a function of the direction in which the field is being applied (illustrated in Figure 7.17).

The most straightforward technique for the measurement of J_c is the *electrical transport method* (illustrated in Figure 7.18). A constriction is made in a film so that the applied current must pass through an area of very small cross

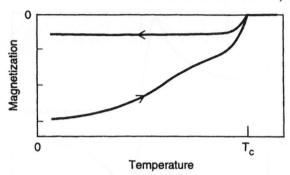

Fig. 7.15. A magnetization curve for a superconducting sample. In the upper curve (indicated by the left-pointing arrow), the sample is cooled in a field while the magnetization is measured to give an indication of flux expulsion or the Meissner effect. In the lower curve (indicated by the right-pointing arrow), a field is applied while the sample is cold, and the magnetization is measured as the sample warms. This provides a measurement of flux exclusion or diamagnetism.

section. The voltage across the sample is measured as the applied current is increased, and the J_c is specified to be the value of current density at which the measured voltage exceeds some small value, such as 1 μV. In its crudest form, this measurement technique requires no patterning, although the edges

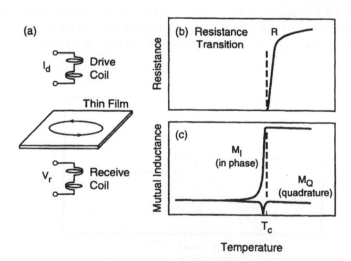

Fig. 7.16. (a) Schematic diagram of the measurement of T_c by the two-coil mutual inductance technique. (b) The resistance transition of an HTS sample. T_c is indicated by the dashed line. (c) The in-phase and quadrature components of the mutual inductance of the same sample.

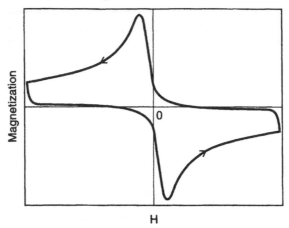

Fig. 7.17. A magnetization loop that could be used to extract J_c in nonzero magnetic field. The temperature of the sample is fixed while the magnetic field is varied and the magnetization is measured. The J_c is related to the change in magnetization at a given field depending on the sign of the change in field.

of the constricted region may be rough, resulting in an uncertain film cross section. The major drawback of this technique is that the J_c criterion is uncertain, both because of thermal propagation from film inhomogeneities and because of possible heating from normal metal leads.

Many of the potential applications of HTS thin films involve operation at high frequencies. At such frequencies, two types of electrons exist in the superconductor: "superelectrons," which do in fact carry current with no

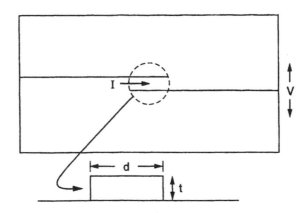

Fig. 7.18. Schematic diagram of the thin-film geometry typically used to measure J_c by the transport method.

resistance, and "normal electrons," which behave like electrons in a normal metal. At 0 K, all electrons that contribute to conduction in a superconductor are superelectrons; above T_c, they are all normal electrons. Between these two temperatures, the fraction of superelectrons contributing to conduction approaches 0 as T_c is approached. This is described by the *two-fluid model*. In a direct current, there is no electric field, so that there is nothing to drive the normal electrons, and hence no normal current. In an alternating field, on the other hand, the electrons are accelerated by the changing field. The inertial mass of the electrons forces them to lag the field. The result is an inductive impedance, causing some current to be carried by normal electrons and leading to the loss that is measured in superconductors at high frequencies. At sufficiently high frequency, a superconductor behaves like a normal metal.

The loss of a superconducting film at high frequency can be measured in a variety of ways, many of which involve patterning of the film. One method of characterization that does not require patterning (and hence avoids the film degradation that could result) is to replace one end of an electromagnetic cavity with the film to be characterized, as described by Drabeck et al. [65]. This technique has several strong points. Since no patterning of the film is required, the danger of film degradation prior to characterization is reduced. The substrate is not actually in the cavity and therefore does not contribute to the losses that are measured. This technique allows the characterization of films grown on lossy substrates such as $SrTiO_3$. The film can be compared directly with Cu. A major disadvantage of the technique, however, is that a film can be characterized over only a limited range of frequencies in a given cavity.

Another way of characterizing the high-frequency properties of an HTS film is to fabricate a meander line (see Figure 7.19) several centimeters long (depending on the frequencies of interest) and to place this in a cavity. A major difficulty of this technique is that a low-loss substrate must be used since the substrate actually resides in the cavity. An advantage is that many resonant frequencies of a given meander line can be probed in a single experiment. Also, contacts to the film are not required, and comparison with Cu lines of the same geometry is straightforward [66]. This technique is typically used in the 1–10 GHz region of the electromagnetic spectrum.

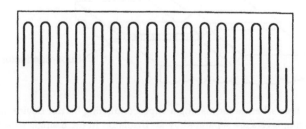

Fig. 7.19. A typical meander line for examining the high-frequency behavior of a superconductor.

For much higher frequencies (up to 1 THz), time-domain characterization is required [67]. One excites an HTS transmission line at one end and then looks for the time delay and amplitude change between the transmitted pulse and a reference pulse at the other end of the line.

Structural and compositional characterization of HTS films is extremely important, especially when commissioning a new growth apparatus or optimizing film properties. A detailed discussion of these techniques is not possible here. Interested readers are referred to many of the fine books on materials characterization techniques such as [68,69].

7.5 Applications of HTS Thin Films

In developing a list of materials needs for the new superconducting electronics, it is first useful to enumerate some of the actual device applications

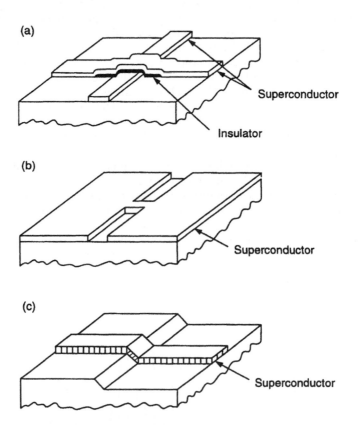

Fig. 7.20. Three schemes for making a superconducting barrier device. In (a), two superconducting layers are separated by an insulator. In (b), a single superconducting film has a constriction to restrict current flow between the two regions. In (c), a superconducting film is grown over a grain boundary or step to cause a grain boundary or other weak link in the superconductor.

of high-temperature superconductors that are contemplated. Each of these potential applications has its own materials needs.

The first application of superconductors to electronics is for *all-superconducting devices*. Nearly all such devices that have been demonstrated to date involve some sort of Josephson junction, or region of restricted current flow (see, e.g., [70,71]). Possible means of accomplishing this are shown in Figure 7.20 and include lithographic patterning of a single superconducting film, incorporation of a grain boundary into a single film, or use of a thin layer of metal or insulator between two superconducting regions, through which superconducting electrons can pass. When the current flowing becomes sufficiently large, this region ceases to be superconducting, resulting in a measurable voltage drop across the device. It is rather obvious that the most important requirement for any process that fabricates such junctions is that it give controllable and stable junction characteristics that are reproducible from device to device. Depending on the device geometry selected, a number of other materials requirements come into play, such as the need for multilayers of superconductor and nonsuperconductor, the necessity of substrate processing before growth to allow the precise location of grain boundaries or other weak-link features, and the need for a suitable barrier material whose properties give good junction characteristics.

A more recent all-superconducting device that has received attention from high-frequency circuit designers dealing with high-temperature superconductors is the superconducting flux-flow transistor (SFFT) [72] sketched in Figure 7.21. In this device a region of a superconducting film is thinned and perforated selectively, resulting in a reduced film cross section. This region is

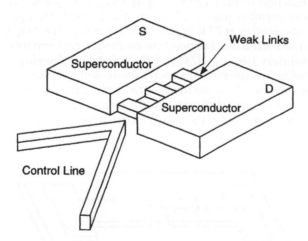

Fig. 7.21. Schematic diagram of a superconducting flux flow transistor (SFFT). The source (S) and drain (D) are made of a single HTS film. These two regions are separated by a region that has been lithographically defined to result in restricted current flow, which forms the gate of the device. The control line is used to induce a magnetic field and cause flux flow in the gate region (and hence a voltage) as described in the text.

located between the source and drain of the device.

A bias current flowing across the device body places the thinned region into the flux-flow regime. A small current path near the thinned region (i.e., the control current) then induces a magnetic field in the "weakened" area that, in turn, causes a rise in source-drain voltage. Small changes in the control current induce measurable changes in the source-drain voltage. One of the appeals of this device is the fact that, in its simplest form, it requires only a single superconducting layer. Other materials requirements that it poses are the need for weak flux pinning and for low-loss substrates, since the devices are typically operated at high frequencies.

Hybrid superconductor-semiconductor devices and systems have been contemplated for many years. Typically, three levels of hybridization are discussed: (1) combining superconductor and semiconductor in a single device; (2) combining individual superconductor devices with semiconductor devices in an integrated circuit; and (3) combining superconducting circuits and semiconducting circuits or chips into a complete system.

An example of a *single device* using both superconducting and semiconducting materials (shown in Figure 7.22) is the semiconductor-coupled Josephson junction (see, e.g., [73]). The semiconductor acts as the weak link for tunneling between two superconducting regions. This sort of device poses stringent materials requirements. The superconductor and semiconductor must be in intimate contact with each other; thus, there may be no interdiffusion or other chemical reaction at this interface. Each class of material must tolerate processing steps that are intended for the other. Depending on the device structure, multilayers of superconducting, semiconducting, and other materials may be required.

The combination of superconducting devices with semiconducting ones in an *integrated circuit* requires, at least for the present, the operation of the semiconducting devices at 77 K or below. This observation begs the question of the impact of such a low temperature on device performance. In general, the effect is salutary (see, e.g., [74] and the references therein). For example, in silicon CMOS (complementary metal oxide–semiconductor) technology, device speed increases because the low field mobilities increase as phonon scattering decreases, and drift velocities increase. Reliability increases and leakage decreases as a result of the decrease of reverse currents and the achieve-

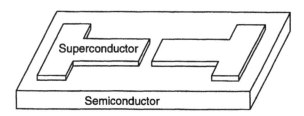

Fig. 7.22. A hybrid superconductor-semiconductor device. The semiconductor forms the barrier between two superconducting regions in a Josephson junction.

ment of sharper I–V characteristics for thermally activated processes. Finally, minority carrier effects are suppressed, giving greater latch-up immunity. The result is VLSI circuit performance that is two to three times better at 77 K than at room temperature. This is not the only semiconductor technology that benefits from low-temperature operation. High-electron mobility transistors (HEMTs), for example, operate only at low temperatures, which, coupled with their high speed, makes them potentially attractive for integration with superconducting devices.

If one is to fabricate integrated circuits using active superconductor and semiconductor devices, one will probably choose to use superconducting elements for those circuit functions that take advantage of the strengths of these materials, such as speed and zero resistance or low loss (see, e.g., [75]). Semiconducting devices are likely to be used for functions requiring other attributes such as high power, mass memory, or specific functions that semiconductors perform well, such as amplification and rectification. Such integrated circuits require that device-quality layers of both superconductors and semiconductors be combined on a single substrate that is compatible with both. As in the case of a single hybrid device, processing steps that can be tolerated by both classes of materials need to be developed. Finally, multilayers are almost certainly necessary to achieve both sorts of devices on the same substrate.

At the *system level*, the greatest interest in combining superconductors with semiconducting circuits has focused on superconducting interconnects, which can increase both bandwidth and wiring density (see, e.g., [76]). In various analyses that have been performed, the relative performance of normal-metal and superconducting interconnects for various functions has been compared. Interconnects between neighboring transistors require switching speeds of 10 to 100 picoseconds over distances on the order of one to several microns. Over this distance, such switching speeds are easily achieved with normal metal, so that there is little motivation to use superconducting interconnects unless the devices themselves are superconducting. Even though the length of the conductor among gate-level circuits is longer than between individual transistors, normal metal can still perform adequately, especially when cooled to 77 K. At the level of integrating system level functions, the possible advantage of superconducting interconnects appears. The switching times are similar to those for gate-level circuits, but the line lengths are considerably greater, so that normal metal interconnects begin to limit the speed of the system. The materials requirements posed by superconducting interconnects include especially the necessity of low surface impedance compared with Cu at 77 K at the relevant frequencies. High current density is required. The substrates that support not only the interconnects but also at least some of the semiconductor circuits must be low loss and compatible with both classes of electronic material. In order to have an impact on the electronics industry, large-area substrates are required. Multilayers of superconducting and dielectric films are required to support a ground plane as well as several interconnected layers of wiring.

Microwave-frequency devices such as filters and delay lines have been tar-

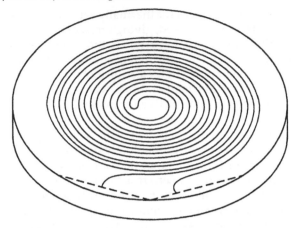

Fig. 7.23. Schematic diagram of a delay line. Large-area, uniform HTS films are required for such a device.

geted by many as potential niches for HTS devices (see, e.g., [77]). In this regime, superconductors offer a number of potential advantages such as dynamic range and bandwidth. In addition, the materials requirements seem perhaps somewhat easier to attain than the requirements posed by active devices. The films must be large area and uniform for many microwave devices such as the delay line shown in Figure 7.23.

The requirement of a ground plane for microwave devices can be satisfied by several geometries, depicted in Figure 7.24. Multilayers or growth on two sides of a thin substrate are required for the stripline and microstrip configurations. The films must have low surface resistance and, at least in some cases, have low dependence of surface resistance on microwave power. More ambitious applications require coupling of superconducting passive elements with active semiconducting elements. Substrates are a major issue in this arena. They must be available with large area and have low dielectric losses (and, in some cases, low dielectric constant). These requirements preclude the use of materials that contain magnetic ions. In order to be able to model the devices accurately, isotropic (or at least predictably anisotropic) substrate materials are needed.

7.6 Status of HTS Thin Films

Because of the widely differing amount of work that has been carried out on thin films of the various HTS materials, the status of each material system is best assessed separately.

Largely because it has received the greatest amount of attention, the state of YBCO films is much more advanced than that of the other HTS systems. Methods have been developed to control the epitaxial orientation of the films, specifically the orientation of the c-axis relative to the growth direction [53]. This development may be particularly important because of the large anisot-

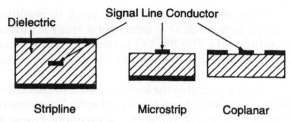

Fig. 7.24. Three geometries for the fabrication of microwave devices. Note that the coplanar geometry requires only a single HTS film, whereas the microstrip geometry requires two HTS layers and the stripline geometry needs three.

ropy of YBCO. While the strength of flux pinning in state-of-the-art films is generally quite high, methods have been developed to achieve low pinning strength for applications requiring this attribute. The surface resistance of the best films is considerably better than that of Cu at 77 K over a wide frequency range, though the question of how much power the films can tolerate is still an issue [78]. Multilayers involving YBCO and many other materials (discussed earlier) have been demonstrated. From the growth standpoint, the most impressive of these is probably the $YBCO/PrBa_2Cu_3O_7$ superlattice [59,60] in which periods as small as one unit cell have been demonstrated. The growth of fifteen layers of several different materials to fabricate a monolithic SQUID magnetometer is also impressive and is likely to be more relevant technologically [79].

Much progress has been made on improving the quality of YBCO film surfaces, but there is still room for improvement. It is fair to say that the atomic level smoothness achievable in many semiconductor systems is still a goal rather than a reality for YBCO. Because of the detrimental effects of grain boundaries and misorientation in YBCO, which are more severe than in the other HTS materials, epitaxy is especially important for YBCO. This has been achieved on a wide variety of substrates and with many sorts of multilayers. Planar or nearly planar methods have been developed to fabricate junctions. The fabrication of a functioning trilayer junction remains a goal, however, which points out the need for continued improvement in the quality of the interface between YBCO and other materials. This is not so important for structures that do not require electrical transport between layers, but is critical for those that do require such transport, especially because of the short coherence length.

The BSCCO system is considerably less studied and characterized, but there are some notable achievements. One major drawback of this system is that all single-phase thin films that have so far been fabricated are of the 2212 phase, which has a relatively low T_c of 84 K. Growth of the higher-T_c 2223 phase has proven a difficult challenge. BSCCO has been found to be more amenable to layer-by-layer growth techniques such as molecular beam epitaxy [80] and atomic layer epitaxy [5] than has YBCO. It is the only HTS system in which artificially layered structures have been fabricated in thin-film form that have not yet been demonstrated in bulk. Work on multilayers has not progressed as far as in the YBCO system, but a trilayer junction has been fabricated successfully [81]. The BSCCO system has much weaker flux pinning than YBCO, and there has not been much success in increasing this in the thin films.

The TBCCO system is probably the least studied and characterized of these three HTS systems. Films have been fabricated with T_c approaching the bulk value of 125 K for the 2223 phase, but with few exceptions these are not single phase. The large density of intergrowths, etc., is probably at least partially responsible for the relative insensitivity of films of this material to deviations in stoichiometry. Because of the volatility of TlO, the *in situ* growth of TBCCO is extremely difficult, and the first reports of such growth, albeit with reduced superconducting properties, are just beginning to appear [7,8]. The lack of a good *in situ* growth process means that multilayer structures have not been fabricated. The weak flux pinning in this material has been used advantageously in devices such as the SFFT.

BKBO is receiving increased attention because of its cubic structure and relatively long coherence length, in spite of its lower T_c. Trilayer junctions have been fabricated successfully [82]. A major issue, however, is film cracking, which arises from the contraction of the film during oxygenation. In general, considerably more characterization (particularly growth optimization) needs to be done in this system.

All of these materials have been grown successfully by a number of different growth techniques, with the caveats mentioned above. Of the *ex situ* growth techniques, only the BaF_2 process for YBCO, MOD for the various materials, and a number of techniques for TBCCO deposition are serious contenders that can grow films at or approaching the state of the art. In contrast, a wide variety of *in situ* growth techniques have been shown to grow high-quality HTS films, including both single- and multiple-target approaches, and such diverse methods as sputtering, pulsed-laser deposition, evaporation, and MOCVD. For physical deposition, the single-target approaches are much simpler than the multiple-target ones and, so far, seem to be giving more reproducible results of high quality. It remains to be seen whether the additional flexibility of multiple source techniques will be an advantage or a liability in the long run.

Some general issues still must be addressed in the thin-film arena. The growth of larger-area films is needed for some device needs in the near term, and will be essential if HTS materials are ever to become part of the semiconductor industry. For large-area devices, as well as for the reproducible fabrication of many small devices on a single substrate, it is necessary to improve film uniformity across a substrate. While these are primarily engineering issues, they will have to be addressed for progress in some areas of superconducting electronics.

Progress in the identification and utilization of substrate materials for HTS films has been substantial, but less impressive than progress in the films themselves. Considerable work is needed, since some of the remaining problems in the films are almost certainly connected with the lack of an ideal substrate on which to grow them. Work continues in this field, and it is likely that much better substrate materials will be available in a year or so than can be obtained today.

Multilayer fabrication has developed quickly. A rather wide variety of insulators have been identified that can be used as buffer layers between the substrate and HTS film or as interlayers. Better dielectric properties, especially lower dielectric constant and/or lower loss, are needed for some applications. An issue that affects device design is the relationship between the properties of the interlayers and those of the bulk material. The identification of appropriate metals for multilayers has been more problematic. Sr_2RuO_4 is the only epitaxial metal that has been identified so far, and it is metallic in only two dimensions. An isotropic metal that is compatible with the HTS materials is needed.

Progress in the fabrication and use of HTS thin films has been most impressive, especially considering the short period of time since the discovery of the high-temperature superconductors and the first report of successful film fabrication. The chemistry and structure of these materials are much more complicated than those of materials that have usually attracted the attention of those interested in thin-film growth. The fact that major issues in the film growth remain to be resolved should not be taken as a discouraging sign. Instead, it is important to view the tremendous progress that has been made as an indication of what can be accomplished when the opportunity and motivation are present to study and optimize a new materials system. It is also a sign of more impressive developments to come, given a continued commitment to unlock the secrets of these fascinating materials.

7.7 References

1. R. J. Cava, Scientific American **262**, 42 (1990).
2. J. Geerk, G. Linker, and O. Meyer, Materials Science Reports **4**, 195 (1989).
3. M. P. Siegal, S. Y. Hou, J. M. Phillips, T. H. Tiefel, and J. H. Marshall, J. Mater. Res. **7**, 2658 (1992).
4. K. Wasa, H. Adachi, K. Hirochi, Y. Ichikawa, and K. Setsune, Mat. Res. Soc. Symp. Proc. **169**, 537 (1990).
5. H. Tabata, O. Murata, T. Kawai, and S. Kawai Appl. Phys. Lett. **56**, 1576 (1990).
6. D. S. Ginley, J. F., Kwak, E. L. Venturini, B. Morosin, and R. J. Baughmann Physica C **172**, 42 (1989).
7. J. Betz, A. Piehler, E. V. Pechen, and K. F. Renk, J. Appl. Phys. **71**, 2478 (1992).
8. D. W. Face and J. P. Nestlerode, Appl. Phys. Lett. **61**, 1838 (1992).
9. E. S. Hellman, E. H. Hartford, and R. M. Fleming Appl. Phys. Lett. **55**, 2120 (1989).
10. P. Chaudhari, R. H. Koch, R. B. Laibowitz, T. R. McGuire, and R. J. Gambino, Phys. Rev. Lett. **58**, 2684 (1987).
11. R. G. Humphreys, J. S. Satchell, N. G. Chew, J. A. Edwards, S. W. Goodyear, S. E. Blenkinsop, O. D. Dosser, and A. G. Cullis, Supercond. Sci. Technol. **3**, 38 (1990).

12. J. R. Kwo, private communication (1991).
13. C. B. Eom, J. Z. Sun, B. M. Lairson, S. K. Streiffer, A. F. Marshall, K. Yamamoto, S. M. Anlage, J. C. Bravman, T. H. Geballe, S. S. Laderman, R. C. Taber, and R. D. Jacowitz, Physica C **171**, 354 (1990).
14. D. J. Carlson, M. P. Siegal, J. M. Phillips, T. H. Tiefel, and J. H. Marshall, J. Mater. Res. **5**, 2797 (1990).
15. P. M. Mankiewich, J. H. Schofield, W. J. Skocpol, R. E. Howard, A. H. Dayem, and E. Good, Appl. Phys. Lett. **51**, 1753 (1987).
16. M. P. Siegal, J. M. Phillips, R. B. van Dover, T. H. Tiefel, and J. H. Marshall, J. Appl. Phys. **68**, 6353 (1990).
17. R. M. Silver, A. B. Berezin, M. Wendman, and A. L. de Lozanne, Appl. Phys. Lett. **52**, 2174 (1988).
18. T. Siegrist, D. A. Mixon, E. Coleman, and T. H. Tiefel, Appl. Phys. Lett. **60**, 2489 (1992).
19. J. R. Kwo, M. Hong, D. J. Trevor, R. M. Fleming, A. E. White, R. C. Farrow, A. R. Kortan, and K. T. Short, Appl. Phys. Lett. **53**, 2683 (1988).
20. A. B. Berezin, C. W. Yuan, and A. L. de Lozanne, Appl. Phys. Lett. **57,** 90 (1990).
21. J. N. Eckstein, I. Bozovic, K. E. von Dessonneck, D. G. Schlom, J. S. Harris, Jr., and S. M. Baumann, Appl. Phys. Lett. **57**, 931 (1990).
22. L. H. Allen, E. J. Cukauskas, P. R. Brousard, and P. K. Van Damme, IEEE Trans. Magnetics **27**, 1406 (1991).
23. M. Kawasaki, S. Nagata, Y. Sato, M. Funabashi, T. Hasegawa, K. Kishio, K. Kitazawa, K. Fuecki, and H. Koinuma, Jpn. J. Appl. Phys. **26**, L738 (1987).
24. Q. Li, X. X. Xi, X. D. Wu, A. Inam, S. Vadlamannati, W. L. McLean, T. Venkatesan, R. Ramesh, D. M. Hwang, J. A. Martinez, and L. Nazar, Phys. Rev. Lett. **64**, 3086 (1990).
25. H. Adachi, K. Hirochi, K. Setsune, M. Kitabatake, and K. Wasa, Appl. Phys. Lett. **51**, 2263 (1989).
26. A. Inam, M. S. Hegde, X. D. Wu, T. Venkatesan, P. England, P. F. Miceli, E. W. Chase, C. C. Chang, J. M. Tarascon, and J. B. Wachtman, Appl. Phys. Lett. **53**, 908 (1988).
27. M. C. Foote, B. B. Jones, B. D. Hunt, J. B. Barner, R. P. Vasquez, and L. J. Bajuk, unpublished manuscript (1992).
28. R. Hiskes, S. A. DiCarolis, J. L. Young, S. S. Laderman, R. D. Jacowitz, and R. C. aber, Appl. Phys. Lett. **59**, 606 (1991).
29. P. C. McIntyre, M. J. Cima, J. A. Smith, R. B. Hallock, M. P. Siegal, and J. M. Phillips, J. Appl. Phys. **71**, 1868 (1992).
30. R. Feenstra, T. D. Lindemer, J. D. Budai, and M. D. Galloway, J. Appl. Phys. **69**, 6569 (1991).
31. A. Mogro-Campero and L. G. Turner, Appl. Phys. Lett. **58**, 417 (1991).
32. S. Y. Hou, Ph.D. thesis, State Unversity of New York, Stony Brook (1993).
33. M. P. Siegal, J. M. Phillips, Y.-F. Hsieh, and J. H. Marshall, Physica C **172**, 282 (1990).
34. J. A. Greer and H. J. van Hook, Mat. Res. Soc. Symp. Proc. **169,** 463 (1990).
35. M. Gurvitch and A. T. Fiory, Appl. Phys. Lett. **51**, 1027 (1987).

36. D. K. Fork, F. A. Ponce, J. C. Tramontana, N. Newman, J. M. Phillips, and T. H. Geballe, Appl. Phys. Lett. **58**, 2432 (1991).

37. Q. Li, O. Meyer, X. X. Xi, J. Geerk, and Linker Appl. Phys. Lett. **55**, 310 (1989).

38. B. H. Moeckly, S. E. Russek, D. K. Lathrop, R. A. Buhrman, J. Li, and J. W. Mayer, Appl. Phys. Lett. **57**, 1687 (1990).

39. A. Mogro-Campero, B. D. Hunt, L. G. Turner, M. C. Burell, and W. E. Balz, Appl. Phys. Lett. **52**, 584 (1988).

40. P. Madakson, J. J. Cuomo, D. S. Yee, R. A. Roy, and G. Scilla, J. Appl. Phys. **63**, 2046 (1988).

41. H. M. O'Bryan, P. K. Gallagher, G. W. Berkstresser, and C. D. Brandle, J. Mater. Res. **5**, 183 (1990).

42. W. G. Lyons, R. S. Withers, J. M. Hamm, A. C. Anderson, D. E. Oates, P. M. Mankiewich, M. L. O'Malley, R. R. Bonetti, A. E. Williams, and N. Newman, *Proc. Fifth Conference on Superconductivity and Applications*, ed. T. H. Kao, H. S. Kwok, and A. E. Kaloyeros, American Institute of Physics, New York, vol. 251 (1992), 639.

43. G. Arivalingam, unpublished manuscript (1991).

44. R. L. Sandstrom, E. A. Giess, W. J. Gallagher, A. Segmuller, E. I. Cooper, M. F. Chisolm, A. Gupta, S. Shinde, and R. B. Laibowitz, Appl. Phys. Lett. **53**, 1874 (1988).

45. R. W. Simon, C. E. Platt, A. E. Lee, K. P. Daly, M. S. Wire, J. W. Luine, and M. Urbanik, Appl. Phys. Lett. **53**, 2677 (1988).

46. G. Koren, A. Gupta, E. A. Giess, A. Segmuller, and R. B. Laibowitz, Appl. Phys. Lett. **54**, 10 (1989).

47. D. K. Fork, K. Nashimoto, and T. H. Geballe, Appl. Phys. Lett. **60**, 1621 (1992).

48. F. Lichtenberg, A. Catana, J. Mannhart, and D. G. Schlom, Appl. Phys. Lett. **60**, 1138 (1992).

49. E. Olsson, A. Gupta, M. D. Thouless, A. Segmuller, and D. R. Clarke, Appl. Phys. Lett. **58**, 1682 (1991).

50. L. J. Schowalter and R. W. Fathauer, CRC Critical Reviews in Solid State and Materials Sciences **15**, 367 (1989).

51. K. Char, M. S. Colclough, L. P. Lee, and G. Zaharchuk, G. Appl. Phys. Lett. **59**, 2177 (1991).

52. A. Walkenhorst, Xi. Doughty, X. X. Xi, S. N. Mao, Q. Li, T. Venkatesan, and R. Ramesh, Appl. Phys. Lett. **60**, 1744 (1992).

53. C. B. Eom, A. F. Marshall, J.-M. Triscone, B. Wilkens, S. S. Laderman, and T. H. Geballe, Science **251**, 780 (1991).

54. S.-W. Chan, E. W. Chase, B. J. Wilkens, and D. L. Hart, Appl. Phys. Lett. **54**, 2032 (1989).

55. A. N.Tiwari, S. Blunier, H. Zogg, Ph. Lerch, F. Marcenat, and P. Marinoli, J. Appl. Phys. **71**, 5095 (1992).

56. M. S. DiIorio, S. Yoshizumi, K.-Y. Yang, J. Zhang, and M. Maung, Appl. Phys. Lett. **58**, 2552 (1991).

57. A. T. Findikoglu, C. Doughty, S. Bhattacharya, Q. Li, X. X. Xi, T. Venkatesan, R. E. Fahey, A. J. Strauss, and J. M. Phillips, Appl. Phys. Lett. **61**, 1718

(1992).
58. C.-B. Eom, R. J. Cava, R. M. Fleming, J. M. Phillips, R. B. van Dover, J. H. Marshall, J. W. P. Hsu, J. J. Krajewski, and W. F. Peck, Jr., Science **258**, 1766 (1992).
59. J.-M. Triscone, M. G. Karkut, O. Brunner, L. Antognazza, A. D. Kent, and O. Fischer, Mat. Res. Soc. Symp. Proc. **169**, 545 (1990).
60. X. D. Wu, M. S. Hegde, X. X. Xi., Q. Li, A. Inam, S. A. Schwarz, J. A. Martinez, B. J. Wilkens, J. B. Barner, C. C. Chang, L. Nazar, C. T. Rogers, and T. Venkatesan, Mat. Res. Soc. Symp. Proc. **169**, 553 (1990).
61. H. C. Montgomery, J. Appl. Phys. **42**, 2971 (1971).
62. A. T. Fiory and A. F. Hebard, Appl. Phys. Lett. **52**, 2165 (1988).
63. C. P. Bean, Phys. Rev. Lett. **62**, 250 (1962).
64. E. M. Gyorgy, R. B. van Dover, K. A. Jackson, L. F. Schneemeyer, and J. V. Waszczak, Appl. Phys. Lett. **55**, 283 (1989).
65. L. Drabeck, G. Grüner, J.-J. Chang, A. Inam, X. D. Wu, L. Nazar, T. Venkatesan, and D. J.Scalapina, Phys. Rev. B **40** (1989) 7350.
66. D. Kalokitis, A. Fathy, V. Pendrick, R. Brown, B. Brycki, E. Belohoubek, L. Nazar, B. Wilkens, T. Venkatesan, A. Inam, and X.-D. Wu, J. Electron Mat. **19**, 117 (1990).
67. M. C. Nuss, P. M. Marikiewich, R. E.. Howard, B. L. Straughn, T. E. Harvey, C. D. Brandle, G. W. Berkstressser, K. W. Goosen, and P. R. Smith, Appl. Phys. Lett. **54**, 2265 (1989).
68. L. C. Feldman and J. W. Mayer, *Fundamentals of Surface and Thin Film Analysis*, North Holland, New York (1986).
69. W.-K. Chu, J. W. Mayer, and M.-A. Nicolet, *Backscattering Spectrometry*, Academic Press, Boston (1978).
70. T. Van Duzer and C. W. Turner, *Principles of Superconductive Devices and Circuits*, Elsevier, New York (1981).
71. T. P. Orlando and K. A. Delin, *Foundations of Applied Superconductivity*, Addison-Wesley, Reading, Mass. (1991).
72. G. K. G. Hohenwarter, J. S. Martens, D. P. McGinnis, J. B. Beyer, J. E. Nordman, and D. S. Ginley, Trans. on Magnetics **25**, 954 (1989) .
73. D. J. Frank, Cryogenics **30**, 996 (1990).
74. W. J. Gallagher, Solid State Technology, 151 (1989).
75. T. Van Duzer, Cryogenics **28**, 527 (1988).
76. R. C. Frye, in *Heterostructures on Silicon: One Step Further with Silicon*, ed. Y. I. Nissim and E. Rosencher, Kluwer Academic Publishers, Dordrecht (1989), 169.
77. T. Van Duzer, Cryogenics **30**, 980 (1990).
78. H. Piel and G. Muller, IEEE Trans. on Magnetics **27**, 854 (1991).
79. L. P. Lee, K. Char, M. S. Colclough, and G. Zaharchuk, Appl. Phys. Lett. **59**, 3051 (1991).
80. M. E. Klausmeier-Brown, G. F. Virshup, I. Bozovic, J. N. Eckstein, and K. S. Ralls, Appl. Phys. Lett. **60**, 2806 (1992).
81. G. F. Virshup, M. E. Klausmeier-Brown, I. Bozovic, and J. N. Eckstein, Appl. Phys. Lett. **60**, 2288 (1992).
82. E. S. Hellman, E. H. Hartford, S. Martin, D. J. , Werder, G. M. Roesler,

and P. M. Tedrow, Physica C **201**, 166 (1992).

7.8 Recommended Readings for Chapter 7

Thin Film Fabrication:

1. *Handbook of Thin Film Technology,* L. I. Maissel and R. Gland, eds., McGraw-Hill, New York, 1970.

Film Characterization:

2. *Introduction to Superconductivity,* 2nd ed., A. C. Rose-Innes and E. H. Rhoderick, Pergamon, Oxford, 1978.

3. *Fundamentals of Surface and Thin Film Analysis,* L. C. Feldman and J. W. Mayer, North-Holland, New York, 1986.

4. *Backscattering Spectrometry,* W.-K. Chu, J. W. Mayer, and M.-A. Nicolet, Academic Press, Boston, 1978.

Applications:

5. *Principles of Superconductive Devices and Circuits,* T. Van Duzer and C. W. Turner, Elsevier, New York, 1981.

6. *Foundations of Applied Superconductivity,* T. P. Orlando and K. A. Delin, Addison-Wesley, Reading, Mass., 1991.

7. *Superconducting Electronics,* H. Weinstock and M. Nisenoff, eds., NATO ASI Series, Vol. 59, Springer-Verlag, Berlin, 1989.

8. *The New Superconducting Electronics,* H. Weinstock and R. W. Ralston, eds., NATO ASI Series, Vol. 251, Kluwer, Dordrecht, 1993.

8

Bulk Processing and Characterization of $YBa_2Cu_3O_{6+x}$

P. McGinn

Despite the vast potential for high-temperature superconductors, in order for them to be commercially useful they have to be processed into appropriate shapes, such as disks or wires. One of the major limitations in applying these materials has been the difficulty in achieving optimized transport properties for bulk applications. The critical current density of these materials depends heavily on the processing method employed. In this chapter various processing methods for preparing powders of $YBa_2Cu_3O_{6+x}$ (Y123) or similar rare-earth superconductors are reviewed. Then we discuss techniques for processing useful shapes such as wires or tapes from powder. In particular, the advantages of melt processing for producing microstructures with a high critical current density are detailed. The properties of melt-textured 123 are reviewed, and avenues for further enhancing the properties are discussed.

8.1 Powder Preparation

We begin with a discussion of the principal processing techniques used to produce Y123 or other rare-earth powders.

8.1.1 Ceramic Method

The most common means of preparing Y123 powder is through the so-called ceramic method, or what is sometimes euphemistically referred to as "shake and bake." In this process all the precursors are powders, generally being metal oxides or carbonates. As typically practiced, the precursor powders are mixed, often being ball milled to enhance mixing and to reduce powder particle sizes. The mixed powders are then heated at high temperature to allow interdiffusion of the cations. The main disadvantages of the ceramic method involve inhomogeneity:

1. The starting mixture is inhomogeneous at the atomic level.
2. The reaction process is very slow because it relies on solid-state diffusion. Multiple cycles of grinding and heating are required, often totaling 48 hours or more. In cases where a liquid phase is present, diffusion through the liquid helps to speed interdiffusion. In the case of Y123, however, the presence of a liquid during processing is undesirable because it limits the critical current density (as will be discussed later.)
3. Even with multiple intermittent grinding steps, a compositionally homogeneous product is difficult to obtain.

Despite these shortcomings, the ceramic method is by far the most widely used process because it is so easy to practice. The only equipment required for the most basic processing is a mortar and pestle and a furnace. Typically, after mixing and grinding, the powders are calcined at 900°–950°C for 8–24 hours and then pulverized, and the process is repeated one or more times. The success of this process depends to some extent on the precursors chosen, and subsequent modifications to the process to enhance decomposition of some of the precursors. The most widely used set of precursors for Y123 are Y_2O_3, CuO, and $BaCO_3$.

$BaCO_3$ is used because it is stable in air and does not tend to pick up moisture (unlike BaO_2). The problem with using $BaCO_3$, however, is that CO_2 released during decomposition can lower the melting point of the low melting point liquid (a eutectic of $BaCuO_2$ and CuO). As a result, to avoid the formation of any liquid phase during calcining or sintering, one is restricted to temperatures below 890°C. By lowering the permissible reaction temperature, the diffusion kinetics are slowed, thereby lengthening the total reaction time required to achieve satisfactory homogeneity. Otherwise, if higher temperatures are used, a liquid phase is present, which leads to rapid grain growth and the presence of frozen liquid at three particle junctions in the final microstructure. (The presence of a liquid phase does offer two advantages, however: increased reaction rates and greater density of the final product.)

One way of avoiding the deleterious effects of CO_2 evolution is to use BaO_2 as the barium precursor [1]. Another way is to use vacuum calcination, a novel technique developed by Balachandran and coworkers [2]. The technique involves using a so-called dynamic vacuum to maintain a reduced CO_2 partial pressure. In this process flowing oxygen is introduced into one end of a tube furnace, while a vacuum pump at the other end maintains a pressure of 2 mm Hg. The CO_2 partial pressure is continuously monitored to maintain it below a level of 2% of the total O_2 pressure. This process permits the use of $BaCO_3$ as the barium precursor without the formation of a liquid phase. In addition, because cation diffusion is faster in a reduced pO_2 environment [3], the reaction times for powder production are reduced with this process. As a result, 1000 grams of phase-pure Y123 can be produced in one 8-hr thermal cycle.

An alternative means of avoiding liquid formation, and also in some cases to speed the reaction time, is to use intermediate precursors. Ruckenstein and his coworkers [4] explored the rates of reaction and the purity of the final product by using a number of different precursors. Their work showed that decomposition of $BaCO_3$ was the limiting factor in the reaction rate to form Y123. Their technique isolates the slow reaction into a first step, allowing the superconducting compound to be obtained in a second, more rapid step. For example, one can prepare $BaCuO_2$ by reacting CuO and $BaCO_3$ in a first step, and then reacting $BaCuO_2$, CuO, and Y_2O_3 in a second step. In this way, the CO_2 evolution is confined to one step, and the second reaction to form the superconducting compound can occur without the presence of CO_2. The $BaCuO_2$ reacts rapidly with the other two precursor materials, so rapid grain growth in the presence of a liquid phase with extended reaction times is not a problem.

8.1.2 Coprecipitation

Another method for preparing superconducting powders is coprecipitation. This process involves a solid containing various ionic species precipitating out of a solution. The resulting precipitates are heated to appropriate temperatures in suitable atmospheres to produce the desired compound. The precipitates offer greatly reduced diffusion distances compared with ceramic processing, resulting in shorter reaction times and more homogeneous reaction products. In addition, finer powder particle sizes can result from the coprecipitation process.

As applied to making Y123, coprecipitation of the component metals from a nitrate solution as a formate, acetate, and oxalate has been studied [5]. Oxalate coprecipitation in particular has been widely reported. In this process an oxalic acid solution is added to an aqueous solution of Y, Ba, and Cu nitrates, and the pH of the solution is adjusted. The resulting oxalate is then heated in air and converted to Y123. The major disadvantage of this process is that control of the final product stoichiometry can be difficult.

8.1.3 Sol-Gel Processing

Sol-gel processing allows one to achieve homogeneous mixing of cations on an atomic scale. Originally developed by Pechini [6], this process permits rapid formation of the final product at reduced temperatures during the solid-state reaction. Sol-gel processing is so named because a concentrated sol (a suspension of colloidal particles) is converted to a semi-rigid gel. The gel is subsequently heated at an appropriate temperature to obtain the final product. Although this technique allows one to produce high-quality powder [7], the resulting bulk solids fabricated from such powder show no increase in J_c over solids prepared from ceramic-processed powder due to the inherent limitations of polycrystalline sintered Y123 (discussed in more detail later). Thus, for bulk applications of Y123, the added cost and limited throughput of this process restrict its use outside of laboratory situations.

8.1.4 Other Processing Methods

Many other processing techniques have been employed to produce Y123 or other rare-earth powders. Noteworthy among these are combustion synthesis, spray drying, freeze drying, and metal precursor routes.

Combustion synthesis, reviewed recently [8], is based on using the heat liberated by a highly exothermic reaction to continue the advance of a reaction front through a sample. Specifically, an energy source (e.g., a pulse from a laser) triggers a reaction at one end of a powder compact, and the heat given off by the reaction allows the reaction front to proceed along the length of the compact in a relatively short time (on the order of seconds). It thus offers a fast, low-energy approach to preparing many inorganic compounds. In the case of Y123, the oxidation of copper metal can act as the fuel, BaO_2 as an oxygen source; when combined with Y_2O_3 and ignited, these allow rapid formation of Y123. Nevertheless, achieving phase-pure material by this technique remains a problem [9].

In *spray drying*, a solution (frequently a nitrate solution) containing the metal cations is sprayed into a hot chamber. The solvent evaporates, leaving behind a powder that, when heated in an oxidizing environment, decomposes to form the desired compound (Y123). A variation on this technique uses an ultrasonic device to atomize very fine liquid droplets. After evaporation, very uniform and chemically homogeneous submicron powder particles can be produced [10]. To date, however, powder production by this process is somewhat limited.

Freeze drying also starts with the reactants in solution. In this case, however, the solution is sprayed into a cold environment (e.g., liquid nitrogen) in order to freeze the droplets. The solvent is removed at low pressure in a vacuum chamber to yield the reactants as a fine powder. This powder is then heated to an elevated temperature to permit solid-state reaction. As in the case of spray drying, the advantage of this route is that diffusion distances required during the solid-state reaction are reduced compared with ceramic processing [11].

Finally, the *metallic precursor method* has also been successfully employed to produce Y123 powder. In this case metal powders are intimately mixed in an inert atmosphere. A high energy mill is effective for this type of mixing. After mixing, controlled oxidation is used to form the superconducting oxide [12]. The advantage of this process is that the metal mixture can be formed into a useful shape (such as a wire through drawing or rolling) prior to oxidation, bypassing some of the limitations of working with brittle fired ceramic shapes.

8.2 Sintering

Once powder of the Y123 superconductor has been prepared, sintering is used to achieve bonding between the particles. Typically, the powder is pressed in a die, usually with a binder to promote cohesion of particles prior to firing, and then fired in a furnace. During firing, the particles bond to one another in order to reduce the total surface area (and thereby the surface energy) of the system. As is the case with the sintering of all powders, with finer starting powder size, the sintering will proceed faster because the driving force is greater. Hence, particle preparation techniques that yield fine particle sizes are advantageous. Techniques such as jet milling can be used to achieve particles of 1 micron. Sintering establishes an electrical link between particles, so that even in porous-sintered Y123, after proper oxygenation a resistivity measurement will give a critical temperature (T_c) on the order of 90 K. Even though the T_c may be high, however, the critical current density J_c of a sintered Y123 solid will typically be relatively low (as will be discussed later).

As is typical of sintering processes, the density of the sintered body can be increased by raising the sintering temperature. For Y123, however, the increased temperature can cause difficulties if there is a liquid phase present. From the Ba_2O_3-CuO phase diagram, a eutectic reaction occurs at approximately 900°C. Sintering above this temperature can lead to liquid-phase formation, leading in turn to more rapid liquid-phase sintering. This results in improved density and increased grain size (because of enhanced grain growth in the liquid), but it also results in an insulating phase along the grain boundaries, thereby decreasing the critical current density.

A dense sample can be formed without resorting to liquid-phase sintering if fine powders (on the order of a micron in size) are used in combination with isostatic pressing (for uniform densification) and long sintering times below the eutectic temperature. Even with a very dense solid, however, the critical current density will still remain low. There are several reasons for this, as will be discussed later. However, one of them, microcracking with oxygenation, is directly related to the grain size that results from sintering. The T_c of Y123 is very dependent on the oxygen content [13]. After sintering at high temperature, Y123 has to be oxygenated at approximately 450°C, below the tetragonal-orthorhombic transition, so that the oxygen stoichiometry changes from 6 to nearly 7. The crystallographic change leads to stress accumulation in the lattice, resulting in the development of cracks in the Y123. This was shown vividly by acoustic measurements by Richardson et al. [14]. By reducing the grain size, crack formation can be avoided. Grain sizes less than 1 micron have been predicted to be sufficient to avoid cracking [15]. Again, the grain size will depend on the sintering conditions, with low temperatures to minimize grain growth being preferred, especially in order to avoid liquid-phase formation which will accelerate grain growth.

8.3 Thick-Film Processing

Many promising electronic applications of superconductors as passive high-frequency components can be realized by using superconducting thick films. For such applications the film is formed by printing an ink containing superconducting particles onto a substrate, and firing to burn off the organic components of the ink and to sinter the particles one another and to the substrate. In forming the ink, the powder preparation is important, because the particle size and distribution will affect the density and uniformity of the final film. Also, the choice of substrate is critical, because deleterious chemical interactions between the substrate and film need to be minimized, and the thermal expansion coefficients of the film and substrate must be matched as closely as possible to minimize cracking of the film. As with thin-film processing, buffer layers can be used to minimize interactions. Fired films have low critical current density, as do sintered pellets, unless the processing involves melt texturing (which is discussed in more detail later). An excellent review of thick-film processing of high-temperature superconducting films has recently been published [16].

Thick films can also be prepared by plasma spraying techniques [17]. In such a process powder is fed (again, the size and distribution are critical for proper flow properties) into a plasma torch, where it is melted and then sprayed onto the substrate. The molten powder then solidifies, and a film is formed. Typically, the film must be reannealed to form the Y123 compound, because of the solidification rates inherent in the process. This technique is suited for coating large surfaces, but results in films with low transport properties.

8.4 Wire Processing

For producing the precursor wires to be used in melt texturing and various applications, several techniques have been employed. Both spinning techniques and extrusion have been reported, with extrusion permitting the production of larger cross section wires. For producing pure 123 extruded wires, the prereacted 123 powder is milled with an organic system consisting of a solvent, dispersant, binder, and a plasticizer. The ratio of additive to 123 powder will depend on the extrusion parameters to be used, the final density desired, etc. After milling, some solvent is allowed to evaporate to yield a plastic mass with the desired viscosity. This mass is then forced through a die at high pressure to produce a green (i.e., unfired) flexible wire. If desired, such a wire can be coiled prior to sintering (the so-called wind-and-react process). For a process such as zone-melt texturing (discussed later), the wire is fired to form a straight rod. From the standpoint of texturing, as long as the wire is relatively pore free, the wire quality does not appear to be a major factor in limiting successful production of textured wire. Large pores, however, can lead to liquid "pooling" and an inhomogeneous microstructure containing regions that are denuded of Y_2BaCuO_5 (211). Pore evolution can be controlled by proper binder burnout during sintering. Density gradients in the extruded wires can also cause problems by making the production of straight wires more difficult, but these can be minimized through constrained sintering (e.g., sintering under a slight applied load). Dorris et al. present an excellent discussion of the extrusion of multilayer coils, including a description of the coextrusion of Y123 wires with a Y211 insulating coating [18].

The addition of externally prepared Y211 powder or other additives to Y123 allows for the extrusion of composite wires for improved mechanical or electrical properties. The process simply involves altering the powder blend that is milled with the organic system.

8.5 Weak Links

In sintered Y123 the critical current density (J_c) is usually rather low (100–1000 A/cm^2 in zero applied magnetic field), being 2–4 orders of magnitude lower than values observed in single crystals or epitaxial thin films. Several factors contribute to the low transport J_c values, including several directly related to the nature of the grain boundaries:

8.5.1 Structural Disorder at Grain Boundaries

The significance of grain boundaries on transport J_c was initially demonstrated by Chaudhari et al. using YBCO thin films grown on polycrystalline $SrTiO_3$ substrates [19]. They found that the current density across a clean grain boundary was always significantly less than that internal to a grain.

Subsequent work by Dimos et al., which evaluated epitaxially grown YBCO thin films (this time using $SrTiO_3$ bicrystals oriented such that the YBCO basal planes were at various angles to each other), supported these findings [20].

Further, their results indicated that a high degree of texture must be achieved both in and normal to the basal plane to obtain high current densities in pure polycrystalline YBCO. They attributed the drop in transport properties with increasing misorientation angle to an observed increase in dislocation density in the grain boundary region. It was proposed that the dislocations might affect the electrical properties by acting either as a barrier to current flow or as an easy path for flux flow. This result was interpreted as showing that grain boundaries are inherently weak linked, presumably because of the structural disorder at the grain boundary. Because of the low coherence length in Y123, even the width of a grain boundary can greatly reduce supercurrent flow, thereby decreasing the J_c.

More recent work by Larbelestier et al. [21] has shown that some specific high-angle grain orientations can carry relatively high critical currents. In that work it was found that during single crystal growth, crystals will rotate to low-energy orientations which support relatively high critical currents.

Work by Zhu et al. [22] shows that low-energy boundaries in Y123 can be described through a constrained coincident-site lattice (CCSL), where two grains will share many lattice sites at the boundary. For a noncubic crystal like Y123, it is necessary to constrain the ratios of the lattice parameters at the boundary. In Y123 it is postulated that this constraint occurs by varying the oxygen content within the boundary region (see below).

8.5.2 Anisotropy in the Material

It has been shown that J_c in Y123 is anisotropic, with the basal plane exhibiting J_c values (J_c^{ab}) higher by a factor of twenty than those along the c-axis (J_c^c) at 4 K [23]. This is a result of the crystal structure of the compound and the intrinsic differences in the coherence lengths for different crystallographic directions in the orthorhombic structure. For Y123, the coherence length is only 4 Å along the c direction and 32 Å in the a-b plane. These small coherence lengths (which are about two to three orders of magnitude smaller than those of conventional superconductors) necessitate that these bulk materials be prepared by special processing techniques to minimize the effect from a second concern, that of grain boundaries. Transport measurements have also been reported on melt-textured 123. Anisotropy ratios ($J_c^{ab} : J_c^c$) of approximately 20 have been measured at 77 K. These values are somewhat lower than have been measured in single crystals, a fact that may reflect improved pinning for transport parallel to the c-axis in textured specimens [24].

8.5.3 Chemical or Structural Variations at Grain Boundaries

If the stoichiometry or structure varies slightly near the grain boundary, then current-carrying capacity will be adversely affected because of the short coherence length of the material. For example, investigations have established that in sintered Y123 the grain boundaries are slightly Cu-rich and Ba- and O-deficient within 5 nm of the boundary [25–28]. The degree to which this contributes to low polycrystalline J_c has not been established. In some studies, amorphous-like regions have been reported at low-angle boundaries between aligned grains, but it is not clear whether this phenomenon is actually an artifact from the specimen-thinning process.

8.5.4 Nonsuperconducting Material along Grain Boundaries

Depending on the method of compound preparation, second-phase, nonsuperconducting material can be found along the grain boundaries in Y123. As previously mentioned, solid-state processing often leads to trace amounts of BaCuO$_2$ and CuO along grain boundaries and at three particle junctions. Alternative synthesis techniques can result in other contaminants or secondary phases being present. The presence of nonsuperconducting phases will limit the effective cross-sectional area through which current can pass. Y$_2$BaCuO$_5$ (211) does not wet the boundaries of Y123, so it is present as discrete particles or precipitates, which intrinsically have a minimal deleterious effect on the current transport. More damaging is the low-melting-point BaCuO$_2$-CuO phase, which appears to wet grain boundaries more effectively, thereby providing a thin insulating layer between grains. This situation is especially important in melt texturing where excess liquid phase may remain because the peritectic reaction is not complete in reforming the 123 during solidification. This is one reason for adding excess 211 to the 123: it helps to tie up excess liquid phase, thereby minimizing the possibility of liquid along grain boundaries. Oxygenation in a CO$_2$-containing environment can lead to local decomposition at grain boundaries, also leading to a decrease in J$_c$. Because the Y123 compound is sensitive to humid environments, sintered samples in particular can show a decrease in J$_c$ with time, as a result of grain boundary degradation.

8.5.5 Microcracking

It has been shown that Y123 crystals shrink anisotropically during cooling. In addition, during oxidation of the tetragonal (T)YBa$_2$Cu$_3$O$_6$ to convert it to the superconducting orthorhombic (O) YBa$_2$Cu$_3$O$_7$, the material also experiences anisotropic dimensional changes. During the T-O transformation the a- and b-axes contract by 0.8% while the c-axis contracts by 3.2% [29]. The excess contraction in the c-axis produces a tensile stress on the (001) boundaries, leading to the formation of microcracks. The formation of microcracks then reduces the effective cross-sectional area for current transport, again reducing J$_c$.

8.6 Melt-Texture Processing

Texture processing was developed as a means to avoid or minimize the effects of many of the weak-link limitations to J$_c$ in polycrystalline Y123. By having a microstructure where the current flow can be directed primarily in the basal plane, and by minimizing the occurrence of high-angle grain boundaries, some of the primary causes of weak link behavior can be minimized, if not eliminated. Formation of texture in bulk Y123 has been achieved by a number of different techniques, including compaction processes, magnetic alignment of particles, and directional solidification.

Numerous groups have shown that *compaction*, both with and without the use of elevated temperatures, can help develop texture in Y123. Although textured microstructures have been produced, however, transport properties comparable to those produced by directional solidification processes have not been achieved [30, 31].

Magnetic alignment has also been successful for developing texture. Particles of Y123 have been aligned in epoxy [32] to simulate single crystal-like behavior, and also in organic solvents [33,34]. After the solvent evaporates, the aligned particles can be sintered, resulting in well-textured bulk samples. Transport properties achieved by this technique, however, are not comparable with those produced through directional solidification (see below) because the alignment produced by the applied field allows for rotation in the nonaligned directions (e.g., a and b directions), producing high-angle twist boundaries between aligned grains. A more effective use of magnetic alignment involves solidification in an applied magnetic field. In this case the 123 crystals tend to rotate and become aligned while in the liquid phase. As a result, growth under the influence of the field results in a well-textured sample with excellent grain-to-grain alignment [35].

Texturing by *directional solidification* from the melt has proven much more successful as a means to achieve texturing from the melt. By this route a microstructure is produced that consists of large grains that are well aligned with respect to each other. Grain boundaries in the direction of current flow can be eliminated by these melt-texturing techniques, so that transport J_c values exceeding 10^4 A/cm^2 in applied fields of several tesla can now be achieved. Because these melt-processing techniques have proven to be so successful, and are the only practical means to achieve high J_c in 123, they will be discussed in more detail below.

Jin et al. used the term "melt texturing" to describe their pioneering work in this area [36–38]. In the case of Y123, melt texturing describes directional solidification from the melt or partially melted state. While there are a number of variations of the texturing techniques, as described below, they all involve

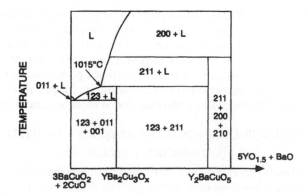

Fig. 8.1. Section of the Y-Ba-Cu-O ternary phase diagram along the tie line between Y$_2$BaCuO$_5$ and YBa$_2$Cu$_3$O$_7$ phase diagram.

heating Y123 above its peritectic decomposition temperature (1015°C in air) and slow cooling to form aligned, generally large, grains of Y123. This can be understood upon examination of the section of the ternary phase diagram shown in Figure 8.1. It can be seen that upon heating above approximately 1015°C, $YBa_2Cu_3O_{6+x}$ will decompose to Y_2BaCuO_5 (211) and a liquid phase consisting of $BaCuO_2$+CuO. Upon slow cooling (e.g., 1°C/hr) from 1015°C to approximately 900°C, the Y123 will nucleate and grow.

8.6.1 The Y123 Pseudo-Binary Phase Diagram

The Y_2O_3-CuO-BaO phase diagram has been the subject of many investigations. Recently a summary of existing phase diagrams for all high-temperature superconductors has been published [39]. Aselage and Keefer [40] identified eleven invariant points on the Y_2O_3-CuO-BaO phase diagram, including the incongruent melting temperatures of 123 and 211. They found that 123 decomposed in air at 1015°C and resulted in the formation of 211 and a barium-rich liquid according to the following relation:

$$2 \text{ Y123} \rightarrow 211 + \text{L}(3 \text{ BaCuO}_2 + 2 \text{ CuO}). \tag{8.1}$$

The decomposition of the 211 phase was found to occur at 1270°C and produced Y_2O_3 and a liquid as follows:

$$211 \rightarrow \text{ Y}_2\text{O}_3 + \text{L}(\text{BaCuO}_2). \tag{8.2}$$

Subsequently, these two sets of peritectic reactions have formed the basis for all YBCO melt-texturing techniques aimed at eliminating the problems associated with both the weak-link character and the poor pinning properties of sintered YBCO. All of the significant melt-processing strategies to date (which will be described later) ultimately involve slowly cooling the required composition from the partial melt (above 1015°C) to below 960°C (either isothermally or in a temperature gradient) to form 123. As with other (metallic) peritectic systems, the reaction of the primary phase (211) with the liquid to form the peritectic product (123) is extremely slow and rarely proceeds to completion to yield phase-pure 123 under normal melt-processing thermal cycles.

For stoichiometric Y123 samples, this incomplete reaction can be exacerbated by a combination of inadequately slow cooling rates and/or liquid phase segregation (i.e., to the sample surface or grain boundaries), which results in a shift in composition to a more 211-rich stoichiometry. Even under optimal conditions, however, the reaction usually will not go to completion because growing 123 will trap unreacted 211, eliminating its further reaction with the liquid. Consequently, on cooling, the reaction is more accurately represented as

$$\text{Y211} + 3\text{BaCuO}_2 + 2\text{CuO} \rightarrow (2\text{-x})\text{Y123} + (x/2)\text{Y211}$$
$$+ (x/2)3\text{BaCuO}_2 + (x/2)2\text{CuO}. \tag{8.3}$$

The typical melt-processed YBCO microstructure consists of aligned platelike 123 grains, which have their most rapid growth direction in the a-b plane. As can be seen in Figure 8.2, the residual 211 phase may be uniformly dispersed throughout the 123 matrix as fine, acicular particles. The size range and distribution of the particles depend on the specific YBCO composition and

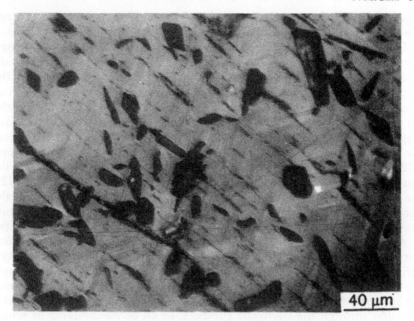

Fig. 8.2. Typical microstructure of melt-processed YBa$_2$Cu$_3$O$_7$, showing aligned 123
platelets and unreacted 211 particles.

melt-texturing process cycle and method.

If the growth of 123 were to proceed by the conventional peritectic mechanism, the primary solid (which would be 211 in this case) would react with
the liquid to form a shell of peritectic product (123) around itself. The reaction
would proceed by the diffusion of species through this shell and would result
in a microstructure containing randomly oriented (123) grains which could
have undissolved primary solid (211) cores. The fact that melt-processed YBCO
does not result in this cored microstructure indicates that its growth from the
melt is not by the conventional peritectic mechanism.

8.6.2 The Y123 Peritectic Reaction

Although a desirable microstructure has been identified that yields high
J$_c$ values, generating such structures so far has necessitated extremely slow
cooling rates (1–3°/h). The slow cooling rate (and therefore the long processing time), which is necessitated by the kinetics of the peritectic reaction, is the
greatest disadvantage of the directional solidification processes as currently
practiced.

Much of the early literature concerning the melt texturing of Y123 assumed
that the formation of Y123 from the melt occurs by the traditional peritectic
reaction in which the Y123 forms a shell around a 211 particle and thus separates the 211 from direct contact with the Ba- and Cu-rich liquid. This Y123
envelope would then slow down any further growth of Y123, since it would

depend on the solid-state diffusion of the reactants through the Y123 layer. The existence of such a shell is, however, not seen in melt-textured Y123.

It has been postulated that in most peritectic systems, the peritectic reaction does not occur because crystallization of the peritectic phase occurs directly from the liquid. In peritectic systems in which the liquidus lines of the properitectic and peritectic phases coincide, however, a peritectic reaction can occur at some temperature below the peritectic temperature. D. H. St. John has recently reexamined the peritectic reaction [41]. In his discussion he distinguishes between the peritectic transformation and the peritectic reaction. The peritectic reaction is the formation of the peritectic or secondary phase (Y123, in this case) by reaction of the primary phase (211 in this case) with the liquid at the peritectic temperature. This is the reaction 211 + liquid → Y123. The peritectic transformation is distinct from this and is the growth of the secondary phase (Y123), which occurs by diffusion through the already-formed secondary phase coating the primary phase. The work by St. John and a number of others has shown that a competition exists between the peritectic transformation, which occurs relatively slowly because of the rate limiting solid-state diffusion step, and nucleation of the secondary phase directly from the melt. The extent to which the peritectic product grows by the peritectic transformation as opposed to nucleation from the melt will depend on the rate of solid-state diffusion through the peritectic product. In all of the systems in which secondary-phase nucleation from the melt have been reported, the secondary phase is not a line compound, as is the case for Y123. In other work [42] St. John has classified peritectic systems into three types depending on the slopes of the α solidus and the solvus line between the α and α +β regions. In Type C systems, in which line compounds form, the extent of the peritectic transformation is observed to be only slight. This suggests that in the formation of Y123, the main mode of formation can be expected to be growth directly from the melt.

8.6.3 Nucleation of Y123

The nucleation and growth of 123 have been the subject of a number of recent studies. Bateman and his coworkers first proposed the possibility of 123 growth directly from the melt [43]. Evidence of this mode of growth is shown in Figure 8.3, which shows a region of a zone-melt-textured sample in which textured regions are separated from one another by large white regions (liquid phase). Within these regions, 211 grains can be seen in direct contact with the melt. If the peritectic transformation were occurring in this sample, one would expect to see a "shell" of Y123 forming around the 211 particles, which is not observed in this photo. Thus, the 211 liquid interface does not seem to be the site of 123 nucleation.

The likely mechanism involves the homogeneous nucleation of 123 in the melt. If a pellet of cooling 123 is quenched from just below the peritectic temperature after sitting at that temperature for some period of time, one observes several independently nucleating 123 grains, as shown in Figure 8.4. The figure shows a 123 grain that is growing in the melt. Further growth of a 123 nucleus requires the diffusion of Y, Ba, and Cu to the solidification

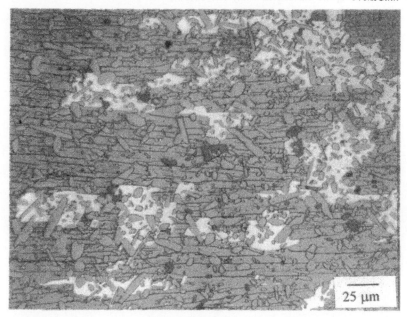

Fig. 8.3. Liquid phase (white) surrounding 211 particles in a textured 123 microstructure.

Fig. 8.4. Growing 123 grains in the frozen liquid, after quenching from 1000°C.

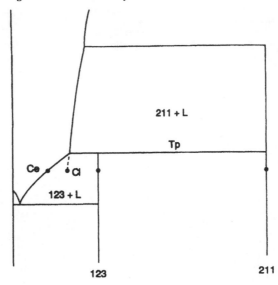

Fig. 8.5. A portion of the phase diagram in Figure 8.1 showing the metastable phase
equilibria during undercooling of Y123 below the peritectic temperature.

front. Ba and Cu are both in plentiful supply, being the primary constituents
of the liquid phase. Y atoms, however, are not so plentiful in the liquid, as the
solubility of Y is approximately 2% [44,45]. As a result, the diffusion of Y to
the growth front appears to be the rate controlling factor for 123 growth.

Bateman et al. discussed the implications of the Y123 pseudo-binary phase
diagram on the growth of Y123 [43]. When the 211 + L mixture is undercooled
below the peritectic temperature, the metastable liquid composition (C_l) falls
within the 123 + L two-phase field, so the liquid is supersaturated with respect to the 123 (see Figure 8.5).

As a result, 123 can precipitate directly from the melt, unlike the traditional case for peritectic transformations. Upon precipitation of 123, the liquid composition shifts to the more Y-deficient equilibrium composition (C_e).
The 211, however, is in equilibrium with liquid of composition C_l; hence, in
an attempt to maintain this liquid composition, the 211 will dissolve to shift
the liquid composition back to C_l. This dissolution of 211 provides the Y
necessary for precipitation and growth of Y123. The dissolution of 211 particles in the liquid will continue until the particles are entrapped by the advancing 123 growth front. Because the particles are dissolving in the melt and
211 reactions with the liquid stop once the particle is engulfed by the 123, the
volume fraction and 211 particle size decrease abruptly at the 123/liquid
interface and remain relatively constant with distance into the crystal.

Recent work by Rodriguez et al. [46] shows real-time SEM images of 123
nucleation at 940°C indicating that 123 nucleates near 211, but not necessarily

on 211 particles. In addition, evidence of a Y gradient in the liquid was also observed, again lending credence to the theory of 211 dissolution and Y diffusion being the rate-controlling step in 123 growth.

8.6.4 Solidification Models

The driving force for the dissolution and solute diffusion flux in the liquid is a concentration gradient. Cima et al. [47] developed a 123 solidification model that suggests that the 211 particles dissolved in the liquid become supercooled close to the growing crystal interface to create the concentration gradient required to drive mass transport and thereby sustain planar growth. This supposition, however, seemingly contradicts classical constitutional supercooling theory, which predicts that the presence of a supercooled region will result in an unstable interface, with perturbations growing and leading to the formation of a cellular or dendritic growth front. Cima et al. resolved the apparent contradiction by suggesting that the surface energy relationships associated with the highly faceted YBCO plates permit interface stability with a limited amount of undercooling. Consequently, it is proposed that the planar growth is limited by the undercooling that will cause the nucleation of new crystals in front of the interface. Further, a quantitative explication of this model yielded the equation

$$R_{max} = D_L / l(c_{sy} - c_{Lp}) \{ (\Delta T_s)_{max} + Gl \} / m_{L\gamma}, \qquad (8.4)$$

which suggests that the maximum YBCO growth rate (R_{max}) at which a planar interface could be sustained is directly proportional to the diffusion coefficient (D_L) of the solute in the liquid and inversely proportional to diffusion distance to the interface (l), which is assumed to be half the 211 particle separation (2l), where the Y211 particle size in the partial melt is l. $(\Delta T_s)_{max}$ is the maximum constitutional supercooling at a distance from the interface x = l, c_{Lp} is the peritectic liquid composition, c_{sy} is the 123 solid composition, and $m_{L\gamma}$ is the slope of the equilibrium 123 liquidus.

Izumi et al. [48] developed an alternative 123 solidification model, also based on the assumption that the rate-controlling step in 123 growth is the diffusion of solute to the interface. They assumed, however, that the concenration gradient in the liquid is due to the change in the chemical potential at the 211/liquid interface caused by the curvature of the 211 particles. The change in the chemical potential was related to the undercooling (ΔT_r) by the Gibbs-Thomson relationship:

$$\Delta T_r = 2\Gamma / r, \qquad (8.5)$$

where Γ is the Gibbs-Thomson coefficient ($\sigma / \Delta s$), σ is the interface energy between the 211 and the liquid, Δs is the volumetric entropy of fusion, and r is the radius of the 211 particle. This relationship was used in conjunction with the 123 phase diagram to determine the resulting composition difference (i.e., yttrium concentration gradient) ΔC between the liquid at the 211 particle and that at the 123 interface. This yielded the following:

$$\Delta C = C^r_L - C^\infty_L = 2\Gamma / rm_L, \qquad (8.6)$$

where C^r_L is the rare-earth concentration at the 211 particle, C^∞_L is the rare-earth concentration in the liquid at the 123 interface, and m_L is the liquidus slope of the 211 phase. The model was based on a mass balance that equated the flux required for 123 growth with that provided from the 211 particles. Given the initial size distribution of 211 particles, the initial volume fraction of 211 phase, and several thermophysical constants (diffusivity, Gibbs-Thomson coefficient, maximum 211 decomposition rate, etc.), the balanced velocity for steady-state directional solidification was estimated. It was found that the theoretical limit of the continuous growth rate (a few millimeters/hour) and the size distribution of 211 particles in the 123 crystal agreed well with experimental results obtained from samples prepared by the Bridgman method.

Related to this model was the finding by Izumi and Shiohara [49] that the final 211 particle size was independent of growth rate, R, which was attributed to only particles greater than a certain critical radius, r_c, being entrapped in the growth front. If the surface energy between a 211 particle (P) and the solid 123 (S), σ_{PS}, is greater than the sum of the surface energy, σ_{PL}, between the particle and the liquid (L) and the surface energy between the liquid and the 123, σ_{LS}, then the material in the fluid surrounding the particle will be transferred to the solidification front, and the particle will be pushed ahead into the liquid. In fact, evidence of such particle pushing has been observed by Varanasi and McGinn [50]. Izumi and Shiohara suggested that the small 211 particles that are pushed by the solidification front will coalesce with other small particles in the liquid and coarsen until a critical radius is reached, at which point the particle is no longer pushed but becomes entrapped in the growth front. This process then yields the growth rate independent of 211 size.

8.6.5 Growth of Y123

One area of continuing research concerns the evolution and the exact character of the melt-textured microstructure within a single domain. The role that 211 particles play in defining the plate structure geometry is still being studied. One model put forth to describe the growth of Y123 platelets involves sympathetic nucleation of additional platelets on existing Y123 as discussed by Bateman et al. [43]. Once a 123 grain nucleates and begins to grow, it will quickly adopt a platelike morphology by virtue of the fact that growth occurs much more rapidly in the a and b directions of the crystal than in the c direction. Growth of the platelet will continue by the reaction between 211 and the melt. According to this model, supersaturation in the liquid as a result of 211 dissolution is quickly removed in front of the grain by the rapid growth in the a-b plane. Because growth in the c direction is slow, however, the local supersaturation builds adjacent to the (001) face of the growing plane. This process, then, provides the driving force for nucleation of a new 123 grain on the (001) face of the existing plane. The new grain assumes a low-energy orientation relative to the existing grain, resulting in a low-angle boundary between the two grains, as are observed in melt-textured Y123 superconductors, yielding the familiar "brick wall" microstructure. According to this model, whenever supersaturation of the liquid exceeds a certain level, a new 123 grain will nucleate. Hence, the observed dependence of plate thickness on 211 size and amount is the result of more frequent supersaturation of the liquid by 211 dissolution. Frequently Y123 platelet growth will extend past a 211 particle before the 211 is completely dissolved, trapping the particle in the platelet structure. The amount of residual 211 will depend on the solubility of 211 in the liquid, the rate of cooling, etc. For example, it is commonly observed that the area around a large pore usually tends to be 211 free as a result of the excess liquid that gathers at the pore [71]. The presence of a large amount of liquid facilitates the dissolution of 211, so that all of the Y from the 211 can combine with Ba and Cu from the melt and can precipitate out as Y123 on platelets in the vicinity.

It is still not clear whether the platelets are distinct crystals separated by low-angle grain boundaries or are actually part of the same crystal. For example, in Figure 8.1, the so-called brick wall structure of the textured Y123 can be easily distinguished. Upon closer examination, however, one notices that the boundary separating adjacent platelets seems to disappear occasionally, reappearing later. In addition, two platelets sometimes appear to merge. Schematically then, the brick wall appears more like what is shown in Figure 8.6. This has given rise to an alternative explanation for 123 growth from the melt, as put forth by Alexander et al. [51].

TEM and convergent beam electron diffraction (CBED) work by Alexander et al. [51] indicates that parallel plates of melt-textured Y123 are physically separated by gaps. Under optical microscopy it is often difficult to distinguish these gaps from cracks caused by thermal contraction. The researchers found that the gap regions frequently contain copper oxides, barium copper oxides, and some Y-rich material as well as some impurity elements such as Si, S, and Mo. The presence of these materials in the gaps indicates that at least some of the gaps are formed during 123 growth (some may be attributable to cracks formed during thermal processing). It was found that the platelets always terminate with an a-b plane parallel to a gap. CBED results on regions on either side of these gaps showed that there was virtually no change in plate orientation between adjacent platelets. The largest misorientation of 1.2° was attributed to local specimen bending. The results of the TEM examination of several adjacent and well-separated platelets indicated that all the "platelets" from a common domain (determined optically) were actually part

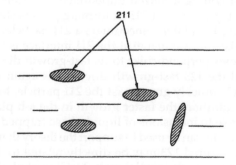

Fig. 8.6. Schematic illustration of modified "brick wall" structure.

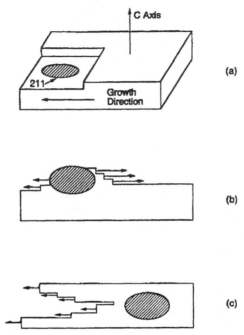

Fig. 8.7. Illustration of model proposed to explain formation of gaps near 211 particles in melt-textured Y123: (a) Y123 approaching 211 particle; (b) nucleation occurs on the 211 particle in the 123-211-liquid interfacial region, allowing for more rapid growth in the c-direction; (c) growth past the 211 particle continues, sometimes resulting in a gap.

of the *same* single crystal and that the gaps narrow and terminate in a stepwise fashion in the 123 matrix. It was proposed that the plate-gap microstructure arises from the anisotropic growth properties of 123 and compositional perturbations associated with the dispersed 211 particles in the partial melt. The presence of 211 particles is believed to directly contribute to the gap formation process by providing localized impediments to uniform 123 growth. The model depicting the role of 211 in forming gaps is shown in Figure 8.7.

After a growing 123 platelet encounters a 211 particle, nucleation on the 123 platelet face (i.e., the a-b plane) at the 211 interface is likely to occur. The 123 plate then grows perpendicular to its fast-growth direction (around the 211 particle) until the 123 fast-growth direction is again unimpeded. Once unimpeded, the 123 quickly grows past the 211 particle, forming the trailing gap structure. In addition, the faster growth in the a-b plane relative to the c direction gives rise to inclusions of liquid phase trapped between platelets [52]. This growth model can be used to support the theory that the plate structure observed in melt-textured 123 may be directly related to the size and distribution of 211 particles in the partial melt. McGinn et al. [53] found that the plate thickness decreased with increasing 211 content. Jin et al. [54] demonstrated a direct correlation between 211 particle size and plate thickness and suggested that the 211 particles both serve as nucleation sites for 123 and limit 123 plate thickness.

Schmitz et al. [55] recently attempted to elaborate on the diffusion controlled growth model of pure 123 proposed by Izumi et al. to explain the direct correlation between particle size and plate thickness. Their study involved evaluating localized growth interfaces found in quenched YBCO pellets that were exposed to a modified MTG thermal cycle. They found that the observable quenched solid-liquid interfaces were not planar but exhibited "bridges" of 123 material that extended from the interface to 211 particles in the partial melt. The bridges are thought to form when the growth interface approaches a high yttrium concentration near a 211 particle. The increased Y concentration gradient results in the localized accelerated growth of 123 towards the 211. Similar to the theory of Alexander et al., it is suggested that a traditional peritectic surface reaction then occurs of 123 growing on the 211 particle which is responsible for the full 211 engulfment. Finally, a negligible peritectic transformation is proposed to occur while the 211 particle is entrapped in the growth interface. Based on this model, it is suggested that the competitive growth of a series of 123 plates impinging on a 211 particle automatically results in a plate thickness corresponding to the mean 211 particle diameter. This is due to an enhanced competitive growth (both parallel and perpendicular to the fast growth direction) of the peripheral plates that are able to grow past the 211 inclusion.

8.6.6 Solidification Processes

A variety of solidification techniques have been reported for texturing of 123, but they can be classified into essentially three types: slow cooling without a temperature gradient, slow cooling in a defined temperature gradient without sample transport, and slow cooling through a temperature gradient by sample transport.

Growth of Y123 single crystals in a flux generally produces thin platelets because of the more rapid growth in the a and b directions than in the c direction of the crystal. This tendency can be used to produce regions of aligned grains (or "domains" of texture) by slow cooling through the peritectic, even in the absence of a temperature gradient. Without a temperature gradient, however, domains tend to nucleate and grow at multiple sites throughout a sample. This approach was first used by Salama et al. [56] and has now been used by many others. Large textured regions can be produced by this technique and can then be sectioned from a sample for characterization and testing. Because of multiple nucleation sites, however, there will be high-angle boundaries and perhaps $BaCuO_2$-CuO–rich phases between domains. These limit the transport of current between domains, although the transport properties of individual domains are extremely good.

The use of a temperature gradient during slow cooling helps to minimize the possibility of multiple nucleation sites for domains along the length of the sample. The size of the imposed temperature gradient is important in determining the resulting microstructure, as is described below. Because no sample (or furnace) transport is involved, only samples of relatively modest dimension can be textured by this technique.

The most versatile melt-texturing process employs the motion of a sample relative to a temperature gradient to achieve texture. Specifically, one either moves a heating source (e.g., a focused IR line heater) along a sample or moves a sample past a heating source. The movement of the sample is preferred from a commercial standpoint because it permits long samples to be processed in a semicontinuous fashion.

8.6.6.1 Melt-Texture Growth (MTG)

The MTG process first demonstrated the feasibility of using melt processing to align grains to overcome both the electrical anisotropy in Y123 materials. With this technique, sintered Y123 was partially melted at about 1100°C and then slowly cooled in a thermal gradient. The resulting microstructure showed preferred orientation for grain growth in the a-b plane and large, nonuniformly distributed spherical 211 particles up to 40 microns in length. The connectivity of the 123 plates was not good and was attributed to nonuniform growth conditions associated with the 211 particle distribution. Although the properties were much improved over those of sintered materials, the J_c was found to drop precipitously in magnetic fields, indicating the absence of adequate flux-pinning sites in the 123 matrix.

After Jin's initial success with the MTG process [36–38], which enhanced J_c two to three orders of magnitude over that of sintered materials (with significantly less field dependence), a variety of melt-texturing methods were developed in an effort to further improve properties and/or facilitate processing. The most frequently applied methods include the modified MTG process, the quench-and-melt-growth (QMG) process, the melt-powder-melt-growth (MPMG) process, the powder-melt process (PMP), the platinum-doped melt-growth (PDMG) process, the Bridgman method, and the floating-zone-melt process. The following presents a brief review of each of these techniques and describes their significance in the evolution of melt texturing.

8.6.6.2 Modified MTG Process

The modified MTG process was developed to homogenize the 211 distribution and minimize the 211 coarsening observed during the MTG process [56]. Here, samples are again quickly heated to approximately 1100°C; but rather than being slow cooled, they are held for only 20 minutes at the maximum temperature before being quickly cooled to the peritectic temperature. The samples are then slow cooled from the peritectic temperature to about 950°C. The initial heating and thermal hold at 1100°C is designed to ensure adequate sample melting (required to minimize pore formation), while the quick cooling to the peritectic temperature minimized 211 coarsening. The final microstructure shows improved 123 connectivity and a more uniform distribution of fine 211 particles ranging from 1 to 5 μm in size.

8.6.6.3 Quench-and-Melt-Growth (QMG) Process

The QMG process was first reported by Murakami and his coworkers [57–59]. This method involves using the 211 peritectic reaction to refine the size of the starting 211 particles. First, a sintered sample (or a mixture of calcined powders of the required stoichiometry) is quickly heated above the 211 decomposition temperature (1270°C) in a platinum crucible. The material is held at 1400°C for 20 minutes and then splat quenched using cold copper plates. The quenched sample consists of Y_2O_3 particles and an amorphous liquid phase. The quenched material is then processed by the modified MTG process described earlier. When heated above 1015°C, the Y_2O_3 reacts with the liquid to form 211, which in turn reacts with the liquid melt on cooling to form 123. The distribution of 211 in the melt is controlled by the distribution of Y_2O_3 in the quenched material. For stoichiometric 123, the QMG process results in a textured material that has little residual 211 phase and J_c values on the order of 2×10^4 A/cm² in zero applied field. The 211 refinement which leads to near complete reaction with the liquid to form 123 has subsequently been attributed to the presence of platinum in the melt (dissolved from the sample crucible).

8.6.6.4 Melt-Powder-Melt-Growth (MPMG) Process

The MPMG process is similar to the QMG process, being usually practiced with a starting powder stoichiometry that is 211 rich [60,61]. The key to this variation on QMG is that the quenched material is powdered before further processing to ensure a homogeneous distribution of Y_2O_3 (which then converts to 211) in both the melt and the final melt textured microstructure. The resulting acicular 211 particles are less than 1 μm in length. Murakami has reported that the properties of MPMG processed materials improve with increasing 211 content and has suggested that 211 acts directly as a flux pinning site.

8.6.6.5 Powder Melting Process (PMP)

The PMP process was developed by Lian and his coworkers [62]. This method uses a starting mixture of $Y211/BaCuO_2/CuO$ powders to secure a uniform distribution of fine 211 in the melt without having to employ a melt-quenching process. The pressed samples are processed either by the MTG or modified MTG thermal profile and yield microstructures identical to those obtained when comparable premixed YBCO stoichiometries are similarly processed. The advantage of this process is that the initial particle size of the 211 can be controlled, for example, by milling or using a sol-gel route to produce the 211. In MTG, by comparison, there is less well defined control over the 211 size. Although not mentioned explicitly by Lian et al., processing can probably be performed below the peritectic temperature because of the precursors used, since the liquid phase will melt below this temperature and begin reacting with the 211.

8.6.6.6 Platinum-Doped Melt-Growth (PDMG) Process

In the PDMG process, small additions of platinum (less than 0.5% by weight) comparable to the Pt contamination levels incurred during the high-temperature melt stage of MPMG are added to the desired stoichiometry prior to melt processing [63]. The 211 size and distribution in the resulting microstructure are similar to that found in MPMG samples, and J_c values as high as 2.5 x 10^4 A/cm^2 at 1 T (H ∥ c) have been reported [64]. Although the reduction in 211 size was originally attributed to an increase in the number of 211 nucleation sites (based on the assumption that 211 nucleated on platinum contaminants), recent studies (discussed later) suggest that the Pt in the melt affects the surface energy of the 211 and reduces its dissolution rate in the liquid.

8.6.6.7 Horizontal Bridgman Method

The Bridgman method is a directional solidification process that involves horizontally transporting a sintered Y123 sample through a preset temperature profile to texture relatively long sections of material. Here, the use of a well-defined temperature gradient helps to eliminate random nucleation and growth of 123 grains throughout the length of the specimen and thereby facilitates the formation of a single, unified planar growth front. The horizontal furnace geometry used with this method, however, requires that the precursor material be supported on a reactive (and contaminating) substrate (often YSZ). Although the substrate/reaction layer can contribute to random nucleation and growth, usually precursor cross-sectional dimensions are great enough to restrict interaction effects to a relatively small, localized volume. This has been employed both with standard Y123 and with PMP processing, utilizing 211 and the liquid phase as the precursors [49]. The primary advantage of the Bridgman method is that it permits the independent control of temperature gradient (G) and crystal growth rate (R). This process control facilitates the study of the effect of G/R and cooling gradient (GR) on the peritectic growth front morphology. The practical application of long YBCO wires critically depends on the ability to reproducibly form aligned microstructure from a planar (or faceted) growth front. According to constitutional supercooling theory, during directional solidification the desired plane front interface becomes stable when the actual temperature gradient in the liquid at the interface (G_L) is larger than the liquidus gradient. This condition, which is represented by Equation (8.7), results in the immediate dissolution and melting of any small protuberance in the growth front since T actual is greater than T liquidus at any point in the liquid (i.e., ahead of the growth front):

$$G_L/R > -(m_L C_o/D_L)((1-K)/K_o) , \qquad (8.7)$$

where R is the growth rate (or pull rate), m_L is the slope of liquidus line, C_o is the alloy composition, D_L is the diffusion coefficient of the solute (211) in the liquid, and K_o is the equilibrium distribution coefficient. Conversely, if G/R is smaller than the limit defined by Equation (8.7), the actual temperature becomes lower than the equilibrium liquidus temperature and a constitutional supercooling region forms in front of the planar interface. The presence of this region results in the preferential growth of plane front perturbations (discontinuous grain growth) and a dendritic or cellular growth front. Izumi and Shiuhara [49] investigated the effect of cooling parameters during Bridgman processing. They held the cooling rate (GR) constant and found an enhancement in magnetic J_c properties with increasing G/R values. They attributed their results to an associated improvement in 123 grain alignment.

Cima et al. [47] performed quench studies on Y211-rich wires that were processed at various growth rates (R) through different thermal gradients (G) by a floating-zone-melt process. The results of the studies highlighted the fact that the growth front was very sensitive to processing speed and relatively *insensitive* to the thermal gradient. At low processing speeds, the growth front was planar. At increasing speeds, the growth front became cellular/dendritic and finally blocky in nature.

8.6.6.8 Floating-Zone-Melt (FZM) Process

The FZM process is a directional solidification technique that involves passing a vertically suspended, sintered YBCO wire through a high-temperature gradient furnace to effect well-textured microstructures. The process requires that a sintered wire be centered in and mechanically supported above a furnace containing a well-defined hot-zone region. The wire is mounted above (and usually outside) the furnace to preclude any chemical interaction between the support and the highly reactive molten precursor material. The process is unique in that only a small section of the wire is melted, and the molten zone is moved from one end of the wire to the other. The furnace hot zone (which defines the wire molten zone) has been produced by a variety of methods including resistive elements [65], halogen lamps, infrared lamps, Nd-YAG lasers [66], and CO_2 lasers [47].

The equipment set can also be constructed in several different ways to permit the relative movement of the wire through the hot zone; however, the simplest and most mechanically stable geometry involves rigidly mounting the wire to a fixed support and moving the furnace up and along the sample to be textured. As the hot zone moves up the wire length, the textured length (below the hot zone region) is supported by the surface tension of the partial melt. For a given wire diameter and processing rate, the surface tension of the partial melt is affected by the hot zone temperature, the furnace design (hot zone width and cooling gradient), and the stoichiometry (i.e., 211 content) of the sintered wire.

Together, these factors influence the extent of liquid-phase segregation, which occurs from the partial melt during zone-melt processing. When the YBCO precursor material is heated into the 211 + liquid region, some liquid migrates towards the cooler section of the wire (i.e., above the hot-zone region), where it ultimately demixes and solidifies. The loss of this liquid shifts the composition of the wire being textured to a more 211-rich stoichiometry and can result in significant residual 211 being left in the melt-textured microstructure of even stoichiometric YBCO. For a given furnace design, the primary processing variables involved in this technique are, again, the temperature gradient (G) and the growth rate (R). The temperature gradient can be adjusted by changing the hot-zone design, changing the maximum hot zone temperature, and installing thermally controllable cooling plates at the bottom (exit end) of the furnace. The growth rate is adjusted by changing the speed at which the furnace moves vertically upward. This processing technique offers the only viable means of fabricating long, well-textured lengths of YBCO superconductor (free of substrate contamination) in a continuous or semicontinuous manner.

When texturing relatively thick sections in a gradient, one of the concerns is whether the lateral temperature gradient in the sample is uniform. This aspect of texture processing was investigated by Selvamanickam et al. [67] They used an auxiliary heater to maintain the outside of the sample hotter than the inside, with a lateral temperature gradient of -15°C/cm. This process minimizes the growth of grains with nonaxial alignment, permitting improved texture to be achieved in thick wires. Similar work by Meng et al. [68] showed that by changing the ratio of the longitudinal gradient to the lateral gradient ((dT/dx)/(dT/dy) in their terminology), the grain orientation could be varied from being aligned along the wire axis to being off the wire axis. In thin samples this is less of a problem because geometrical constraints encourage growth parallel to the axis of the wire.

One complicating feature that arises in processing Y123 in a thermal gradient is demixing. Migration of liquid occurs in the thermal gradient, producing a concentration gradient along the texturing direction [69]. This liquid (BaCuO-rich) migrates down the temperature gradient and freezes, producing a "hump" in a wire. Researchers have reported different temperatures corresponding to the freezing point of the liquid. In zone melting, the liquid will initially migrate both up and down the wire. Thus the material in the hot zone will become deficient in Ba and Cu and will not yield stoichiometric Y123 on freezing. If zone melting proceeds by the movement of the furnace up the wire, the hump on the top end of the wire will continue to be pushed out ahead of the hot zone as the wire slowly moves into the hot zone. Thus a steady-state condition will be achieved.

The occurrence of this migration is important in a process such as zone melting, however, because the Y123 sintered "feed" material coming into the hot zone experiences this excess liquid phase in the hump prior to entering the hot zone. This hump can be 1–2 mm in length; hence, assuming a 2-mm/hr travel speed, the sintered Y123 will be in the liquid for up to one hour. This liquid will promote grain growth of the Y123 at the least, and perhaps decomposition of the Y123, depending on the composition of the liquid. If grain growth of the Y123 results, the final textured microstructure for pure Y123 will have larger 211 than a sample that experienced no grain growth [70]. If decomposition occurs, for example, by the reaction of 123 + CuO → 211 + L, then excess 211 will be produced in a liquid which can then begin to coarsen.

In fact, a change in 211 size with distance along the wire has been observed [71]. This size change might also be due to Y enrichment along the wire, as reported by Meng et al. They observed the Y content to increase along the wire during zone melting, a process they attributed to the zone refining that occurs during solidification [68]. It is not clear, however, whether this is, in fact, the cause, since zone refining is usually associated with the transport of low-melting-point material along with the hot zone, not high-melting-point compounds such as 211. Research by Jiang et al. has shown that the contact angle of the liquid phase on the Y123 varies with atmosphere [72]. In their work, O_2 pressures above 1 atm resulted in decreased liquid flow away from the hot zone.

8.6.7 Effect of 211 on the Properties of 123

Many reports in the literature address the beneficial effects of 211 particles in the melt-textured 123 microstructure. The 211 particles have been observed to assist in blunting the progress of cracks in the 123 matrix [73], and 211 particles were also reported to enhance the fracture toughness of 123 [74]. Perhaps the most important effect of 211 particles, however, is to influence the flux-pinning characteristics of 123 either directly or indirectly (by acting as sources of crystal defects) [75–77].

Most of the significant enhancements in the properties of bulk Y123 materials have evolved from process modifications that exploited the unique properties of the Y123 phase diagram. Although the development of various melt-texturing techniques has effectively demonstrated the benefits of both eliminating grain boundary weak links and aligning the strongly superconducting CuO (a-b) planes, the electrical properties (J_c) of bulk melt-textured materials, although improved, still remain two orders of magnitude lower than those of 123 thin films. The discrepancy in J_c values between bulk-processed and thin-film samples is attributed to inherent differences in the materials' flux-pinning characteristics. Consequently, a significant amount of research is now being focused on elucidating the effectiveness of the various extrinsic pinning sites in melt-textured Y123 with the hope of defining and introducing improved pinning mechanisms in the future. Although the irradiation of bulk samples with neutrons [78–80] and protons [81] has been demonstrated to significantly improve properties, the mechanism by which the enhancement occurs, and therefore the practical extension of the technology, is still unclear. It is for this reason that the most studied practical consideration involves the role of residual 211 particles in flux pinning and J_c enhancement.

Using pellets processed by the MTG method in which the size and volume of 211 were varied, Jin et al. [82] concluded that additions of the residual phase yielded no significant improvement in J_c. Other results [83] based on zone-melt-textured wires indicated that transport and magnetic Jc values actually decreased with increasing RE211 content. In turn, these results seemingly contradicted those of others [73] who processed pellets by the QMG method and found that J_c values increased with increasing 211 content. The significant factors that were varied in these sets of experiments were the stoichiometry and processing method, which ultimately determined the size, distribution, and volume of 211 in the 123 matrix.

Subsequently, Lee et al. [84] performed a study to directly address the effect of 211 particle size and content on transport J_c properties. Using a modified MTG method, they controlled the 211 size and content by varying the rate at which the samples were cooled from 1025°C. Samples cooled at the fastest rate of 4 degrees per hour had the greatest volume and finest dispersion of 211, while those cooled at 1 degree per hour had the smallest total 211 volume and largest average 211 particle sizes. Lee et al. attempted to reconcile those of previous investigators by suggesting that the effect of 211 content on J_c could be divided into three regimes. In the first regime, 211 content is low, particle sizes are small, and J_c increases with increasing 211 content. In the second regime, an intermediate amount of 211 has competing influences on the measured J_c. On the one hand, the J_c is increased with improved pinning; on the other hand, the J_c is decreased as a result of a drop in superconducting volume. Samples representative of this regime showed little or no net improvement in J_c depending on 211 particle size and the strength of the applied field. In the third regime, the J_c monotonically decreases with increasing 211 content because the benefit from flux pinning cannot overcome the reduction in the currents percolation path. It follows that, for a given amount of added 211, different J_c may result if the samples being compared have 211 inclusions of different average size. Lee et al. suggested that the optimum second-phase content should vary with particle size and shift upward with decreasing average 211 sizes. This supposition is directly linked to a second debate regarding 211, which involves the means by which the presence of 211 particles might enhance pinning.

Since the size of residual 211 particles are two to three orders of magnitude greater than the coherence length in 123, many researchers have suggested that 211 acts as an indirect pinning site by providing a pinning defect structure in the 123 at the 123/211 interface. The findings of Murakami et al. that the critical current density scales with the surface area of 211 trapped in the 123 are regarded as among the best supporting evidence for this theory [85]. They suggest that since there is no energy difference inside the 211 particles to drive flux motion, flux pinning occurs at the 211 interface. Although the effectiveness of 211 flux pinning per unit volume is not as great as with small pinning centers, a 1-μm sized 211 particle results in pinning potentials that are at least two orders of magnitude greater than those associated with typical (coherence length) pinning sites. Based on the enhanced pinning potential, Murakami et al. believe the 211 may provide significant collective flux-pinning capability.

In an effort to elucidate the enhanced pinning phenomena observed in melt-processed samples, several research groups have investigated the 123/211 interface using electron microscopy techniques. A common finding is a significant increase in the number of dislocations and stacking faults in the 123 near the interface. Wang et al. studied a variety of samples prepared by different melt-texturing methods and in all cases found that disc-shaped stacking faults (having diameters ranging from a few to 30 nm) were inhomogeneously distributed around the 211 and seemed to preferentially form perpendicular to the c direction in the 123 [86]. The density of the stacking faults was greater than $10^{15}/cm^3$. Wang et al. [87] reported that the density of stacking faults in the interface region increased as the radius of curvature of the 211 particle decreased. Calculations by Wang et al. suggest that these stacking faults may act as effective flux pinners for fields directed both parallel and perpendicular to the basal plane of the aligned 123. Consequently, these faults may be responsible for the observed increase in J_c with increasing 211 content for QMG-processed samples. Further, EDS scans near the 123/211 interface regions containing few stacking faults indicated a significant enrichment of Y and a deficiency in Ba in the 123 [88]. It was also proposed that this local cation nonstoichiometry may result in a region of high point defects which also contributes to flux pinning.

Further evidence of the importance of dislocations and stacking faults in affecting the pinning properties was shown in deformation studies on melt-textured 123 performed by Salama and his coworkers [89]. Hot deformation at 45° to the slip plane (001) and the slip directions [100]/[010] resulted in an increase in the defect density. It was observed that this caused a dramatic increase in the critical current in the a-b plane for H $\parallel$ c, such that the measured values were as high as those for H $\parallel$ a-b. In isostatically deformed samples, point pinning by dislocations was seen, leading to improvements in J_c for H $\parallel$ c magnetic orientations [90].

8.6.8 Controlling 211 Particle Size

As noted, Y_2BaCuO_5 (211) particles have been seen to lead to improvements in the critical current density as long as they are below a critical size [84]. While this critical size has not been clearly identified, it is clearly of great importance to be able to control the size of 211 particles. A number of different strategies have been employed in bulk Y123 processing to produce fine 211. These include the following:

1. Additions of Pt and PtO_2 have been found to lead to decreases in 211 size and also to change the morphology of the 211 into particles with a very high aspect ratio by reducing the coarsening rate by decreasing the energy of the 211/liquid phase interface [63,91–96]. This has been widely used to reduce 211 size in bulk melt-texturing techniques, but has not been reported for wire fabrication. In techniques such as QMG, the Pt additions can result from corrosion of the Pt crucible by the melt, whereas in other techniques discrete additions are required.
2. Rapid heating of Y123 during processing above the peritectic has been found to result in refined 211. The faster the Y123 is heated above the peritectic (to 1050°C, for example) the more unstable the Y123 becomes. This, in turn, provides a greater free energy change with decomposition, promoting greater nucleation of 211, and resulting in the production of finer 211 [97]. Thus, in normal melt-texture growth, a more rapid heating rate will yield finer 211 in the textured microstructure, assuming the cooling rate is kept constant.
3. The addition of heterogeneous nucleation sites in the form of excess 211 or other second-phase additions has been found to act as nucleation sites for the peritectically produced 211, thereby decreasing the average 211 particle size [98–99]. For example, Jin et al. added aerosol-processed 211 particles that initially were 0.2 μm in size and, after texturing, were 0.7 μm in size in the Y123 matrix [54]. Figure 8.8 shows a zone-melted Y123 wire with Er211 additions. The fact that most of the 211 particles contain Er-rich cores (the bright centers) indicates that the added 211 acts as a heterogeneous nucleation site for the Y211 during Y123 decomposition. The inability to resolve bright cores in all the 211 particles is due to polishing angle effects.

Other additions have been employed to reduce the 211, but it is not clear that any of them actually act as heterogeneous nucleation sites. $BaTiO_3$, $BaSnO_3$, $BaZrO_3$, SiC, Al_2O_3, CeO_2, and related compounds have been employed in efforts to improve the flux pinning. Some, such as $BaSnO_3$, promote 211 dissolution, resulting in finer 211 which reacts with the melt and ties up Y in a Y-Ba-Sn-O compound, producing refined 211 along with excess liquid phase [100]. Others, like CeO_2, act like Pt in that they refine the 211 size and change the particle morphology to being more acicular [101]. No clear evidence exists of any particle acting as a heterogeneous nucleation site (with the exception of 211), although many authors have offered it as an explanation for observations of 211 refinement.

Although in many instances the additions were made with the hope that they would act as fine pinning centers, the additions were often observed to modify the morphology and size of the 211 phase. Improvements in pinning noted with some of these additions are, in fact, probably due to their effect on the 211 volume, size, and distribution.

Coarsening of 211 particles in the presence of second phases can be considerably different from that which occurs in the absence of any additions. Morphological differences in 211 caused by second-phase additions can also affect the coarsening rate. For example, high-energy interfaces will dissolve faster than lower-energy ones. Thus, it can be expected that additions that affect the morphology (by changing the interfacial energy) will alter the coarsening rate. The coarsening rate can also be affected by

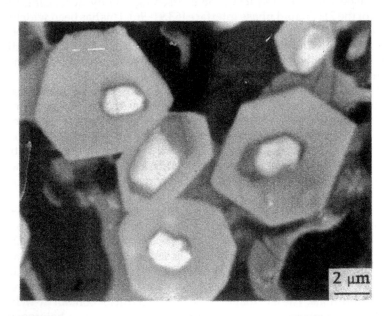

Fig. 8.8. Textured Y123 with added Er 211 particles. Er-rich cores in 211 particles after texturing result from Er211 acting as a heterogeneous nucleation site.

the size distribution of the initial peritectically formed 211; and anything, such as second-phase additions, that affects the distribution will change the coarsening rate.

4. Use of alternative precursors results in the formation of refined 211 particles in the Y123 matrix. For example, one can begin with 211 + liquid phase (PMP) [62], Y_2O_3 + liquid, produced as a result of a quench process (QMG or MPMG) [57–61], or well-dispersed Y_2O_3 + liquid produced by a solid-state route (SLMG) [102,77].

Since much of the melt-texture processing relies on the use of sintered Y123 as the precursor, research has also been performed to determine whether the grain size of the Y123 affects the size of the 211 that forms during decomposition [70]. This work indicates that the nucleation and growth of 211, in the absence of any second-phase additions which might act as heterogeneous nucleation sites, depend on the Y123 grain size. This effect is expected to be less important, however, when different heating rates are employed or there are significant second-phase additions. From the perspective of a texturing process such as zone melting, the sharpness of the gradient on the entrance side of the hot zone, along with the travel speed, will control the heating rate. Because the speed is a greater concern from the point of controlling the solidification, however, altering the speed to increase the heating rate is not a practical option. The only option is to select a furnace design or heating technique that will yield a sharp gradient. Just as for generating a sharp gradient on cooling, a laser source is probably the best option in this regard.

8.7 Critical Current Density Measurements

Many of the envisioned bulk applications for 123 generally involve either monolithic structures such as flywheels or bearings, or wire applications such a magnets or other long-length conductors. In both of these areas, among the most important characteristics of the superconductors are their critical current density and their ability to pin magnetic flux. Thus, much of the characterization of the superconductors is aimed at assessing these properties. Figure 8.9 shows magnetic hysteresis loops at 77 K for a zone-melt-textured 123 wire and a sintered-123 pellet acquired with a vibrating sample magnetometer (VSM). The great difference in the magnetic properties is illustrated here where the hysteresis loop for the sintered sample is so small that it appears as a flat line in comparison with the loop for the textured sample. One of the most useful aspects of the hysteresis curve is that it can be used to approximate the critical current density by applying Bean's model, where the critical current density is described by $J_c = C \Delta M/d$, where C is a geometric constant that depends on the sample shape, ΔM is the width of the hysteresis loop, and d is the diameter of the supercurrent loop. Modifications to the model are necessary for various sample geometries, but it allows the J_c to be characterized in a contactless fashion with a VSM.

Measuring J_c by a four-point electrical measurement requires the application of low-resistance electrical contacts, usually formed by firing silver paint at 500°C, followed by soldering with a low-melting-point solder. Poor contacts will inhibit the ability to fully characterize the material, because at high currents the contacts will lead to sample heating, thereby yielding results that are too low. One way to avoid this difficulty is to characterize the sample in high magnetic fields, where the J_c will be lower, so the applied currents will be reduced. Otherwise, the sample cross section must be reduced in order to lower the current necessary to achieve a certain current density level. A recent review discusses in great detail the preparation of contacts to high-temperature superconductors [103].

One of the precautions that must be kept in mind in using the magnetic J_c measurements is that they are relatively insensitive to the effect of cracks and insulating layers at grain boundaries. Thus, a sample that might carry zero transport current because of a crack may still exhibit very good magnetic properties and pin flux extremely well. This is not a problem in considering materials for flux-trapping applications such as frictionless bearings or flywheels, but it is a consideration in characterizing wires.

8.8 Mechanical Properties

Assuming that sufficiently high J_c values do not prevent the use of Y123 superconductors, the other major obstacle to their widespread application is poor mechanical strength. The Y123 compound is brittle, and, as mentioned earlier, Y123 is prone to microcracking during cooling through the tetragonal-orthorhombic phase transformation. In addition, thermal cycling from room temperature to 77 K has also been shown to lead to degradation from thermal fatigue. The most practical means to improve the mechanical properties is by using Y123 as one component in a composite, in combination with either a metal or another ceramic.

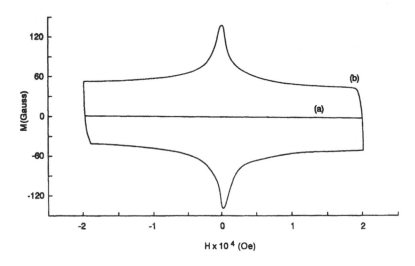

Figure 8.9. Hysteresis loops at 77 K for (a) a sintered Y123 pellet and (b) a zone-melt-textured Y123 wire showing the large increase in magnetization due to texturing.

Fortuitously, the toughness of 123 has been found to be improved by the addition of fine 211 particles. Thus, the very particles that improve the critical current density of 123 can also improve its mechanical toughness. As with 211, fine CeO_2 particles have also been observed to increase fracture toughness by impeding crack growth through the 123 [104].

Metal additions have been widely used to improve the mechanical properties of 123. Silver is by far the most popular metal employed for making metal-123 composite structures. It results in little, if any degradation of the 123, unlike most metals which, when sintered with 123, tend to reduce T_c. In sintered 123, by adding enough silver a sample with modest ductility that is superconducting can be produced, but the J_c will be quite low. Thus, texturing of 123-silver composites is necessary to achieve reasonably high J_c values. Several studies have considered the effect of silver additions on the properties of melt textured Y123 [105–109]. Silver additions do not alter the textured microstructure significantly, but do tend to prevent crack formation and propagation, and improve the mechanical strength of the Y123. Silver does not appear to enhance the J_c in textured Y123, as it does in sintered Y123; but mixed results have been reported on whether it degrades J_c as compared with textured Y123 with no silver additions.

One consideration in using silver is that the melting point is below that of the peritectic of Y123 in air. Thus, when texturing in air, the silver melts and silver particles coarsen considerably, making it difficult to produce a microstructure with fine, well-dispersed Ag particles. Two potential solutions to avoid melting have been offered. By alloying Ag with approximately 10% Pd, the melting point of the alloy is raised above the peritectic of the Y123. At this doping level the amount of Pd is still low enough to avoid any substantial degradation of the Y123, although there is some reaction of the melt with the Pd, leading to the formation of some intermetallic compounds. The other solution is to texture in a reduced pO_2 atmosphere to lower the peritectic temperature of the Y123 below the silver melting point. This issue is currently the subject of several investigations.

Only one study has looked at the compatibility of a range of metals with molten Y123 [110]. The results show that complete recovery of the 90 K transition temperature can be achieved by melt processing of transition-metal-doped Y123. When such doped Y123 is sintered, the T_c is degraded as a result of the incorporation of the dopant into the Y123 lattice. With melt processing, however, the Y123 that forms during solidification is dopant-free, with the transition metal being incorporated into a second-phase solidification byproduct. This result provides hope that a variety of possible additions can be used to increase the mechanical strength of melt-textured Y123. Whichever metal is chosen for use, it clearly will be necessary to use textured 123 as part of a composite for many bulk applications because of its poor mechanical strength.

8.9 Summary

A variety of techniques have been used to make Y123 powder. Many are aimed at producing as fine a particle size as possible, or achieving homogeneity on an atomic level. Despite the various advantages the different techniques have (e.g., fine particle size, well-controlled stoichiometry), when sintered the result is a sample with low critical current density and poor flux trapping at 77 K. This is a result of the weak links in the form of high-angle grain boundaries in polycrystalline samples. By processing the sample to produce an aligned microstructure with low-angle grain boundaries, however, the properties can be improved tremendously. Textured microstructures can be produced by directional solidification processes that exhibit transport J_c values in excess of 10^5 A/cm^2.

The textured microstructure does not develop by the usual peritectic transformation, in this case the growth of 123 directly on 211 particles, but instead evolves by 123 nucleating homogeneously from the melt. Solidification models have been developed that indicate that the rate controlling step for the formation of 123 is 211 dissolution and diffusion of Y to the solidification front. Because commercialization of texturing is hindered by the required slow cooling rates, strategies must be devised to speed the 211 dissolution.

Although relatively high critical current densities have been achieved in textured structures, higher J_c and flux-trapping properties are desirable for certain applications. Efforts in this regard are directed at identifying effective flux-pinning sites in the textured structure and implementing processing routes to produce such sites. In particular, fine 211 particles have shown to be effective pinning sites. Certain processing routes (QMG, PMP) have proven to be well suited to producing fine 211 particles, yielding samples with improved flux-pinning properties.

For producing monolithic structures for applications such as frictionless bearings or energy storage in flywheels, texturing processes such as melt-texture-growth or quench-melt-growth are suitable. In order to produce textured wires, processes such as zone melting or Bridgman methods are necessary. All have been used with success in generating textured structures.

The widespread commercial applicability of Y123 depends on the development of processing methods to produce nonweak-linked, strongly flux-pinning microstructures in reasonable periods of time. If such processing improvements cannot be devised, the material will be confined to more specialized applications, and systems where texture can be more quickly achieved (such as the Bi-based high-T_c superconductors) will occupy a more commercially significant role.

8.10 References

1.   M. Leskela, C. Mueller, J. K. Truman, and P. Holloway, *Mater. Res. Bull.* **23**, 1469 (1988).
2.   U. Balachandran, R. Poeppel, J. Emerson, S. Johnson, M. Lanagan, C. Youngdahl, K. Goretta, and N. Eror, *Mater. Lett.* **8**, 454 (1989).
3.   N. Chen, D. Shi, and K. Goretta, *J. Appl. Phys.* **66**, 2485 (1989).
4.   E. Ruckenstein, S. Narain, and N. L. Wu, *J. Mater. Res.* **4**, 267 (1989).
5.   P. Barboux, J. Tarascon, L. Greene, G. Hull, B. Bagley, *J. Appl. Phys.* **63**, 2725 (1988).
6.   M. P. Pechini, U.S. Patent 3,330,697 (July 11, 1967).
7.   E. Hayri, M. Greenblatt, K. Ramanajachary, M. Nagano, J. Oliver, M. Miceli

and R. Gerhardt, *J. Mater. Res.* **4**, 1099 (1989).

8. J. Subrahmanyam and M. Vijayakumar, *J. Mater. Sci.* 6249 (1992).

9. J. Lebrat and A. Varma, *Physica C* **184**, 220 (1991) 220.

10. T. Kodas. E. Engler, V. Lee, R. Jacowitz, T. Baum, K. Roche, S. Parkin, W. Young, S. Hughes, J. Kleder, and W. Auser, *Appl. Phys. Lett.* **52**, 1622 (1988).

11. Y. Kimura, T. Ito, H. Yoshikawa, and A. Hiraki, *Jap. J. Appl. Phys.* **30**, L798 (1991).

12. G. J. Yurek, J. Vander Sande, W. Wang, D. Rudman, Y. Zhang, and M. Matthiesen, *Met. Trans.* **18A**, 1813 (1987).

13. H. Poulsen, N. Andersen, J. Andersen, and H. Bohr, O. Mouritsen, *Nature* **349**, 594 (1991).

14. T. Richardson, L. De Johnge, *J. Mater. Res.* **5**, 2066 (1990).

15. D. Clarke, T. Shaw, and D. Dimos, *J. Am. Ceram. Soc.* **72**, 1103 (1989).

16. A. Bailey, G. J. Russell, and K. N. R. Taylor, in *Studies of High Temperature Superconductors*, ed. by A. Narlikar, Vol. 9, Nova Science Publishers, New York (1992), 145.

17. P. McGinn, N. Jain., and D. Lee, *Surface and Coatings Technology* **37**, 359 (1989).

18. S. Dorris, J. Dusek, M. Lanagan, J. Picciolo, J. Singh, J. Creech, and R. Poeppel, *Ceram. Bull.* **70**, 722 (1991).

19. P. Chaudhari, J. Mannhart, D. Dimos, C. Tsuei, J. Chi, M. Oprysko, and M. Scheuermann, *Phys. Rev. Lett.* **60**, 1653 (1988).

20. D. Dimos, P. Chaudhari, J. Mannhart, and F. LeGoues, *Phys. Rev. Lett.* **61**, 211 (1989).

21. D. Larbalestier, S. Babcock, X. Cai, M. Field, Y. Gao, N. Heinig, D. Kaiser, K. Merkle, L. Williams, and N. Zhang, *Physica C* **185–189**, 315 (1991).

22. Y. Zhu, H. Zhang, H. Wang, and M. Suenaga, *J. Mater. Res.* **12**, 2507 (1991).

23. T. Dinger, T. Worthington, W. Gallagher, and R. Sandstrom, *Phys Rev. Lett.* **58**, 22 (1987).

24. V. Selvamanickam and K. Salama, *Appl. Phys. Lett.* **57**, 1575 (1990).

25. S. Babcock, T. Kelly, P. Lee, J. Seuntjens, L. Lavanier, and D. Larbalestier, *Physica C* **152**, 25 (1988).

26. S. Babcock and D. Larbalestier, *Appl. Phys. Lett.* **55**, 393 (1989).

27. S. Babcock and D. Larbalestier, *J. Mater. Res.* **5**, 919 (1990).

28. K. B. Alexander, D. M. Kroeger, and J. Bentley, J. Brynestad, *Physica C* **180**, 337 (1991).

29. H. M. O'Bryan and P. K. Gallagher, *Adv. Ceram. Mater.* **2**, 632 (1987).

30. I.W. Chen, X. Wu, S. Keating, C. Keating, P. Johnson, and T.Y. Tien, *J. Am. Ceram. Soc.* **70**, C388 (1987).

31. K. Matsuzaki, A. Inoue, and T. Masumoto, *Jap. J. Appl. Phys.* **29**, L1789 (1990).

32. D. Farrell, B. Chandrasekhar, M. DeGuire, M. Fang, V. Kogan, J. Clem, and D. Finnemore, *Phys. Rev. B* **36**, 4025 (1987).

33. R. Arendt, A. Gaddipati, F. Luborsky, and L. Schilling, in *High Temperature Superconductors*, ed. by M. R. Brodsky, R. Dynes , K. Kitazawa, and H. Tuller. Materials Research Society, Pittsburgh (1988), 203.

34. J. E. Tkaczyk and K. W. Lay, *J. Mater. Res.* **5**, 1368 (1990).

35. M. Lees, D. Bourgault, D. Braithwaite, P. de Rango, P. Lejay, A. Sulpice, and R. Tournier, *Physica C* **191**, 414 (1992).
36. S. Jin, T.Tiefel, R.Sherwood, R. van Dover, M. Davis, G. Kammlott, and R. Fastnacht, *Phys. Rev. B* **37**, 7850 (1988) 7850
37. S. Jin, T. Tiefel, R. Sherwood, M. Davis, R. van Dover, G. Kammlott, R. Fastnacht, and H. Keith, *Appl. Phys. Lett.* **52**, 2074 (1988).
38. S. Jin, R. Sherwood, E. Gyorgy, T. Tiefel, R. van Dover, S. Nakahara, L. Schneemeyer, R. Fastnacht, and M. Davis, *Appl. Phys. Lett.* **54**, 584 (1989).
39. J. Whitler and R. Roth, *Phase Diagrams for High Tc Superconductors*, The American Ceramic Society, Westerville, Ohio (1991).
40. T. Aselage and K. Keefer, *J. Mater. Res.* **3**, 1279 (1988).
41. D. H. St. John, *Acta Metall. Mater.* **38**, 631 (1990).
42. D. H. St. John, L. M. Logan, *Acta Metall. Mater.* **35**, 171 (1987).
43. C. A. Bateman, L. Zhang, H. M. Chan, and M. P. Harmer, *J. Amer. Ceram. Soc.* **75**, 1281 (1992).
44. K. Oka, K. Nakane, M. Ito, M. Saito, and H. Unoki, *Jap. J. Appl. Phys.* **27**, L1065 (1988).
45. M. Maeda, M. Kadoi, and T. Ikeda, *Jap. J. Appl. Phys.* **28**, 1417 (1989).
46. M. Rodriguez, B. J. Chen, and R. Snyder, *Physica C* **195**, 185 (1992).
47. M. Cima, M. Flemings, A. Figueredo, M. Nakade, H. Ishii, H. Brody, and J. Haggerty, *J. Appl. Phys.* **72**, 179 (1992).
48. T. Izumi, Y. Nakamura, and Y. Shiohara, *J. Mater. Res.* **7**, 1621 (1992).
49. T. Izumi and Y. Shiohara, *J. Mater. Res.* **7**, 16 (1992).
50. C. Varanasi and P. J. McGinn, *Physica C* **207**, 79 (1993).
51. K. Alexander, A. Goyal, D. Kroeger, *Phys. Rev. B* **45**, 5622 (1992).
52. A. Goyal, K. Alexander, D. Kroeger, P. Funkenbush, and S. Burns, *Physica C* **210**, 197 (1993).
53. P. McGinn, N. Zhu, W. Chen, S. Sengupta, and T. Li, *Physica C* **176**, 203 (1991).
54. S. Jin, G. W. Kammlott, T. H. Tiefel, T. T. Kodas, T. L. Ward, and D. M. Kroeger, *Physica C* **181**, 57 (1991).
55. G. Schmitz, J. Laakmann, C. Wolters, S. Rex, W. Gawalek, T. Habisreuther, G. Bruchlos, and P. Gornert, *J. Mater. Res.* **8**, 2774 (1993).
56. K. Salama, V. Selvamanickam, L. Gao, and K. Sun, *Appl. Phys. Lett.* **54**, 2352 (1989).
57. M. Murakami, M. Morita, and N. Koyama, *Jap. J. Appl. Phys.* **28**, L1125 (1989).
58. M. Murakami, M. Morita, K. Doi, and K. Miyamoto, *Jap. J. Appl. Phys.* **28**, 1189 (1989).
59. M. Murakami, M. Morita, K. Doi, K. Miyamoto, and H. Hamada, *Jap. J. Appl. Phys.* **28**, 399 (1989).
60. H. Fujimoto, M. Murakami, S. Gotoh, T. Oyama, Y. Shiohara, N. Koshizuka, and S. Tanaka, in *Advances in Superconductivity II.* Springer-Verlag, Tokyo (1990), 85.
61. M. Murakami, *Mod. Phys. Lett.* **4**, 163 (1990).
62. Z. Lian, Z. Pingxiang, J. Ping, W. Keguang, W. Jingrong, and W. Xiaozu, *Supercond. Sci. Tech.* **3**, 490 (1990).

63. N. Ogawa, I. Hirabayashi, and S. Tanaka, *Physica C* **177**, 101 (1991).
64. M. Morita, M. Tanaka, S. Takebayashi, K. Kimura, K. Mayamoto, and K. Sawano, *Jap. J. Appl. Phys.* **30**, L813 (1991).
65. P. McGinn, W. Chen, N. Zhu, U. Balachandran, and M. Lanagan, *Physica C* **165**, 480 (1990).
66. S. Nagaya, M. Miyajima, I Hirabayashi, Y. Shiohara, and S. Tanaka, *IEEE Trans. Mag.* **27**, 1487 (1991).
67. V. Selvamanickam, C. Partsinevelos, A. McGuire, and K. Salama, *Appl. Phys. Lett.* **60**, 3313 (1992).
68. R. L. Meng, Y. Sun, P. Hor, and C. W. Chu, *Physica C* **179**, 149 (1991).
69. D. Dube, B. Arsenault, C. Gelinas, and P. Lambert, *Mater. Lett.* **15**, 1 (1992).
70. C. Varanasi and P. J. McGinn, *Mater. Lett.* **17**, 205 (1993).
71. D. Balkin, unpublished.
72. X. P. Jiang, M. Cima, H. Brody, J. Haggerty, M. Flemings, *Proc. International Workshop on Supercond.*, MRS, Pittsburgh (1992), 259.
73. M. Murakami, S. Gotoh, N. Koshizuka, S. Tanaka, T. Matsushita, S. Kambe, and K. Kitozawa, *Cryogenics* **30**, 390 (1990).
74. H. Fujimoto, M. Murakami, and N. Koshizuka, *Physica C* **203**, 103 (1992).
75. D. Lee, V. Selvamanickam, and K. Salama, *Physica C* **202**, 83 (1992).
76. S. Sengupta, D.Shi, Z. Wang, A. Biondo, U. Balachandran, K. Goretta, *Physica C* **199**, 43 (1992).
77. C. Varanasi, S. Sengupta, P. J. McGinn, and D. Shi, *Appl. Supercond.*, in press (1994).
78. P. Hor, J. Bechtold, Y. Xue, C. Chu, E. Hungerford, X. Maruyama, H. Backe, F. Buskirk, S. Connors, Y. Jean, and J. Farmer, *Physica C* **185**, 2311 (1991).
79. J. Bechtold, Y. Xue, Z. Huang, E. Hungerford, P. Hor, C. Chu, X. Maruyama, H. Backe, F. Buskirke, S. Connors, D. Snyder, Y. Jean, and J. Farmer, *Physica C* **191**, 199 (1992).
80. B. Weaver, M. Reeves, G. Summers, R. Soulen, W. Olson, M. Eddy, T. James, and E. Smith, *Appl. Phys. Lett.* **59** (1991) 2600
81. Y. Zhao, W. Chu, M. Davis, J. Wolfe, S. Deshmukh, D. Economou, and A. McGuire, *Physica C* **184**, 144 (1991).
82. S. Jin, T. H. Tiefel, and G. W. Kammlott, *Appl. Phys. Lett.* **59**, 540 (1991).
83. P. McGinn, N. Zhu, W. Chen, S. Sengupta, and T. Li, *Physica C* **176**, 203 (1991).
84. D. F. Lee, V. Selvamanickam, and K. Salama, *Physica C* **202**, 83 (1992).
85. M. Murakami, H. Fujimoto, S. Gotoh, K. Yamaguchi, N. Koshizuka, and S. Tanaka, *Physica C* **185–89**, 321 (1991).
86. Z. Wang, A. Goyal, D. Kroeger, and T. Armstrong, in *Layered Superconductors: Fabrication, Properties and Applications*, ed. by D. T. Shaw, C. C. Tsuei, T. R. Schneider, and Y. Shiohara.*Materials Research Society Proceedings*, Vol. 275 (1992), 181.
87. Z. Wang, R. Kontra, A. Goyal, and D. Kroeger, *Matls. Sci. Forum* **129**, 1 (1993).
88. Z. Wang, A. Goyal, and D. Kroeger, *Phys. Rev. B.* 1812 (1992).
89. V. Selvamanickam, M. Mironova, and K. Salama, *J. Mater. Res.* **8**, 249 (1993).
90. V. Selvamanickam, M. Mironova, S. Son, and K. Salama, *Physica C* **208**,

238 (1993).
91. C. Varanasi and P. J. McGinn, *Physica C* **207**, 79 (1993).
92. S. Gauss, S. Elschner, and H. Bestgen, *Cryogenics* **32**, 965 (1992).
93. M. Morita, M. Tanaka, S. Takebayashi, K. Kimura, K. Miyamoto, and K. Sawano, *Jap. J. Appl. Phys.* **30**, L813 (1991).
94. C. Varanasi, P. J. McGinn, V. Pavate, and E. Kvam, preprint.
95. T. Izumi, Y. Nakamura, T. H. Sung, and Y. Shiohara, *J. Mater. Res.* **7**, 801 (1992).
96. N. Ogawa, M. Yoshida, I. Hirabayashi, S. Tanaka, Supercond. *Sci. Technol.* **5**, S89 (1992).
97. J. Rignalda, X. Yao, D. G. McCartney, C. J. Kiely, G.J. Tatlock, *Mater. Lett.* **13**, 357 (1992).
98. C. Varanasi, D. Balkin, and P. McGinn, *Appl. Superconductivity* **1**, 71–80 (1993).
99. D. Balkin, C. Varanasi, P. McGinn, in *Layered Superconductors: Fabrication, Properties and Applications*, ed. D. T. Shaw, C. C. Tsuei, T. R. Schneider, and Y. Shiohara,*Materials Research Society Proceedings*, Vol. 275 (1992), 207.
100. C. Varanasi and P. McGinn, *Supercond. Sci. Tech.*, in press (1994).
101. C. J. Kim, preprint
102. D. Shi, S. Sengupta, J. S. Lou, C. Varanasi, and P. McGinn, *Physica C* **213**, 179 (1993).
103. J. W. Ekin, in *Processing and Properties of High Tc Superconductors*, Vol 1, ed. by S. Jin, World Scientific, Singapore (1993), 371.
104. C. J. Kim, K. B. Kim, G. W. Hong, D. Y. Won, B. H. Kim, C. T. Kim, H. C. Moon, and D. S. Suhr, *J. Mater. Res.* **7**, 2349 (1992).
105. P. McGinn, N. Zhu, W. Chen, M. Lanagan, and U. Balachandran, *Physica C* **167**, 343 (1990).
106. D. F. Lee, X. Chaud, and K. Salama, *Physica C* **181**, 81 (1991).
107. A. Goyal, P. Funkenbusch, D. Kroeger, and S. Burns, *Physica C* **182**, 203 (1992).
108. D. F. Lee and K. Salama, *Jap. J. Appl. Phys.* **29**, L2017 (1990).
109. G. Kozlowski, S. Rele, D. F. Lee, and K. Salama, *J. Mater. Sci.* **26**, 1056 (1991).
110. G. Kammlott, T. Tiefel, and S. Jin, *Appl. Phys. Lett.* **56**, 2456 (1990).

8.11 Recommended Readings for Chapter 8

1. *Processing and Properties of High T_c Superconductors*, Vol. 1, ed. S. Jin, World Scientific, Singapore, 1993. (Deals with processing and characterization of high-T_c superconductors.)

2. *Melt Processed High-Temperature Superconductors*, ed. M. Murakami, World Scientific, Singapore, 1992. (Deals exclusively with melt processing and characterization of Y123.)

3. *Techniques of Melt Crystallization*, G. J. Sloan and A. R. McGhie, John Wiley and Sons, New York, 1988. (General text on directional solidification, zone refining, phase diagrams, etc.)

4. *Chemical Synthesis of Advanced Ceramic Materials*, D. Segal, Cambridge University Press, New York, 1989. (Introductory treatment of general ceramics processing techniques.)

9

Processing Bi-Based High-T$_c$ Super-conducting Tapes, Wires, and Thick Films for Conductor Applications

E. E. Hellstrom

This chapter is written with two goals in mind. One is to discuss the underlying materials science issues so the reader understands the chemical and microstructural development that occurs as BSCCO (Bi-Sr-Ca-Cu-O) conductors are fabricated. The main focus is on $Bi_2Sr_2CaCu_2O_8$ (2212) conductors, where there exists a more detailed understanding of the fundamental mechanisms that occur during processing compared with $Bi_2Sr_2Ca_2Cu_3O_{10}$ (2223) conductors. It provides a snapshot of the technology currently used to make 2212 conductors. The other goal is to provide sufficient experimental information for senior or graduate-level students to make 2212 conductors.

The technology in this field is advancing quickly. Hence, the processing details described here, particularly the heating schedules, may be outdated in a few years. One may wonder whether it is worthwhile reading this chapter, particularly if the processing has changed. The answer is yes. Although this chapter discusses specific processing technologies used when it was written, more important, in it we have extracted the basic underlying materials science issues that are generic to all these processes. This basic understanding is vital as it forms the basis from which new processes are developed.

Note: Throughout this chapter the terms conductor, wire, tape, and film are used. They have various meanings in the literature. Here *conductor* is used as a generic term for wires, tapes, and films. *Wires* are round (or square), and *tapes* are flat with high aspect ratios; both are sheathed in Ag. *Films* are superconducting layers (> 1 μm thick) on Ag foil.

9.1 Introduction

The majority of work around the world developing practical high-temperature superconductors focuses on the BSCCO systems. Why? Figure 9.1 provides much of the answer. It shows that the critical current density (J_c) in BSCCO conductors does not decrease precipitously with small magnetic fields as it does for Y123; thus BSCCO conductors do not suffer from the weak-link problem that plagues Y123 conductors. Also, BSCCO conductors have higher J_c in high magnetic fields (> 15 T) than conventional low-temperature superconductors such as Nb-Ti and Nb_3Sn, so they offer the possibility of building higher-field magnets than can be made with conventional materials. The BSCCO systems are not the ideal systems for conductor

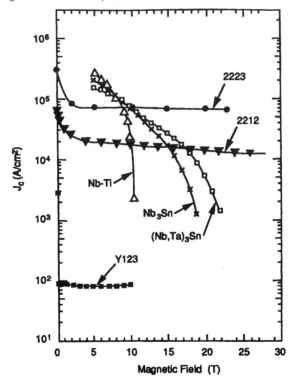

Fig. 9.1. J_c as a function of applied field for 2212 wire and 2223 tape at 4 K. Data for 2212 wire, Nb₃Sn, Nb-Ti, (Nb,Ta)₃Sn, and Y123 (YBa₂Cu₃O₇) from [1] and for 2223 wire from [2]. This plot shows the original data for melt-processed 2212 wire.

applications, however, because of their low irreversibility line (Figure 9.2). Since Y123 ($YBa_2Cu_3O_{7-x}$) and Tl-1223 ($TlBa_2Ca_2Cu_3O_{9-x}$) have higher irreversibility lines than Bi-2212 and Bi-2223, they could be used at higher temperatures than BSCCO conductors. But because the BSCCO conductors do not suffer from the weak link problems of Y123 and are much easier to process than Tl-based superconductors, they are the systems that are being developed.

The BSCCO system comprises three superconductors. They have nominal compositions of $Bi_2Sr_2CuO_6$ (2201; T_c = 7–22 K), $Bi_2Sr_2CaCu_2O_8$ (2212; T_c = 75–95 K), and $Bi_2Sr_2Ca_2Cu_3O_{10}$ (2223; T_c = 105–110 K). They all have two-dimensional crystal structures in which the supercurrent flows in the (001) Cu-O planes. Conductors are made from 2212 and 2223, both of which exist over a range of stoichiometry that does not include the nominal composition. Majewski, Sunghigh, and Hettich [5] and Müller et al. [6] have mapped out the range of stoichiometry for the 2212 phase, which is shown in Figure 9.3. The 2223 system is even more complicated than 2212, because one typically substitutes Pb for part of the Bi [7] to enhance the 2223 formation and pos-

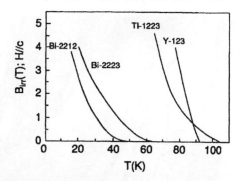

Fig. 9.2. Irreversibility lines for 2212, 2223, Y123 [3], and Tl-1223 [4]. Plot courtesy of D. K. Christen of Oak Ridge National Laboratory.

sibly increase the stability of the 2223 phase. For 2223, the general region in which the phase forms [8] has been determined, but its stoichiometry range is not known with the same detail as that for 2212. Phase-pure, polycrystalline 2212 or 2223 has not been synthesized yet.

In order to attain high J_c in BSCCO conductors, the microstructure must be homogeneous. Ideally, this means that the conductor must contain only a

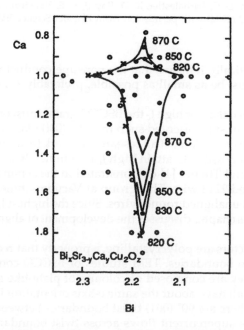

Fig. 9.3. One-phase region for the 2212 phase [5].

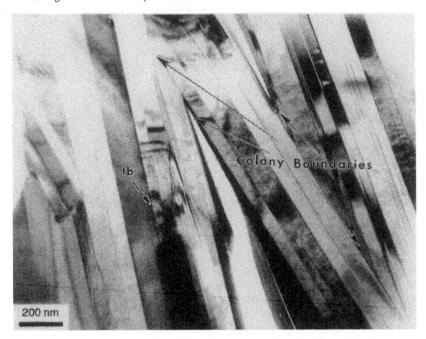

Fig. 9.4. TEM micrograph of 2212 colonies showing grains 100–200 nm thick within the colonies separated by (001) twist boundaries (tb)[9]. (Reprinted from Y. Feng, K. E. Hautanen, Y. E. High, D. C. Larbalestier, R. D. Ray II, E. E. Hellstrom and S. E. Babcock, *Physica C* **192**, 293 (1992) with permission from Elsevier Science BV, Amsterdam.)

single phase. Practically, it means that the nonsuperconducting phases in the microstructure must be as small as possible, preferably < 1–2 μm, and uniformly dispersed.

It is also assumed that for high J_c the BSCCO grains must be highly aligned, that is, the grains must be oriented such that the (00l) planes in all the grains are parallel to each other. Almost all researchers work to align the microstructure; however, it is possible to attain high J_c in completely unaligned 2212, as Heine, Tenbrink, and Thöner [1] demonstrated in the seminal report of high J_c in melt-processed 2212 wire. This group at Vacuumschmelze is one of the few that works on unaligned round wires. Since the highest J_c conductors are highly aligned, this chapter discusses the development of alignment in BSCCO conductors.

BSCCO conductors are polycrystalline, a property that requires supercurrent to cross grain boundaries. TEM studies of BSCCO conductors (Figure 9.4) reveal that they are composed of colonies of plate-like grains each 100–200 nm thick that all have about the same c-axis orientation in the colony [9]. Within a colony, there are 90° (001) twist boundaries between the individual grains. Hence, the supercurrent flows across twist boundaries within colonies and general grain boundaries between colonies.

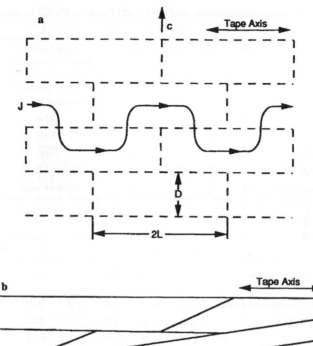

Fig. 9.5. Microstructure for the (a) brick wall [10] and (b) railroad switchyard [11] models showing a superconducting current path. In (a) D and 2L are the width and length of the 2212 "bricks". The c direction is shown. In (b) the center lines of the BSCCO colonies are shown. The colonies intersect one anoter at small-angle c-axis boundaries. The c direction is perpendicular to the lines and is in the plane of the drawing.

Two competing theories have been proposed to describe how the supercurrent transports across the boundaries. These models permit slightly different microstructures, as shown in Figure 9.5. In one model, popularly called the *brick wall model* [10], supercurrent flows between grains and colonies in the c direction, which requires the microstructure to be as highly aligned as possible to maximize the area over which current can flow in the c direction (Figure 9.5a). In the other model, known as the *railroad switchyard model* [11],

Table 9.1. Nominal compositions used for (*a*) 2212 and (*b*) Pb-2223 conductors.

(*a*)

Organization	Bi	Sr	Ca	Cu	Conductor form/comments	Ref.
Kobe Steel	2.1	2	1	1.9	Tape, powder contains 0.1 Ag	12
Mitsubishi Cable	2	2	0.64	1.64	Tape	13
NRIM	2	2	0.95	2	Film	14
Showa Electric	2.05	2	1	1.95	Tape	15
Sumitomo Metals	2	2.3	0.85	2	Tape, 1% O_2	16
Vacuumschmelze	2	2	1	2	Round wire	28

(*b*)

Organization	Bi	Pb	Sr	Ca	Cu	Reference
American Superconductor Corp.	1.8	0.3	1.9	2.0	3.1	17
Intermagnetics General Corp.	1.8	0.4	2.0	2.2	3.0	18
Sumitomo Electric	1.8	0.4	2.0	2.2	3.0	19
Toshiba	1.72	0.34	1.83	1.97	3.13	107

the current flows from (00l) planes in one colony to (00l) planes in an adjacent colony through small-angle c-axis grain boundaries between the colonies (Figure 9.5b), which permits a small amount of misalignment in the structure. Which transport mechanism occurs is not known, and it may be some time before one can achieve the high degree of alignment in 2212 or 2223 tapes that will be needed to distinguish unambiguously between them. In the meantime, work continues on increasing the alignment as much as possible.

Some general comments on the compositions used for 2212 and 2223 conductors are in order here. At present, the community uses a variety of compositions for 2212 and 2223 (Table 9.1). To make a single-phase conductor, one needs to use a composition within the 2212 or 2223 one-phase field. This suggests that the community should be working only with compositions reported to be within the one-phase regions for 2212 and 2223. Using these compositions does not, however, assure the formation of single-phase samples. The problem stems from the basic chemical reactions by which 2212 and 2223 form in the conductors. Both phases form *in situ* in the conductor from a mixture of phases, and remnants of the reactant phases exist in fully processed conductors—even those with the highest J_C [20,21]. These remnant phases may be thermodynamically unstable in the fully processed conductor, but they are kinetically stabilized. Until processing schemes are developed that eliminate these remnant phases, using an overall composition in the one-phase region will not necessarily produce a single-phase conductor.

There are similarities and differences in fabricating 2212 and 2223 conductors. The powder processing, conductor assembly, and mechanical deformation are similar for both systems. The thermal processing of these two conductors, however, is very different. As a preview, the highest J_c 2212 conductors are made by melt processing. Here all the 2212 phase initially present in the conductor melts, and new grains of 2212 form from the melt during cooling. In contrast, the highest-J_c 2223 conductors are formed by an *in situ* reaction from a mixture of 2212 and nonsuperconducting phases during a series of heating and deformation cycles. Here only a small amount of liquid develops that assists in forming 2223.

This chapter focuses on the reactions involved in development of homogeneous, highly aligned microstructures. It contains a discussion of powder processing and conductor fabrication common to 2212 and 2223. It then discusses a generic melt-processing schedule for 2212 and actual schedules used by various groups. We examine the chemical reactions that occur in 2223 conductors. Throughout, the emphasis is on developing a basic understanding of the reactions that occur and how the microstructure develops during thermal processing.

The presentation is most detailed for 2212, where there is greater understanding of how the 2212 conductor evolves during melt processing than there is of how 2223 forms. The reason 2212 is better understood is that when 2212 melts, it always forms the same phase assemblage. Thus there is a common starting point when melt processing 2212, and the subsequent phase evolution during melt processing is the same for all groups. In contrast, the 2223 conductors form from a mixture of phases that typically includes 2212; however, the other phases that must be present in this assemblage can vary widely. So, unlike 2212, one has no common starting point to begin to understand 2223, and the wide scatter in phase and microstructure evolution between groups has not led to a unified understanding of the 2223 system.

9.2 Powder Processing

What is the ideal powder to make BSCCO conductors? Of course it must contain the correct overall composition, but what about such issues as what constitutes the best phase assemblage for the powder, the grain size and shape in the powder, and the chemical purity of the powder? No adequate answers to most of these questions exist at present; in fact, these are probably not all the questions that should be asked about the powder. Nevertheless, acceptable BSCCO conductors are currently being made, and they can provide guidance on the powder characteristics that are needed. Specific powder requirements are discussed in the following sections.

A wide variety of methods have been used to prepare 2212 and 2223 powder. Several of these are listed in Table 9.2. Two common features underlie these diverse synthesis techniques: all the processes mix the cations on a microscopic scale and the phase of interest is formed at elevated temperature. Advantages and disadvantages of the methods are also listed in Table 9.2.

Solid-state synthesis, which is the simplest of the methods listed in Table 9.2, yields adequate powder. It is used here to illustrate how to prepare 2212

Table 9.2. Synthesis techniques to make 2212 and 2223 powder. References are given for the less common techniques used to make these powders.

Method	Description	Advantages	Disadvantages
Solid state reaction	Mix oxides, peroxides, carbonates, or nitrates of Bi, (Pb), Sr, Ca, and Cu. React at elevated temperature where no melting occurs. Grind sample and refire. Repeat until reaction is complete.	Simple, inexpensive technique.	Large grain size of reactants can cause slow reactions. Can have large grain size product. Can introduce impurities during grinding.
Coprecipitation	Dissolve Bi, (Pb), Sr, Ca and Cu compounds in acid. Add base to precipitate cations. Fire precipitate to yield the desired phase.	Intimate mixing of cations.	Not all cations may precipitate out at the same pH, causing segregation. Initial composition and precipitate composition may be different.
Aerosol Spray Pyrolysis	Make solution containing cations. Produce a fine mist of the solution and pass it through a hot furnace to form a powder of mixed oxides. Fire mixed powder to yield the desired phase.	Intimate mixing of cations. Product has very fine grain size (1-2μm). Product can have low C content (using nitrates).	Can lose species, particularly Pb, during pyrolysis. Powder formed in pyrolysis is not fully reacted to the desired phase.
Burn technique [103]	Form nitrate solution of cations. Add organic species, such as sugar, to solution. Heat solution to remove water, then heat powder at elevated temperature. The sugar (fuel) and	Intimate mixing of cations. Powder can have fine grain size.	Can lose species, particularly Pb, during burn process. Powder formed in burn process is not fully reacted to the desired phase. Product may contain C.

Method	Procedure		Comments
	nitrate ion (oxidant) react (i.e. burn) at elevated temperature yielding a high temperature that forms mixed oxides. Fire this powder to yield the desired phase.	Intimate mixing of cations. Powder can have low C content.	Cations may demix during freeze drying if the temperature is not carefully controlled. Nitrates present after freeze drying may melt during firing leading to large grains of nonsuperconducting phases.
Freeze drying	Spray aqueous nitrate solution of Bi, (Pb), Sr, Ca and Cu into liquid nitrogen. Collect frozen droplets and freeze dry them to remove water. Fire dried powder to yield the desired phase.		
Liquid mix method [104]	Form nitrate solution of cations then add glycol or citric acid. Heat to remove water and form polymerized gel, then heat to elevated temperature to yield the desired phase.	Intimate mixing of cations. Powder can have fine grain size.	Product may contain C.
Microemulsion [105]	Form suspension of microdroplets of aqueous nitrate solution of Bi, (Pb), Sr, Ca and Cu in oil. Add base to form precipitates. Separate precipitate from oil by washing in solvent. Fire precipitate to yield the desired phase.	Intimate mixing of cations. Powder can have fine grain size.	Product may contain C.
Sol Gel [106]	Form alkoxide solution of cations. Add water or alcohol to cross link molecules, forming gel through polymerization and condensation reactions. Heat to elevated temperature to burn off the organics and yield the desired phase.	Intimate mixing of cations. Powder can have fine grain size.	Method is better suited to making films than bulk powders.

powder. One typically begins by mixing together Bi_2O_3, $SrCO_3$, $CaCO_3$, and CuO in the desired stoichiometry. The carbonates of Sr and Ca are used because the oxides (SrO and CaO) are difficult to handle, as they pick up water or CO_2 from the atmosphere and become hydroxides or carbonates. These powders, which must have a small particle size (1–5 µm) to achieve fast reaction kinetics, are thoroughly mixed, then calcined at elevated temperature to remove the CO_2 and begin the solid-state reaction. Normally the powders are heated in air. During heating it is important to prevent local melting in the powder resulting from local variations in composition. To prevent this, one carries out the reaction in stages, beginning at low temperature (750°C). After several hours of heating, the powder is cooled, ground using a mortar and pestle, and refired at a higher temperature. This process is repeated several times, with the maximum temperature being limited to about 850°C. At this point the powder is mainly the 2212 phase. The synthesis can be done with loose powder, which gives a slower reaction but makes it easier to grind the powder between heating cycles, or with compacted powder, which accelerates the reaction kinetics but requires more grinding between heating steps that can introduce more impurities into the powder.

It is important to remove the CO_2 from the powder during the solid state synthesis. Practical experience shows that the C content in the powder used for OPIT conductors should be 0.04 wt% or lower. (The effects of higher C contents are described later in this chapter.) To remove the CO_2 during synthesis, one must have a furnace arrangement that allows the CO_2 to escape, such as a tube furnace with open ends or through which gas, typically air or O_2, can be flowed. In contrast, a typical box furnace is rather well enclosed, impeding the loss of CO_2. Solid-state synthesis of 2212 in a typical box furnace can take much longer than in a tube furnace, unless gas is flowed into the box furnace to sweep out the evolving CO_2. (Methods for removing C from reacted powders are discussed later.) One must be aware that Bi (and Pb for 2223) can vaporize from the powder at elevated temperature, thereby changing the overall composition. (Note that Pb loss is much greater than Bi loss [22].)

Balachandran et al. [23] have used low-pressure synthesis to increase the rate of solid-state reactions using carbonates. They synthesize the powder in a reduced-pressure reactor with flowing O_2 ($P_{tot} = pO_2 = 3$ torr for the BSCCO system [24]). The advantages with this synthesis are that CO_2 can be removed from the carbonates at lower temperatures than in air (which contains 0.033 vol% CO_2), the gas flow sweeps the CO_2 from the reaction chamber, allowing the reaction to continue, and the low pressure pulls the evolving CO_2 gas out of the interior of the mass of powder where it forms.

At the highest temperature used to melt process 2212, no trace of the 2212 phase remains in the melt. Thus, one may wonder whether it is necessary to react the powders completely to 2212. Or is it adequate just to remove the carbonates and have a melange of phases in the powder? From a thermodynamic point of view, one would expect that for a given overall composition, the melt state consists of an equilibrium mixture of liquid and crystalline phases independent of the initial phase assemblage in the powder. Hase et al., however, have found that J_c in their melt-processed OPIT tapes increased with increasing reaction time used for the solid-state reaction [12]. They attribute this phenomenon to having a more complete conversion to the 2212 phase in the powder with increasing time, but they have not offered a microstructural explanation for this effect. Clearly, more experiments are needed in this area.

It is best to work with fully reacted 2212 powder for several reasons. Synthesizing single-phase 2212 powder assures that the reaction has gone to completion and that no bismuth oxycarbonates (such as $Bi_2Sr_4Cu_2CO_3O_8$, which is superconducting at 30 K [25]) are present in the powder. Fully reacted 2212 powder also eliminates segregation and mixing problems that may can occur if a multiphase powder is used. Also, in a single-phase powder all of the grains have the same mechanical properties, so they deform identically, which is advantageous for the mechanical deformation of tape and wire.

9.3 Fabricating BSCCO Conductors

In this section we discuss the fabrication of BSCCO conductors and describe the characteristics required of the substrate and sheath materials.

9.3.1 Oxide Powder-in-Tube Wire and Tape

The oxide powder-in-tube (OPIT) method for making conductors, reviewed by Sandhage et al. [26], is shown schematically in Figure 9.6. The powder is packed in a tube, typically made of Ag, which is sealed and mechanically worked into the desired conductor form. Usually the Ag tube is drawn to a small diameter and then rolled into a flat tape. The tube need not be drawn,

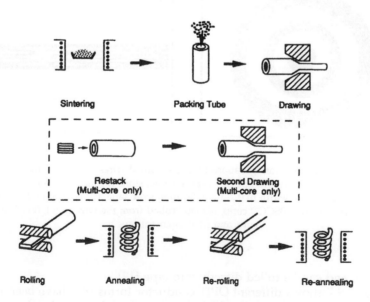

Fig. 9.6. Schematic diagram of the oxide-powder-in-tube (OPIT) method to make wires and tapes [27]. (Reprinted from L. R. Motowidlo, P. Haldar, S. Jin and N. D. Spencer, *IEEE Trans. Appl. Supercond.* **3**, 942 (1993), © 1993 IEEE.)

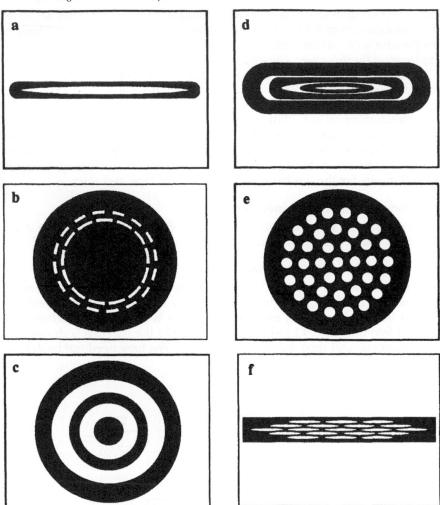

Fig. 9.7. Representative configurations for Ag-sheathed wires and tapes. Black regions are Ag, and white regions are BSCCO. Typical dimensions for tapes are a total thickness of 100–250 μm and a width of 0.2–1 cm: (a) monocore tape, (b) coaxial multifilament wire, (c) wire with two BSCCO cores, (d) tape rolled from the wire in (c), (e) 37 filament wire, and (f) 19 filament tape.

however, and can be rolled directly into tape [12].

Figure 9.7 shows different OPIT conductor forms that have been made. Configurations are shown in which the outer Ag tube contains the powder plus a solid Ag rod, or multiple concentric Ag tubes filled with powder. Figure 9.7 also shows multifilamentary round and rectangular wires, as well as

Fig. 9.8. Longitudinal cross section of a BSCCO tape showing undulations in the Ag/ oxide interface. This is known as sausaging.

flat tapes. These are made by restacking small diameter OPIT wire into a Ag tube and mechanically working this assemblage. The rectangular wire is made using a turk's head die or a series of rectangular dies.

The mechanical deformation of the Ag/oxide powder composite is an important step in fabricating OPIT conductors, but it is not well understood. As

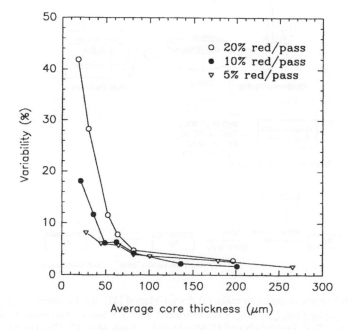

Fig. 9.9. Variation in the uniformity of the core thickness as a function of core thickness for a 2223 tape rolled with different reductions in thickness per pass [29]. The sausaging becomes significant when the core thickness is reduced below 50–80 μm.

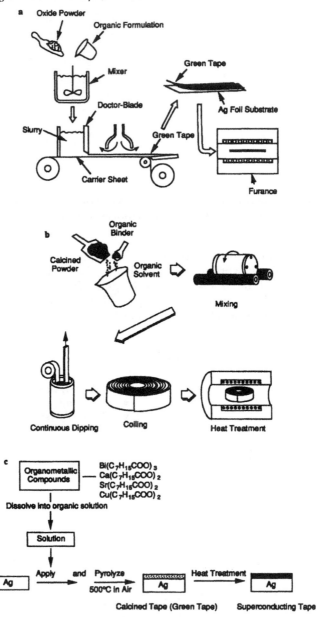

Fig. 9.10. Schematic diagram to make (a) doctor-bladed [31], (b) dip-coated [32], and (c) organic precursor films [33]. (Fig. 9.10a reprinted from J. Kase, T. Morimoto, K. Togano, H. Kumakura, D. R. Dietderich and H. Maeda, *IEEE Trans. Mag.* **27**, 1254 (1991), © IEEE. Fig. 9.10c reprinted from T. Hasegawa, T. Kitamura, H. Kobayashi, F. Takesha, H. Kumakura, H. Kitaguchi and K. Togano, *Physica C* **190**, 81 (1991) with permission from Elsevier Science BV, Amsterdam.)

Table 9.3. Formulation for the slurry used to make doctor-bladed films [34].

Component	Material	Fraction (wt%)
Oxide powder	2212	22
Solvent	trichloroethylene	68
Binder	polyvinyl butyral	6
Dispersant	sorbitan trioleate	5

discussed later, to induce high alignment in BSCCO tapes, the oxide core in the tape should be thin (40–50 μm or less); hence, the powder must have small particle size (1–3 μm). As mentioned earlier, ideally the powder should contain a single phase so that each grain has the same mechanical properties and the powder deforms uniformly. For 2212 one can use a single-phase powder, but with 2223, a multiphase powder mixture is required. The powder must be uniformly packed in the Ag tube, but it is not clear how densely the powder should be packed into the tube initially. Tenbrink et al. [28] compared the final oxide density in wires made from 2212 powder that was hand packed into a Ag tube (low packing density) and a rod of 2212 powder that had been isostatically pressed, then loaded into a Ag tube (high packing density).

After drawing, the oxide core for both wires was 70% dense. Although the final core density was the same for both wires, the Ag:superconductor volume ratio was higher for loosely packed powder; hence, this difference may have translated into a thicker Ag sheath in wire made from the loosely packed powder. This difference in sheath thickness has important implications for conductor design where the Ag:superconductor ratio is a critical parameter.

The Ag sheath is softer than the oxide core, thus causing difficulties with the deformation of the Ag/oxide composite. A serious problem that can occur in the latter stages of rolling a tape is forming an undulating oxide/Ag interface, which is known as *sausaging*. Figure 9.8 shows a longitudinal cross section of a sausaged Ag-sheathed tape. It has been found empirically that after the oxide core is reduced to some minimum thickness during rolling, additional rolling causes it to sausage. A hypothesis is that sausaging occurs when the oxide powder in the core starts to deform as blocks of oxide rather than individual grains. At this point, the rolling stress shears the core into smaller blocks that deform independently in the softer Ag sheath causing the undulations in the Ag/oxide interface. Figure 9.9 shows that for 2223 tapes, the variation in core thickness, which indicates the onset of sausaging, began to increase after the core thickness was reduced to 50–80 μm [29]. The minimum core thickness that appears to be achievable without significant sausaging is about 40–60 μm. As more is understood about the deformation process, it is anticipated that optimizing such parameters as the particle size in the powder, powder packing density, and deformation parameters (e.g., roll size, roll speed, reduction per pass) will lead to thinner cores without sausaging.

During rolling, the plate-like BSCCO grains, present in 2212 and 2223 tapes, can be mechanically aligned with their (001) basal planes parallel to the rolling direction. As discussed below, this mechanical alignment is lost when the 2212 tapes are melt processed, but it is retained during the processing of 2223 tapes.

9.3.2 2212 Films on Ag Foil

Melt-processed 2212 films on Ag foil have the highest J_c ($2.6 \bullet 10^5$ A/cm^2, at 4 K and 12 T) of any of the 2212 conductors [20]. (2223 films are not made because the PbO is lost during the thermal processing [30].) Figure 9.10 shows the processes used to make doctor-bladed [31], dip-coated [32], and organic precursor [33] films. The powder for doctor-bladed and dip-coated films can be made using any of the processes listed in Table 9.2. This powder is added to a mixture of organic compounds, an example of which is listed in Table 9.3, to make a slurry.

In the simplest form of the doctor-blade process, one makes a green film (i.e., film before heat treatment) from the organic/powder mixture by pouring a pool of the slurry on a flat surface, such as a piece of glass, then leveling the slurry with a straight-edged blade that is held above the flat surface at the desired film thickness. (Using a blade to control the film thickness is called the doctor-blade or tape-casting process. Sophisticated doctor-blade systems are commercially available. In a laboratory, however, one can simply use a piece of plate glass, two thin wires 100–150 μm dia. laid parallel to one another on the glass to control the thickness of the film, and the edge of a microscope slide for the straight-edged blade.) The film is allowed to dry, cut into strips that are placed on the Ag foil, then melt processed.

For *dip-coated films*, the Ag foil is passed through the organic/powder mixture, which adheres to the foil [32]. The thickness of the film is controlled by changing the organic compounds, modifying their proportions in the mixture, and adjusting the solids loading in the mixture. After passing through the organic/powder mixture, the coating is dried, the organics are burned out, and the film is melt processed.

For the *organic precursor films*, a solution of organometallic compounds (metal octylates) of Bi, Sr, Ca, and Cu is deposited on the Ag foil, the solvent is burned out, and the process is repeated until the desired layer thickness is built up [33]. The film is then melt processed.

A 2212 film can also be made by painting 2212 powder onto a Ag foil [35]. Here the powder is mixed with an organic liquid with a high vapor pressure, such as butanol, plus a dispersant. This slurry is brushed onto the foil, then melt processed.

9.3.3 Sheath and Substrate Material

Figure 9.7 shows that the Ag sheath is an integral part of the conductor configuration. The Ag provides a normal-conducting current path in case the superconductivity is lost. In tapes and wires the Ag also protects and strengthens the brittle oxide superconductor, and in films it is the supporting substrate on which the film is formed.

The sheath material used for the BSCCO conductors must fulfill several

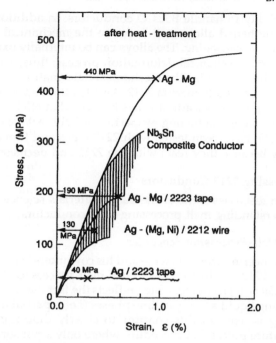

Fig. 9.11. Strength of fully processed Ag-sheathed BSCCO conductors: 2223 tapes with pure Ag and Ag-Mg alloy sheaths [107] and Ag-sheathed 2212 wire [36]. The strength of a Nb₃Sn composite conductor is shown for comparison. The data for 2212 wire (from [36]) were plotted using a 0.1% offset from the modulus for Ag-Mg (from [107].)

important chemical requirements. It must not react with the BSCCO during the high-temperature heat treatment, since such a reaction could lead to a decrease in the superconducting properties. It should not oxidize during the heat treatment. And since the oxygen content of BSCCO varies with temperature, oxygen must be able to pass through the sheath. At present, only Ag fulfills all these requirements. It is not the ideal material, however, because its cost is relatively high, its strength is low (making it difficult to handle fully processed conductors without damaging them), and its coefficient of thermal expansion does not match that of the BSCCO phases.

It is well known that the strength of Ag can be increased by solid solution strengthening and precipitation hardening. Tenbrink et al. [36] first reported making 2212 wire using Ag alloys containing small amounts of Ni and Mg. The Ni and Mg provide solid solution strengthening for the alloy; and when the BSCCO conductor is heat treated in an oxidizing atmosphere, these alloying elements oxidize internally, forming oxide precipitates that strengthen the alloy. Motowidlo, Galinski, and Haldar [37] used Ag-Al alloys, which have the same strengthening mechanisms. The strength of fully processed tapes and wires made with these alloys is comparable with conventional Cu-sheathed Nb₃Sn conductors, as shown in Figure 9.11. This strengthening significantly

improves the ability to handle BSCCO conductors. In addition, it is anticipated that strengthened alloys will improve the mechanical deformation,
helping to minimize sausaging. The alloys can be internally oxidized before,
during, or after the mechanical deformation process; thus, a wide range of
options exist to help optimize the mechanical deformation.

Since MgO is essentially inert in 2212, Ag-Mg alloys appear to be a good
choice for a strengthened sheath. It is not known what effect Al from Ag-Al
alloys has on the superconducting properties of 2212. Alloys of Ag-Mn [38]
and Ag-Cu [38,39] have been tested with 2212 tapes and films, but they are
not satisfactory because they react with the 2212 and decrease the J_c.

9.4 Melt-Processing 2212 Conductors

This section discusses some of the basic materials science issues fundamental to understanding melt processing 2212 conductors.

9.4.1 Generic Melt-Processing Schedules

Since the pioneering work of Heine and his coworkers [1] on 2212 conductors, almost all 2212 conductors have been melt processed. This method is
also called *partial melt processing*, which reflects the fact that 2212 melts incongruently, forming liquid and crystalline phases (i.e., a partial melt). The term
melt processing is used in this chapter to clearly distinguish it from a
lower-temperature processing procedure where only a portion of the 2212 is
melted, which is called *liquid-assisted processing* [40]. In general, it is difficult
to make single-phase samples when solidifying an incongruently melting phase
because of the sluggishness of the diffusion and chemical reactions that must
occur to form the phase of interest on cooling. Much of the discussion on
processing 2212 conductors deals with the phases from which 2212 forms on
cooling, the problems they cause, and the methods that have been devised to
alleviate them.

Many different heating schedules are currently used to melt process 2212.
The essential features of all these, however, are captured in the generic
melt-processing schedules for 2212 tapes and wires, and for films such as that
shown in Figure 9.12. To aid in understanding the processes that occur during melt processing, we have divided this schedule into four regions [41]. The
development of the phase assemblage and microstructure within each region
will be discussed. The processes that occur in each region are for the most
part generic to 2212 tapes, wires, and films; however, differences in the behavior of tapes, wires, and films during melt processing will be noted. Since
most melt processing is currently done in air, the discussion below empha-
sizes the processes that occur in air.

The following is a synopsis of melt processing. In Region 1 the 2212 powder melts incongruently, forming liquid and nonsuperconducting crystalline
phases. Region 2 begins when 2212 starts to form from the melt during cooling. It encompasses the formation of 2212 from the melt, growth and alignment of the 2212 grains, and reaction of the nonsuperconducting phases. In
Region 3 the formation and alignment of 2212 are maximized, and in Region
4 the conductor is cooled to room temperature. Regions 1 and 2, which are

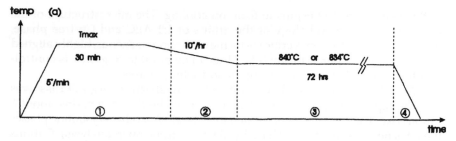

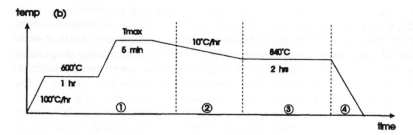

Fig. 9.12. Generic 2212 melt-processing schedules used for (a) tapes and wires and (b) films. The schedules have been divided into four regions that are described in the text.

intimately connected and are most critical for forming a homogeneous highly aligned microstructure, are discussed in the most detail below.

At the elevated temperatures used for melt processing, Bi can evaporate from the 2212 melt as a bismuth oxide [22]. In tapes and wires, the Ag sheath prevents the Bi from evaporating, but in films, the Bi can readily evaporate. To minimize Bi loss during melting, Shimoyama et al. [20] placed the film in a semi-sealed alumina container with a mixture of $Bi_2Al_4O_9$-Al_2O_3 powders, from which bismuth oxide vaporized, setting the bismuth oxide partial pressure in the system.

9.4.1.1 Region 1 – Melting

In Region 1 the conductor is heated above the melting point of 2212, held at this temperature for a short time, then cooled. Region 1 ends when 2212 begins to form from the melt on cooling. For films, the organic species used to form the film must be burned out at 400–600°C (Figure 9.12b). On melting, the 2212 powder releases oxygen, which is not a problem for films that are open to the atmosphere, but can be a problem for tapes and wires, where it can cause the Ag sheath to bubble. Methods to alleviate such bubbling are discussed later.

In air, 2212 powder in the conductor melts incongruently at about 870°C into liquid and $(Sr,Ca)CuO_2$, hereafter called 1:1 alkaline earth cuprate (1:1 AEC), and a Cu-free phase of nominal composition $Bi_2(Sr,Ca)_4O_x$. (The presence of Ag depresses the melting point of 2212.) These same phases are present in the melt when 2212 begins to form on cooling. The microstructure of the melt, that is, the morphology of the grains of 1:1 AEC and Cu-free phase, when 2212 begins to form is critical to achieve the homogeneous, highly aligned microstructure required for high J_c. Thus a key point in Region 1 is controlling the morphology of the nonsuperconducting phases.

When 2212 melts, a small amount of the Ag sheath or Ag foil dissolves into the liquid. Quench studies [42] show that the liquid contains about 5 cation% Ag. The Ag is not incorporated into the 2212 or other crystalline phases (i.e., it is not detected in the phase by electron probe microanalysis). Cations that dissolve into the liquid from a Ag alloy can be incorporated in the 2212, however, and may adversely affect its superconducting properties.

Table 9.4 lists the phases present in the melt as a function of temperature in air [43]. These were determined from oil-quenched tape samples that were analyzed at room temperature by using X-ray diffraction, energy dispersive spectroscopy on an SEM, and wavelength dispersive spectroscopy with an electron probe microanalyzer. Fast quenching is needed to preserve the high-temperature phase assemblage. If the quench is too slow (e.g., using air cooling), 2201 forms on cooling [42]. The phase assemblages in Table 9.4 have also been observed in quenched films [44] and in films using high-temperature x-ray diffraction [45,46]. The 1:1 AEC and Cu-free phase that form when 2212 melts are replaced in the liquid by $(Sr,Ca)_2CuO_x$ (2:1 AEC), and CaO with increasing temperature. The maximum temperature in these studies was limited to 925°C, because Ag melts in air at about 930°C. Since the 2212 composition lies in the primary phase field for CaO, it is the last crystalline phase present in the melt on heating [47].

In order to develop a homogeneous, highly aligned microstructure, the crystalline phases in the melt (1:1 AEC and the Cu-free phase) should be as small as possible when 2212 begins to form. Throughout this chapter references to small and large grains of 1:1 AEC or Cu-free phase refer to the grain size relative to the thickness of the oxide core in the tape (40–60 μm) or film (< 25 μm). Small grains are typically less than 10 μm, and large grains are a few tens to hundreds of micrometers.

Figure 9.13 shows that J_c depends on the maximum processing temperature, with the optimum maximum temperature ranging from 890°C to 900°C for tapes and films [20,48,49]. Hellstrom, Ray, and Zhang [43] have correlated this optimum processing temperature with variations in the microstructure of the melt at the temperature where 2212 begins to form on cooling (870°C). Specifically, they found that on cooling to 870°C the grain size of the Cu-free phase varied with the maximum processing temperature. Table 9.4 shows that the Cu-free phase is no longer present in the melt above 905°C. This disappearance is not sudden; Zhang, Ray, and Hellstrom [50] reported that the amount of Cu-free phase in the melt decreased with increasing temperature and was completely gone above 905°C. If the Cu-free phase is present in the melt at the maximum processing temperature (< 905°C), the grains of Cu-free phase grow in size by a small amount on cooling.

As Figure 9.14a shows, however, they are still relatively small at 870°C. In contrast, when the maximum processing temperature is high enough that the

Table 9.4. Phases present in 2212 tape oil quenched from various temperatures during heating. The samples were heated to the indicated temperature and held for 30 min before being quenched. Samples were heated in air [43]. (Reprinted from E. E. Hellstrom, R. D. Day II and W. Zhang, *Applied Superconductivity* 1, 1535 (1993) with permission from Elsevier Science Ltd, UK.)

Quench Temp. (°C)	2212	14:24 AEC[1]	Liquid[2]	1:1 AEC[3]	Cu-free[4]	2:1 AEC[5]	CaO
<856	X	X					
864	X	X	X	X(1)[6]			
869			X	X	X		
875			X	X	X		
878			X	X	X		
880			X	X	X		
885			X	X	X	X(2)[6]	
890			X	X	X	X	
895[7]			X	X	X	X	X(5)[6]
900[7]			X	X	X	X	X
905			X	X		X	X
910			X	X		X(3)[6]	X
915			X			X	X
920			X				X

[1]14:24 AEC = $(Sr,Ca)_{14}Cu_{24}O_x$. [2]The quenched liquid contained Ag. [3]1:1 AEC = $(Sr,Ca)CuO_2$. [4]Cu-free = $Bi(Sr,Ca)_2O_x$. [5]2:1 AEC = $(Sr,Ca)_2CuO_3$. [6]Phases that have a number in parenthesis were only occasionally found in the tapes. The number indicates the number of grains of that phase that were observed. The tapes contained many crystals of the other phases, and at least ten crystals of each phase were analyzed. [7]The number of phases at 895 and 900°C violates the phase rule. This may have occurred because the system had not come to complete thermodynamic equilibrium in 30 min. However, in this temperature range, the amount of Cu-free phase diminished with increasing temperature, whereas the amount of 2:1 AEC increased with increasing temperature.

Cu-free phase is not present in the melt (> 905°C), then the grains of Cu-free phase nucleate on cooling and grow to be very large, as shown in Figure 9.14b. This behavior has also been observed in tapes and films [44]. Since having smaller grains of the Cu-free phase in the melt when 2212 forms is more conducive to developing a homogeneous, highly aligned microstructure, these experiments show that the maximum processing temperature should be limited to about 895–900°C. This prevents the formation of large grains of the Cu-free phase and a subsequent decrease in J_c because the microstructure is not homogeneous or highly aligned.

In contrast to the Cu-free phase, the grain size of the 1:1 AEC on reaching 870°C is not affected by the maximum processing temperature (Figure 9.14). Figure 9.15 shows the microstructure of a 2212 tape as a function of time at 890°C. This series of micrographs graphically illustrates the relative growth

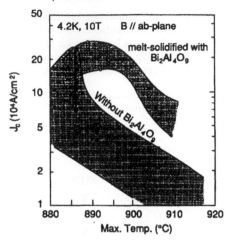

Fig. 9.13. J_c in 2212 films as a function of maximum processing temperature [20]. Data are shown for films processed with and without $Bi_2Al_4O_9$-Al_2O_3 powder, which is used to prevent Bi loss.

rates of 1:1 AEC and the Cu-free phase. It shows that in the four minutes needed to heat (5°C/min) from 870°C, where 2212 melted forming liquid 1:1 AEC and the Cu-free phase, to 890°C, where the tape was quenched, the grains of 1:1 AEC had grown very large. They continued to grow with time at 890°C. In contrast, the grains of Cu-free phase were relatively small even after 60 min at 890°C.

Figure 9.16 shows that the size of the 1:1 AEC is smaller in films than in tapes that have had identical heat treatments, indicating different behavior for this phase in the two types of conductors. What causes this is not known, but the difference in behavior is reflected in the melt processing used for tapes and films (Figure 9.12). (This behavior is discussed further below.)

Many melt-processing schedules in current use minimize the time the conductor spends in the melt state, by holding for a short time at the maximum processing temperature, then rapidly cooling to just above the temperature where 2212 begins to form. From the discussion of the processes that occur in Region 1, one can understand that the short time in the melt state is meant to minimize grain growth of 1:1 AEC. Since 1:1 AEC is large even after very short times in the melt state (Figure 9.15), however, the short time in the melt state does not prevent the formation of large grains of 1:1 AEC, but it does prevent excess growth of this phase. For films, an additional reason to minimize the time in the melt state is to limit Bi loss, even when using the $Bi_2Al_4O_9$-Al_2O_3 powder.

In summary, small grains of the Cu-free phase are present in the melt when the maximum processing temperature in air is limited to < 900°C. At present, however, it is not possible to achieve small grain size of the 1:1 AEC in the melt in tapes and wires. In films, the 1:1 AEC grains are naturally smaller

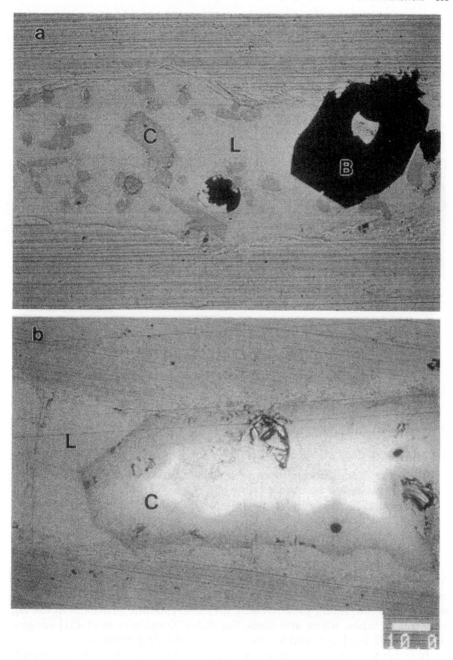

Fig. 9.14. The microstructure of 2212 tape that was heated to (a) 910°C and (b) 890°C for 10 min, cooled at 10°C/hr to 870°C, then quenched in oil. B = 1:1 AEC, C = Cu-free phase, L = quenched liquid.

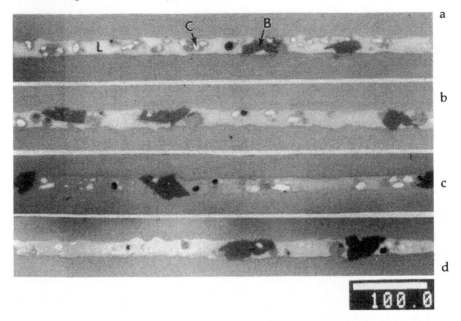

Fig. 9.15. Microstructure of 2212 tapes quenched after various times at 890°C. The 1:1 AEC grains (B) are large at 0 min and grew exensively with time, whereas the Cu-free grains (C) grew only a small amount. Tapes were quenched into oil after (a) 0, (b) 10, (c) 30, and (d) 60 min at 890°C. Heating rate was 5°C/min.

than in tapes and films.

9.4.1.2 Region 2 – Forming and Aligning 2212

In Region 2, 2212 forms from the melt and becomes aligned. These processes are discussed separately below.

Forming 2212. Region 2 begins when 2212 starts to form from the melt. Since 2212 melts incongruently, on cooling it should form by a reaction between the liquid, 1:1 AEC, and Cu-free phase. However, micrographs of samples quenched just after 2212 begins to form (e.g., Figure 9.17), show no evidence of 2212 forming by a liquid/solid reaction. Rather, they indicate that 2212 forms directly from the liquid. Hence, the 1:1 AEC and Cu-free phase must be consumed in the latter stages of cooling. Furthermore, there is no indication of 2212 nucleating preferentially on the crystalline phases or at the Ag interface in tapes, wires, and films. For films, Kumakura et al. [51] suggest that the 2212 nucleates and aligns from the oxide/air interface.

In order to form a homogeneous, highly aligned microstructure, the melt must contain small rather than large grains of 1:1 AEC and Cu-free phase when 2212 begins to form, for the following reasons: (1) small grains can be uniformly distributed throughout the conductor on a micrometer scale, so that the average diffusion distance between grains of 1:1 AEC, the Cu-free

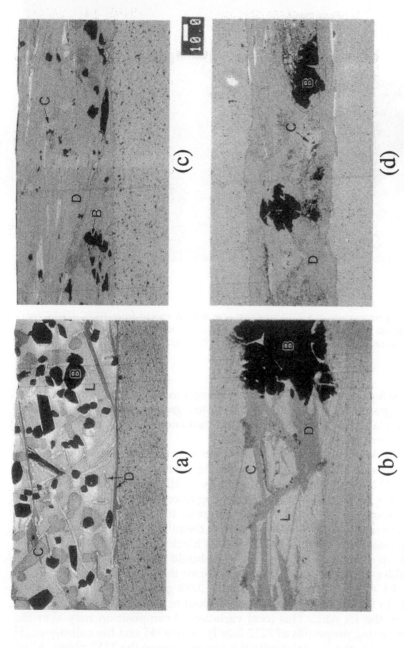

Fig. 9.16. Microstructure of 2212 tapes and doctor-bladed films that had identical melt processing and were oil quenched during cooling. (Processing: 890°C for 10 min, cool at 10°C/hr in air.) (a) film quenched at 870°C, (b) tape quenched at 870°C, (c) film quenched at 840°C, and (d) tape quenched at 840°C. B = 1:1 AEC, C = Cu-free phase, D = 2212, L = quenched liquid.

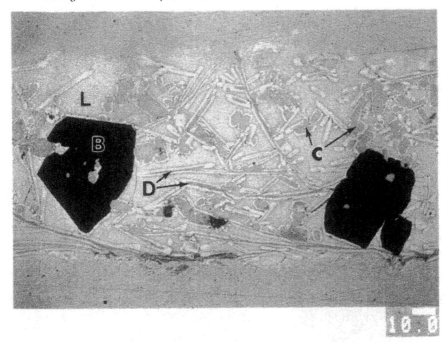

Fig. 9.17. Microstructure of a 2212 tape that was rapidly cooled from 890°C (240°C/hr) and quenched into oil from 868°C. The 2212 grains are randomly oriented throughout the core. B = 1:1 AEC, C = Cu-free phase, D = 2212, L = quenched liquid.

phase, and liquid, which must ultimately react to form 2212, is shorter than with large grains; (2) small grains can react to a greater extent in a given time period than large grains; and (3) small grains do not interrupt the developing 2212 alignment as much as large grains.

1:1 AEC does not coexist in thermodynamic equilibrium with the 2212 phase in the solid state. Thus, remnant 1:1 AEC that is always present in fully processed 2212 conductors should have chemically reacted to form the 2212 phase during cooling. Therefore, the cooling rates used to melt process 2212 conductors are too fast for the 1:1 AEC to react to form 2212. The problems with having nonsuperconducting phases in the fully processed conductor are that the phases are too large to pin flux, they block the supercurrent path, and they lock up a portion of the cations that should have reacted to form 2212 [9]. Consider a fully processed 2212 conductor that contains large grains of remnant 1:1 AEC. The composition of the 2212 grains may vary with their proximity to the 1:1 AEC grains, being Bi-poor near the 1:1 AEC and Bi-rich away from the 1:1 AEC. This local variation in composition may alter the superconducting properties of 2212 locally: Majewski and his colleagues [5] have shown that the T_c varies with the composition of the 2212 phase. Alternatively, the system may compensate for the cations tied up in the 1:1 AEC

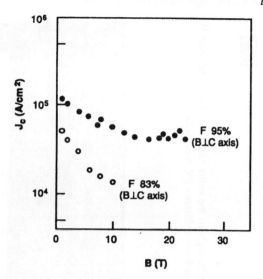

Fig. 9.18. J_c as a function of alignment in 2212 tape [52]. The f-factor (Lotgering factor) [53] is a measure of the alignment determined from X-ray peak intensities in powder diffraction patterns: 0 = completely random, 1 = completely aligned.

by forming other phases, such as 2201, that would not exist in the system if the 1:1 AEC were not present.

Even though the melt from which 2212 forms contains both 1:1 AEC and the Cu-free phase, the dominant remnant phase in fully processed 2212 conductors is 1:1 AEC; usually very little Cu-free phase is present. This situation shows that the reaction kinetics for 1:1 AEC are slower than for the Cu-free phase, and it also suggests there are two reaction paths involving the consumption of the Cu-free phase. One reaction must involve 1:1 AEC in the formation of 2212, whereas the other does not involve 1:1 AEC.

Surprisingly, the reaction kinetics for 1:1 AEC are much faster in films than in wires and tapes. Figure 9.16 shows the microstructure of tape and film with the same powder and identical processing. In the tape the 1:1 AEC remained large, whereas in the film it was smaller initially and decreased in size during cooling. This difference may be due to oxygen effects, possibly the inability to supply an adequate amount of oxygen to the oxide core in Ag-sheathed tape and wire compared with films that are open to the atmosphere.

Aligning 2212. The original work on melt-processed 2212 wire by Heine and coworkers [1] used round wire 1.5 mm in diameter, in which there was no evidence of 2212 alignment. This group has continued to work on unaligned wires. For tapes and films, however, it has been found that slow cooling (< 10°C/hr) through Region 2 aligns the 2212 grains, and J_c increases with increasing alignment, as illustrated in Figure 9.18. In two-dimensional configurations, such as tapes and films, the aligned grains have their c-axis perpen-

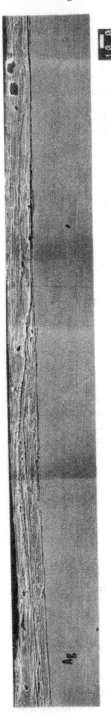

Fig. 9.19. Cross section of a thin (20 μm) 2212 doctor-bladed film showing high alignment throughout the entire film [44]. (Reprinted from W. Zhang and E. E. Hellstrom, *Physica C* **218**, 141 (1993) with permission from Elsevier Science BV, Amsterdam.)

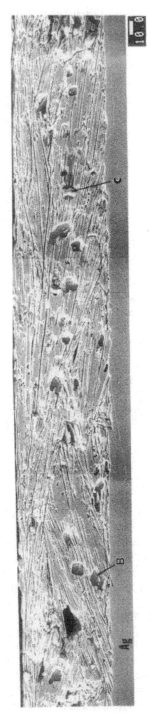

Fig. 9.20. Cross section of a thick 2212 doctor-bladed film showing a well-aligned layer at the oxide/air interface and a poorly aligned layer at the Ag/oxide interface [44]. B = 1:1 AEC, C = Cu-free phase. (Reprinted from W. Zhang and E. E. Hellstrom, *Physica C* **218**, 141 (1993) with permission from Elsevier Science BV, Amsterdam.)

dicular to the plane of the tape or film. In one-dimensional round wires, they have their c-axis radially aligned perpendicular to the long axis of the wire. This section examines the alignment mechanism and methods to achieve high alignment.

Figure 9.19 shows the cross section of a highly aligned doctor-bladed film on Ag foil. The 2212 was etched to show the 2212 colonies. The (001) planes of 2212 are parallel to the plane of the Ag foil.

Plate-like 2212 grains grow from the melt because growth of 2212 is faster in the a and b directions than in the c direction. This two-dimensional growth is critical for the alignment that develops during cooling. When melt processing a tape or film, one can easily imagine that the plate-like 2212 grains begin to align at the flat Ag/oxide interface and the alignment proceeds from this interface into the oxide layer as the grains grow. The microstructure of tapes and films quenched during the initial stages of 2212 growth shows, however, that it does not begin in an orderly fashion at this interface. Rather, 2212 appears to nucleate and initially grow in completely random directions, as shown in Figure 9.17. There is no preferred 2212 growth or alignment along the Ag/oxide interface. Figure 9.20 shows that the alignment at the Ag interface in fully processed thick films (75–100 µm) can be very poor, a result that supports the observation that the alignment does not begin at the Ag interface.

It is known experimentally that the cooling rate and thickness of the oxide melt affect the alignment: slower cooling and thinner oxide yield higher alignment. Although the alignment mechanism is not known in detail, researchers believe that an essential feature of the alignment process is the selective growth of properly aligned 2212 grains [54–56]. Thus, the 2212 grains that initially nucleate and grow with their basal planes nearly parallel to the plane of the tape or film can grow to large size, whereas grains aligned at other orientations cannot grow large. (This is called the constrained-volume model for alignment [54].) For a grain to grow to a specified length, for example, 100 µm, geometry specifies the maximum allowable angle (misorientation angle) between the plane of the tape or film and the (001) planes of the 2212 grain. The misorientation angle for a given grain size decreases with decreasing oxide thickness. As the cooling rate decreases, fewer 2212 nuclei form per unit time, so each grain can grow to a larger size. Thus, for a given cooling rate, the 2212 grains will be larger and more highly aligned with decreasing oxide thickness. This same alignment mechanism is thought to operate in round wires, where the one-dimensional geometry allows grains oriented with their (001) planes nearly parallel to the axis of the wire to grow to large size. The result is a microstructure where the grains have their c-axes radially aligned normal to the wire axis. (More information about the development of 2212 alignment is presented later in this chapter.)

The alignment mechanism just described requires that the large, properly oriented grains grow at the expense of the smaller, misoriented grains. Soylu et al. [57] have reported that this growth mechanism occurs in samples in which they used MgO fibers to align the growing 2212.

When one observes a fully processed 2212 tape, there are often regions of highly aligned 2212 at the Ag interface. These highly aligned regions formed not during the initial stages of 2212 growth, but later in the process. In other regions the 2212 is poorly aligned at the interface. High-resolution TEM studies of fully processed tape show that the 2212 (001) planes are parallel to the Ag, irrespective of the macroscopic alignment of the 2212 grain with respect to the plane of the tape [58]. In addition, a half cell of 2201 is always found between the Ag and 2212 [58].

Films that are < 20–25 μm thick align easily. In thicker films, the alignment is usually not uniform throughout the film, being higher close to the free surface than near the Ag interface. It has been suggested that in films the 2212 growth and alignment begin at the free surface and proceed into the oxide layer [51]. Micrographs of quenched films show that the free surface plays an important role in the alignment process. But they do not show conclusive evidence that 2212 nucleates at the free surface or that aligned 2212 grows in from this interface. Rather, the micrographs suggest that the 2212 initially nucleates and grows randomly in the film, just as it does in tape (Figure 9.17). Since films, like tapes, are two dimensional, one would expect a higher alignment in thinner films because of the smaller misorientation angle a grain could have and still grow to a given length. In addition, as the aligned grains grow in films, they may rotate misaligned grains into alignment, which, having nothing to attach to at the free surface, are less well anchored than in Ag-sheathed conductors.

The free surface in films is also important in film alignment. As seen in Figure 9.20, thick films (50–100 μm) can have a 20–25 μm thick layer of aligned 2212 at the free surface, with poorly aligned 2212 below this layer [44]. The aligned layer at the surface in the thick films shows that the surface induces alignment; however, since this 20–25 μm layer at the surface is not as well aligned as a 20–25 μm thick film, the alignment induced by the free surface is not the dominant alignment mechanism in thin (< 25 μm) films.

In general, the alignment and J_c are higher in films than tapes. The reasons are thought to be the smaller size of the crystalline phases in the melt in films, the faster reactivity of 1:1 AEC in films (which allows more of the 1:1 AEC to be consumed forming more 2212 during cooling), and the combination of constrained volume alignment and surface induced alignment in films.

Ray has developed a method for attaining high alignment in 2212 tape, wires, and films (> 50 μm) that also allows much of the large 1:1 AEC in tapes and wires to react [55]. This method, called *step solidification melt processing*, is based on the understanding of the processes that occur in Regions 1 and 2. It is presented later.

As the discussion on alignment suggests, the Ag does not induce 2212 to align. Indeed, 2212 has been aligned on Au foil and MgO single crystals by raising the maximum processing temperature in Region 1 to 905°C [44].

9.4.1.3 Region 3 – Maximizing the Formation and Alignment of 2212

The purpose of Region 3 (Figure 9.12) is to maximize the formation and alignment of 2212. The temperature used ranges from 820–850°C, and times for tapes and wires range from a few hours to one hundred hours. In tapes and wires that contain large grains of 1:1 AEC, annealing for times up to 100 hr at ≤ 850°C does not significantly reduce the size of 1:1 AEC. Thus the re-action to form 2212 cannot go to completion. Ray and Hellstrom [59] studied tapes that were fast cooled (240°C/hr) from about 920 to 840°C and found that there was little alignment when the tape initially reached 840°C, but the amount of 2212 and its alignment increased during the extended anneal at 840°C. Tapes that were slow cooled (< 10°C/hr) over this same temperature range, however, were well aligned on reaching 840°C; and during the extended anneal at 840°C, little additional 2212 formed, and the alignment did not increase significantly.

The processes that occur in Region 3 have not been studied in detail yet. It is known that the 2212 grains contain intergrowths of the 2201 phase, and Heeb et al. [60] reported that the number of 2201 intergrowths in pellet samples decreased with extended annealing. Such a decrease in 2201 intergrowths may occur in Region 3.

9.4.1.4 Region 4 – Cooling to Room Temperature

In Region 4 the conductor is cooled to room temperature. This cooling is critical. Figure 9.21 illustrates that high J_c requires fast cooling, with a rate of 1200°C/hr yielding the highest J_c in films [61]. This same enhancement of J_c using fast cooling rates has also been observed with tapes and wires [62,63]. At very high cooling rates the conductors may crack, leading to low J_c [61]. The 2212 is reported to decompose, forming 2201, with slow cooling rates (< 300°C/hr) [61]. The details of the decomposition mechanism and the tem-perature range over which it occurs are not well defined, however. The ne-cessity for fast cooling may present technical difficulties, because large coils needed for practical applications may be difficult to cool uniformly at high rates.

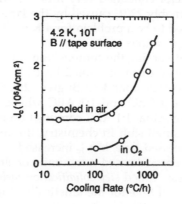

Fig. 9.21. J_c in doctor-bladed 2212 film as a function of cooling rate in air and 1 atm O_2 [61].

During cooling, 2212 can also pick up additional oxygen, depressing T_c [64,65]. Region 4 may also include a low-temperature anneal (400–650°C) in inert gas (not shown in Figure 9.12) that increases T_c by reducing the oxygen content in the 2212 [64]. An example of such an anneal for film is 650°C in N_2 for 2 hr [61]. However, the 2212 phase can decompose during such an extended low-temperature anneal in inert gas [65,66]. Noji et al. [67] observed the growth of 2:1 AEC on the surface of 2212 films after annealing for 15 hr at 500°C in N_2.

By comparison, work on single crystals of 2212 annealed in pure O_2 at 450–650°C showed they also decomposed [68], but surprisingly, the AC susceptibility measured after annealing in O_2 showed that the T_c transition was narrow ($\Delta T_c \sim 2.2$ K) and the T_c values were high (85–88 K). This response on annealing 2212 single crystals in O_2 was opposite that observed in polycrystalline 2212 conductors [64,65], which illustrates that there is much that is not understood about the 2212 system.

9.5 Toward Phase-Pure 2212 Conductors – Methods to Minimize or Eliminate 1:1 AEC

The 2212 phase does not form at the ideal 2:2:1:2 (Bi:Sr:Ca:Cu) composition. At this composition the phases that coexist in the solid state (850°C in air) are 2212, 2:1 AEC, 14:24 AEC, and $Bi_2(Sr,Ca)_3O_6$ [5], which is a different Cu-free phase from that forming in the liquid; 2212 does not coexist with 1:1 AEC in the solid state.

One approach to deal with the troublesome 1:1 AEC is to modify the overall composition so it is poor in the elements that form 1:1 AEC. Sawada et al. [13] at Mitsubishi Cable used this approach, modifying the 2212 composition according to 2:2:1-x:2-x (Bi:Sr:Ca:Cu), which corresponds to eliminating 1:1 AEC of composition $CaCuO_2$. The actual 1:1 AEC composition [69] in fully processed 2212, however, is approximately $(Sr_{0.6}Ca_{0.4})CuO_2$. For x = 0.36, their fully processed conductor contained very little 1:1 AEC, but it contained a different phase, presumably 2201, caused by the large shift in composition. The 2201 is not expected to be a problem for 4 K applications, but would be a problem for 20 K applications.

Another example of varying the composition is the work of the group at Kobe Steel [12] that uses the composition 2.1:2:1:1.9:(0.1), which includes 0.1 Ag. Normalizing this composition to 2 Bi gives 2:1.90:0.95:1.81:(0.1), which corresponds to removing CuO and 1:1 AEC whose composition (Sr:Ca = 0.67:0.33) is close to the actual 1:1 AEC composition in fully processed 2212 conductors. With this small shift in chemistry, the researchers still had 1:1 AEC in their fully processed tape, but J_c increased.

Other methods have been devised to minimize or eliminate 1:1 AEC during processing. One of these, called *step solidification melt processing* (discussed later), provides additional time for the kinetically sluggish 1:1 AEC to react in Region 2. Another method is to melt process in pure O_2, where $(Sr,Ca)_{14}Cu_{24}O_x$ (14:24 AEC) rather than 1:1 AEC forms in the melt. The 14:24 AEC phase does not grow to large size in the melt, and it reacts faster than 1:1 AEC [70].

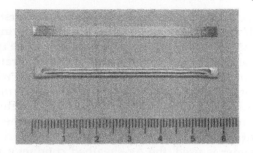

Fig. 9.22. Sections of Ag-sheathed 2212 tape that are bubbled (upper) and unbubbled (lower). Bubbling was caused by the O_2 released when the 2212 melted.

9.6 Bubbling in 2212 Tapes and Wires

Figure 9.22 shows a bubbled tape. The bubbles are bulges in the Ag sheath in tapes and wires, caused by high internal gas pressure that develops at elevated temperature. Bubbles are a serious problem because they cause dimensional nonuniformity in the conductor and J_c in the bubbled region may be severely degraded or even lost. Bubbling occurs when a condensed species transforms to a gas, building up a pressure that deforms the Ag sheath. If the gas pressure that develops is not high enough to create a large bulge in the sheath, it may deform the Ag sheath only slightly and form a pore in the melt. Here, too, J_c in the region of the pore can be severely degraded or even lost.

Three chemical species can cause bubbling: water, CO_2, and O_2. Water can enter the OPIT conductor as an absorbed species on the surface of the powder or as $Ca(OH)_2$ or $Sr(OH)_2$. The CO_2 probably enters as a carbonate, either from incomplete reaction of a carbon-containing precursor used to make the powder or by reaction between the powder and CO_2 in the atmosphere. Water and CO_2 are relatively easy to eliminate by heating the powder in flowing O_2 or inert gas at about 800°C, or in vacuum at 500–600°C, for times ranging from 8 to 48 hr. The flowing atmospheres are most useful for loose powder. The time needed depends on how well the evolving gases can be removed from the powder. Heating the powder in a flow system with a reduced total pressure, in a fluidized bed where gas is forced through the powder, or in a rotary kiln speeds the removal of the evolving gases. On a laboratory scale, one can heat the powder in a crucible in a system with flowing gas and stir the powder occasionally.

Bubbling from water typically occurs at 400–600°C, from CO_2 at or just below the 2212 melting point, and from O_2 at the 2212 melting point. The O_2 release that occurs on melting is thought to be caused by the Cu being reduced in the melt. Although oxygen readily diffuses through Ag, if a tape or wire is rapidly heated through the melting point of 2212, O_2 may evolve so fast that it builds up a high pressure before it can diffuse through the Ag.

Several different approaches have been used to prevent bubbling due to

O_2 release. These can be grouped in four categories. The first is a mechanical method in which the tape is pierced to release the evolving gas [12]. This method can be messy, because liquid may flow out of the openings during processing. The second method is to reduce the heating rate when passing through the 2212 melting point, which decreases the rate at which O_2 evolves. Although this slow heating rate may prevent bubbling, it may allow the 1:1 AEC to grow excessively large. The third method is to remove the oxygen from 2212 before melting it. This can be done by heating the powder close to the melting point in air [71] or at a lower temperature in vacuum. If this oxygen-poor powder is used, the Ag-sheathed conductor must be heated quickly to prevent the powder from picking up oxygen before it is melted. Another way to remove oxygen before melting the 2212 is to melt process in reduced pO_2 (e.g., 1% O_2) and hold the tape below the melting temperature, allowing oxygen to diffuse out through the Ag sheath before melting the 2212. The fourth method is melt processing in 1 atm O_2. Nomura et al. [72] observed no mass loss when 2212 tape was melted and heated to 900°C. Using 1 atm O_2 may decrease Cu reduction in the liquid; however, Yoshida [73] reported bubbling when processing long tapes in pure O_2.

9.7 Carbon Effects in 2212 Tapes and Wires

The BSCCO powder used to make conductors often contains C, which is

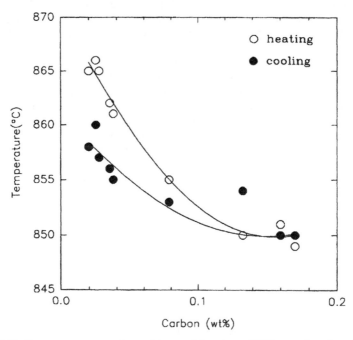

Fig. 9.23. Onset temperature for melting and freezing of 2212 as a function of carbon content [74]. The temperatures were determined from DTA runs on 2212 tapes heated and cooled at 5°C/min in air. The curves are drawn to guide the eye.

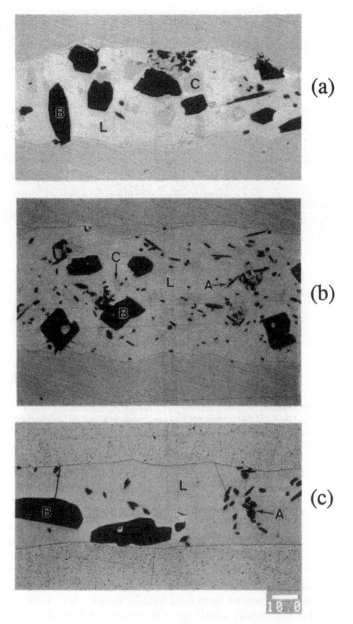

Fig. 9.24. Phases present in 2212 tapes that have different C contents. Tapes were heated to 890°C in air for 10 min, cooled at 10°C/hr to 880°C, then quenched in oil. The carbon contents are (a) 0.027 wt%, (b) 0.078 wt%, and (c) 0.16 wt%. A = 2:1 AEC, B = 1:1 AEC, C = Cu-free phase, L = quenched liquid. *Note*: A Sr-rich phase was also seen in the sample that contained 0.16 wt% C, but this phase is not present in the micrograph shown here.

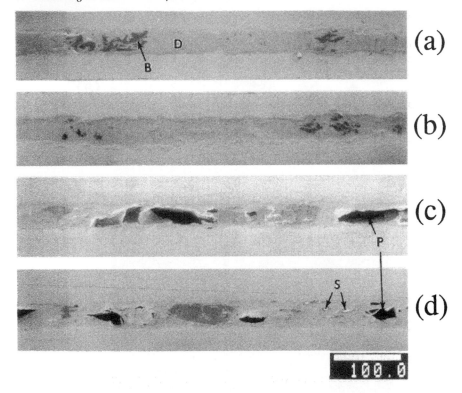

Fig. 9.25. Microstructure of fully processed 2212 tape that contains 0.027 wt% C [(a) and (b)] and 0.16 wt% C [(c) and (d)]. The maximum processing temperature was 905°C for the tapes in (a) and (c) and 890°C for the tapes in (b) and (d). (890°C or 905°C for 10 min, 10°C/hr to 820°C, hold 12 hr, furnace cool). B = 1:1 AEC, D = 2212, P = porosity, S = Sr-rich phase.

probably present as $SrCO_3$, the most stable carbonate in the system. In films, C is present in the organic species that are burned off during heating, and it may form carbonates if the CO_2 is not flushed out of the furnace during burnoff. CO_2 can cause bubbling in tapes and wires, as mentioned above. In addition, C in tapes and wires affects the melting and freezing point of 2212 [74]. Figure 9.23 shows the onset of 2212 formation in tape decreased from 866°C to 849°C with an increase in C content from 0.02 to 0.17 weight % C. One should not be misled by the apparently small weight percent of C in the samples, because 0.17 wt% C corresponds to 1.7 cation% C.

Zhang and Hellstrom [74] found that the C content of 2212 powder synthesized in air could not be reduced below about 0.03–0.04 wt%, which is thought to be due to equilibration with CO_2 in the atmosphere. By treating the powder in pure flowing O_2 at 800°C or in vacuum at 600°C, however, they were able to reduce the C content to about 0.01 wt%. Tenbrink et al. [28] reported the C content in their powder to be as low as 0.02 wt% C, but did not say what atmosphere had been used.

The phase assemblage in the melt changes with C content [74]. The phase data in Table 9.4 were determined on a powder with about 0.04 wt% C. This same phase assemblage was also found with powder having lower C content. With increasing C content in the powder, a Sr-rich phase appears, and the amount of Cu-free phase in the melt decreases. At 0.78 wt% C the Sr-rich phase is present in the liquid, and at 0.13 wt% C there is no Cu-free phase in the liquid. Figure 9.24 shows the progression of microstructures in the melt with increasing C content. Figure 9.25 shows that the microstructure of fully processed tapes varies with C content. The fully processed tape with the higher C content contained the Sr-rich phase, which was also present in the melt, and the 1:1 AEC grains and amount of porosity were larger.

C is not expected to be a problem in films so long as there is adequate gas flow to flush the CO_2 out of the furnace as the organics are burned off.

We are not certain how much C can be tolerated in powder used for tapes and wires. At present, it seems prudent to keep the C content as low as possible. The 0.03–0.04 wt% C present in solid-state powder currently used to make tapes and wires can be taken as the upper bound for the C content. Large amounts of C affect the microstructure of fully processed 2212 conductors, and therefore J_c is expected to decrease. The effect of small amounts of C on J_c is not known; however. carbon in Y123 [75] and 2223 [76] degrades the superconducting properties, so C would also be expected to degrade the superconductivity in 2212.

9.8 Step Solidification Melt Processing of 2212 Conductors

Ray developed a process called step solidification melt processing that attains high 2212 alignment and allows the 1:1 AEC to be consumed during

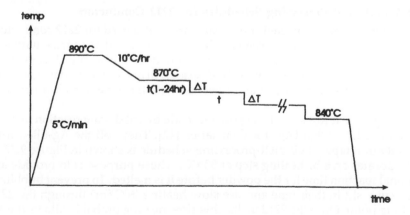

Fig. 9.26. Schematic diagram of the heating schedule used for step solidification melt processing.

cooling [55]. The process, shown in Figure 9.26, replaces the continuous cooling in Region 2 with a series of steps.

The first step must be just below (< 5°C) the temperature at which 2212 begins to form. At this temperature a small number of 2212 nuclei form, and the constant temperature of the step provides time for these to grow into a few large, well-aligned 2212 grains. If the temperature is too high, then 2212 does not form; instead, extremely large 1:1 AEC grains can form in the conductor. These will not be fully consumed during the later stages of step solidification, and the remnant 1:1 AEC will be large.

It has been observed that 2212 initially nucleates and grows in random directions (Figure 9.17). Researchers believe that the large, well-aligned 2212 grains that grow in the first step of step solidification melt processing grow at the expense of the small, misoriented grains (see earlier). The entire oxide core does not transform to 2212 during the first step: the temperature must be decreased to form more 2212.

During the first step, when 2212 starts to form, one could imagine that grains of 1:1 AEC would either shrink by reacting to form 2212 or grow as they did at higher temperatures in the melt state. Ray found, however, that the grains of 1:1 AEC remain about the same size during the first step. Only during lower temperature steps (about 860°C) does the grain size of the 1:1 AEC begin to decrease significantly. In addition, during the lower temperature steps, the composition of the 1:1 AEC begins to change, possibly moving towards 14:24 AEC. Thus, in the high-temperature steps the 2212 phase forms and aligns, and the lower-temperature steps provide time for the 1:1 AEC to react.

Step solidification has been used [77] to align thick (75–100 μm) 2212 films on Ag. It has also been used to align multifilamentary round wires in which neutron diffraction pole figures have shown the wires have radial c-axis alignment [78].

9.9 Actual Melt-Processing Schedules for 2212 Conductors

In this section, actual melt-processing schedules used for 2212 conductors and resulting J_c data are presented. These were selected as representative of processes that are currently being used; they show the diversity of melt-processing schedules that can be devised. All are understandable in terms of the mechanisms discussed in connection with the generic melt-processing schedule.

The group at Kobe Steel use powder made by solid-state reaction of composition 2.1:2:1:1.9:(0.1Ag) for their tapes [12]. They roll the Ag-filled tube directly into tape. Their melt-processing schedule is shown in Figure 9.27. It incorporates an 8-hr heating step at 835°C, whose purpose is to provide additional reaction time for the powder before it is melted. To prevent bubbling, they cut a slit in their tape and use slow heating (6°C/hr) through the 2212 melting point. Once the 2212 melts, this slow heating probably allows the 1:1 AEC to grow very large in size. They cool the tape from 885°C to 875°C at 50°C/hr to reduce the time in the melt state, then use slow cooling while forming 2212. Their process does not incorporate Region 3. They cool the samples [62]

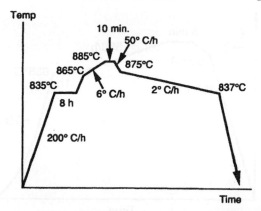

Fig. 9.27. Melt-processing schedule used by the Kobe Steel group to make coils with 2212 tape [12]. (Reprinted from T. Hase, T. Egi, K. Shibutani, S. Hayashi, Y. Masuda, I. Shigaki, R. Ogawa and Y. Kawate, *Proc. 6th Int. Symp. on Supercond.*, in press, © 1994 Springer-Verlag.)

to room temperature at 700°C/hr. J_c in short sections of tape made with this procedure was 90,000 A/cm^2 (4 K, 0 T) and 17,000 A/cm^2 (20 K, 0 T) in a coil [12] made from tapes whose total length was 1500 m.

The heating schedule used by the Vacuumschmelze group is shown in Figure 9.28. They work with round mono- and multifilamentary wire. The cooling rate in Region 2 is very fast, providing little time for the 2212 to grow and align or for the 1:1 AEC and Cu-free phase to react. Their processing relies on forming 2212 in Region 3. With this processing schedule there is no detectable alignment of the superconductor in their wires. It is questionable, however, whether much alignment could be induced even with slow cooling

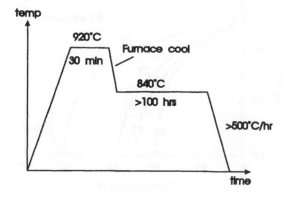

Fig. 9.28. Melt-processing schedule used by the Vacuumschmelze group to make coils with 2212 wire [63,79].

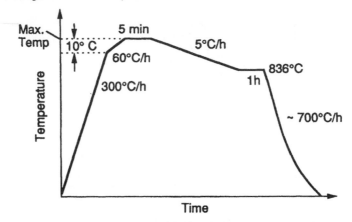

Fig. 9.29. Melt-processing schedule used by the NRIM-led consortium to make coils with 2212 dip-coated film [20].

or step solidification because of the large diameter (> 100 μm) of the filaments they typically use. The J_c for short sections of Vacuumschmelze round wire is as high as 20,000 A/cm^2 (4 K, > 20 T) [28].

A collaboration between the National Research Institute for Metals (Japan) (NRIM), Ashahi Glass, and Hitachi Cable uses the heating schedule shown in Figure 9.29 to make dip-coated films for coils. The composition is 2.0:2.0:0.95:2.0. The coated Ag foil is wound into a loose coil, placed edge up on a bed of $Bi_2Al_4O_9$-Al_2O_3 powder on an alumina setter plate, heated in air to burn out the organics, cooled, covered with an alumina pan, and then melt

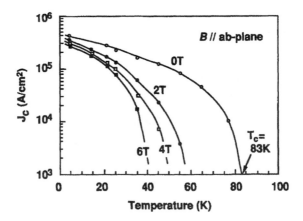

Fig. 9.30. J_c as a function of applied magnetic field for doctor-bladed film processed using the heating schedule shown in Figure 9.29 [20].

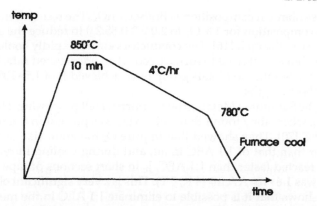

Fig. 9.31. Melt-processing schedule used by the Sumitomo Metals group to make coils using 2212 tape in 1% O_2 [80].

processed. Even with the protective atmosphere, the time at the maximum temperature is short to minimize Bi loss. The coil is cooled to room temperature at about 700°C/hr. When heat treating the coil, the layers of Ag foil must be kept from touching one another, or they stick together, ruining the coil. At the maximum processing temperature, the Ag foil is very weak, so it occasionally sags, allowing the layers to touch. To alleviate this problem, the researchers are investigating Ag alloys to strengthen the foil [38]. The J_c values that this collaboration attain with doctor-bladed 2212 using the process described above are shown in Figure 9.30.

The heating schedule the group at Sumitomo Metals uses is shown in Figure 9.31. It was developed for an atmosphere of 1% O_2 in Ar, which they use to prevent bubbling problems in long tapes used to make coils. In 1% O_2, the phases present in the melt when 2212 begins to form are 1:1 AEC and a Cu-free

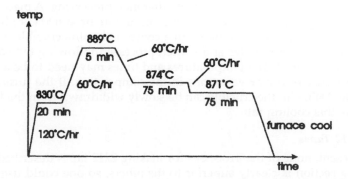

Fig. 9.32. Melt-processing schedule used by the group at the State University of New York-Buffalo to process coils made with Ag-sheathed 2212 tape [81].

phase whose nominal composition is $Bi_2(Sr,Ca)_3O_x$. The researchers modified the overall composition for 1% O_2 to 2.0:2.3:0.85:2.0 to reduce the size of this Cu-free phase in the melt [16]. The conductors are not rapidly cooled to room temperature but are allowed to furnace cool at an unspecified rate. Using this procedure, the Sumtiomo Metals group have achieved J_c of $1.5 \bullet 10^5$ A/cm^2 (4 K, 0 T) in short lengths.

Earlier, the Sumitomo Metals group reported melt processing 2212 (2:2:1:2) in 1 atm O_2, where they found that 14:24 AEC was present in the melt rather than 1:1 AEC [70]. They showed that in pure O_2 the grain size of 14:24 AEC was smaller than that of 1:1 AEC in air, and during cooling in region 2, the 14:24 AEC reacted faster than 1:1 AEC. J_c in short sections of tape processed in pure O_2 was $1.6 \bullet 10^5$ A/cm^2 (4 K, 0 T). This is a very significant observation because it shows that it is possible to eliminate 1:1 AEC in the melt. In spite of having eliminated 1:1 AEC in the melt by using pure O_2, however, they switched to 1% O_2 to prevent the bubbling they encountered while melt processing long tapes in pure O_2 [73].

The heating schedule used by the group at the State University of New York-Buffalo for tape, painted films, and jelly rolls (a film that is rolled up) is shown in Figure 9.32. This schedule heats the sample relatively slowly and minimizes the time in the melt state on cooling. This process is similar to step solidification, but does not continue down to the temperatures where the 1:1 AEC phase was observed to be consumed [56]. Using this procedure, the researchers measured J_c of 32,000–54,000 A/cm^2 for 110 m tapes wound into coils [81].

9.10 Suggestions for Making 2212 Conductors

Here we discuss processes for making 2212 films and tapes.

9.10.1 2212 Films

A good place to start making 2212 conductors is with doctor-bladed films, which seem to be the easiest 2212 conductors to fabricate. Powder of overall composition 2:2:0.95:2.0 can be made by solid-state synthesis, then mixed with the organics listed in Table 9.3 to make a slurry. With this slurry, a dried 50-μm thick green film reduces to about 15 μm after melt processing. A piece of film is placed on Ag foil (50–75 μm thick) and heated in air at 600°C to burn off the organics. The film is then placed in a covered alumina crucible that contains $Bi_2Al_4O_9$-Al_2O_3 powder and is melt processed following the schedule shown in Figure 9.29. The temperatures and times may need to be adjusted slightly because of minor variations in the composition. If the furnace will not cool at 700°C/hr, the crucible can be slowly withdrawn from the furnace to achieve this cooling rate.

9.10.2 2212 Tapes

At present, none of the processes for making 2212 tapes described in the preceding section is clearly superior to the others, so one could use any of these processes and its nominal 2212 composition. One could also use the 2:2:1:2 composition. Solid-state powder that has been fully reacted to 2212 can be used to fabricate 2212 tapes. Commercially available Ag tubing (6 mm OD x 4 mm ID) can be used. A 10-cm long tube is relatively easy to handle and requires less than 10 g of powder. The tube is sealed at one end with a Ag plug and the 2212 powder is hand packed into the tube. The tube, which still has one open end, is then heated in vacuum at about 600°C for 12–24 hr, cooled in vacuum, and backfilled with O_2 at room temperature. The open end of the tube is then plugged. If wire-drawing facilities are available, the tube can be drawn until it is 1.8–2.0 mm in diameter, then rolled to form a 100–200 μm thick tape. Or the sealed tube can be rolled directly into tape. During drawing, if the wire begins to break, it can be annealed (200–400°C for 1 hr in air) to relieve the cold work. The reduction in thickness during rolling should be about 5–10% per pass.

If a specific process from the preceding section is being followed to make 2212 tapes, then its heating schedule should be followed. If one is experimenting with melt-processing schedules, however, a starting point is given below. For a tape in which the powder was heated in vacuum in the Ag tube prior to being rolled into tape, a suggested melt-processing schedule is to quickly heat (300°C/min) to 890–895°C in air, hold for about 10 min, quickly cool (as fast as the furnace can cool) to 875°C, which is just above the temperature where 2212 begins to form, then slowly cool (3–5°C/hr) to 840°C, hold for 100 hr, then quickly cool (about 700°C/hr) to room temperature.

9.11 Future Directions for Melt-Processing 2212 Conductors

The melt-processing parameters one can vary are time, temperature, heating and cooling rates, and the pO_2. Our discussion thus far has concentrated on the current technology, mainly melt processing in air, which has serious problems with 1:1 AEC in the melt. It appears imperative to eliminate this troublesome phase to continue to improve the performance of tapes. Changing the atmosphere to pure O_2 appears to be the only possible way to accomplish this, as in pure O_2 14:24 AEC replaces 1:1 AEC in the melt [70]. The 14:24 AEC does not appear to grow large, as 1:1 AEC does, and it has more favorable reaction kinetics to form 2212. Thus, processing in pure O_2 provides a clear possibility for improving the homogeneity and therefore the performance of 2212 tapes. Future work is expected to concentrate on understanding and optimizing melt processing in pure O_2, and eliminating the bubbling problem.

Step solidification melt processing appears as though it will be a more robust process in pure O_2 than in air. Recall that in air, if the temperature for the first step is above the point where 2212 forms, then 2212 does not form, but the 1:1 AEC grains grow excessively large and then cannot fully react during the latter stages of step solidification. When using pure O_2, however, if the temperature of the first step is too high for 2212 to form, the 14:24 AEC in the melt does not grow, but remains relatively small. Future studies that investigate step solidification melt processing in pure O_2 are expected.

The fast cooling to room temperature (700–1200°C/hr in Region 4) may be very difficult to achieve when processing large coils. Future work is expected to examine methods to use slow cooling to room temperature and attain high

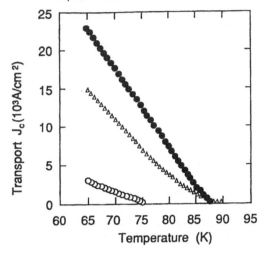

Fig. 9.33. J_c as a function of temperature for 2212 films that were quenched (triangle), slow cooled in air (open circle), and slow cooled in controlled pO_2 (solid circle) [82].

J_c. Here, controlling the pO_2 during cooling appears to offer the possibility to use slow cooling. For example, Yoshida and Endo [16] used furnace cooling (the cooling rate was not specified) when processing in 1% O_2 and achieved high J_c. Noji et al. [82] have studied J_c in film cooled to room temperature under controlled pO_2. They described their experimental procedure as decreasing the oxygen content from 21 to 10^{-4}% O_2 while cooling from 850°C to 500°C in such a manner that the pO_2 at all temperatures corresponded to the pO_2 that gave the maximum T_c measured in 2212 samples quenched from that temperature [64]. Although they did not specify their cooling rate, it is presumably much slower than 700°C/hr. Their controlled cooling with decreasing pO_2 increased the J_c around 77 K, as shown in Figure 9.33. They did not show how the controlled pO_2 cooling affected J_c at 4 K.

9.12 2223 Conductors

In this section, we discuss the processing to form the 2223 phase. Recent experiments with 2223 reaction pathways are described, and novel metal precursor 2223 tapes are briefly reviewed.

9.12.1 Forming 2223

The 2223 phase is difficult to form. It does not form directly by using any of the synthesis techniques in Table 9.2, but by a multistep reaction in which the 2212 phase reacts with other phases to form 2223. Thus, several significant differences exist between the processing of 2212 and 2223 conductors. Specifically, 2223 tapes are not melt processed, since 2212 plus 2223 form from the melt on cooling [83], and extensive additional heat treatment is needed to convert the 2212 to 2223. Rather 2223 tapes are heated at lower tempera-

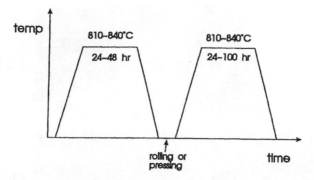

Fig. 9.34. Schematic diagram of the thermomechanical heat treatment used for 2223 tapes. The tape is rolled or pressed between each heat treatment.

tures where only a small amount of liquid is thought to form. They are made using precursor powder that is a mixture of phases that react *in situ* during the heat treatment to form 2223. Fully reacted 2223 powder is not used because J_c in these tapes is lower than in tape made from a mixture of powders that react *in situ* to form 2223 [84].

2223 conductors are made by packing a Ag tube with powder consisting of a mixture of phases that typically contains the 2212 phase, which is called the 2223 precursor powder, fabricating a tape, then heat treating the tape (Figure 9.34) to form 2223 *in situ*. As Table 9.1 shows, part of the Bi is replaced with Pb, to aid in forming the 2223 phase [7]. The heat treatment of 2223 conductors, shown schematically in Figure 9.34, consists of two or more heat treatment steps separated by a mechanical deformation step (rolling or uniaxial pressing). The 2223 phase forms during the heat treatment steps. The mechanical deformation step densifies the core, it helps break up large unreacted grains allowing the reactions to proceed in the subsequent heat treatment, and it is thought to improve the alignment of the 2223 phase. Only 2223 tapes can be fabricated because of the need to roll or press the conductor between heat treatments.

The temperature window in which 2223 is stable is rather small, with 2223 forming between 830°C and 870°C in air [86]. Endo, Koyama, and Kawai [86] showed that this window was widest when 2223 was formed in about 7.7% O_2. Many groups process 2223 tapes in 7–8% O_2, although processing is also done in air [87].

The presence of nonsuperconducting phases causes the same types of problems in 2223 conductors as in 2212. Here, too, reducing the nonsuperconducting phases in fully processed 2223 tape increases J_c [19]. Unlike 2212, one particular phase does not cause the majority of problems in 2223 tapes; rather, several nonsuperconducting phases including $(Sr,Ca)_2CuO_3$ (2:1 AEC), 14:24 AEC, Ca_2PbO_4, SrO, CuO, and a Bi-rich phase can be present in fully processed 2223 conductors [88]. Figure 9.35 shows the microstructure of 2223 tapes made with 2223 precursor powders that all had the same nominal composition.

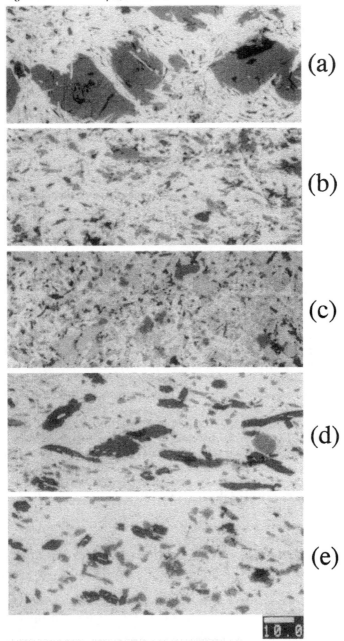

Fig. 9.35. Micrographs of 2223 tapes that were processed together, showing different microstructures after the first heat treatment. The powder used to make each tape had the same nominal composition but had been synthesized differently. Powders were synthesized using the following techniques: (a) microemulsion, (b) coprecipitation, (c) solid-state reaction, (d) aerosol spray pyrolysis, and (e) freeze drying.

Although these tapes were processed together, they developed very different microstructures. A significant difference was that the precursor powders were synthesized by different techniques.

The standard approach to make a 2223 precursor powder is to mix all of the reactants together with the required composition, then use one of the techniques in Table 9.2 to synthesize the powder. This is what was done to synthesize the precursor powders used to make the tapes in Figure 9.35. Dorris et al. [24] call this "single-powder" synthesis. The powder is reacted until the X-ray pattern is dominated by diffraction peaks for 2212. The X-ray pattern also contains peaks from the nonsuperconducting phases that are present in the powder. But these peaks are much weaker than those from 2212, and consequently they are very difficult to use to identify the minor nonsuperconducting phases in the powder. (However, one can often identify peaks in the pattern from Ca_2PbO_4 and CuO.)

Figure 9.36 shows an X-ray pattern for a mixture of Pb-doped 2212, Ca_2CuO_3, and CuO in a 1:0.5:0.5 mole ratio. This mixture contains the maximum possible amounts of Ca_2CuO_3 and CuO relative to the 2212 to form 2223. The pattern shows that the peaks from Ca_2CuO_3 and CuO are fewer in number and much smaller than for 2212. One could turn to SEM/EDX to identify the nonsuperconducting phases in the precursor powder. But because of the typically small grain size in the powder, these phase cannot be identified by this

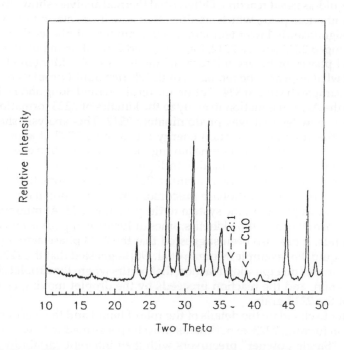

Fig. 9.36. X-ray diffraction pattern (Cu K_α) for a mixture of Pb-2212, Ca_2CuO_y, and CuO in a 1.0:0.5:0.5 mole ratio. The arrows mark the strongest peaks for 2:1 AEC and CuO.

technique. The point is that the nonsuperconducting phases in "single-powder" precursors are difficult to identify and therefore usually not well known. It is now thought that such phases play a key role in determining the reaction pathway to form the 2223 phase, and the wide range of microstructures in Figure 9.35 is believed to result from differences in the nonsuperconducting phases present in the precursor powders.

Note: For the remainder of this chapter, Pb-2201, Pb-2212, and Pb-2223 will be used to designate the respective phase with part of the Bi replaced with Pb.

If the nonsuperconducting phases in the 2223 precursor can vary widely, how have groups managed to make such significant advances with 2223 tapes? The key is to make highly reproducible powder and then to optimize the thermal processing for this powder to achieve high J_c. This requires careful control of all facets of powder synthesis. Since each group in the community synthesizes its own precursor powder using different methods, significant variations may occur in the nonsuperconducting phases in these precursor powders. Thus, specific times and temperatures one group uses to process its 2223 tape may be valid only for its precursor powder and may not be transferable to powders other groups use. This variation in the precursor powder may also explain why it has not been possible for different groups to reproduce each other's high J_c values.

Strong evidence exists that the 2223 phase forms from the precursor powder in a liquid-assisted reaction. Differential thermal analysis shows the onset of an endothermic peak, which signifies that a melt has started to form, occurs at a significantly lower temperature in a mixture of phases that react to form Pb-doped 2223 (such a 2212, Ca_2PbO_4, and CuO) than in powders of the individual phases or binary mixtures of the phases [89]. Morgan et al. [90] observed what appear to be remnants of Bi/Pb-rich liquid droplets on grains of 2212 in samples heated at 834°C in air. Luo et al. [91] and Sung and Hellstrom [92] used the Avrami equation to analyze the kinetics of 2223 formation from a precursor powder that was predominately 2212. This analysis showed a change in the apparent activation energy for forming 2223 at about 820°C, with the higher temperature regime having the lower activation energy. They interpreted this change as being due to liquid forming above about 820°C.

What role does liquid play in forming 2223? The reactions that occur on forming 2223 from the multiphase precursor powder require mass transport from the various phases in the system to the growing 2223. At minimum, the liquid increases the formation kinetics, since diffusion in liquids is much faster than in solids. It has also been suggested that the 2223 phase actually forms from the liquid. For example, Morgan et al. [90] suggested that the 2223 forms by a dissolution/precipitation reaction that occurs where the droplet contacts the 2212 grain, and the reaction proceeds by the droplet moving across the surface of the 2212 grain.

In order to elucidate the details of the role of liquid and the basic chemical reactions in forming 2223, experiments must be performed with well-defined reactants. "Single-powder" precursors with their inherent variability are not adequate. Thus, a key point is to design experiments in which all the reactant phases are well defined and controllable. To do this, one writes a chemical reaction to form 2223, such as

$$0.9Bi_2Sr_2CaCu_2O_x + 0.4PbO + 0.55CuO$$
$$+ 0.65(Sr_{0.15}Ca_{0.85})_2CuO_3 = Bi_{1.8}Pb_{0.4}Sr_2Ca_2Cu_3O_y.$$

Here, each reactant phase can be synthesized separately, thoroughly characterized, then mixed together to make a well-defined precursor powder. This precursor powder can be made into a pellet or tape, then reacted and characterized. Several examples using this approach are given below.

An early example of this approach was the work of Fukushima [93], who studied the following reactions:

$$Bi_{1.75}Pb_{0.25}Sr_2CuO_x + Ca_2CuO_3 + CuO$$
$$= Bi_{1.75}Pb_{0.25}Sr_2Ca_2Cu_3O_x$$

and

$$Bi_{1.75}Pb_{0.25}Sr_2CaCu_2O_x + 0.5Ca_2CuO_3 + 0.5CuO$$
$$= Bi_{1.75}Pb_{0.25}Sr_2Ca_2Cu_3O_x.$$

Fukushima synthesized all of the reactants separately, then mixed them together in the correct proportions, pressed pellets, and reacted them. Pb-2223 formed with both reaction schemes, but the reacted pellets contained Pb-2212 and unidentified Sr-Ca-O phases along with the Pb-2223. With their process (860–870°C, 48 hr, air) the grains of Sr-Ca-O were quite large (> 50 μm) when they used Pb-2212 as a reactant.

More recently, Arendt, Garbauskas, and Bednarczyk [94] used this approach to study the following reaction:

$$2212 + 0.3PbO + (1+x)"CaCuO_2" = Pb\text{-}2223 + x"CaCuO_2" + 0.15Bi_2O_3,$$

where x ranged from 0 to 1. Here $CaCuO_2$ is placed in quotes to show that it is not a single phase but is a mixture of Ca_2CuO_3 and CuO. Arendt and his coworkers synthesized 2212 and "CaCuO_2" separately, then mixed them with PbO. This reaction was designed to evolve Bi_2O_3 during the reaction and to have unreacted "CaCuO_2" for x > 0. They found faster, more complete conversion to 2223 in their pellet samples using this technique than in pellet samples made with conventional "single-powder" precursor.

More recently, Dorris et al. [24] investigated the formation of Pb-2223 based on the following reaction:

$$Bi_{1.8}Pb_{0.4}Sr_{2-x}Ca_{1+x}Cu_2O_y + (Sr_xCa_{1-x})CuO_2$$
$$= Bi_{1.8}Pb_{0.4}Sr_2Ca_2Cu_3O_z,$$

where x ranged from 0 to 1. This is a schematic representation of the reaction and phases present in the experiment, because 1:1 AEC decomposes to 2:1 AEC and CuO at the reaction temperature, and the stoichiometry range for 2212, and presumably Pb-2212, does not extend over the entire range they investigated (see Figure 9.3). They call this a "two-powder" process. Their highest J_c values were for x = 0, corresponding to starting powders of "CaCuO_2" and $Bi_{1.8}Pb_{0.4}Sr_2CaCu_2O_y$. Their Pb-2212 contained some Ca_2PbO_4. A particularly significant result of this study was the small size of the nonsuperconducting phases in the fully processed tapes (< 2–3 μm [see Figure 7 in [24]]).

In a subsequent study, Dorris et al. [95] investigated how adding the Pb in different phases in the precursor powder affected the formation of Pb-2223.

They used the following reaction schemes.

$$Bi_{1.8}Pb_zSr_2CaCu_2O_x + (0.4-z)Ca_2PbO_4 + (0.1+z)Ca_2CuO_3$$
$$+ (0.9-z)CuO = Bi_{1.8}Pb_{0.4}Sr_2Ca_2Cu_3O_x$$

and

$$Bi_{1.8}Sr_2CaCu_2O_x + 0.4PbO + 0.5Ca_2CuO_3 + 0.5CuO$$
$$= Bi_{1.8}Pb_{0.4}Sr_2Ca_2Cu_3O_x.$$

The Pb was added in Pb-2212, Ca_2PbO_4, or PbO. They found the highest J_c in fully reacted tape when the Pb was added in the Pb-2212. With Pb-2212 the number and size of the nonsuperconducting alkaline earth cuprates in their fully processed tape was smallest (< 5 µm), whereas these phases were more numerous and larger when Pb was added as Ca_2PbO_4 (5–10 µm) or PbO (10–25 µm). This result agrees with the second step of the following reaction sequence, which was suggested by Flükiger et al. [96]:

2212 + second phases = Pb-2212 + second phases = Pb-2223.

Concurrent with the work of Dorris and coworkers, Sung and Hellstrom [98] were using reaction couples to study the following reactions:

$$Bi_{1.8}Sr_{1.9}Ca_{1.2}Cu_2O_x + 0.4Ca_2PbO_4 + CuO$$
$$= Bi_{1.8}Pb_{0.4}Sr_{1.9}Ca_2Cu_3O_y$$

and

$$Bi_2Sr_2CaCu_2O_x + 0.5Ca_2PbO_4 + CuO = Bi_2Pb_{0.5}Sr_2Ca_2Cu_3O_y.$$

Reaction couples are useful as they limit the reaction to the interface between the two pellets. Figure 9.37 shows a schematic diagram of their reaction couple and the actual interface after reaction. Here Pb-2223 had formed on the 2212 side of the interface, and 2:1 AEC plus liquid formed on the Ca_2PbO_4 plus CuO side. From these reaction-couple studies Sung and Hellstrom concluded that Ca_2PbO_4 plays two roles in the reaction to form 2223, one positive, one negative. Ca_2PbO_4 decomposes at about 820°C (in 7.5% O_2), which is triggered by the presence of 2212 in the system (presumably by the Bi in the 2212), releasing CaO, PbO, and O_2. On the positive side, part of the PbO reacts with the 2212 forming Pb-2212 and part of it reacts with CuO, forming a liquid. Some of the CaO that is released reacts with the Pb-2212 and CuO, forming 2223. The Pb-2212 and liquid enhance the 2223 formation kinetics. On the negative side, part of the CaO reacts with the CuO, forming 2:1 AEC, which grew to large size in the reaction couple (Figure 9.37b). Once large, the 2:1 AEC grains react slowly, impeding the formation of 2223. It appears that the formation of 2:1 AEC is kinetically favored, since its apparent activation energy for formation is about 300 kJ/mole [88], whereas the energy associated with forming 2223 derived from kinetic studies ranges from 0.46 to 1.9 MJ/mole [91,92,97]. This very high energy is probably a combination of the activation energy to form 2223 and the energy associated with another process, such as the formatin of the liquid.

Since 2:1 AEC is often present in fully processed 2223 tapes and it grew large in reaction couples, Sung and Hellstrom [98] suggested that Ca_2PbO_4 should be eliminated from the 2223 precursor powders. This observation is consistent with that of Dorris et al. [95] that the best microstructures and highest J_c were obtained when the Pb was present only in Pb-2212. Sung and Hellstrom

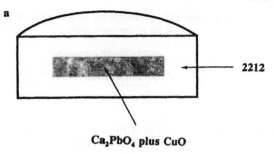

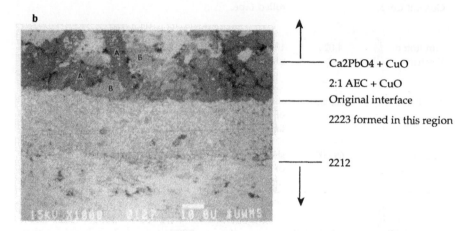

Fig. 9.37. (a) Schematic diagram of a reaction couple to study the formation of 2223; (b) interface between the 2212 and the Ca_2PbO_4 plus CuO pellets after 120hr at 825°C in 7.5% O_2.

also concluded that the final step in the formation of 2223 involves a liquid-assisted reaction between Pb-2212, 2:1 AEC, and CuO [98]. This conclusion is consistent with the results of Dorris et al. [24] and the suggestion of Majewski and coworkers [5] that 2223 formed from 2212, Ca_2CuO_3 and CuO. Majewski and his colleagues, however, were able to form only ~50% 2223 (Pb-free 2223) from these phases in ~100 hr after raising the reaction temperature for the pellets to 880–890°C.

How is the 2212 phase converted to the 2223 phase? Two different classes of mechanism have been proposed. The first is a *precipitation reaction* in which 2212 dissolves into a liquid and 2223 precipitates from the liquid, which was mentioned above. The other is an *insertion process* in which the Ca, Cu, and O diffuse into the 2212 phase and it converts to 2223. (See [97] and references there.) There is evidence for both of these mechanism. Matsubara et al. [97] have converted 2212 whiskers into 2223 by burying the whiskers in a Bi-Pb-Sr-Ca-Cu-O powder, then analyzing the individual whiskers after the

Table 9.5. J_c for 2223 tapes. Data include results for long lengths of rolled tape and short lengths of pressed tape.

Group	J_c (A/cm²) 77 K, self field	Type of sample	Voltage criterion	Ref.
American Superconductor Corp.	7,000	650 m of 19 filament rolled tape	1^{-11} Ωcm	17
Intermagnetics General Corp.	20,000	70 m of monocore rolled tape	1 μV/cm	18
Sumitomo Electric	4,020	1080 m of 61 filament rolled tape	10^{-11} Ωcm	19
Toshiba	66,000	short section of pressed monocore tape	1 μV/cm	20
Li et al.	69,000	short section of pressed monocore tape	1 μV/cm	21

reaction. They found that the conversion to 2223 occurred only when they had a powder that partially melted. The whiskers retained their original shape after conversion, but the initially sharp edges of the whiskers became rounded. This result is thought to be due to the 2212 dissolving and 2223 nucleating from the liquid at the interface between the 2212 whisker and the liquid. One can envision forming 2223 from the liquid at the surface of the whisker. But forming 2223 in the interior of the whisker appears to require an insertion type of mechanism where the Ca, Cu, and O diffuse into the 2212 forming 2223.

In summary, our understanding of the basic reactions that occur on forming 2223 is quite limited. Recent experiments with controlled-phase powders, however, are beginning to elucidate the details of the 2223 reaction pathways. If Pb-2212, Ca_2CuO_3, and CuO are the final phase assemblage from which Pb-2223 forms, precursor powder made from these phases can provide a common starting point to study Pb-2223 reaction mechanisms.

9.12.2 Performance of 2223 Tapes

Earlier, we presented details about melt processing 2212 tapes and films. A parallel section on 2223 tapes is not appropriate here because the processing is very sensitive to the precursor powder that is used, and only scant information exists in the literature on the details of synthesizing the precursor powder and the heat treatment conditions for high J_c tape. In the future, when the phase assemblage in the precursor powder is under better control, it should be much easier to fabricate reproducible 2223 tape.

The J_c of 2223 tapes made by various groups is listed in Table 9.5. J_c in long tapes (up to 1km) is steadily improving. However, since J_c in long tapes is only a fraction of the highest J_c in short tapes, there is substantial room to improve the processing of 2223 tapes.

9.12.3 Metal Precursor 2223 Tapes

This section presents a novel method to make Pb-2223 tape by internally oxidizing a metal precursor in a Ag sheath. Although the details of the chemical reactions that occur in the process are not well documented in the literature, the process is mentioned here to highlight its similarities and differences compared to processing OPIT tape. Otto et al. [100] reported fabricating long lengths of multifilamentary Pb-2223 tape using this technique. They produced a mechanically alloyed powder of the metals (Bi,Pb,Sr,Ca,Cu), packed this powder into a Ag tube, extruded the tube to form wire, then used standard wire working techniques to make a multifilamentary conductor. The tape was then heated in an oxidizing atmosphere using an undisclosed heating/deformation process, forming Pb-2223 *in situ* in the tape.

An important advantage of metal precursors over OPIT tape is the ductility of the metallic core. This allows the Ag/metal precursor composite to be easily drawn to fine filament diameter and to achieve high filament count. Pb-2223 tape with 9853 filaments (5 µm thick) have been made with the metal precursor technique. In contrast, the highest number of filaments reported for a 2223 OPIT tape [101] is 1296. (American Superconductor has made a Y124 (YBa$_2$Cu$_4$O$_8$) metal precursor with almost 1,000,000 filaments [102].) The J_c in Pb-2223 metal precursor multifilament tape was reported to be 7,500 A/cm^2 (77 K, 0 T) over 40-cm lengths [100].

9.13 Summary

This is an exciting time in the development of BSCCO conductors. Steady progress is being made in improving J_c in long lengths of tape, particularly 2223. Tapes are being fabricated into coils and their practical capabilities are being demonstrated. Much headroom remains, however, to improve the performance of these conductors. The continuing challenge is to understand the basic chemical and microstructural processes that occur on forming 2212 and 2223, and to improve the performance of these materials. This requires carefully designed experiments to elucidate the fundamental processes in these complicated systems and ingenuity to apply this basic understanding to engineer improved processing techniques.

The information contained in this chapter is unique as it has attempted to summarize our current understanding of the chemical and microstructural development in 2212 and 2223 conductors. It is certainly not the final statement on how these conductors form. There is much in the literature on processing 2212 and 2223, but unfortunately, we are unaware of any concise review articles that would neatly augment this chapter. In selecting what to include in this chapter, much work was not included, and some was undoubtedly overlooked. For further, more detailed information on processing 2212 and 2223 conductors, a good beginning point is the references in this chapter and the references in these articles.

9.14 References

1. K. Heine, J. Tenbrink, and M. Thöner, Appl. Phys. Lett. **55**, 2441 (1989).
2. N. Shibuta, M. Ueyama, H. Mukai, and K. Sato, Jpn. J. App. Phys. **30**, L2083 (1991).
3. D. K. Christen, J. R. Thompson, H. R. Kerchner, L. Civale, A. C. Marwick, and F. Holtzberg, in AIP Conf. Proc., Superconductivity and Its Applications **273**, 24 (1993).
4. J. E. Tkaczyk, J. A. DeLuca, P. L. Karas, P. J. Bednarczyk, D. K. Christen, C. E. Klabunde, and H. E. Kerchner, Appl. Phys. Lett. **62**, 3031 (1993).
5. P. Majewski, H. L. Sunghigh, and B. Hettich, Adv. Mat. **4**, 508 (1992).
6. R. Müller, Th. Schweizer, P. Bohac, R. O. Suzuki, and L. J. Gauckler, Physica C **203**, 299 (1992).
7. M. Takano, J. Takada, K. Oda, H. Kitaguchi, Y. Miura, Y. Ikeda, Y. Tomii, and H. Mazaki, Jpn. J. Appl. Phys. **27**, L1041 (1988).
8. S. Koyama, U. Endo, and T. Kawai, Jpn. J. Appl. Phys. **27**, L1861 (1988).
9. Y. Feng, K. E. Hautanen, Y. E. High, D. C. Larbalestier, R. D. Ray II, E. E. Hellstrom, and S. E. Babcock, Physica C **192**, 293 (1992) .
10. L. N. Bulaevskii, J. R. Clem, L. I. Glazman, and A. P. Malozemoff, Phys. Rev. B **45**, 2545 (1992).
11. B. Hensel, J.-C. Grivel, A. Jeremie, A. Perin, A. Pollini, and R. Flükiger, Physica C **205**, 329 (1993).
12. T. Hase, T. Egi, K. Shibutani, S. Hayashi, Y. Masuda, I. Shigaki, R. Ogawa, and Y. Kawate, *Proc. 6th Int. Symp. on Supercond.*, to appear (1994).
13. K. Sawada, M. Hiraoka, J. Kai, A. Okuhara, and M. Nakamura, *Proc. International Workshop on Superconductivity*, ISTEC and MRS, Honolulu (1992), 205.
14. J. Shimoyama, N. Tomita, T. Morimoto, H. Kitaguchi, H. Kumakura, K. Togano, H. Maeda, and K. Nomura, *Proc. 5th Int. Symp. on Supercond.*, ed. Y. Brando and H. Yamaguchi, Springer-Verlag, Berlin (1993), 697.
15. T. Hasegawa, H. Kobayashi, H. Kumakura, and K. Togano, preprint (1993).
16. M. Yoshida and A. Endo, Jpn. J. Appl. Phys. **32**, L1509 (1993).
17. G. N. Riley, Jr., J. J. Gannon, Jr., P. K. Miles, and D. R. Parker, preprint (1994).
18. A. N. Iyer, U. Balachandran, P. Haldar, J. G. Hoehn, Jr., and L. R. Motowidlo, in *Processing and Properties of Long Length Suerpconductors*, ed. U. Balachandran, E. Collings, and A. Goyal, to appear (1994).
19. M. Ueyama, T. Hikata, T. Kato, and K. Sato, Jpn. J. Appl. Phys. **30**, L1384 (1991).
20. J. Shimoyama, N. Tomita, T. Morimoto, H. Kitaguchi, H. Kumakura, K. Togano, H. Maeda, K. Nomura, and M. Seido, Jpn. J. Appl. Phys. **31**, L1328 (1992).
21. Q. Li, K. Brodersen, H. A. Hjuler, and T. Freltoft, Physica C **217**, 360 (1993).
22. T. Sata, K. Sakai, and S. Tashiro, J. Am. Cer. Soc. **75**, 805 (1992).

23. U. Balachandran, R. B. Poeppel, J. E. Emerson, S. A. Johnson, M. T. Lanagan, C. A. Youngdahl, D. Shi, K. C. Goretta, and N. G. Eror, Mater. Lett. **8**, 454 (1989).

24. S. E. Dorris, B. C. Prorok, M. T. Lanagan, S. Sinha, and R. B. Poeppel, Physica C **212**, 66 (1993).

25. D. Pelloquin, M. Caldes, A. Maignan, C. Michel, M. Hervieu, and B. Raveau, Physica C **208**, 121 (1993).

26. K. H. Sandhage, G. N. Riley, Jr., and W. L. Carter, J. Met. **43**, 21 (1991).

27. L. R. Motowidlo, P. Haldar, S. Jin, and N. D. Spencer, IEEE Trans. Appl. Supercond. **3**, 942 (1993).

28. J. Tenbrink, M. Wilhelm, K. Heine, and H. Krauth, IEEE Trans. Mag. **27**, 1239 (1991).

29. W. L. Starch, D. Slauson, J. Ponty, and D. C. Larbalestier, unpublished results, University of Wisconsin-Madison (1993).

30. T. H. Tiefel, S. Jin, G. W. Kammlott, J. E. Graebner, R. B. van Dover, and N. D. Spencer, Appl. Phys. Lett. **58**, 1917 (1991).

31. J. Kase, T. Morimoto, K. Togano, H. Kumakura, D. R. Dietderich, and H. Maeda, IEEE Trans. Mag. **27**, 1254 (1991).

32. K. Togano, H. Kumakura, K. Kadowaki, H. Kitaguchi, H. Maeda, J. Kase, J. Shimoyama, and K. Nomura, Adv. Cryo. Eng. (Materials) **38**, 1081 (1992).

33. T. Hasegawa, T. Kitamura, H. Kobayashi, F. Takeshita, H. Kumakura, H. Kitaguchi, and K. Togano, Physica C **190**, 81 (1991).

34. W. Zhang, A. Pashitski, and E. E. Hellstrom, Am. Inst. Phys. Proc., Superconductivity and Its Applications **273**, 599 (1993).

35. J. Ye, S. Hwa, S. Patel, and D. T. Shaw, in AIP Conf. Proc., Superconductivity and Its Applications **219**, 524 (1991).

36. J. Tenbrink, M. Wilhelm, K. Heine, and H. Krauth, IEEE Trans. Appl. Supercond. **3**, 1123 (1993).

37. L. R. Motowidlo, G. Galinski, and P. Haldar, presented at the Materials Research Society Spring Meeting, San Francisco (1993).

38. K. Nomura, J. Sato, S. Kuma, H. Kumajura, K. Togano, and N. Tomita, preprint (1993).

39. Y. Tanaka, T. Asano, T. Yanagiya, M. Fukutomi, K. Komori, and H. Maeda, Jpn. J. Appl. Phys. **31**, L235 (1992).

40. E. E. Hellstrom and R. D. Ray II, *HTS Materials, Bulk Processing and Bulk Applications*, World Scientific, Singapore (1992), 354.

41. R. D. Ray II and E. E. Hellstrom, manuscript in preparation from R. D. Ray's Ph.D. thesis, University of Wisconsin-Madison (1993).

42. R. D. Ray II and E. E. Hellstrom, Physica C **175**, 255 (1991).

43. E. E. Hellstrom, R. D. Ray II, and W. Zhang, Applied Superconductivity **1**, 1535 (1993).

44. W. Zhang and E. E. Hellstrom, Physica C **218**, 141 (1993).

45. T. Hasegawa, T. Kitamura, H. Kobayashi, H. Kumakura, H. Kitaguchi, and K. Togano, Appl. Phys. Lett. **60**, 2692 (1992).

46. J. Polanka, M. Xu, Q. Li, A. I. Goldman, and D. K. Finnemore, Appl. Phys. Lett. **59**, 3640 (1991).

47. S. T. Misture, D. P. Matheis, and R. L. Snyder, in AIP Conf. Proc., Super-

conductivity and Its Applications **279**, 582 (1993).

48. H. Noji and A. Oota, Supercond. Sci. Tech. **5**, 269 (1992).
49. K. Shibutani, T. Egi, S. Hayashi, R. Ogawa, and Y. Kawate, Jpn. J. Appl. Phys. **30**, 3371 (1991).
50. W. Zhang, R. D. Ray II, and E. E. Hellstrom, Adv. Cryo. Eng. (Materials) **40**, to appear (1994).
51. H. Kumakura, T. Hasegawa, H. Kobayashi, H. Kiaguchi, K. Togano, and H. Maeda, Adv. Cryo. Eng. (Materials) **40**, to appear (1994).
52. N. Enomoto, H. Kikuchi, N. Uno, H. Kumakura, K. Togano, and K. Watanabe, Jpn. J. App. Phys. **29**, L447 (1990).
53. F. K. Lotgering, J. Inorg. Nucl. Chem. **9**, 113 (1959).
54. E. E. Hellstrom, J. Met. **44**, 48 (1992).
55. R. D. Ray II, Ph.D. thesis, University of Wisconsin-Madison (1993).
56. T. D. Aksenova, P. V. Bratukhim, S. V. Shavkin, V. L. Melnikov, E. V. Antipova, N. E. Khlebova, and A. K. Shikov, Physica C **205**, 271 (1993).
57. B. Soylu, N. Adamopoulos, W. J. Clegg, D. M. Glowacka, and J. E. Evetts, IEEE Trans. Appl. Supercond. **3**, 1131 (1993).
58. Y. Feng, D. C. Larbalestier, S. E. Babcock, J. B. vander Sande, Appl. Phys. Lett. **61**, 1234 (1992).
59. R. D. Ray II and E. E. Hellstrom, Appl. Phys. Lett. **57**, 2948 (1990).
60. B. Heeb, L. J. Gauckler, H. Heinrich, and G. Kostorz, J. Mater. Res. **8**, 2170 (1993).
61. J. Shimoyama, J. Kase, T. Morimoto, H. Kitaguchi, H. Kumakura, K. Togano, and H. Maeda, Jpn. J. Appl. Phys. **31**, L1167 (1992).
62. K. Shibutani, T. Egi, S. Hayashi, Y. Fukumoto, I. Shigaki, Y. Masuda, R. Ogawa, and Y. Kawate, IEEE Trans. Appl. Supercond. **3**, 935 (1993).
63. J. Tenbrink and H. Krauth, Adv. Cryo. Eng. (Materials) **40**, to appear (1994).
64. G. Triscone, J.-Y. Genoud, T. Graf, A. Junod, and J. Muller, Physica C **176**, 247 (1991).
65. T. Morimoto, J. Shimoyama, J. Kase, and E. Yanagisawa, Supercond. Sci. and Tech. **5**, S328 (1992).
66. L. M. Rubin, T. P. Orlando, J. B. vander Sande, G. Gorman, R. Savoy, R. Swope, and R. Beyers, Physica C **217**, 227 (1993).
67. H. Noji, W. Zhou, B.A. Glowacki, and A. Oota, Physica C **205**, 397 (1993).
68. W. Wu, L. Wang, X. G. Li, G. Zhou, Y. Qian, Q. Qin, and Y. Zhang, J. Appl. Phys. **74**, 7388 (1993).
69. R. D. Ray II and E. E. Hellstrom, Physica C **172**, 435 (1991).
70. A. Endo and S. Nishikida, IEEE Trans. Appl. Supercond. **3**, 931 (1993).
71. T. Kanai and T. Kamo, Supercond. Sci. Tech. **6**, 510 (1993).
72. K. Nomura, M. Seido, H. Kitaguchi, H. Kumakura, K. Togano, and H. Maeda, Appl. Phys. Lett. **62**, 2131 (1993).
73. M. Yoshida, private communication, 6th ISS, Hiroshima, Japan (1993).
74. W. Zhang and E. E. Hellstrom, manuscript in preparation.
75. Y. Masuda, R. Ogawa, Y. Kawate, K. Matsuba, T. Tateishi, and S. Sakka, J. Mater. Res. **8**, 693 (1993).
76. R. Flükiger, A. Jeremie, B. Hensel, E. Seibt, J. Q. Xu, and Y. Yamada, Adv. Cryo. Eng. (Materials) **38**, 1073 (1992).

77. E. E. Hellstrom, presented at ICMC 1993, Albuquerque (1993).
78. E. E. Hellstrom, W. Zhang, A. C. Larson, L. R. Motowidlo, and J. Tenbrink, manuscript in preparation.
79. J. Tenbrink, K. Hein, H. Krauth, M. Szulczyk, and M. Thöner, VDI Berichte **733**, 399 (1989).
80. M. Yoshida, A. Endo, and N. Hara, in *Proc. 6th Int. Symp. on Supercond.*, Springer-Verlag, Berlin, to appear (1994).
81. T. Haugan, M. Pitsakis, S. S. Li, S. Patel, and D. T. Shaw, *Processing and Properties of Long Length Superconductors*, ed. U. Balachandran, E. Collings, and A. Goyal, to appear (1994).
82. H. Noji, W. Zhou, B. A. Glowacki, and A. Oota, Appl. Phys. Lett. **63**, 833 (1993).
83. Y. Yamada, T. Graf, E. Seibt, and R. Flükiger, IEEE Trans. Mag. **27**, 1495 (1991).
84. M. Däumling, A. Jeremie, and R. Flükiger, Supercond. Sci. Tech. **6**, 721 (1993).
85. S. S. Oh and K. Osamura, Supercond. Sci. Tech. **4**, 239 (1991).
86. U. Endo, S. Koyama, and T. Kawai, Jpn. J. Appl. Phys. **27**, L1476 (1988).
87. G. Grasso, A. Perin, B. Hensel, and R. Flükiger, Physica C **217**, 335 (1993).
88. Y. E. High, Y. Feng, Y. S. Sung, E. E. Hellstrom, and D. C. Larbalestier, Physica C **220**, 81 (1994).
89. Y. T. Huang, W. N. Wang, S. F. Wu, C. Y. Shei, W. M. Hurng, W. H. Lee, and P. T. Wu, J. Am. Cer. Soc. **73**, 3507 (1990).
90. P. E. D. Morgan, R. M. Housely, J. R. Porter, and J. J. Ratto, Physica C **176**, 279 (1991).
91. J. S. Luo, N. Merchant, V. A. Maroni, D. M. Gruen, B. S. Tani, W. L. Carter, G. N. Riley, Jr., and K. H. Sandhage, J. Appl. Phys. **72**, 2385 (1992).
92. Y. S. Sung and E. E. Hellstrom, presented at the Spring 1992 meeting of the Am. Cer. Soc., Minneapolis (1992).
93. K. Fukushima, Jpn. J. Appl. Phys. **29**, L1295 (1990).
94. R. H. Arendt, M. F. Garbauskas, and P. J. Bednarczyk, Physica C **176**, 126 (1991).
95. S. E. Dorris, B. C. Prorok, M. T. Lanagan, N. B. Browning, M. R. Hagan, J. A. Parrell, Y. Feng, A. Umezawa, and D. C. Larbalestier, Physica C, to appear (1994).
96. R. Flükiger, B. Hensel, A. Pollini, J.-C. Grivel, A. Perin, and A. Jeremie, *Proc. 5th Int. Symp. on Supercond.*, 17 (1992), 17.
97. I. Matsubara, R. Funahashi, T. Ogura, H. Yamashita, H. Uzawa, K. Tanizoe, and T. Kawai, Physica C **218**, 181 (1993).
98. Y. S. Sung and E. E. Hellstrom, manuscript in preparation.
99. K. Ohkura, H. Mukai, T. Hakita, M. Ueyama, T. Kato, J. Fujikami, K. Murakana, and K. Sato, *Proc. 6th Symp. on Supercond.*, Springer-Verlag, Berlin, to appear (1994).
100. A. Otto, L. J. Masur, J. Gannon, E. Podtburg, D. Daly, G. J. Yurek, and A. P. Malozemoff, IEEE Trans. Appl. Supercond. **3**, 915 (1993).
101. K. Sato, T. Hikata, H. Mukai, M. Ueyama, N. Shibuta, T. Kato, T. Masuda, M. Nagata, K. Iwata, and T. Mitsui, IEEE Trans. Mag. **27**, 1231 (1991).
102. L. J. Masur, presented at the Fall 1993 meeting of the Mater. Res. Soc.,

Boston (1993).

103. I. Aksay, C. Han, G. D. Maupin, C. B. Martin, R. P. Kurosky, and G. C. Stangle, U.S. Pat. No. 5,061,682, July 25, 1991.

104. M. Pechini, U.S. Pat. No. 3,330,697, July 11, 1967.

105. P. Kumar, V. Pillai, and D. O. Shah, Appl. Phys. Lett. **62**, 765 (1993).

106. C. J. Brinker and G. W. Sherer, *Sol Gel Science: The Pysics and Chemistry of Sol-Gel Processing*, Academic Press, Boston (1990).

107. Y. Yamada, M. Satou, T. Masegi, S. Nomura, S. Murase, T. Koizumi, and Y. Kamisada, in *Proc. 6th Int. Symp. on Supercond.*, Springer-Verlag, Berlin, to appear (1994).

9.15 Recommended Readings for Chapter 9

The articles referenced in this chapter and the references in those articles provide a good starting point for further reading. No comprehensive surveys or reviews of this area are currently available.

10

Applications for High-Temperature Superconductors

Z. J. J. Stekly and E. Gregory

10.1 Introduction

When the phenomenon of superconductivity was first discovered by Onnes in 1911, scientists envisioned numerous industrial applications [1]. Not until fifty years later, however, was the first practical high-field magnet realized [2]. This 8-T magnet was wound from wire made of Nb_3Sn a mechanically brittle compound with a critical temperature of 18K.

Since then, other alloys based on Nb have been developed and used in superconductivity applications. In particular, a solid-state alloy of NbTi has received considerable attention because of its ductility and ease of fabrication. This alloy with a critical temperature of 9.5K is suitable primarily for low and intermediate fields (<10 T). For higher fields and for temperatures above 20 K, other materials are needed. Until the discovery of high temperature superconductors (HTS) in 1986, the most used high temperature and high field material remained Nb_3Sn. Nb_3Sn belongs to the A15 class of compounds which includes Nb_3Al, Nb_3Ge, and V_3Ga which have promising superconductor properties.

Superconductors have been used in a variety of applications (see Table 10.1 for a representative list). Coils of multifilamentary conductors wound into the required geometry have been used in high-field test facilities, hybrid magnet facilities, and magnetic fusion research. We emphasize, however, that these applications all involve conventional, *low-temperature* superconductors.

In 1986, a new world in superconductivity was opened by Bednorz and Müller, who carried out experiments with *high-temperature* ceramic superconductors [3]. Within less than a decade, high-temperature superconductivity has become a forefront research and development area in materials science and condensed-matter physics, with potentially significant energy-saving technology applications and considerable economic importance. While the recent reports of superconducting materials with critical temperatures approaching room temperature T_c=250K [4,5] are very promising for the future, currently available conductors using HTS materials offer the ability to operate with critical temperatures approaching 120 K, far higher than the 18–25 K range of the A15 superconductors. Moreover, these HTS superconductors appear capable of carrying more current than A15 materials at fields over 14 T.

The larger scale applications of high temperature superconductors are still embryonic. The availability of the superconductor with the required technical, and later economic, characteristics in large enough quantities is a prerequisite for broad application.

Applications now using low temperature superconductors (LTS) are a natural place to consider using HTS materials. It is for this reason that selected superconductor applications have been chosen for discussion in this chapter. The applications have been selected to highlight the broad range of applications. No attempt is made to be inclusive of all applications.

The material presented relies significantly on material presented in references 6 and 7.

The principal reason that, until recently, we have not seen more applications is that the current-carrying capacity of HTS was unacceptably low and their cost high. In the early 1990s, however while the cost is still high, the performance of these materials has increased dramatically, as shown in Figure 10.1. The figure compares the current-carrying capability, at various temperatures, of Nb_3Sn, NbTi, and short samples and coils of HTS made of bismuth-strontium-calcium-copper oxide (BSCCO).

Based on this improvement, scientists are optimistic that commercial applications of HTS are not far away. The principles are already developed: the coil winding technology widely used for A15 conductors should be di-

Table 10.1. Representative applications of superconductivity.

Current

Commercial Applications
      Nuclear Magnetic Resonance
      Magnetic Resonance Imaging

Future Potential Commercial Application

Propulsion/Levitation
      Maglev
      MHD Ship Propulsion
      Marine Electric Motors and Generators
Power Applications
      Energy Storage
      Power Transmission
      Motors and Generators
      Magnetic Fusion

Non-Commercial

Particle Accelerators
      Beam Transport Magnets
      Radio-Frequency Cavities

---

rectly applicable to HTS materials. Thus, as the availability and performance of HTS materials improves, and their cost decreases, we can expect to see these materials become the superconductors of choice, whenever appropriate.

In the next section, we discuss some HTS conductor and coil performance, followed by applications.

10.2 HTS Conductors/Coils

Many unsuccessful attempts to make useable conductors by various techniques using YBCO pointed out the difficulty of using this material for conductors. It was only when the BSCCO/Ag system was used that progress in conductors was made.

Most of the early development work on conductors for coil applications was on tape configurations. The properties that have been recently achieved for such BSCCO/Ag materials are shown in Table 10.2.

Pressed tapes are superior to rolled samples, but obviously, the produc-

Table 10.2. Critical current densities of short and long tapes at 77K.

	I_c (A)	Core J_c (A/cm^2)	Overall J_c (A/cm^2)	SC (%)
SHORT SAMPLE (PRESSED)	51	−45,000	9,000	20
SHORT SAMPLE (ROLLED)	33	−21,000	5,500	24
LONG LENGTH (34m)	16	−11,000	2,500	24
LONG LENGTH (70m)	23	−15,000	3,500	24
SHORT SAMPLE (ROLLED)	51	−29,000	7,800	27

These data were taken in zero external magnetic field.

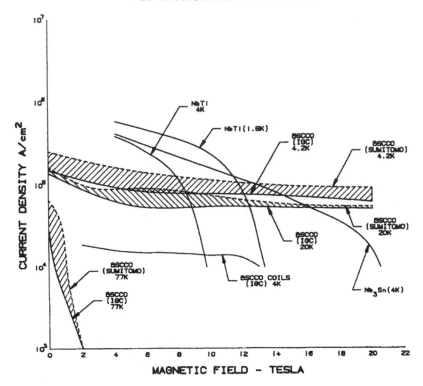

Fig. 10.1. Comparison of the superconductor current-carrying capacity at various temperatures of LTS and HTS. (Courtesy of Intermagnetics.)

Fig. 10.2. Photo of two coils using BSCCO tapes developed by Intermagnetics with partial DOE funding. (Courtesy of Intermagnetics.)

MIT MEASUREMENTS

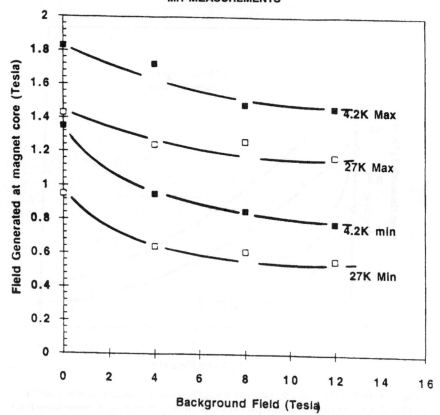

Fig. 10.3. Performance of a 10 pancake BSCCO tape coil assembly at 4.2K and 27K developed by Intermagnetics with partial DOE funding. (Courtesy of Intermagnetics.)

tion of long lengths by a pressing method is not as straight forward as that by rolling. The Table also shows that the long lengths exhibit somewhat reduced properties [14].

Perhaps more importantly, tapes show property anisotropies of some significant magnitude limiting the flexibility of use in coils.

A 2.6T magnet made with ten tape pancake coils is shown in Figure 10.2. Its performance is shown in Figures 10.3 and 10.4 at 4.2K and 27K as well as at 64K and 77K. The details of the coil are shown in Table 10.3.

The mimimum and maximum values represent the performance range from when the first pancake coil begins to become resistive to the point where the last pancake coil begins its transition.

Since the development of the tapes started some four or five years ago, there has been a considerable interest in developing a wire which would facilitate layer winding and reduce the anisotropic properties. At first, circu-

MIT MEASUREMENTS

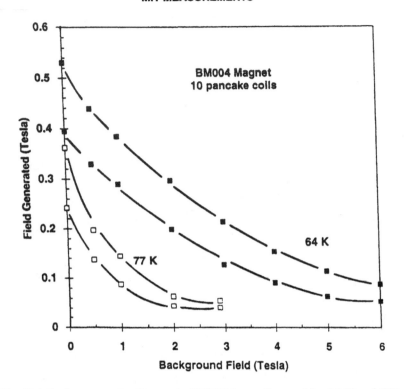

Fig. 10.4. Performance of a 10 pancake BSCCO tape coil assembly at 64K and 77K developed by Intermagnetics with partial DOE funding. (Courtesy of Intermagnetics.)

lar wires had markedly inferior properties to those of tapes, but a recent paper [15] shows that some significant success has now been achieved in this area in wires such as those shown in Figure 10.5. The highest J_c (@ 1μV/cm) of 165,000 A/cm^2 at 4.2K in zero field in the superconductor, was achieved in wires with 11 μm size filaments. The corresponding I_c was 45 amperes in a 0.368 mm o.d. wire which gave an overall J_c of 42,308 A/cm^2. A plot of the critical current density (J_c) versus the filament size is shown in Figure 10.6.

The J_c dependence on temperature and applied magnetic field on these round multifilament wires has also been measured and is given in the above reference.

10.3 Commercial Applications

Superconductors provide the foundation for a wide range of commercial applications. Existing commercial applications range from small, high field magnets used for laboratory experiments or as part of an analytic instrument such as in Nuclear Magnetic Resonance to superconducting magnets for

Table 10.3. HTS Magnet Data.

Winding inner diameter (cm)	2.50
Winding outer diameter (cm)	11.30
Coil height (cm)	6.35
No. of co-wound tapes per pancake	3
Total length of tape in magnet (m)	480
Total no. of turns in the magnet	700
Overall winding cross-section (cm^2)	0.0363
No. of pancake coils	10
Magnet constant (Guass/amp)	111.2

77K (measured)

I_c max	32A	B_o max	0.36 T
I_c min	19A	B_o min	0.21 T

4.2K (measured)

I_c max	234A	B_o max	2.60 T
I_c min	140A	B_o min	1.56 T

27K (extrapolated)

I_c max	160A	B_o max	1.80 T

Magnetic Resonance Imaging, a medical diagnostic technology that enjoys a major international market.

Applications in propulsion/levitation and in power, with the exception of small energy storage systems (SMES) have so far been limited to one-of-a-kind prototypes and have not reached full commercial fruition.

Because HTS coils appear at 4K to have high current densities at high field, their use as a high field insert in small diameter high field superconducting magnet systems appears promising.

Progressively higher magnetic fields have been required for Nuclear Magnetic Resonance and HTS should enable further increases.

10.3.1 Nuclear Magnetic Resonance

Nuclear magnetic resonance (NMR) spectrometry is widely employed in chemistry and biochemistry laboratories. The principle is as follows: Using

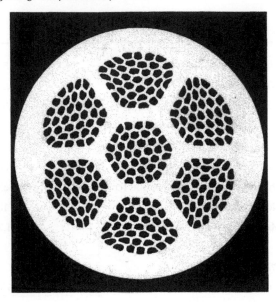

Fig. 10.5. A cross-section of 259 filament BSCCO-2212 round wire at 0.518 mm O.D. and average filament diameter of 16μ developed by Intermagnetics with partial funding from W.P. A.F. Base. (Courtesy of Intermagnetics.)

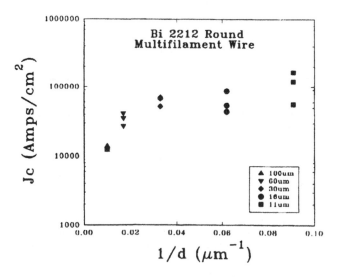

Fig. 10.6. Dependence of the critical current density on final filament diameter for filaments ranging from 100μm to 11μm developed by Intermagnetics with partial funding from W.P.A.F. Base. (Courtesy of Intermagnetics.)

Fig. 10.7a. Developmental Nb₃Sn Model Coils for a 900Mhz (21T) NMR Spectrometer. These operate in a 10.7T background. (Courtesy of Intermagnetics.)

Fig. 10.7b. High Field Coils Assembly. (Courtesy of Intermagnetics.)

resonance techniques, one can identify chemical and nuclear species based on their spin. The frequency of the spin is proportional to the magnetic field. To achieve a high field (> 400 MHz, or 9 T), scientists have traditionally relied on superconducting coils combining Nb-Ti and Nb_3Sn.

In addition to high magnetic field, NMR applications require high stability in time. Such stability is achieved by energizing the magnet and then operating the coils in a "persistent" mode; i.e., a switch is used that short-circuits the coil terminals, producing a closed circuit.

Early systems used Nb_3Sn tape coils operating in a liquid helium bath. Such systems were rated at about 600 MHz (14 T) [16]. In the past decade, however, commercial models have been developed that achieve progressively higher frequencies [17]. Figure 10.7, for example, shows three Nb_3Sn coils for a model high-field NMR spectrometer [900Mhz, 21T]; the model coils operate successfully in a background field of 10.7 T [18]. Even more promising is the work planned at the Pacific Northwest Laboratory, which is developing a 1-GHz (23.5-T) system. A distinctive feature of this system is that it will include a magnetic enclosure to confine the large magnetic fields generated [19].

Commercial application of NMR has seen advances in the past few years. The successful development of HTS materials for NMR would provide further increments in available field. It is important to realize that one of the most demanding applications for a superconductor is *persistent* mode of operation. For this, the perfection of wire and the joints must be extremely high. HTS technology has not yet developed to a stage suitable for the construction of high quantity NMR magnets. However, it should be very suitable for high field magnets for other uses.

10.3.2 Magnetic Resonance Imaging

Similar to nuclear magnetic resonance is a medical diagnostic technique known as Magnetic Resonance Imaging (MRI). MRI produces an image by modulating the spin frequency of the protons. A magnetic field provided by the magnet must be extremely stable (1 part in 10^7 per hour). Also required is extreme uniformity (10–50 parts per million in a sample volume of 30–50 cm dia.). These requirements are met in two ways: (1) by using shim coils and/or small pieces of magnetic material, and (2) by operating a closed superconducting circuit (persistent).

More than two thousand MRI systems are being used for medical diagnosis worldwide. All of these are currently based on NbTi alloys, that is, low-temperature superconducting technology. While the systems have proven remarkably reliable, they still have one major undesirable feature: the use of NbTi, requires operation in liquid helium. An approach has been proposed to address this situation: design the system in such a way that most of the coil windings are not in contact with the liquid helium, but are cooled by conduction.

One variation of conduction cooling, suggested by Laskaris [20] in 1992, uses Nb_3Sn wires or tapes cooled by conduction. In this design, the superconductor windings are impregnated with epoxy (and interleaved with a high thermal conducting metal) and are placed directly in the vacuum space. The windings are then cooled from contact with the low temperature state of a refrigerator. Van der Laan et al. [21] report that such a conduction-cooled magnet has been produced for MRI, and many other research groups are experimenting with conduction-cooled low-temperature Nb_3Sn materials.

Because MRI technology is costly, many hospitals share equipment through the use of mobile vehicles. Unshielded magnets, with a large external magnetic field, must be discharged before being moved. To overcome this inconvenience, scientists have developed shielded MRI designs such as the one shown in Figure 10.8. Developed by Intermagnetics General Corporation for mobile applications, this shielded superconducting magnet has a significantly reduced external magnetic field does not need to be discharged between moves from site to site.

HTS radio frequency receivers have the potential of increasing the MRI signal-to-noise ratio which is a major factor in image quality. These are being developed and utilized on an experimental basis in imaging systems [22,23]

To reduce the refrigeration required to a minimum, HTS dysprosium-based leads have been incorporated in a novel open MRI configuration. This system uses two separate donut-shaped magnet/cryostat assemblies to create a cen-

Fig. 10.8. Shielded MRI superconducting magnet made by Intermagnetics General Corp. (Courtesy of Intermagnetics.)

tral region that is open allowing direct access to the patient [24,25,26]. This system utilizes an Nb_3Sn tape for the main magnet windings.

Use of HTS tapes in magnets of this type has already been considered and as such, they are a natural next step in technology evolution.

Several challenges still limit the widespread application of HTS superconductors to MRI coils themselves. In addition to meeting the requirements for uniformity and stability already discussed, any new magnet system must produce sharp images. Most important, new HTS superconducting systems must be proven reliable and economical before they receive daily acceptance as a medical diagnostic tool.

10.4 Propulsion/Levitation

Superconducting technology is being studied for propulsion of vehicles such as ships (by electric motor/generator and electromagnetics) and high-speed trains (by magnetic levitation/propulsion). The higher magnetic fields and potentially higher operating temperatures possible with HTS make them attractive for such applications. Progress has been achieved by building more advanced and higher performance prototypes and experimental facilities. These applications have not yet resulted in full implementation for commercial use or full-scale implementation for other applications.

One of the earliest applications of superconductors to ship propulsion using rotating machines is shown in Figure 10.9. A turbogenerator generates electricity for an 3000 hp electric homopolar motor that drives the propeller shaft [27].

The superconducting magnets in the homopolar machine are stationary and provide the magnetizing current for an iron magnetic circuit. Room temperature cylinders are caused to rotate by the interaction of the dc current from the generator and the magnetic flux in the iron circuit.

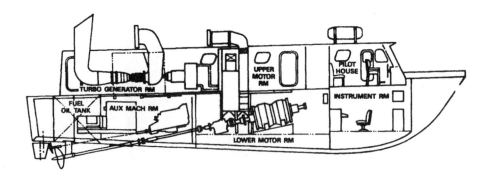

Fig. 10.9. David Taylor Naval Ship Research and Development Center (DTNSRDC) Electric Drive Test Craft - Jupiter II. (Courtesy of U.S. Navy.)

Fig. 10.10. Cryomagnetic systems for naval ship propulsion motors. These provide magnetomotive force for a 3,000 HP homopolor motor. Dimensions of the cryostat are 53 cm diameter x 97 cm long. (Courtesy of the U.S. Navy.)

Intermagnetics supplied the original cryomagnetic modules using Nb-Ti LTS shown in Figure 10.10 [28].

The current plan is to replace these with modules using HTS. The coils are solenoidal and not much different from those depicted in Figure 10.2.

Laskaris [20] discusses a superconducting machine with a stationary field winding which produces magnetic field in a central hole [much like an MRI magnet]. A room temperature armature rotates in the central room temperature space. The field winding could be either Nb_3Sn or Hi T_c depending on availability.

10.4.1 Magnetic Levitation

Magnetic levitation (popularly known as maglev) using superconductors has been under investigation for at least twenty-five years. Most of the effort has been in two countries—Japan and the United States.

Although the technology was invented in the U.S., Japan has clearly held the foremost role in the development of high-speed maglev trains [29]. In the Japanese design, oval superconducting coils are placed on the vehicle. These coils interact with guideway conductors to provide suspension, guidance, and propulsion. Prototype Japanese maglev trains have been tested extensively at the Miyazaki Maglev Test Track shown in Figure 10.11. Current plans are to explore commercialization of maglev vehicles, using a new track being built in Yamanashi. This 42 km length will be part of a future Tokyo to Osaka line.

After a period of diminished interest, the United States is now reentering the field. Under a National Maglev Initiative, U.S. Federal Railroad Administration's efforts have focused on an alternative design. The technical challenge here is to ensure that the conductors remain stable and superconducting throughout acceleration, deceleration, and cruise operation of the

Fig. 10.11a. Experimental maglev vehicle at the Miyazaki Maglev Test Track in Japan.
(Courtesy of the Railway Technical Research Institute.)

Fig. 10.11b. Details of the track. (Courtesy of the Railway Technical Research Institute.)

vehicle, which can weigh up to five tons.   In most cases, the guidance and propulsion forces are exerted directly on the superconducting conductors.

As part of the Federal initiative, U.S. researchers are also exploring the use of Nb^{3}Sn alloys in a variety of configurations, including a Nb$_3$Sn cable in a conduit [30] and Nb$_3$Sn conduction-cooled windings in a vacuum [31]. The former configuration uses supercritical helium adjacent to the superconductor. The latter configuration uses a refrigerated heat sink to cool the windings by conduction. Both configurations offer potentially higher operating temperatures (8–10 K compared with the 4 K achieved by the Japanese design using NbTi) as well as increased thermal stability.

A system configuration which alleviates the problem of large forces on the conductor was proposed in the study by the Grumman/Intermagnetics team [32] shown in Figure 10.12.

Superconducting coils are used to magnetize a laminated iron core which is attracted to a laminated iron rail. Ambient temperature control coils with power supplies are used to maintain a constant gap between the iron core and the rail as the vehicle moves. Most of the forces are exerted on the rail and the iron core. The forces on the superconductors are very much reduced.

The study recommended development of HTS coils as an alternate to the baseline design using LTS.

10.4.2 Magnetohydrodynamic Propulsion

Magnetohydrodynamic (MHD) technology is potentially important for naval and civilian applications because it replaces the propeller and requires no moving parts. The technique involves producing a thrust by passing current through a conducting fluid (seawater) in the presence of a magnetic field. High fields are especially desirable because the efficiency increases with increasing magnetic field (the magnetic forces increase with increasing magnetic field, while the resistive seawater losses remain essentially the same). To produce these high fields, engineers use a superconducting system which, when energized, requires only minimal power for refrigeration.

Three basic configurations have been explored. The simplest involves a *single transverse field magnet* (normally referred to as a dipole magnet); (Figure 10.13a). The principal disadvantage of this design is that it creates relatively large external magnetic fields. The second configuration seeks to overcome this disadvantage by using *multiple magnets* in a cluster [33]. Each magnet has a slightly different field direction (Figure 10.13b). In addition to reducing the total external field of a single dipole, this design offers the advantage that it is modular. The third configuration is a *toroidal magnet*, in which the magnetic field is in an annulus (Figure 10.14). Windings at the inner and outer radii maximize the high field region and produce little external magnetic field. The U.S. Navy has explored several toroidal magnet designs [34].

By far, the largest accomplishment in this area is the YAMATO I, an MHD-propelled surface ship commissioned in 1993 [35] using two clusters of magnets (Figure 10.15). A follow-on effort is planned that has as yet not been defined. However, higher fields would increase the propulsive efficiency.

The U.S. Navy has set up a closed loop system with a superconducting magnet to study this technology [36].

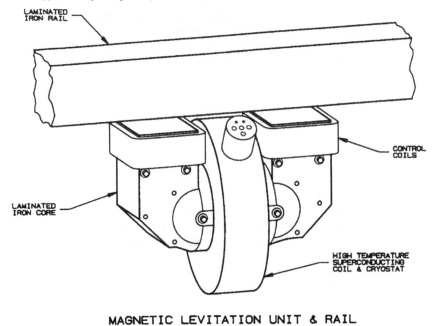

MAGNETIC LEVITATION UNIT & RAIL

Fig. 10.12. A proposed magnetic levitation unit using HTS. (Courtesy of Intermagnetics.)

While the magnets for this facility are made with LTS, preliminary concepts and fabrication techniques for a test facility scale toroidal magnet using HTS have been proposed by Intermagnetics.

10.5 Particle Accelerators

The largest *noncommercial* application of superconductivity has been in charged-particle accelerators used for basic scientific research. Superconductors are a necesary enabling technology for these. Such accelerators involve three steps: (1) very high energy is produced by several stages of acceleration, (2) the high-energy beams are then collided with each other or with a target; and (3) a complex particle-diagnosis is conducted of post-collision fragments.

Superconducting technology is used in particle accelerators in several ways. *Beam transport magnets* bend and focus the beam; *radio-frequency* cavities add energy to the beam and large low field magnets are used as detectors for diagnostics.

10.5.1 Beam Transport Magnets

Since the early 1980s, Fermi National Accelerator in Illinois has operated a charged-particle accelerator known as the Tevatron. The facility comprises more than one thousand superconducting magnets using multifilament NbTi cable. The dipole magnets bend the beam into a roughly circular orbit, in a field of 4.4 T.

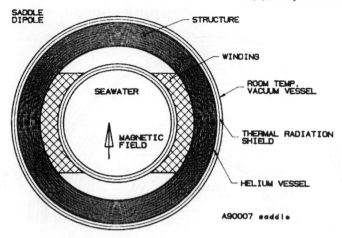

Fig. 10.13a. MHD single magnet configuration. (Courtesy of Intermagnetics.)

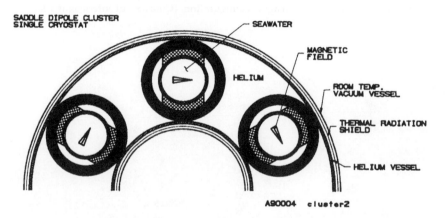

Fig. 10.13b. MHD multiple magnet configuration. (Courtesy of Intermagnetics.)

Another example of an extremely large accelerator is the Superconduct-ing Super Collider (SSC). Although funding for the project has been discon-tinued, the SSC design is worthy of note (see Figure 10.16). The plans called for more than 12,000 beam transport magnets. Typically these have a field of 6.7 T in a 5-cm bore approximately 15 m long [37]. Such a design demands extreme magnetic field accuracy, which in turn requires very small (< 6 mi-crons) superconductor filaments to minimize distortions caused by the su-perconductor, and high current densities. Conductor development work car-ried out before the project was canceled resulted in a superconductor current density of 3×10^5 A/cm^2, compared with 1.8×10^5 A/cm^2 used in magnets for

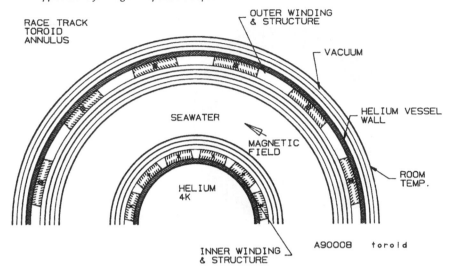

Fig. 10.14. MHD toroidal magnet configuration. (Courtesy of Intermagnetics.)

Fig. 10.15 The YAMATO I, a 185 ton MHD propelled ship 30 m long, 10 m wide with a thrust of 8kN. (Courtesy of the Ship and Ocean Foundation, Japan.)

the Tevatron particle accelerator.

One interesting study during the SSC project involved comparison of the use of higher temperature superconductors with conventional low-temperature materials. Because the design called for a 4.2 K magnet bore in order to cryopump any impurities in the beam tube, it was concluded that low-temperature cryopumping would still be required even if superconducting windings operated at 77 K. Thus, the net effect was minimal: the decrease in refrigeration cost was countered by an increased cost in insulation, conductor, and cryopump system [38].

Most of the beam transport magnets which operate at about 7kA are connected in series at low temperature. However, high temperature superconducting power leads have been proposed for the corrector magnets which are used to fine-tune the high energy beam. These coils operate at 100A and require power connections to room temperature for each magnet.

One other particle accelerator, the Large Hadron Collider (LHC), deserves mention. Under development at CERN (the European Organization for Nuclear Research), the LHC will utilize a 10T dipole magnet 9 m long [39]. Two superconducting technologies are being considered: NbTi at 2 K, and Nb_3Sn at 4.3 K. The latter alloy is brittle, therefore, researchers are focusing on the use of NbTi for this application.

For future accelerators, fields even higher may be required where NbTi cannot be used.

A high field dipole magnet is being built to answer questions relating to the effect of stresses on the current-carrying capability of the alloy strands and cable [40]. The current design requires four layers of cable integrated in a 13-T dipole to be built at the Lawrence Berkeley Laboratory [40]. Coils will be impregnated with epoxy, supported by a stainless steel collar, and surrounded by a large yoke area to return the large magnetic flux generated; see Figure 10.17.

The various designs of particle accelerators illustrate the need for a tradeoff among three factors: the achievable field, the cost of the magnets, and the cost of the tunnel in which the accelerator is placed. If cost-effective high current density HTS become available and the mechanical stress problems associated with higher field magnets are solved, a smaller number of magnets are required, with a consequent reduction in tunneling size and cost.

10.5.2 Radio-Frequency Cavities

The Continuous Electron Beam Accelerator Facility (CEBAF), scheduled to begin operation in 1994, will utilize more than 330 superconducting cavities to accelerate an electron beam to 5 GeV [42]. The facility will operate at 1.5 GHz. To ensure low resistance under such conditions, pure niobium is being used for the cavities. Pure niobium presents several problems, however. In particular, it limits operation to 2 K and requires a refrigeration heat load of 4.8 kW—the largest for operations at this temperature. An alternative material for the superconducting cavities is Nb_3Sn, which can operate at higher temperatures and thus can reduce the complexity of the cryogenic system. Using a niobium cavity coated with a thin film of this alloy, Campisi et al.

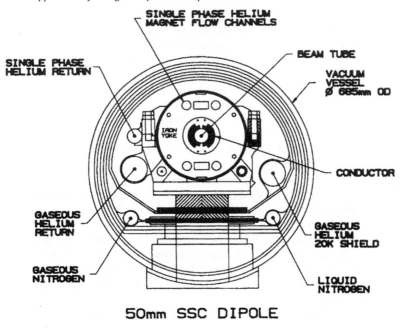

50mm SSC DIPOLE

Fig. 10.16. SSC dipole magnet cross section. (Courtesy of the SSC.)

have obtained performance comparable to Nb up to 11 K [43]. Dasbach et al. also report excellent results at 4.2 K with a 0.6-mm layer of Nb_3Sn deposited by a vapor diffusion process on a Nb cavity [44].

HTS films appear promising in being able to raise the operating temperature above that achievable with Nb_3Sn. Small area thin films of Yttrium Barium Copper Oxide (YBCO) have been shown to have surface resistances that would make them suitable candidates for future devices [45,46]. Scale up to larger areas of film, while maintaining the performance obtained in small films, is the challenge.

10.6 Power Applications

Because of their ability to carry high current densities, superconductors are excellent candidates for use in compact electrical power equipment.

While many full-scale devices have been considered or even built, full-scale commercialization has been limited to small magnetic energy storage (SMES).

Funded by the U.S. Department of Energy (DOE), conductors for power applications using HTS, have been worked on for at least five years in National laboratories and industry.

As part of their continuing effort, the DOE's Superconductivity Partnership Initiative (SPI) has started funding development of electric motors and generators as well as fault current interruptors.

As is well known, superconductors have no resistance to the flow of elec-

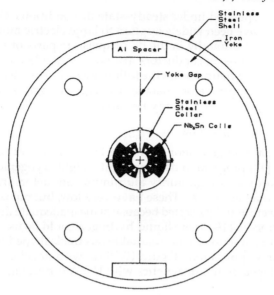

Fig. 10.17a. Developmental 13T, 50-mm dipole magnet for accelerator applications. (Courtesy of Lawrence Berkeley Laboratory.)

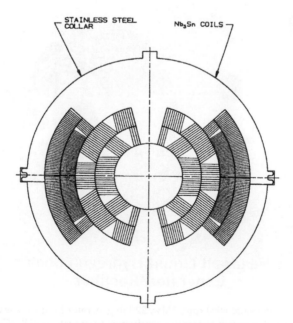

Fig. 10.17b. Detail of Winding. (Courtesy of Lawrence Berkeley Laboratory.)

tric current when operated under steady-state dc conditions. Consequently, superconductors have been widely applied to large electric motors and synchronous generators primarily in the dc or quasi dc parts of the machines. However, very low loss ac conductors are now available and can be used throughout. Here we briefly discuss the following power-related applications of superconductivity: generators, magnetic energy storage, ac power transmission, and magnetic fusion energy research.

10.6.1 Light Weight Generators

The 1Mw, 6000 rpm generator shown in Figure 10.18 was developed as part of a USAF [47] program to develop light weight cryogenic electric machinery. It makes use of high purity aluminum conductors (not superconducting) which operate at 21K. These have very low, but not zero, electrical resistance, and were initially selected because of the inadequate thermal margin of superconductors at 21K where liquid hydrogen could be used for cooling.

Now that HTS is a serious option, suitable racetrack-shaped coils are being developed for this application. Figure 10.19 shows a developmentally HTS coil being developed by Intermagnetics with U.S.A.F. funding for this application [48].

The coil characteristics and performance are summarized in Table 10.4.

1 Megawatt Liquid Hydrogen Cooled
Generator/Rectifier

Fig. 10.18. Developmental 6000 rpm, 1Mw, 220 lb generator built by Westinghouse for U.S.A.F. with high purity aluminum conductors to operate at 21K. (Courtesy of Westinghouse Electric Corp. and the U.S. Air Force.)

The coil as shown has about 30,000 Ampere turns, about 1/3 of that required for one pole of the generator.

Rotating superconducting field windings using NbTi and Nb_3Sn have been the approach taken for synchronous generators. While rotating low temperature field windings are not trivial, these have been demonstrated.

Once the cryogenic and reliability problems associated with continuous rotation of a HTS field winding are dealt with, the current state-of-the-art appears to be adequate for this application.

With funding from the DOE Superconductivity Partnership Initiative, a GE, Niagara Mohawk, Intermagnetics, in collaboration with DOE National Laboratories, will design and develop components for a 100 MVA electric generator utilizing HTS windings.

10.6.2 Magnetic Energy Storage

The superconducting coil has the ability to store and later release magnetic energy in a compact and efficient manner.

On a small scale, 1 to 10MJ coils are used to rapidly supply electric power for a short time in cases where main electric power is interrupted. These systems are available commercially at the 1MJ level. Figure 10.20 shows a

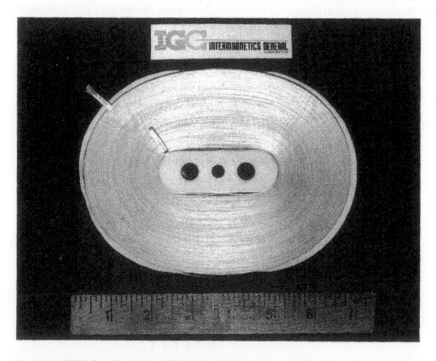

Fig. 10.19. HTS development coil built by Intermagnetics with U.S.A.F. funding for potential use in the 1Mw generator - about 30,000 A-turns. (Courtesy of the U. S. Air Force.)

Table 10.4. Characteristics of double race-track BSCCO pancake coil.

Critical Current at 4.2K	193.5A
Turns	170
Amp Turns: 4K	32,895
27K	29,500
77K	5,270
Outer winding	12.7 cm x 16.83 cm
Inner winding	2.54 cm x 7.46 cm

proposed mobile system which would provide megawatts of power for a few seconds [49].

On a larger scale, a 20Mw hr system has been under development by the Department of Defense for several years Figure 10.21 [50,51].

The objective of this larger device is twofold: for utility applications, it enables the storage of energy at times of low demand for later discharge in periods of high use. For military applications, it also provides the high electrical power for large lasers. While considerable effort has been expended in development and design, it currently appears that there will be no funding for this project. It still has the potential, however, for being applied commercially in even larger sizes.

The large conductor currents, 50kA in one version and 200kA in the other,

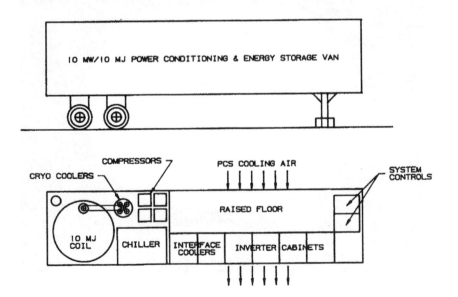

Fig. 10.20. Proposed mobile superconducting energy storage system (SMES). (Courtesy of Intermagnetics.)

Fig. 10.21a. Artist's rendering of Superconducting Magnetic Energy Storage developed by Raytheon Ebasco with Department of Defense funding. (Courtesy of Raytheon Ebasco.)

Fig. 10.21b. Cut away detail showing vacuum vessel, supports, low temperature vacuum vessel and windings developed by Raytheon EBASCO with Department of Defense funding. (Courtesy of Raytheon Ebasco.)

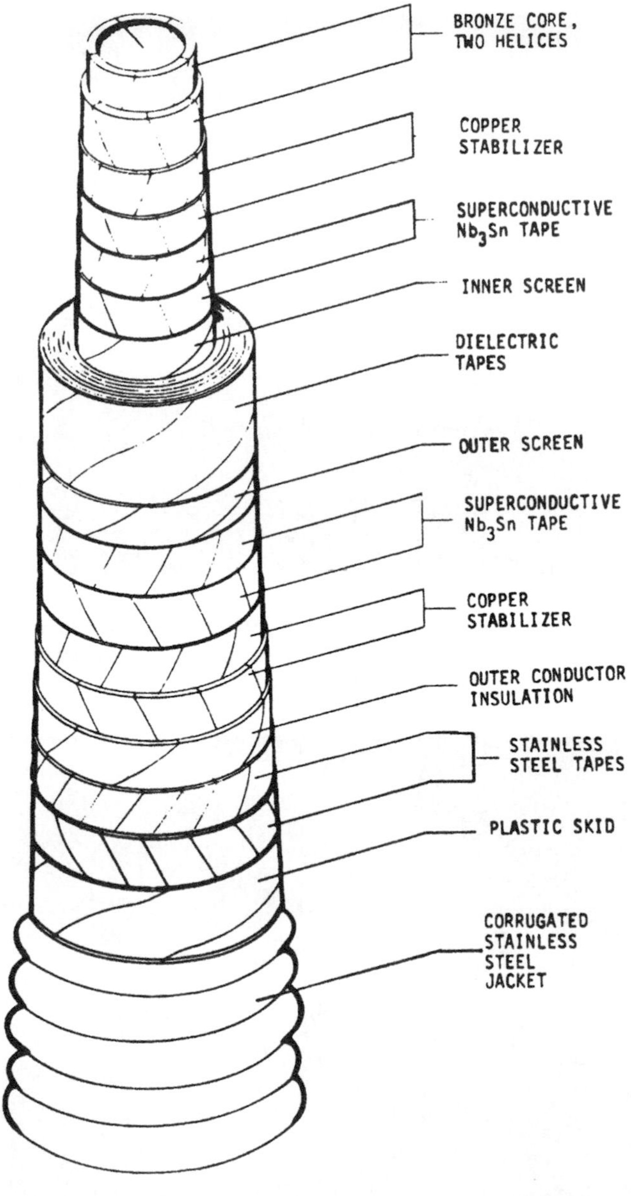

BRONZE CORE, TWO HELICES

COPPER STABILIZER

SUPERCONDUCTIVE Nb_3Sn TAPE

INNER SCREEN

DIELECTRIC TAPES

OUTER SCREEN

SUPERCONDUCTIVE Nb_3Sn TAPE

COPPER STABILIZER

OUTER CONDUCTOR INSULATION

STAINLESS STEEL TAPES

PLASTIC SKID

CORRUGATED STAINLESS STEEL JACKET

Fig. 10.22. Schematic of an Nb_3Sn tape experimental 1000 MVA transmission line built at Brookhaven National Laboratory. (Courtesy of Brookhaven National Laboratory.)

result in a large refrigeration requirement using conventional power lead technology. High temperature superconducting leads that connect the LTS coil to room temperature have been considered seriously for this application.

Further, it has been estimated that use of HTS could reduce the cost of the project by 8–10% [7].

10.6.3 ac Power Transmission

Using superconductivity for ac applications may seem puzzling at first: after all, under ac conditions, superconductors *do* exhibit losses and *do not* have zero resistance. As insight into the mechanism of conductor loss has been gained, superconductors have been developed that can operate successfully under transient or steady-state ac conditions. One such full-scale developmental facility was built at Brookhaven National Laboratory [52] in the late 1980s. Nb^{3}Sn conductor tape 6 mm wide was used for the 100-m three-phase cable. A schematic of the tape is shown in Figure 10.22. The facility, which operated at 1000 MVA, demonstrated the successful application of superconducting power transmission. The cost and complexity of low temperature underground cryogenic systems necessitate the transmission of large amounts of power for the application to be cost-efficient (conventional overhead power transmission is more economical). High-temperature ceramic superconductors, on the other hand, offer particular promise. Once again, the principal technical challenge is the development of materials with suitable current density.

Economic and technical studies of HTS transmission have recently been

Fig. 10.23. Nb$_3$Sn coil made by Westinghouse for the LCT with Department of Energy, Oak Ridge National Laboratory funding. (Courtesy of Westinghouse Electric Corp.)

carried out [53,54]. In general, these indicate that current densities of 30k A/ cm^2 or higher would make a HTS underground replacement cable in the 200 to 600 MVA range attractive.

At power levels of 1 to 3 GVA, current densities greater than 120 KA/cm^2 are necessary. At this current density, the HTS cable costs are equal to that copper cable of the same rating.

10.6.4 Magnetic Fusion

The objective of magnetic fusion is to produce energy from a hot, electrically conducting plasma. In order to confine, control, and induce current in the plasma, superconducting coils are used. As the need for higher fields grew, these coils have been made of A15 materials, because of their proven ability to carry current at high fields.

The first and largest application of A15 superconductors in magnetic fusion research was for the Large Coil Task (LCT). This was a collaborative effort by Japan, the United States, and Europe to demonstrate large toroidal superconducting technology. Six toroidal coils were designed, in the shape of a D to minimize mechanical stress [55]. Five of the coils used NbTi; the sixth coil was wound with a Nb$_3$Sn cable in a conduit (see Figure 10.23). A cable of 500 Nb$_3$Sn strands was formed and then sheathed in a vacuum-tight tube through which supercritical helium was forced in the voids between the strands. The conductor was wound into slots machined in aluminum plates and then assembled to form the integrated coil [56].

Since the LCT, a high field test facility to test large conductors has been built. The Fusion Engineering International Experimental Magnet Facility, FENIX, as it is popularly known, uses a pair of Nb$_3$Sn coils originally designed for a mirror fusion test facility. Each coil is wound with a rectangular conductor with 145,000 five-micron filaments. Using both coils plus iron [57], magnetic fields up to 14 T have been achieved. Short lengths of full-size conductors for applications up to 40 kA [58] have successfully tested.

This facility has been suggested as a good place to initially test and then install HTS power leads to provide the 40kA required for conductor tests.

The *International Thermonuclear Experimental Reactor* (ITER) project, a collaborative effort with the United States, the European Community, Japan, and Russia, is intended to lead to the design of a large tokamak by 1997 [59]. The project will use 16 D-shaped coils to generate a magnetic field of 11.2 T and confine the plasma (see Figure 10.24). At present, the magnet configuration remains undecided, as does the choice of conductor for the coil systems. While Nb$_3$Sn offers the advantage of higher current-carrying capability, Nb$_3$Al is being tested because of its lower sensitivity to mechanical stress [60].

In parallel with ITER, a smaller Tokamak Physics Experiment (TPX) is being built in the U.S. to provide data on compact, continuously operating lower cost fusion reactors. It also has 16 D-shaped coils (smaller than the ones for ITER) and which generate a field of 8.5T. The plans call for Nb$_3$Sn cable in conduit conductor with an operating current of 33 kA.

Application of HTS to fusion so far appears to be in the power lead area. Conductor availability and cost have been a major factor. As conductors become available and confidence is built up on operating their reliability, there is every reason to expect that they will compete with LTS.

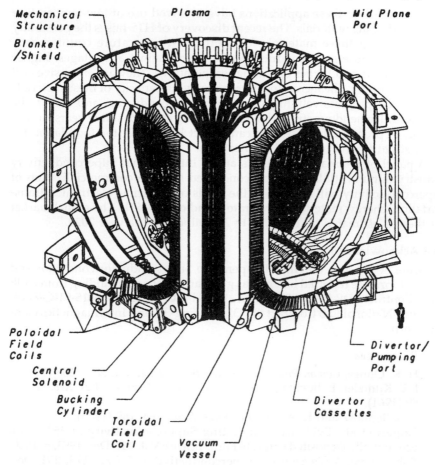

Mechanical Structure

Blanket /Shield

Plasma

Mid Plane Port

Poloidal Field Coils

Central Solenoid

Bucking Cylinder

Toroidal Field Coil

Vacuum Vessel

Divertor/ Pumping Port

Divertor Cassettes

Fig. 10.24. Cutaway view of ITER system. (Courtesy of ITER.)

Because Nb_3Sn is a brittle material, and special manufacturing techniques are necessary to insure conductor performance is not degraded during fabrication, much of the technology should be applicable to HTS materials which are also brittle.

10.7 The Future

Superconducting materials have been widely used in both scientific and commercial applications. At one end of the spectrum are very large high-field installations such as the Tevatron at Fermilab, comprising 1000 superconducting magnets supplied with liquid helium coolant. At the other end of the spectrum is MRI technology, involving lower field superconducting magnets that have 1 m bores and that have extremely low liquid helium consumption.

To date, all of these applications have involved use of low-temperature superconducting materials. The recent discovery of HTS raises the possibility of substituting these materials for their low-temperature counterparts. High-temperature superconductors offer two distinct advantages: higher magnetic fields and higher operating temperatures. They are expensive and in limited supply. The increasing current-carrying capacity of these materials has enabled progressively higher performance conductors and coils to be produced. Tremendous strides have been made in the past five years, and as the rate of progress increases, we can expect HTS to lay the foundation for major scientific and industrial applications.

A panel of international experts at the International Superconductivity Industry Summitt (ISIS) in Japan concluded that the commercialization of superconductivity will develop in the fields of medicine, electronics, transportation, energy, materials, accelerators and others [61]. The global market for these is estimated to reach 150 to 200 billion by the year 2020.

10.8 Acknowledgement

Some of the data on High Temperature Superconductor conductor and coil performance reported herein was generated by partial funding from DOE SBIR Contract #DE-FG02-92ER81461, WPAFB Contract #F33615-91-C-2149, Argonne National Lab Contract #W-31-109-Eng-38, and Intermagnetics General Corp.

10.9 References

1.   H. K. Onnes, *Comm. Phys. Lab.*, University of Leiden, No. 120b (1911).
2.   J. E. Kuntzler, E. Buehler, F. S. L. Itsu, and J. Wernick, *Phys. Rev. Lett.* **6**, 89 (1961).
3.   J. G. Bednorz and K. A. Müller, *Phys. Rev. B* **64**, 189 (1986).
4.   Laques et al. Evidence Suggesting Superconductivity at 250K in a Sequentially Deposited Cuprate Film. *Science* **Vol. 262**, Dec. 1993, p. 1850.
5.   Robert Pool, A Big Step for Superconductivity? *Science* **Vol. 262**, Dec. 1993, p. 1816.
6.   Stekly, Z.J.J. and Gregory, E. Chapter on Applications of A-15 S.C., *Intermetallic Compounds Principles and Practice*, to be published by J. Wiley & Sons 1994, J.H. Westbrook and R.L. Fleisher, editors.
7.   Z. J. J. Stekly, *Proc. Intersociety Energy Conversion Engineering Conf.* (1990).
8.   Motowidlo, L.R., Haldaar, P., Hoehn, J., Rice, J., and Walker, M.S. (1992), In *Proc. 3rd World Congress on Superconductivity*, Munich, September.
9.   Motowidlo, L.R. (1992), In *Proc. HTS Wire Development Workshop*, United States Department of Energy, Conference 920286, Washington, DC, p. 519.
10.   Haldaar, P. and Motowidlo, L.R. (1992), *Journal of Met.*, October.
11.   Sato, K. (1991), *J. Appl. Phys.*, 70(10), 6484.
12.   Sato, K. (1992), *J. Appl. Phys.*, 61(6), 714.
13.   Walker, M.S., Hazelton, D.W., Haldaar, P., Rice, J.A., Hoehn, J.G., Motowidlo, L.R., and Gregory, E. (1992), In *Proc. Applied Superconductivity Conference*, Chicago, IL, August, IEEE, N.Y.
14.   Haldaar, P., verbal communication.

15. Motowidlo et al, To be published in *Applied Phys. Letter* 1994, Dependence of J_c on Fil. dia. in round Multifilamentary wires.
16. Dadok, J.J., and Bothner By, A.A. (1979), In *Proc. Symposium in Honor of M. Cohen in NMR and Biochemistry* (eds S.J. Opaila and P. Lu). Dekker, New York, p. 169.
17. Williams, J.B.C., Pourrahimi, S., Iwasa, Y., and Neuringer, L.J. (1989), *IEEE Trans. Magn.*, **25(2)**, 1767.
18. Markiewicz, W.D., Clancetta, G.M., Grandin, A.W., Hazelton, D.W., et al. (1992b). In *Advances in Cryogenic Engineering*, **Vol. 37, Part A**. Plenum Press, New York, p. 361.
19. Battelle Pacific Northwest. 1000 MHz NMR Magnet System Specification. Battelle Pacific Northwest Laboratories, Richland, Washington (1992).
20. Laskaris, T., *Proc. HTS Wire Development Workshop.*, U.S. Department of Energy Conf. 920286, Washington, D.C. (1992), 2.3.
21. M. T. G. Van der Laan, R. B. Tax, H. H. ten Kate, and L. J.. M. van de Klundert, *IEEE Trans. Magn.* **28**, 633 (1992).
22. Withers, R.S., Ling, C.A., Cole, B.F., and Johansson, M., Thin Film HTS Probe for Magnetic Resonance Imaging, *Trans. on Appl. Superconductivity*, **Vol. 3(1)**, p. 2450, IEEE, New York (1993).
23. Black, R.D., Early, T.A., Roemer, P.B., Mueller, O.M., Mogro-Campero, A., Turner, L.G., and Johnson, G.A., A High Temperature Superconducting Receiver for Nuclear Magnetic Resonance Microscopy, *Science*, **Vol. 259**, Feb. (1993), p. 793.
24. Herd, Kenneth, G., and Laskaris, E.T., Refrigerated High-Tc Superconducting Devices, *AIP Conference Proceedings* **273**, Buffalo, N.Y., American Institute of Physics, 1992.
25. Herd, K.G., Dorri, B., Laskaris, E.T., Tkaczyk, J.E., and Lay, K.W., Grain-Aligned YBCO Superconducting Current Leads for Conduction-Cooled Applications, *Trans. on Appl. Superconductivity*, IEEE, New York, **Vol. 3**, No. 1., March 1993.
26. G.E. preparing to announce extraordinary new MRI system, *Superconductivity News*, **Vol. 6**, No. 16, Jan. 1994, published by Superconductivity Publications, Inc., Somerset, N.J., editor C. Jim Russell.skaris,.
27. Superczynski, M. Navy Motor Program, presented at DOD/DOE Workshop on Cryogenic Rotating Machinery, March 15, 1994, Dayton, OH.
28. *IGC Newsletter*, Spring, 1982.
29. Tanaka, H. (1992b), *Q. Rep Railway Technical Research Institute*, **33(3)**, 160.
30. R. J. Thome, D. B. Montgomery, J. V. Minervini, and J. R. Hale, *Proc. Appl. Superconductivity Conf.*, IEEE, New York (1992).
31. E. Vermilyea and C. Minas, *Proc. Appl. Superconductivity Conf.*, IEEE, N.Y. (1992).
32. Grumman Aerospace Corp. System Concept Definition of a Superconducting Maglev Electromagnetic System, National Maglev Initiative Report, Sept. (1992).
33. S. Motoro et al., 16th Meeting U.S.–Japan Marine Facilities Panel Conf. (1989).
34. J. C. S. Meng, Workshop on Superconducting Electromagnetic Thruster

for Marine Applications, Newport, R.I. (1989).

35. Motoro, S., Imaichi, K, Nakato, M., and Takezawa, S., An Outline of the R&D Project on Superconducting MHD Ship Propulsion in Japan, in *Proc. Ship and Ocean Foundation International Symposium on Superconducting Magnetohydrodynamic Ship Propulsion*, (1991), Kobe, Japan.

36. Meng, J., Hrubes, J., Hendricks, P, Thivierge, D, and Henoch, C, Experimental Studies of a Superconducting Electromagnetic Thruster for Seawater Propulsion, in *Proc. Ship and Ocean Foundation International Symposium on Superconducting Magnetohydrodynamic Ship Propulsion*, (1991), Kobe, Japan.

37. S. Wolff, *IEEE Trans. Magn.* **28**, 96 (1992).

38. P. Liton, Superconductors in Magnetics Commercialization Workshop, Sunnyvale, Calif. (1988).

39. R. Perin, *IEEE Trans. Magn.* **27**, 1735 (1991).

40. J. M. Van Oort, R. Scanlon, H. W. Weijers, S. Wessel, and H. H. J. ten Kate, *Proc. Appl. Superconductivity Conf.*, IEEE, New York (1992).

41. D. dell'Orco, R. Scanlon, and C. E. Taylor, *Proc. Appl. Superconductivity Conf.*, IEEE, New York (1992).

42. I. E. Campisi, R. Ahlman, M. Augustine, K. Crawford, M. Drury, X. Jordan, P. Kelley, T. Lee, J. Marshall, J. Preble, J. Robb, W. Schneider, J. Sista, J. Van Dyke, and M. Wiseman, *IEEE Trans. Magn.* **27** (1991).

43. I. E. Campisi, *IEEE Trans. Magn.* **21**, 134 (1985).

44. D. Dasbach, G. Muller, M. Peiniger, H. Piel, and R. W. Roth, *IEEE Trans. Magn.* **25**, 1862 (1989).

45. Hein, M., Hill, F., Mullen, G., Piel, H., Schneider, H.P., and Strupp, M., Potential of Polyclystalline YBCO Layers for Applications, *Trans. on Appl. Superconductivity*, IEEE, New York, Vol. 3, No. 1, March (1993) p.1745.

46. Oates, D.E., Nguyen, P.P., Dresselhaud, G., Dresselhaud, M.S., and Chin, C.C., Non-linear surface resistance in $Y-Ba_2Cu_3O_{7-x}$ Thin Films, *Trans. on Appl. Superconductivity*, IEEE, New York, Vol 3, No. 1, March (1993) p. 1114.

47. Oberly, C. and Long, L., Air Force Cryogenic Generators presented at DOD/DOE Workshop on Cryogenic Rotating Machinery, March 15, 1994, Dayton, OH.

48. Motowidlo, L., personal communication.

49. Intermagnetics presentation at SMES Utility Interest Group Meeting April 1993, Washington, DC.

50. Bechtel presentation at SMES Utility Group Meeting April 1993, Washington, DC.

51. EBASCO presentation at SMES Utility Interest Group Meeting April 1993, Washington, DC.

52. Forsythe, E.B., *Science*, **Vol. 242**, pp391-399, Oct. 1988.

53. Engelhardth. J.S., Von Dollen, D., and Samm, R. *Application Considerations for HTSC Power Transmission Cables* (1992) published by American Institute of Physics, p. 692

54 Ashworth, S.P., Metra, P., Pirelli, C., and Slaughter, R.J., The Technical and Economic Feasibility of High Temperature Superconducting Power

Transmission Cables, *Procedures of EUCAS 93 Conference*, Gottingen, Germany (Oct. 1993).

55. C. J. Heyne, J. L. Young, and D. T. Hackworth, *Proc. 7th Symposium on Engineering Problems of Fusion Research*, eds M. S. Lubell and C. Whitmore, Jr, IEEE, New York (1977), 935.

56. P. Sanger, E. Adam, W. Marancik, and S. Poulsen, *Proc. 7th Symposium on Engineering Problems of Fusion Research*, eds M. S. Lubell and C. Whitmore, Jr, IEEE, New York (1977), 948.

57. S. S. Slack, F. E. Patrick and J. R. Miller, *IEEE Trans. Magn.* **27**, 1835 (1991).

58. P. Bruzzone, N. Mitchell, H. Katheder, E. Salpietro, et al., *Proc. Appl. Superconductivity Conf.*, IEEE, New York (1992).

59. R. W. Conn, V. Cliuyanov, N. Inoue, and D. Sweetman, *Sci. Amer.* **266** (1992).

60. T. Ando, Y. Takakashi, M. Sugimoto, M. Nishi, and H. Tsuji, *Proc. Appl. Superconductivity Conf.*, IEEE, New York (1992).

61. Intermagnetics General Corporation's Annual Report, Latham, NY (1993)

10.10 Recommended Readings for Chapter 10

1. *1993 Proceedings Annual Review Meeting for U.S. Department of Energy Superconductivity Technology Program for Electric Power Systems* - July 28-29, 1993 [yearly review].

2. *Advances in Cryogenic Engineering* published by Plenum Press, NY (every two years).

3. *ASM Handbook, Vol. 2, Properties and Selection: Non Ferrous Alloy & Special Purpose Materials*, p. 1025-1085, published by ASM International (1992).

4. *DESY Workshop on High Field Dipoles beyond NbTi*, July (1990), published by Deutche Electron Synchrotron, Hamburg, Germany.

5. *Proceedings of Applied Superconductivity Conference* published by IEEE, NY Trans. MAG (every two years).

6. Proceedings of International Conferences on Magnet Technology published by IEEE Trans. MAG (every two years).

7. Proceedings of the International Symposium on ac Superconductors, Smolenice, CSFR ISBN 80-900506-0-3, (1991), Czechoslovakia Federal Republic.

8. Several authors in "Filamentary A-15 Superconductors" eds M. Suenaga and A.F. Clark, Plenum Press, New York (1980).

Index

Printed and bound by CPI Group (UK) Ltd, Croydon, CR0 4YY
03/10/2024

01040418-0010